Mickey C. Smith
Albert I. Wertheimer
Editors

Social and Behavioral Aspects of Pharmaceutical Care

More pre-publication
REVIEWS, COMMENTARIES, EVALUATIONS . . .

"This book should be required reading for pharmacy students and practitioners in meeting the challenges of today's dynamic health care environment. The authors offer an excellent and comprehensive overview on the social-economic aspect of pharmaceutical care. Theoretical models and practical examples are used to elaborate the pharmacist's role in identifying patients' non-compliant behavior and managing drug-related problems. Various chapters address both the interaction between health care providers as well as the communication with patients that are essential to the pharmacist's provision of cognitive services. The approach taken in discussing the quality of life issues is particular appealing. This valuable resource, which includes clinical, economic, and humanistic considerations, should be of great interest to those who are striving to achieve optimal therapeutic outcomes for their patients."

W. Kenneth Wu, PhD
Assistant Professor,
College of Pharmacy,
St. John's University

"Since pharmaceutical care has become an essential part of pharmacy practice and the demand for and accountability of pharmacists to provide this care is growing, it is essential that pharmacists understand it and develop standard care in their practices. This comprehensive book presents a broad spectrum of pharmaceutical care. This well-organized book integrates pharmaceutical care concepts drawn from such diverse disciplines as patients' various behaviors regarding health and sickness, pharmacists' roles in providing care in the process of drug therapies, pharmacists' monitoring and counseling to different types of patients, outcomes assessment, and current issues related to providing pharmaceutical care. It also provides information about the concepts and skills necessary for the development and implementation of pharmaceutical care. This book is highly recommended as an excellent text and reference for both pharmacy students and pharmacists."

Dong-Churl Suh, PhD, MBA
Assistant Professor,
Rutgers–The State University
of New Jersey

"This is a self-described textbook complete with a useful Instructor's Manual. The book is organized into three parts, the core of which is Part I: The Sickness Career. Part I is subdivided further into six sections and 17 chapters. These provide a logical progression from concepts of health and illness through symptoms, care-seeking behavior, prescribing and OTC selection decisions, compliance and, finally, outcomes. Most of the chapters in Part I are thoroughly researched and well-written, and they provide an enlightening description of the antecedents, activities, and consequence of pharmaceutical care delivery.

Part I of this book could stand alone as its own text. The editors, however, have seen fit to round out the material by adding two other sections that suggest the complexity of applying principles of pharmaceutical care delivery to real patients and that locate care delivery in its larger context. Part II (Special Classes of Patients) comprises eight chapters that cover the major patient age groups, major patients sites (e.g., hospital), two especially complex patient groups (the terminally ill and the mentally ill), and cultural factors. Part III (Review of Useful Concepts/Models) provides chapters that triangulate on pharmaceutical care delivery from the perspectives of public policy; reimbursement for cognitive services; behavioral medicine; professional expectations, education, and technology; and ethical concerns in drug research. Parts II and III amplify the meaning of Part I by suggesting that the application of pharmaceutical care delivery takes place in a larger context that both influences and is influenced by that delivery.

In summary, this book would be extremely useful as a text for courses that explore pharmaceutical care delivery. The Instructor's Manual includes exercises and sample test questions for 26 of the 30 chapters in the book. The contributions of the various authors to the Instructor's Manual show a thoughtful and earnest desire to engage students in the content of each chapter. In addition, this book would make a useful reference source, because it provides summaries of important topic areas in the social and behavioral sciences applied to pharmacy (e.g., prescribing and compliance behavior).

Thomas J. Reutzel, PhD
Associate Professor of Pharmacy Administration,
Chicago College of Pharmacy,
Midwestern University

Social and Behavioral Aspects
of Pharmaceutical Care

PHARMACEUTICAL PRODUCTS PRESS
Pharmaceutical Sciences
Mickey C. Smith, PhD
Executive Editor

Social and Behavioral Aspects of Pharmaceutical Care

Mickey C. Smith
Albert I. Wertheimer
Editors

Pharmaceutical Products Press
An Imprint of the Haworth Press, Inc.
New York • London

Published by

Pharmaceutical Products Press, an imprint of The Haworth Press, Inc., 10 Alice Street, Binghamton, NY 13904-1580

Library of Congress Cataloging-in-Publication Data

Social and behavioral aspects of pharmaceutical care / Mickey C. Smith, Albert I. Wertheimer, editors.
    p.   cm.
Includes bibliographical references and index.
ISBN 1-56024-952-8 (acid-free paper)
    1. Pharmacy–Social aspects. 2. Pharmacy management. 3. Pharmacist and patient. I. Smith, Mickey C. II. Wertheimer, Albert I.
RS92.S638 1996
362.1′782–dc20   96-14088
                                                                                                    CIP

CONTENTS

ABOUT THE EDITORS

Mickey C. Smith, RPh, PhD, is F.A.P. Barnard Distinguished Professor at the University of Mississippi School of Pharmacy. He has published more than 300 research and professional articles in more than 100 different journals. He has published eight books and presented more than 150 papers at professional meetings. Smith is editor of the *Journal of Pharmaceutical Marketing & Management* and the *Journal of Research in Pharmaceutical Economics* and Executive Editor of the Pharmaceutical Products Press. His research interests include the determinants of medication use, medication compliance, and medication use in the elderly.

Albert I. Wertheimer, PhD, is Vice President of Managed Care Pharmacy at First Health Services Corporation. He also holds a professorship at the School of Pharmacy of the Medical College of Virginia, in Richmond. Previously, he was Professor of Pharmacy Administration, Dean, and Vice President at the Philadelphia College of Pharmacy and Science. He earned a BS in Pharmacy at the University of Buffalo (1965), an MBA from the State University of New York at Buffalo (1967), and a PhD in Pharmacy Administration from Purdue University (1969). His research interests involve the application of social science and managerial principles to the drug use process and its regulation. He is on the editorial boards of ten journals and is Vice President of the International Pharmaceutical Federation (F.I.P.).

CONTRIBUTORS

Lisa Ruby Basara, RPh, MBA, is Research Assistant in the Research Institute of Pharmaceutical Sciences and a doctoral candidate in the Department of Pharmacy Administration at the University of Mississippi. Her research and publications have concentrated on the evaluation of consumer-oriented pharmaceutical marketing strategies, such as direct-to-consumer advertising and patient education programs, as well as the development of attitudinal and demographic profiles of consumers who are likely to respond to direct pharmaceutical marketing efforts. Additionally, Basara has studied the quality of patient education materials and the attitudes of health professionals toward the use and impact of these materials. Basara is a member of the American Pharmaceutical Association, the Drug Information Association, and the American Marketing Association.

Joni I. Berry, RPh, MS, is Consultant Pharmacist at Hospice of Wake County in Raleigh, North Carolina and Clinical Assistant Professor in the School of Pharmacy, University of North Carolina at Chapel Hill. Berry was a founding member of the Board of Directors of Hospice of Wake County. She has published on pharmaceutical services in hospice care and in pain and symptom control in terminal disease. Berry's research interests include outcome measurements of interventions in palliative care and physicians' knowledge of symptom management issues.

Royce A. Burruss, RPh, MBA, is General Manager at HILLMED Home Medical Systems in Ashland, Virginia. He is also a doctoral candidate in Pharmacy Administration at the School of Pharmacy, Medical College of Virginia, Virginia Commonwealth University, Richmond, Virginia.

Patricia J. Bush, PhD, is Adjunct Professor in the Department of Family Medicine, Georgetown University School of Medicine, Washington, DC. As a pharmacosociologist, she has long been interested in the medicine use process and, especially, the socialization of children into the use of medicines. In 1995, as Visiting Scholar at the United States Pharmacopoeial Convention, Inc., she facilitated the development of guidelines for educating children about medicines. She is the author of more than 70 books, book chapters, and journal articles on the medicine use process and children's health behaviors.

Norman V. Carroll, RPh, PhD, is Professor of Pharmacy Administration at the Medical College of Virginia, Virginia Commonwealth University. His research and teaching focus on financial management, pharmaceutical marketing, and pharmacoeconomics. Dr. Carroll's research has been published in a variety of journals, including *Medical Care, Journal of Clinical Anesthesia, Journal of Health Care Marketing,* and *Journal of Research in Pharmaceutical Economics.* His textbook entitled *Financial Management for Pharmacists: A Decision Making Approach* is widely used in schools of pharmacy. Dr. Carroll received a BS degree in pharmacy and MS and PhD degrees in pharmacy administration from the University of North Carolina at Chapel Hill.

Dale B. Christensen, RPh, PhD, is Associate Professor of Pharmacy at the University of Washington School of Pharmacy and School of Public Health (adjunct). He has authored numerous articles, book chapters, and reports on the subjects of drug utilization, drug policy, drug-taking compliance, pharmacoeconomics, management, and computer applications in pharmacy. He has a particular interest in pharmacy practice in ambulatory and managed care settings. He currently serves as a consultant on drug use review to the state Medicaid program and to private third-party management of drug benefit programs. His research interests are in the areas of drug-taking compliance, drug use review evaluation, and compensation issues related to the provision of value-added pharmaceutical care services by pharmacists.

Stephen Joel Coons, RPh, PhD, is Associate Professor, Department of Pharmacy Practice and Director, Center for Pharmaceutical Economics, University of Arizona College of Pharmacy. He also holds an appointment as Assistant Clinical Professor, Division of Health Care Sciences, Department of Family and Preventive Medicine, School of Medicine, University of California, San Diego. Coons has worked as a pharmacist in community pharmacy practice, in institutional pharmacy practice, and in a drug abuse treatment facility. His scholarly interests include enhancing patient compliance, the pharmacist's role in self-care/self-medication, and assessment of outcomes associated with the use of pharmaceuticals.

Stephanie Y. Crawford, PhD, is Assistant Professor, Department of Pharmacy Administration, the University of Illinois at Chicago. Her research interests include the evaluation of pharmaceutical services and delivery systems, automation in health systems, and medication error outcomes and prevention initiatives.

Charles E. Daniels, PhD, is Chief of the Pharmacy Department at the National Institutes of Health Clinical Center and Clinical Assistant Professor at the University of Minnesota College of Pharmacy. He is a graduate of the University of Arizona College of Pharmacy and earned a PhD in Social and Administrative Pharmacy from the University of Minnesota. Daniels has published and worked in the fields of clinical pharmacy services, pharmacy purchasing, quality improvement, cost management, and pharmacy systems.

Richard Dehn, MPA, PA-C, is Assistant Director of the University of Iowa Physician Assistant Program and has a clinical appointment in the Department of Family Practice, College of Medicine. His major interest is in health care for underserved and rural patients. His research focuses on documenting the practice of physician assistants.

Donna E. Dolinsky, PhD, is Professor of Pharmacy Administration at the Arnold and Marie Schwartz College of Pharmacy and Health Sciences, Long Island University, Brooklyn, New York. Her experience includes teaching, research, and presentations in applications of social and behavioral sciences to pharmacy practice. She has a PhD in educational psychology and has spent the last 16 years teaching at A&M Schwartz College of Pharmacy and at the University of Illinois at Chicago College of Pharmacy. She is the author of more than 30 articles and five book chapters. She has given over 50 presentations and continuing education seminars. Her areas of interest are patient and health professional decision making, behavior, and communications.

Steven Erickson, PharmD, is an Assistant Professor of Pharmacy Practice at Wayne State University. He practices as an ambulatory care clinical pharmacist specializing in geriatrics. His research has addressed patient counseling, medication compliance, and the delivery of pharmacy services to older persons.

Jack E. Fincham, RPh, PhD, serves as Dean and Professor of Pharmacy Practice at the University of Kansas School of Pharmacy. Fincham is a graduate of the University of Nebraska Medical Center College of Pharmacy. Fincham was a Kellogg Pharmaceutical Clinical Scientist Fellow at the University of Minnesota from 1980-1983. He obtained his PhD in Social and Administrative Pharmacy in 1983. Fincham has researched issues pertaining to the drug use process, including the outcomes of drug therapy and pharmacist interventions, adverse drug reactions, patient compliance, drug use in the elderly, and smoking cessation therapies and counseling. He serves as editor of the *Journal of Pharmacoepidemiology,* is an associated editor of the *Journal of Pharmacy Teaching,* and is on the editorial board of the *Journal of Geriatric Drug Therapy.*

Dennis A. Frate, PhD, is Research Professor in the Research Institute of Pharmaceutical Sciences, Professor of Pharmacy Administration, and Professor of Anthropology at the University of Mississippi. He has conducted health behavior research since 1971, including 15 years of research on rural, biracial populations in Mississippi. One of his main areas of interest has been individual health care seeking behaviors. His research has been supported mainly by the National Institutes of Health. He has co-authored over 30 journal articles, including publications in *Medical Care, Hypertension, Women and Health,* the *Journal of Rural Health,* and the *British Medical Journal.* He has also coauthored and/or edited three books.

Juliet B. Frate, BA, is a Research Associate in the Research Institute of Pharmaceutical Sciences, University of Mississippi. She has conducted health behavior research since 1978, 13 of those years on rural, biracial populations in Mississippi. She has co-authored three journal articles and over 20 papers presented at scientific meetings.

Jennifer A. Gowan, PhC, Grad Dip Comm. Pharm., FACPP, is a practicing pharmacist and the Director of Training and Development of the Pharmaceutical Society (Victorian Branch) responsible for the training of preregistration pharmacy students, plus the development of new continuing professional education courses. Her courses emphasize the importance of patient medication counseling and effective communication with patients, caregivers, and health care professionals. She has published a number of pharmacy practice papers.

Amy M. Haddad, PhD, is an Associate Professor and Assistant Dean for Administration and Research at the School of Pharmacy and Allied Health Professions at Creighton University. She is also a Faculty Associate of Creighton University's Center for Health Policy and Ethics. Her teaching and research interests include pharmacy ethics, the contextual dimensions of ethical decision making, and narrative ethics. Haddad has published several books and numerous articles in the area of applied ethics and clinical practice in a variety of health science disciplines. She recently chaired the Council of Faculties Committee on Ethics Content in the Curriculum and was Co-Chair of the Ethics Special Interest Group of the American Association of Colleges of Pharmacy for three years.

David A. Holdford, RPh, MS, is a PhD candidate at the University of South Carolina. He has 11 years of hospital pharmacy experience in clinical and managerial positions. His main areas of interest are the evaluation of pharmaceutical service quality and the influence of pharmaceutical services and programs on patient clinical and economic outcomes.

Peter D. Hurd, PhD, is Professor of Pharmacy Administration at the St. Louis College of Pharmacy. He is also Director of the Division of Pharmaceutical and Administrative Sciences and Director of Graduate Studies. His research includes the application of social psychological principles to the use of medications, health promotion in children and seniors, and the application of management techniques to pharmacy. He has been studying the medication use of national senior athletes, as well as their health-related behaviors, since 1989.

Jeffrey A. Johnson, MSc, received his BSP (1988) and MSc (1994) from the University of Saskatchewan College of Pharmacy, Canada. He also completed a hospital pharmacy residency and was employed as a staff pharmacist at the Royal University Hospital in Saskatoon, Canada. He is currently a Research Assistant at the

Center for Pharmaceutical Economics and enrolled in a PhD program in pharmaceutical economics at the College of Pharmacy, University of Arizona. His current research interests include measuring health-related quality of life in diabetic patients and assessing the impact of enhanced pharmacy services on patient outcomes.

Sheldon Xiaodong Kong, PhD, is Assistant Professor in the Department of Pharmacy Administration, the University of Illinois at Chicago. His research interests are in the areas of pharmacoeconomics and quality of life studies. He also conducts research on social support and pharmacists' work-related attitudes (e.g., organizational and career commitment).

Lon N. Larson, RPh, PhD, is Associate Professor of Social and Administrative Pharmacy at Drake University in Des Moines, Iowa. He also serves as Director of the Drake Center for Health Issues, whose mission is to promote and enhance public participation in health ethics and health policy issues. Prior to his 1991 appointment at Drake, he was on the faculty at the University of Arizona. Larson received a BS in Pharmacy from Drake and an MS and PhD in Health Care Administration from the University of Mississippi. He worked in regional health planning and health insurance before beginning his academic career. Larson has an ongoing interest in managed care pharmacy, evaluation research, pharmacoeconomics, patient satisfaction with pharmacy services, and community health planning. He has presented and published several papers in these areas.

Pedro J. Lecca, RPh, PhD, is Professor and Director of Health Care Specialization at the University of Texas at Arlington and the Director of the Southern Association of School and Colleges on Reaccreditation for U. T. Arlington. He has more than 15 years of experience in teaching, scholarship, research, and academic administration in higher education. Lecca received his BS from Fordham University, his MS from Long Island University, and his PhD from the University of Mississippi. He completed a postdoctorate at Baruch College of Business/Mt. Sinai Medical School in Health Services Research. Lecca has also completed a fellowship in higher education from the American Council on Education. He has special training in academic administration, undergraduate and graduate education, planning, governance, research, fund raising, health/mental health research, substance abuse training, AIDS research, grantsmanship, and resource development. He has been associated with several institutions, including the University of Illinois at Chicago, New York University, and St. John's University. Prior to his appointment at the University of Texas, he held an appointment as Assistant Commissioner of Mental Health/Mental Retardation and Alcoholism Services for the City of New York. Lecca has written over 20 books including monographs and more than a score of articles.

Nancy J. W. Lewis, PharmD, MPH, is President of Lewis Consulting Group, Inc., a Delaware-based corporation that specializes in pharmaceutical economics and outcomes research. She has served as a consultant to pharmaceutical companies, pharmacy trade associations, pharmacy benefit management companies, and government agencies. Previously, Lewis was Assistant Professor of Pharmacy Practice and a clinical pharmacist in Family Medicine at Wayne State University, a health policy analyst for the Greater Detroit Area Health Council, and the Associate Executive Director for the Institute for Pharmaceutical Economics at the Philadelphia College of Pharmacy and Science.

Arthur G. Lipman, RPh, PharmD, is Professor of Clinical Pharmacy in the Department of Pharmacy Practice, College of Pharmacy and at the Pain Management Center, University Hospital, University of Utah Health Sciences Center in Salt Lake City. He was an investigator in the National Cancer Institute Demonstration Project on Hospice Care in New Haven, Connecticut in the 1970s. Lipman is a former president and pharmacological consultant to Hospice of Salt Lake. He is the author of over 80 articles, chapters, and monographs and over 200 reviews and editorials. His research is in pharmacological management of pain and in symptom control of advanced disease. Lipman is the founding editor of the *Journal of Pharmaceutical Care in Pain & Symptom Control.*

Alistair Lloyd, PhC, FPS, has been the Registrar of the Pharmacy Board of Victoria and the Director of the Victorian Branch of the Pharmaceutical Society of Australia since 1982. Prior to this he was in community practice as a pharmacist for 25 years, during which time he served on the Council and as President of the Pharmaceutical Society at state level, and on the Council and as President of what was then called the Pharmaceutical Society of Australia and New Zealand. At present he is actively involved in administering a national project to have pharmaceutical care adopted as the accepted standard of practice by all Australian pharmacists.

Michael Montagne, RPh, PhD, is Associate Professor at Massachusetts College of Pharmacy. His research efforts have focused on the social, cultural, and historical aspects of drug use, the pharmacoepidemiology of drug use problems, images of drugs and drug use in mass media, the occurrence of drug epidemics in society, and the development and testing of drug education programming. Montagne's teaching activities include the social and behavioral considerations of medical and nonmedical drug taking. He has published extensively on drug development and drug use, the occurrence of drug-use problems, and consumer involvement in health care. Montagne belongs to several professional associations, including the American Association of Colleges of Pharmacy, the American Public Health Association, and the International Pharmaceutical Federation (F.I.P.).

Brian G. Ortmeier, PharmD, PhD, is Director of Health Economics & Outcomes Management of MedXL, a Hoechst-Roussel Health Care Company. He is responsible for economic evaluation and disease state modeling and implementation. Ortmeier jointed Hoechst-Roussel Pharmaceuticals, Inc. (HRPI) in June 1994 as a manager, pharmacoeconomics in market development for managed health care. He has been responsible for incorporating economic and outcome evaluations into protocols during Phases 2-3B and 4 for HRPI products. He is a member of the National Pharmaceutical Council Task Force on Compliance and Chair-Elect of the Drug Policy Section/Medical Care of the American Public Health Association. Previously, Ortmeier was Assistant Professor of Pharmacy Practice, Division of Behavioral and Administrative Sciences and Faculty Associate, Center for Pharmaceutical Economics, at the University of Arizona. Ortmeier received his PharmD from Creighton University School of Pharmacy and Allied Health Professions in Omaha, Nebraska, and his PhD in Pharmacy with emphasis in marketing and economics from the University of Georgia.

Andrew M. Peterson, PharmD, is Assistant Director of Pharmacy/ Clinical Services at Thomas Jefferson University Hospital and Clinical Assistant Professor at the Philadelphia College of Pharmacy and Science, Philadelphia, Pennsylvania. Peterson received his BS in Pharmacy from Rutgers College of Pharmacy and his PharmD from the Medical College of Virginia. He also completed a hospital pharmacy residency at Rush-Presbyterian St. Luke's Medical Center and an advanced administrative residency at Thomas Jefferson. Peterson has an interest in developing new practice roles and models for providing pharmaceutical care.

Paul G. Pierpaoli, MS, is Director of Pharmacy Services at Rush-Presbyterian St. Luke's Medical Center and Professor in the Department of Pharmacology, Rush Medical College, Chicago, Illinois. He also holds the rank of Assistant Professor, Health Systems Management, Rush University School of Health Sciences. Over the last 30 years, he has been a director of pharmacy services in a number of major U.S. hospitals and has been active in developing and evaluating innovative pharmacy service programs. He has held leadership positions in a number of national professional organizations and has served as President of the American Society of Hospital Pharmacists. Pierpaoli is the author of numerous articles in the pharmacy and health care literature.

Tracy S. Portner, RPh, PhD, is Assistant Professor in the Department of Clinical Pharmacy, University of Tennessee College of Pharmacy. Her research interests include evaluation of the provision and outcomes of pharmaceutical care in ambulatory care environments, economic evaluations of community and institutional pharmacy services, and the organization and provision of medication information to patients. A specific area of interest is pharmacists' activities in patient care and mechanisms of obtaining compensation for those services. Portner teaches students in the areas of pharmacoeconomics, managed care, and community pharmacy practice management.

Dennis W. Raisch, RPh, PhD, is the Associate Chief of the Veterans Affairs Cooperative Studies Program Clinical Research Pharmacy Coordinating Center and a Research Assistant Professor at the University of New Mexico. He has conducted research on programs designed to influence and improve prescribing, pharmacist interactions with physicians and patients, and pharmaceutical education. His primary research interests are evaluating methods for improving drug prescribing, assessing pharmacy services, and pharmacoeconomics. He has authored 14 and co-authored five research publications in peer-reviewed pharmacy journals. He is currently serving on the *Annals of Pharmacotherapy* editorial advisory panel for pharmacoeconomics.

Louis Roller, PhD, is Associate Professor in the Department of Pharmacy Practice, Victorian College of Pharmacy, Monash University, Melbourne, Australia. He is also Associate Dean (Teaching) and Student Counsellor of the College and a Senior Associate of the Medical Faculty, University of Melbourne. He is a practicing pharmacist, has been involved in professional continuing education for almost three decades, and has lectured regularly interstate and overseas. He is a member of the Pharmacy Board of Victoria with a strong involvement in the educational aspects of the board. He is on the editorial boards of two international journals of pharmacy practice, and has published widely in the area of pharmacy practice.

Richard M. Schulz, RPh, PhD, is Associate Professor of Pharmacy Practice at the University of South Carolina. He received a BS in Pharmacy from the University of Maryland in 1976 and an MS and PhD in Pharmacy Administration from the University of North Carolina in 1980 and 1983, respectively. Prior to joining the University of South Carolina faculty in 1986, he was on the faculty at West Virginia University School of Pharmacy from 1982 to 1986. Schulz's research focuses on quality of life and patient compliance with therapeutic regimens. His work has been published in the pharmacy, health care, communication, and business literature.

Bernard Sorofman, RPh, PhD, is Associate Professor of Pharmacy in the Clinical and Administrative Pharmacy Division, College of Pharmacy, University of Iowa. His research focuses on issues surrounding the patient's use of the health care system, cultural issues in care, and delivery of pharmaceutical care to the rural elderly. Recent publications have appeared in the *Annals of Pharmacotherapy, Social Science and Medicine,* and *International Journal of Cross-Cultural Gerontology.*

Kenneth A. Speranza Sr., PhD, is Associate Professor of Pharmacy Administration at the University of Connecticut School of Pharmacy. His areas of teaching and research specialization are pharmacy law, pharmacy and research ethics, and human resources development in managed care systems. Speranza has published extensively in educational, scientific, and professional journals and has made numerous presentations to professional and lay audiences. He is a past Chairman of the Council of Faculties and Section of Teachers of Social and Administrative Sciences, and a past member of the Board of Directors of the American Association of Colleges of Pharmacy.

Kent H. Summers, RPh, PhD, is the Vice President of Drug Utilization and Pharmacoeconomics, Prudential Pharmacy Management. He is responsible for drug utilization reviews and interventions, as well as pharmacoeconomic research. Prior to this, Summers was an Assistant Professor at the University of Mississippi School of Pharmacy. He has authored and coauthored over 40 articles in peer-reviewed and trade journals. His research interests include investigations of patients' and physicians' decision making, as well as pharmacoeconomic and outcomes assessments of pharmaceutical therapy. Summers received a BS in Pharmacy (1978) and a PhD in Pharmaceutical Sciences (1989) from the University of Missouri-Kansas City School of Pharmacy. Summers has worked in chain drugstores and in health maintenance organizations. In addition, he has over ten years of experience in the pharmaceutical industry.

Sheryl L. Szeinbach, RPh, PhD, is currently Associate Professor in the Department of Pharmacy Administration and Research Associate Professor in the Research Institute of Pharmaceutical Sciences at the University of Mississippi. She received her BS in Pharmacy from the University of Texas at Austin (1980), MS in Pharmacology from the University of Kentucky (1983), and PhD in Pharmacy Administration from Purdue University in 1989. Her research interests include the distribution of pharmaceutical products and services, decision-making processes and risk management, technology (automation), and pharmacoeconomic techniques used in utility assessment. She has published over 60 papers in peer-reviewed journals, including the *Journal of Structural Equation Modeling, American Journal of Hospital Pharmacy, American Journal of Pharmaceutical Education, Journal of Applied Business Research, Journal of Health Care Marketing, Journal of Pharmaceutical Marketing & Management, Journal of Research in Pharmaceutical Economics*, and *The Consultant Pharmacist.*

Toni Tripp-Reimer, RN, PhD, is Professor of Nursing and Anthropology at the University of Iowa. She is Director of the Office of Nursing Research, Development and Utilization; Director of the Gerontological Nursing Research Center; and Director of the Iowa-Veterans Affairs Research Consortium. Her research focuses on self-care behavior in cultural context with an emphasis on ethnogerontology. She has over 200 published scholarly articles, book chapters, and books.

Julie Magno Zito, PhD, is Associate Professor of Pharmacy Practice and Science and Research Associate Professor of Psychiatry at the University of Maryland at Baltimore. Her research area is pharmacoepidemiology and drug treatment refusal among those with mental disorders. During the past ten years, these studies have been supported by funding from the National Institutes of Mental Health and by a state mental health agency. She is the editor and major author of the *Psychotherapeutic Drug Manual*, a handbook on drug therapy for mental disorders published by Wiley in 1994. She has co-authored more than 20 journal articles in journals such as the *American Journal of Psychiatry* and written book chapters on adult and child psychiatry for books in the fields of pharmacoepidemiology and child psychiatry.

Mary Zwygart-Stauffacher, RN, PhD, is Assistant Professor of Nursing and Head of the Gerontological Nurse Practitioner Program at the University of Iowa. She maintains a primary care practice in rural Iowa and Wisconsin. Her research is focused on long-term care issues with emphasis on medication management.

Foreword

Things are just not as they seem to be. First impressions provide stark reminders of that truism as one explores in greater depth the phenomena from which the first impression is made. More often than not, upon more reflective study and scrutiny, what seemed simple and mundane becomes complex and difficult to understand. That is eminently true with human behavior. It is equally true in foreign affairs, the life cycle of a cell, the goings-on in an organization, and a host of other examples in life. So why should it not be true in that aspect of life that is perhaps more personal and simultaneously affects the entire planet's population: namely, illness, sickness, disease, healing, health practitioners, and the systems that attempt to bring them all together. The latter we call the health care system; however, most of us would agree that there is little relationship between the concepts of system and the why's, wherefore's, where's, and how's of the care rendered to sick individuals.

Indeed, the relationships among the individual who is sick (by whatever criteria that may be determined), the individual who is sought out to determine a resolution to the sickness, the institutions where the "cure" might be obtained, and the overarching aspects of financing and assuring appropriateness of care form one of the most complex webs of human and institutional interactions in society, a web which few people understand. Social and behavioral scientists have only recently begun to study systematically the intricacies of the web and its components. Through these studies have come more focused approaches to caring for the patient, educating future health care professionals, and developing dialogues with policymakers and the general public.

That is what this book is all about. In its 30 chapters, the diverse authors attempt to take pieces of the complex web of pharmaceutical care, describe the known microcosmic components of such care, and then relate the pieces back to the integrity of the web. The scrutinizing reader of this collection of important essays will find that the behavior of the patient, the prescriber, the systems that allow for these interactions, and ultimately, the outcomes of medication use are, in fact, not as simple as they appear.

Beginning with the decision to seek care, it will be noted that a variety of perceptions of health and illness exist which are drawn from various individual, cultural, and social interpretations. Some people interpret their illness from the perspective of stoic tolerance, while others perceive it as minor inconvenience and simply move on. Still others will define it as a turning point in life and consequently seek out the best and brightest minds to find the cure. Why is that? Some of the answers lie in the pages you are about to read.

On the other end of the spectrum lie the issues surrounding the outcomes of illness and the care that has been provided. Specifically with regard to medication use, are the desired outcomes being achieved and if not, why not? We know the answers to the latter questions are also not easily determined. Interwoven in medication use are cultural and social interpretations of chemical treatment, including the assurances related to compliance with therapy; the decision structures that are in place to monitor drug therapy; and the difficulties associated with defining and measuring outcomes.

The adventure upon which you are about to embark as you read the varied chapters of this book will likely evoke an equally varied set of responses. Students in the health professions typically do not like the gray zones of the social and behavioral sciences. They find more comfort in the black and white so often attributed to the physical and biological sciences. For readers who find themselves in that frame of mind, this text may prove to be frustrating reading. They may find themselves reading words on many pages and not finding the answers they expect. They will also miss the profoundness of human behavior.

On the other hand, the reader who approaches the reading of this text with a real desire to understand better how behavior links to the complexities of an individual's or social group's actions and deeds will find an exhilarating series in the following pages. This reader will find that microcosms of the web are as fascinating as the web itself. To be sure, there will be few definitive answers; rather, there will be extensive description of phenomena known to be true but which are all subject to change when new variables are introduced. That is what makes inquiry into the social and behavioral aspects of pharmaceutical care both challenging and frustrating.

A better understanding of and appreciation for pharmaceutical care and its behavioral underpinnings will assuredly better prepare pharmacists for contemporary and future roles that more closely bind them to their patients and the prescribing community. As the functional role of the pharmacist as direct caregiver comes to be better understood by the patient, the pharmacist can more effectively influence the behavior of the patient toward achieving better outcomes with drug therapy. On the other side of the coin, as the pharmaceutical caregiving pharmacist comes to understand the patient better from a behavioral perspective, the pharmacist can create a more effective partnership as care is rendered.

Enjoy your journey and find comfort in the gray zones.

Henri R. Manasse Jr.

PART I:
THE SICKNESS CAREER

SECTION A:
CONCEPTS OF HEALTH AND ILLNESS

Chapter 1

Definitions and Meaning of Health

Richard M. Schulz
David A. Holdford

THE PARADOX OF HEALTH

If you were to survey every person in your professional pharmacy practice about the meaning of health, you would find agreement about many aspects of health. First, there would be significant agreement that health is important. Evidence of the importance of health is pervasive. Health care reform has dominated our national dialogue. Fitness and nutrition centers are part of our communities. Bookstores dedicate entire sections to the topic of health. Health has become a vital part of our individual and collective psyche. We are a health-conscious people. Second, the survey would indicate that health is highly valued. Our language is filled with phrases to indicate the value we place on health: "Healthy, wealthy, and wise," and "If you don't have your health, you don't have anything." You could also measure the extent to which we value health by the amount of money we spend to maintain health or to return to a healthy state. Although people might complain about the cost of medical services, pharmaceuticals, and insurance, they still pay it if possible. The $900 billion spent on health care in 1993 reflects the value we place on health.

Within the same survey, however, you would find that people do not agree on other aspects of health. There would be disagreement over how best to achieve health. Some would argue that making frequent visits to traditional health care providers is the surest path to maintaining health. Others would contend that nontraditional providers are better suited to provide guidance and advice. Still others believe that minimizing contact with health care providers altogether improves their chances of being healthy. Within our practice, we encounter some people who believe that compliance with prescribed medication is necessary to maintain or improve health, whereas others believe that such medicines do as much harm as good.

One reason for this lack of consensus over the means to health is the absence of a universal definition and meaning of health. At the core of this dilemma is the realization that health is difficult to measure and to assess. Like intelligence and happiness, it is intangible. It cannot be seen, heard, touched, or felt. Health does not lend itself to assessment by our senses, meaning that its measurement can only be indirect, at best. And as with any indirect, value-laden measure, we can never be completely sure that our assessment is correct.

The paradox about health, therefore, is that while it is important and valuable, we are not sure what it is or how best to achieve it. For health professionals, this paradox poses a fundamental dilemma. Your patients may agree with you that their health is important and valuable, but their concept of health and vision of achieving it may differ significantly from yours. Expanding our understanding of the meaning of health may broaden our capabilities to assist our patients in reaching higher levels of health.

THE MEANING OF HEALTH

The meaning of health cannot be context free. Our views are bound by culture, experience, circumstance, and perspective. An exploration of several different views of the meaning of health may help in clarifying the beliefs we might hold.

Eastern cultures view health as an integrated system of body, mind, spirit, and environment interacting with each other to produce a harmonious balance. When one achieves this balance, he or she is considered "healthy." When any of these factors falls out of balance, health is not possible. Imbalance can be caused by excesses or shortages of necessary components in one's life. The purpose of healing, therefore, would be the process of returning to a state of balance within oneself and with the larger cosmic force. The religious and philosophical connotation to this holistic view of health is rooted in the ancient religions of that region. Today, this view of health can be found in Western society as "holistic health," whose advocates emphasize the mind-body connection in health and illness.[1]

The spiritual aspect of health and disease is prevalent in American Indian culture as well. American Indians believed that disease and death, and therefore health, was brought on them by spirits. Evil spirits would visit the body and bring disease to an individual or community. The art of medicine, therefore, was the skill and technique of ridding the body of these spirits. The person who performed these feats was the medicine man, or shaman. The shaman was granted power from the spirits and served as a "middleman"–a medium between the everyday world of natural occurrence and the supernatural world of gods and spirits.

Although the ancient Greeks had their version of shamans and shamanism as part of Greek mythology, this approach to health and illness took a dramatic turn with the emergence of Hippocrates. Hippocrates espoused the nontraditional view that illness was a natural process owing more to natural phenomenon than to magic and supernatural powers. Interestingly, he and other Greek philosophers believed that health was attained through a balance of the four humors–blood, phlegm, yellow choler, and melancholy–and four qualities of dryness, dampness, heat, and cold. As health was viewed as a balance of these humors and qualities, sickness was the result of their imbalance. In this context, Hippocrates caused a rethinking of the meaning of health, disease, and healing.

The belief that health and disease had natural causes was evident during the period of the Roman Empire. The elaborate aqueducts and subterranean water systems reflected the recognition of the need for clean water and adequate sanitation. Attention to these environmental issues and concern over the quality of general living conditions prompted Cicero to write, "The well-being of the people should be the supreme law."[2]

This view, however, changed with the emergence of Christianity and the gradual infiltration of the Roman empire by various tribes of barbarians. The scientific view of health and disease stagnated as health, disease, and healing became once again the domain of religious or spirituals who summoned the supernatural. As a result, health matters were largely ignored up through the Middle Ages, a time when the world suffered catastrophic epidemics.[1,2]

Although the current Western view of health and disease may be traced back to Hippocrates, it became our dominant view only as recently as the seventeenth century, with William Harvey's discovery that the heart pumps blood through the body in an uninterrupted circulation. Since that time, science and medicine have collaborated in forming the dominant view today of what constitutes health and disease.

This short history of alternative views of health and disease may help in understanding the various beliefs held by patients whom we as pharmacists serve. This is especially important if our own beliefs are inconsistent with those held by others. Patients may place their health in the hands of God and ignore the science-based recommendations of health care providers. By doing this, they are invoking the long-held view that a supernatural power can purge the body of evil. Alternatively, patients may look to medicine and pharmaceuticals as a way to achieve health, but they ignore their responsibilities to maintain balance of mind and body. By doing this, they are reflecting the view, formulated by Descartes in the seventeenth century, that the body is a machine that can be fixed by removing the broken parts.[3] Both types of people are typical of those found in any pharmacy practice. They can frustrate pharmacists and other health care providers because they will not follow our "correct" advice. It is important to note that these views are not transient or superficial. They will not be easily discarded or changed by health care professionals. These ideas have developed over centuries and are often deeply rooted in a person's belief system.

DEFINITIONS OF HEALTH

The various meanings of health held by our patients have been in the making for centuries. Understanding these meanings helps us understand patients' actions and behaviors. However, left unstated at this point is a succinct definition of health. The task of defining health is not as easy as it seems. A major reason for the difficulty lies in the various meanings of health previously discussed. For the patients described above, one definition of health would be insufficient. Another reason for the problem is that any definition must be related necessarily to a context or perspective. For example, in addition to medical definitions of health, there are sociological, epidemiological, health planning, and psychological definitions as well.

Anderson has summarized epidemiological and health planning definitions into five categories: (1) health as a product or outcome (the result of adequate planning and utilization of resources), (2) health as a potential or capacity to achieve goals, (3) health as an ever-changing dynamic process (the interaction between agent, host, and environment), (4) health as something experienced by individuals, and (5) health as an attribute of an individual (similar to characteristics such as fitness or emotions).[4] In addition, various sociological definitions have been offered. Parsons argued that health was a prerequisite for social systems to be maintained in that health maximized individuals' capacity to perform social roles.[5] Dreitzel proposes that health is defined as the capacity to produce surplus in society.[6] These sociological views suggest that health is defined as any state in which individual or societal capacity is maximized. Psychological definitions of health focus on the individual. Some psychologists cite the critical role of needs and need fulfillment. Having one's needs fulfilled, or progressing successfully through stages of development, represents health. Maslow's Needs Hierarchy and Erikson's Developmental Stages are examples of this view of health.[7,8] A medical definition for health may be the absence of disease or the maintenance of physiological parameters within accepted norms (e.g., blood pressure, cholesterol, etc.). A person would be considered healthy if blood pressure and cholesterol readings were within normal limits.

All of the above-mentioned definitions of health present a single limited perspective. Each, although interesting and insightful, is bounded by the discipline from which it came. Perhaps the most fully integrative and widely used definition of health was developed by the World Health Organization. This definition purports that health is the state of complete physical, mental, and social well-being.[9] It specifically does not take a negative perspective in defining health as the "absence" of something, but views health as a positive state. Well-being may be determined subjectively by the individual and may include aspects of functioning and satisfaction.

HEALTH, PHARMACEUTICAL CARE, AND QUALITY OF CARE ASSESSMENT

Patient health and well-being are the sole justification for pharmacy and all other health professions. Pharmaceutical care–the responsible provision of drug therapy for the purpose of achieving specific outcomes that improve a patient's quality of life–is a means to achieve the goal of patient health.[10] Pharmaceutical care serves as a philosophy of practice and has become a part of the mission statements of colleges of pharmacy and the guiding principles of pharmacy organizations.

The development of pharmaceutical care cannot be seen as an isolated event. Pharmaceutical care is the logical consequence of the development of both pharmacy practice and quality of care assessment. These two fields have run parallel courses leading undeniably to health as the ultimate patient outcome. Understanding how and why quality of care assessment progressed will provide insight into the development of pharmaceutical care and its relation to health.

Although efforts to evaluate the quality of health care date back to the turn of the century, a modern era dawned with the conceptualization of health care being composed of structure, process, and outcome components.[11] This view survives to the present. One could assess the quality of care provided by examining the resources (structure) considered necessary for high-quality care to be rendered. Within pharmacy, such resources may include laminar flow hoods (necessary to provide highest quality parenterals) and computer systems integrated with medical records (necessary to conduct drug utilization review). The implication is that quality of care provided by the pharmacist could be accomplished only with those resources present. These resources (structure) would be considered necessary for quality care to be received. "Process" refers to the activities health care providers perform that result in quality care being received. For pharmacists, process refers to the many activities pharmacists perform that are considered a part of quality of care. Staying with the parenteral and computer examples, process would include the use of aseptic technique while making IVs or the conducting of prospective drug utilization review for each prescription dispensed. Like structure variables, process variables are not direct measures of quality of care; rather, they are variables that are believed to be linked to quality of care. For such care to be received, pharmacists would need to engage in these activities. "Outcomes" refer to the experiences of the patient who has received care. The relationship between outcomes and quality of care is that high levels of care must have been provided if the patient experienced desirable outcomes. Conversely, poor patient outcomes suggest that lower levels of care have been rendered. Typically, the outcomes of death, disease, disability, discomfort, and dissatisfaction are measured.[12]

Structure, process, and outcome are intended to be parts of an integrative model of quality of care assessment. In practice, however, their use for the stated purpose has been essentially sequential. As quality of care became an issue of increasing importance, systems were first evaluated on structure, or resource, variables. Hospitals vied for CAT scans and sought affiliation with medical groups because these were the means by which a hospital could demonstrate the level of care it provided. A hospital with a CAT scanner or affiliation with certain physicians was presumed to render a higher quality of care than a hospital with neither. This was the first method of care assessment because it was the easiest to do. Assessment shifted to a process orientation as the methods of assessment evolved. A system was evaluated not on the type of equipment it possessed, but on the number and type of transplants it conducted. The presumption was that a hospital performing transplants was giving better care–across all levels of patients–than a hospital that did not. Recently, the focus has shifted to patient outcomes as a measure of quality of care. Evidence of this emphasis can be found in the publication of mortality rates of different hospitals or patient satisfaction surveys used by hospitals in their own marketing efforts.

These stages of quality of care assessment help us understand how and why pharmacy, specifically pharmaceutical care, is focused on health. Pharmacy practice itself has evolved along the lines of structure, process, and outcome. What may be considered the modern era for pharmacy began at about the midpoint of the twentieth century with the rapid development of safe and effective products. These products are analogous to structure within the framework. With the emergence of clinical pharmacy in the 1960s, pharmacists engaged in the process of providing information to patients through patient information leaflets and verbal counseling. Within the framework, pharmacy had evolved to a process orientation. As pharmacy continued its evolution, it could develop in only one way: namely, to a true orientation toward patient outcomes. Pharmaceutical care is the logical consequence of the evolution of the profession of pharmacy. It has run a track that is parallel to the development of quality of care assessment. It is in step with the broader trend toward outcomes assessment found in all other health-related disciplines. This convergence means that pharmacists who practice pharmaceutical care orient their practice not toward themselves and what they do, but toward the patient and the patient's health. A pharmacist successfully practicing pharmaceutical care is one who maintains the patient's physical, mental, and social well-being by preventing hospitalizations caused by adverse drug interactions. A pharmacist successfully practicing pharmaceutical care maintains a patient's physical, mental, and social well-being by keeping that person in his or her job because side effects of a prescription medication were avoided through appropriate counseling. In effect, pharmaceutical care–the responsible provision of drug therapy for the purpose of achieving definite outcomes that improve a patient's quality of life–is the mechanism by which pharmacy can influence health. Furthermore, as patient health becomes the focus of our practice, it necessarily becomes the only means by which we can evaluate the effectiveness of our individual or collective action.

HEALTH BELIEFS

Pharmacists' ability to influence patient health through the provision of pharmaceutical care relies on the active participation of pharmacists–but says little about the role of patients. In fact, patients exercise free will at every decision-making point. Free will is at the heart of the apparent contradictions in patients' health behaviors that pharmacists observe every day in their practices. An examination of the reasons for patient decisions to engage in health behaviors follows.

Health behavior is any action taken by a healthy person for the purpose of remaining healthy or of remaining in an asymptomatic state. It differs from illness behavior, which is developed further in Chapter 2. Examples of health behavior may include brushing teeth, wearing a seat belt, avoiding tobacco products, exercising regularly, and avoiding high-risk behaviors. People engage in these behaviors for several reasons, including habit, fear, attraction, and health. They develop these behaviors and associated beliefs about them from a variety of sources. Parents teach their children not to touch the stove and to look both ways before crossing a street. Parents' own behaviors concerning health serve as a powerful source of information for children. Teachers serve as another source of information. Peers and the media provide a constant source of information as children get older. These sources, like all others, can provide either positive or negative influences. Perhaps more powerful than any of these sources is one's own experience. The child or adolescent who has been told of the dangers of particular behaviors may experiment and discover that the threatened outcomes did not materialize. Such experience reinforces the person's emerging health beliefs.

Pharmacists, like other health professionals, confront daily behaviors of their patients that are unhealthy. Such behaviors are puzzling to us, especially when we provide information that clearly states the benefits of healthy behavior and the dangers of unhealthy behavior. We do not understand why people continue to smoke even when their respiratory systems are compromised. We do not understand why individuals will engage in destructive eating behaviors when they have been warned of the consequences. Our frustration can easily spill over into anger and ultimately lead to removing ourselves psychologically from a position of positively influencing the person's health.

The frustration described above need not be reason to isolate ourselves from the patient. The predicament is best understood and resolved by recognizing that health behavior is the behavioral extension of health beliefs and motives. Health behavior is unavoidably irrational and inconsistent because our motives are irrational and our sources are inconsistent. A person will continue to smoke in spite of the presence of a respiratory tract infection because smoking is a habitual behavior. It is triggered by cues of situation and circumstance. A person may know intellectually that smoking is not healthy but will continue to smoke because he or she is not thinking of health and is not motivated by health considerations at the time of smoking. Likewise, people may be anorexic and refuse to consume sufficient calories to sustain themselves because of their unrealistic image of the ideal body shape and weight. Furthermore, their friends who have preceded them in this behavior may not have suffered any consequences.

Certainly, there are health professionals specifically trained to handle these potentially dangerous situations. However, pharmacists, because of their accessibility within the community and opportunity for long-term relationships with patients, may have a unique opportunity to influence patient health by fostering health behavior. Effectiveness comes from recognizing that a patient's motives and sources may differ from yours. Targeting an intervention to the motive that generates the behavior or the source of information upon which the behavior is based allows the pharmacist and patient to explore in a nonthreatening way why such behavior exists. The pharmacist who is unaware of a patient's motive yet provides a standard response based on health motivation (eating well is necessary for good health) or fear (if you do not eat better, you will wind up in the hospital) will be hopelessly ineffectual at altering patient behavior and improving patient health. Pharmacy's challenge in the face of irrational and contradictory motives and information is to create practice environments in which pharmacist-patient dialog can occur and health motives and information sources can be explored.

A MODEL OF HEALTH BEHAVIOR

Recognizing that health behavior is rooted in one's beliefs and sources of information is a helpful first step toward understanding and influencing health behavior. However, the realities of practice do not allow us the luxury of extended periods of time with every patient. We need to identify certain disease or patient characteristics that influence health behavior. Pharmacy, like other health professions, needs to rely on certain unifying frameworks that enable it to be efficient as well as effective. These frameworks guide our efforts to influence a patient's own health behavior.

One such framework is the Health Belief Model.[13] The model was developed to increase health professionals' understanding of why and under what conditions people take preventive health action (health behavior). The model proposes that health behavior depends primarily on three classes of variables: (1) the individual's psychological state of readiness to take specific action, (2) the degree to which a particular course of action is believed to have a net beneficial effect in reducing a health threat, and (3) a cue to action that may trigger appropriate action within the individual. Two dimensions define whether a state of action exists. First is the degree to which a person feels vulnerable or susceptible to a particular health condition. Second is the seriousness of the condition itself. Likewise, two dimensions define the perception that a course of action would be beneficial. First is the degree to which the behavior is viewed as beneficial. The second related dimension is the barriers or costs incurred while engaging in such behavior. The three classes of variables influence behavior in the following manner: A person who has a high psychological state of readiness by virtue of high perceived susceptibility or high perceived seriousness of the condition will likely engage in a health behavior. Similarly, one who sees high benefits and minimal costs would be more inclined to act. The cue to action can be internal or external and is of sufficient strength to motivate a person to action. Conversely, low levels of these variables would predict a lower likelihood that an individual would engage in any health behavior, even if the threat to health is present.

Pharmacists can use this model to better understand patient behavior. The model can also serve as a guide to effective and efficient intervention. "Routine" pharmacist interaction with patients can be directed toward assessing patients' psychological state of readiness to act, or their perception of the benefit of such action. For example, pharmacists could reinforce the seriousness of elevated cholesterol to a person not following dietary recommendations. The patient may have stated that "no one in his family has ever had a heart attack," even though there is a known history of coronary artery disease. Such an action on the part of the pharmacist is consistent with the model, is directed and focused, and is in keeping with pharmacists' orientation toward patient health.

Christensen, Fincham, and Wertheimer have applied the Health Belief Model to the issue of patient compliance with therapeutic regimens.[14,15] Compliance, especially for chronic conditions, can be seen as a health behavior done for the purpose of remaining in an asymptomatic state. Christensen proposed that compliance with drug therapy is a dynamic process in which the patient continually reassesses the decision to comply. This decision is best understood within the context of the Health Belief Model. Fincham and Wertheimer used the Health Belief Model to predict initial drug defaulting–patients not picking up the filled prescription from the pharmacy. Using the model, they successfully categorized 69% of the patients into groups that complied or did not comply with the initial prescription.

Janz and Becker reviewed the literature on the Health Belief Model to determine which variable was most effective.[16] They concluded that "perceived barriers" was the most powerful of the model's dimensions. Pharmacists who use the model as a framework for their own professional behavior may improve their effectiveness in influencing patient health behaviors by thoroughly exploring the costs and barriers patients experience while trying to remain compliant. Costs can be economic in nature. Pharmacists should explore ways to reduce the financial burden on patients for whom prescription cost might be a barrier. Lack of access might pose a significant barrier to a segment of the population. Pharmacists could identify patients who have difficulty with transportation. Assistance might include delivery service, mailing, or identifying neighbors who frequent the pharmacy and may be able to deliver medicines to a neighbor. Barriers can be psychological in nature.

The pharmacy can be an intimidating place, and discussion of disease and treatment with the pharmacist can be uncomfortable for some. Pharmacists can check their own verbal and nonverbal behaviors to determine if their behaviors encourage discussion and are nonthreatening. Do pharmacists create a private space where personal discussions about medical conditions and their treatment can take place? While these pharmacist behaviors appear mundane, within the context of the model, they are focused and directed toward improving compliance and fostering healthy behavior among patients.

WELLNESS AND HEALTH

Wellness is both a concept and a movement that is directly related to health. Wellness is defined as an integrated method of functioning that is oriented toward maximizing the potential of which the individual is capable, within the environment where he or she is functioning.[17] Wellness involves the total person. Mind, body, and spirit are inseparable and constantly interact to determine one's experience and behavior.[18] Wellness is seen as one's potential for wholeness and well-being and is strongly influenced by personal choice and environmental factors.[19] Wellness is possible for everyone, is individualized, and is each person's responsibility.

Wellness is mentioned as an example of a current movement rooted in the distant past. The need for balance between mind and body has been voiced for thousands of years. The interaction between person and environment has strong epidemiological underpinnings that can be traced for centuries. Perhaps the greatest contribution of the wellness movement is the focus on individual responsibility. Wellness is action-oriented. It requires the person to take an active part in searching for and maintaining health.

HEALTH PROMOTION AND DISEASE PREVENTION

Healthy People 2000 is a statement of America's public health objectives to be reached by the year 2000.[20] Developed by the U.S.

Department of Health and Human Services' Public Health Service, it argues forcefully that we as a people can control our health destinies in significant ways. The report documents the benefits of appropriate health behavior in terms of lives and dollars saved. The report offers a vision for a more healthy America in which disease and disability are greatly reduced and quality of life is enhanced for all Americans.

Pharmacy is mentioned specifically in one of the priority areas (Food and Drug Safety), and indirectly in other areas (e.g., Alcohol and Other Drugs, and Educational and Community-Based Programs). Pharmacy's role in food and drug safety is delineated as an information source that can prevent adverse drug reactions and minimize the negative effects of polypharmacy. This task is accomplished through the use of linked computerized medication systems by which pharmacists check prescriptions against patient medication profiles for possible therapeutic problems. Furthermore, by linking the pharmacy to other pharmacies and health care sites, pharmacists can assess the drug-taking history of patients regardless of the source of previous prescriptions. While this system reduces the chances of adverse reactions, it also allows the pharmacist to function at a public health level. Clearly, the report sees pharmacists' professional activities in this area as promoting health, preventing disease or adverse events, and enhancing patients' quality of life. This view demonstrates the significance of pharmacists within the community and suggests that our own perspective on health can and should be broadened. Just as we have a responsibility to the patients we serve, we also have a responsibility to the community through our role in drug use control, an idea originally developed by Brodie approximately 30 years ago.[21] *Healthy People 2000* documents our role as agents for change in the community and solidifies our position as public health professionals.

Opportunities to fulfill this role abound, and many pharmacists have seen public health as part of their responsibility and incorporate it in their practices. Pharmacists serve specific and unique populations within the community. Pharmacists have assumed responsibility for health rooms and medicine chests for high school and college sports teams and for entire athletic conferences. Pharmacists carry their profession and professional expertise with them as they serve on community boards. Pharmacists embrace this broadened view of health by participating in community health fairs.

Although it is gratifying that pharmacy is included in our country's formal statement of health objectives, it is less than satisfying that pharmacy's role appears to be defined so narrowly. Adverse drug reactions and drugs of abuse are important public health issues. However, the suboptimal use of prescribed medicines resulting in unnecessary hospitalizations, physician visits, and additional medications is a problem of greater proportions. Polypharmacy, noncompliance, and cognitive or physical impairment of special populations are some of the situations associated with suboptimal use of prescribed medicines. Pharmacists can influence significantly the health of their communities by identifying specific groups of patients who may need assistance in managing their drug therapy.

CONCLUSION

Every action taken by pharmacists is performed to influence in some positive manner the health of our patients. Yet, for an issue that is so critical, there is little we know with certainty. The many definitions of and perspectives on health suggest that we are far from consensus about its meaning. However, what we do know is that patients' views of health may differ from ours, and their motives may conflict with our own. As a result, patient behavior may challenge even the most altruistic among us.

Our response to the challenge is both empathic and strategic. Empathy allows us to see patient behavior through their eyes. It requires us to set aside, as best we can, our own philosophy and expectations, and recognize the validity of our patients' viewpoints. Our challenge is also strategic. Influencing health behavior requires interventions that are appropriately targeted to match patients' motives and beliefs. In so doing, we improve the likelihood that our patients will experience positive outcomes resulting in higher levels of health.

REFERENCES

1. Powers of healing. Alexandria, VA: Time-Life Books, 1989.

2. Harper AC, Lambert LJ. The health of populations. New York: Springer Publishing Company, 1994.

3. Tillich P. The meaning of health. Richmond, CA: North Atlantic Books, 1981.

4. Anderson R. Health promotion: an overview. Eur Monogr Health Educ Res 1984;6:4-119.

5. Parsons T. The social system. Glencoe, IL: Free Press, 1951.

6. Dreitzel HP, comp. The social organization of health. New York: Macmillan, 1971.

7. Maslow AH. Motivation and personality. 2nd ed. New York: Harper and Row, 1970.

8. Erikson E. Childhood and society. 2nd ed. New York: Norton, 1963.

9. World Health Organization. Constitution of the World Health Organization. Annex I. In: The first ten years of the World Health Organization. Geneva: World Health Organization, 1958.

10. Hepler CD, Strand LM. Opportunities and responsibilities in pharmaceutical care. Am J Pharm Educ 1989;53:7S-15S.

11. Donabedian A. Evaluating the quality of medical care. Milbank Mem Fund Q 1966;44:166.

12. Lohr KN. Outcome assessment: concepts and questions. Inquiry 1988;25:37-50.

13. Rosenstock IM. Why people use health services. Milbank Mem Fund Q 1966;44:94.

14. Christensen DB. Drug-taking compliance: a review and synthesis. Health Serv Res 1978;13:171-87.

15. Fincham JE, Wertheimer AI. Using the health belief model to predict initial drug therapy defaulting. Soc Sci Med 1985;20:101-5.

16. Janz NK, Becker MH. The health belief model: a decade later. Health Educ Q 1984;11:1-47.

17. Dunn HL. What high level wellness means. Can J Public Health 1959;50:447.

18. Swinford PA, Webster JA. Promoting wellness. Rockville, MD: Aspen Publishers, Inc., 1989.

19. Webster JA. The wellness model: feeling good about yourself. Assoc Operating Room Nurses J 1985;41:713.

20. U.S. Department of Health and Human Services. Public Health Service. Healthy people 2000. Boston: Jones and Bartlett, 1992.

21. Brodie DC. Drug use control: keystones to pharmaceutical service. Drug Intell Clin Pharm 1967;1:63-5.

ADDITIONAL READINGS

1. Ardell DB. The history and future of wellness. Dubuque, IA: Kendall/Hunt Publishing Company, 1984.

2. Calnan M. Health and illness: the lay perspective. New York: Tavistock Publications, 1987.

3. Hochbaum GM. Health behavior. Belmont, CA: Wadsworth Publishing Company, Inc., 1970.

4. Nordenfelt L. On the nature of health: an action-theoretic approach. Boston: D. Reidel Publishing Company, 1987.

5. Simon HB. Staying well: your complete guide to disease prevention. New York: Houghton Mifflin Company, 1992.

Chapter 2

Illness, Sickness, and Disease

Mickey C. Smith

INTRODUCTION

William Osler once wrote that it is as important to know what kind of person is taking a medication as it is to know what kind of medicines a person is taking. Pharmacists, like it or not, must be prepared to be *social*, as well as physical and biological scientists if they are to practice their profession completely. Eliot Friedson, an eminent medical sociologist, writes:

> One is immediately obligated to distinguish between illness as a purely biophysical state and illness as a human, societal state. Illness as a biophysical state involves changes in bone, tissue, vital fluids, or the like in living organisms, human or not. Illness as a social state involves changes in behavior that occur only among humans and that vary with culture and other organized sources of symbolic meaning.

In this chapter, we hope to demonstrate that social-psychological-behavioral issues are an integral part of any successful treatment or prevention program. Certainly, we cannot touch on every aspect of this broad and complex topic, but we hope to engage the reader in thinking about the *human* aspects of illness.

Both health and sickness mean something different to different people. Their meaning is influenced by prior experience, age, education, and a variety of other personal social factors. These same factors will also influence the patient's perceptions of his experience with pharmacy services.

It is common knowledge that all individual aspirations are not necessarily attainable in the health care system as it exists today. Some major barriers to total availability of health services are built into the current system. Others, however, are at least in part a function of the variations in individual patients and their behavior. The pharmacist, particularly the community pharmacist, is in a unique position to aid in removing barriers such as limitations in education and communication that prevent some people from finding a point of entry into the system. Indeed, the potential exists for making the pharmacist a major point of entry. To do so will require a better understanding on the part of the pharmacist of the importance of individual behavior patterns.

DEFINITIONS AND DETERMINANTS OF ILL HEALTH

If one is to approach an understanding of the behavior of people seeking health care, and in so doing serve them better, one must recognize certain distinctions. One of these is the conventional distinction between illness and disease. *Illness* is defined by laymen—often imprecisely—as a reaction to a perceived biological alteration. It has both physical and social connotations and is highly individual, depending on state of mind and cultural beliefs, as well as physiological and psychological stimuli.

Disease is professionally defined, most often by physicians, and for that reason is perceived to be a more precise term. The definition of disease and its identification (diagnosis) in a given patient form the basis for much of the practice of medicine, including the choice of therapy. Beyond that, disease has become the essential framework for the organization of the health care system and often for the organization of resources within that system.

Health professionals should recognize that precision in the definition of disease is, in fact, often illusory. Physicians disagree both in diagnosis of physical and mental disorders and in the relationship between the two. It should be obvious that imprecision in the *definition of disease* is highly likely to lead to imprecision in *therapy*. It is necessary to go a step further and point out that by these constructions:

1. A person may have a disease and not be ill.
2. A person may be ill and not have a disease.
3. Both disease and illness may be present.

An easy example is hypertension, a "disease" that has been precisely defined—by combination—as a combination diastolic and systolic blood pressure outside "normal" limits. This is, of course, precision by *definition*, with the fact remaining that hypertension has multiple causes, some of them not understood. In any case, a person with the disease of hypertension may be asymptomatic, and consequently not ill. Not being ill, that person may not seek care. In contrast, a person who experiences dizziness or has headaches may perceive himself or herself as ill, seek care, and be diagnosed as disease free. These seemingly inconsequential differences may have serious consequences: failure to receive needed care in the first instance and possible waste of medical resources in the second.

A third definition that warrants brief discussion is that of *sickness*. In the context of this chapter, we will refer to sickness as a condition that is *socially* defined, that is, a social status conferred on an individual by others. This is the approach taken by sociologists, and it will be important in our discussion of the sick role.

Finally, it is valuable to mention *health* or wellness. The World Health Organization's definition of health is the most widely quoted one: "Health is a state of complete physical, mental, and social well-being, and not merely the absence of disease or infirmity." As this definition would conceivably exclude the existence of dental caries or a mild case of athlete's foot, it would similarly exclude virtually all of the world's population. It may be helpful, then, to think of health as an ideal state. Health, or wellness, becomes a goal or destination that is approached but never reached. Some prefer the term "wellness" to denote such a goal, which is actively sought through positive action (good nutrition, exercise) and not merely passively sought through avoidance of contact with disease-causing agents.

With these definitions in hand, perhaps we can examine the determinants of disease, illness, and (conversely) health. Bezold has argued that our state of health is determined by the interaction of four sets of interrelated variables:

1. Biology (i.e., genetic determinants).
2. Behavior (e.g., smoking, drug abuse, eating habits).
3. Pre- and postnatal environments (including physical, biological, economic, and social).
4. The health care system.[1]

Unquestionably, all of these have an influence on our individual and collective health. Inarguably, it is as difficult to assign credit for improved health status to any one of these as it is to find a single cause for most cases of hypertension.

We do know that insofar as *disease* processes are concerned, certain pathophysiological processes are involved. These are vascular, inflammatory, neoplastic, toxic, metabolic, and degenerative. They are modified in their manifestations, however, by factors in the individual such as age, immunological status, ingestion of medication, existence of other disease, and psychological factors. We know further that the passage of time and the accompanying social and technical changes have brought about changes in the ways we die and the diseases from which we suffer.

BEHAVIOR IN HEALTH AND SICKNESS

Three types of behavior are relevant to anyone serving the needs of patients or potential patients. These are, with definitions:

1. Health behavior: "Any activity undertaken by a person who believes himself to be healthy, for the purpose of preventing disease or detecting disease in an asymptomatic stage."
2. Illness behavior: "Any activity undertaken by a person who feels ill, for the purpose of defining the state of his health and of discovering suitable remedy."
3. Sick role behavior: "Activity undertaken by those who consider themselves ill for the purpose of getting well."[2]

The pharmacist has been most involved in the second of these, to a lesser degree in the third, and only minimally in the first. We will look at them in reverse order.

The Sick Role in Acute and Chronic Illness

Understanding the "sick role" or behavior expected of a person defined as sick is made easier by placing it in historical perspective. Treatment, both medical and social, of the sick person has changed with the level of civilization. At the most primitive level, the member who was ill was left to fend for himself or die, with no obligation placed on his neighbors to come to his aid. As civilization advanced, illness was frequently ascribed to evil spirits, and methods of assistance were limited to incantations and sometimes magic potions. The Old Testament, while changing the frame of reference somewhat, still placed illness in a religious context and is rife with references to illness as punishment for sins—either the patient's own or those of his family.

The New Testament brought about a sharp change in attitudes, even to the point of allocating "grace" to those who associated with or aided the sick. By the eighteenth and nineteenth centuries, secular authorities had become influential in health care, and as it became obvious that health care contributed to the common good, "contributions to the care of the sick grew larger and larger, finding in the course of time, social security as its most striking expression."[3]

In spite of this progress, vestiges of the old attitudes remain. We still have difficulty adopting a wholesome attitude, for example, toward mental illness, and a subconscious uncomfortable reaction to crippled people is still widespread. All of this background is necessary to an understanding of the behavior pattern that society expects of those who are officially "sick." Sociologist Talcott Parsons has characterized the sick role as consisting of two rights and two duties:[4]

> *Rights:* Freedom from blame for illness
> Exemption from normal roles and tasks.

The rights are bestowed conditionally, however, and are appropriate only if the patient fulfills his duties:

> *Duties:* To do everything possible to recover
> To seek technically competent help.

Obviously, there are deviations from this model, but when such deviance occurs, society's approval is usually withheld.

The sick role gives the individual a reasonable excuse for making claims on others for care. People with symptoms (i.e., who are "ill") can, with confirmation, adopt this special social role. A person can enter the sick role if a doctor confirms that the person is ill or if the family or friends of the individual are willing to accept the status of "sick." Thus illness (individually defined) becomes sickness (socially defined), especially if a physician confirms the existence of disease. Indeed, the prescription may be thought of as a "ticket" to occupy the sick role.

Considerable evidence exists to suggest that four factors play an important part in determining whether one is allowed to be "sick":

1. Legitimization by physician: Someone is under a doctor's care. (Prescriptions are important evidence of this.)
2. Symptoms: Pains, discomfort, or other manifestations that indicate a change in health.
3. Functional incapacity: The inability of persons to perform normal work activities.
4. Prognosis: The expected outcome of the illness, i.e., probably get worse, get better, stabilize, uncertain, etc.

As valuable as the sick role model is for understanding patient behavior, it should be clear that it does not apply to all cases. How, for example, is the chronically ill patient to "recover?" Indeed, such people cannot, but they can adopt a *chronic illness* role. Pharmacists must know that this role is frequently a difficult one and that medication may be part of the solution but also part of the problem. The benefits of drug therapy that works are most easily understood as the elimination of problems.

Chronic illness is a twentieth-century phenomenon. Prior to this time, illness was generally acute in nature and limited in duration. Advances in sanitation, as well as other advances in medical knowledge and management of illness, resulted in impressive gains over the infectious and parasitic diseases. Regimens for curing or preventing previously irreversible or fatal diseases are currently available that could not have been imagined a century ago. These gains have not been entirely free of negative consequences. The successful treatment of acute life-threatening illnesses has resulted in an increase in the number of individuals with residual limitations and chronic physical or emotional problems. The difficulties attending many chronic conditions continue long after the acute stage of the illness has been successfully managed.

The way ill persons define chronic illness depends on the extent of abnormality of biological structure and function: the nature and severity of symptoms, the competence and skill to manage or control symptoms, and the values, norms, and expectations of others. Except for the anatomical or structural changes caused by the disease process itself, most of the factors that determine the meaning of illness are related in some way to the sociocultural world of the ill person.

The world of the chronically ill is made up of family, friends, and health care professionals, each with a personal perspective on the meaning of the illness and each, therefore, with a framework or rationale for responding to the ill person. Differences in perspectives and expectations between and among these sets of individuals create ambiguities, confusion, tension, and sometimes distress for the individual with chronic illness.

Professionals frequently have treatment goals that are at odds with those of patients. Professionals may define illness only, or primarily, in terms of physiological deviations from normal (elevated blood gases, diminished breath sounds, sugar in the urine, low hemoglobin) and plan treatment to manage these abnormalities with little concern for broader aspects of the patient's life. The individual with a chronic illness may, on the other hand, define the situation primarily in terms of his or her quality of life and establish goals related primarily to social functioning (eating out, entertaining friends, spending time with children, maintaining some level of gainful employment).

The achievement of both sets of goals (professional and personal) may be difficult, and the potential for tension and misunderstanding between patient and caregiver is great. Unless steps are taken to change or modify goals, the final outcome may be discouragement, disappointment, and perhaps giving up on the part of both the professional and the patient. What is needed is communication and negotiation to achieve a compromised, but common, goal.

Sharing a common goal, professionals and patients can establish reasonable criteria for judging the success of treatment. It may be that the most appropriate and humane criteria for measuring success are related to the clients' ability to function at some acceptable level of social intercourse and not near normal body chemistries.

Some case studies by various contributors to Anselm Strauss's book, *Chronic Illness and the Quality of Life*, will provide understanding for the reader and should also demonstrate the value of patient contacts to assess the negative effects of illness and the positive (and negative) effects of the therapy.[5]

Rheumatoid Arthritis (Carolyn W. Wiener)

Eliciting help decreases the arthritic's potential for covering-up and keeping-up. A case in point: the woman who took a leave of absence from work because she could no longer perform to her own satisfaction–she could not lift the heavy robes on her saleswoman job and could not stand having other workers do it for her since: "after all, they're being paid the same thing as I." Awkward or embarrassing situations may occur when eliciting help and these only serve to highlight dependency. One man, for example, was forced to ask a stranger in a public toilet to zip his pants up; his fingers are closed to the palms of his hands, and he had left his trusty buttonhook at home.

Arthritics may be put on strict drug regimens: the drugs hopefully provide control–to help the arthritic normalize. Patients with long histories of frequent flare-ups often undergo sequential trials of potent antirheumatic drugs, all of which have adverse side effects. Some have a difficulty in recalling the sequence of these trials; frequently they were not told what was in their injections and did not ask. For them, the balancing was weighted in favor of relief at any cost: "When you're hurting like that you have to do something."

Ulcerative Colitis (Laura Rif)

Both the symptoms and the medical regimen generate many problems of a technical and physical nature: the individual faces the task of coping with the technically difficult proce-

dures, physical discomfort, pain, fatigue, various sorts of physiological disfunction, and even life-threatening episodes. However, many of the most important ramifications of this illness are personal and social in character. For much of the time, these latter problems are the foremost concern of persons chronically ill with ulcerative colitis.

Indeed, the two major concerns of the sick persons are perhaps: first, the personal and social consequences of the odor and excrement associated with their illness; and second, how the illness and its accompanying regimen restrict their use of time.

The symptoms and regimens associated with ulcerative colitis complicate the management of time because they consume time, pre-empt time, and interfere with the structuring of time. It is important to note that medical regimens are used quite selectively. That is, far from "buying" the treatment package outright, the sick person implements medical recommendations when they are effective for facilitating daily activities. In addition to judging the regimen in terms of its efficacy for improving the illness itself, he evaluates treatment procedures, medications, and other medical interventions according to their costs and benefits for enabling participation in valued activities and insuring attention to high-priority goals.

Emphysema (Shizuko Fagerhaugh)

Possibly the main problem of people who suffer from emphysema is the management of scarce energy. They may be able to increase their energy through a proper regimen. Primarily, however, they must allocate their energy to those activities which they must do or wish most to do. Hence, two key issues for them are symptom control (energy loss) and the balancing of regimen versus other considerations. . . .

Whether a patient does or does not comply to his prescribed regimen and how he complies are determined in part by how the regimen interferes with his life-style and mobility needs. In part, it also depends upon how he comprehends or miscomprehends the uses and effects of therapy.

Sometimes drugs via aerosol spray may be prescribed to immediately relieve periodic respiratory distress. These drugs

*may be ordered at most three/six times a day because of unde-
sirable side effects. When an activity requires extended oxygen
expenditure, there is likelihood of drug overdose, since these
sprays provide extra needed energy to complete an activity.
One patient patted his portable nebulizer, stating it was his
"life saver" as it gave him the needed extra spurt of energy in
difficult routing situations. Although he had been warned by
the physician about overdosing, unfortunately he did not know
what were the symptoms of overdose.*

*Also, although bronchodilator drugs help breathing, these
frequently cause untoward side effects such as nervousness
and sleeplessness. Sometimes sedatives are required to counteract these side effects, but such drugs may leave the patient
groggy. Not infrequently, balancing of the two drugs can
become a problem. An imbalance may mean a patient has no
energy to be mobile because he lacks sleep, or is too groggy to
get around. When health personnel are not attentive to the
patient's drug balance problems, he may doubt both the usefulness of the therapy and the competence of his physician.*

Besides the exception of the chronically ill from the classic sick
role, it is important to note that there is some evidence of unwilling-
ness to grant the exemption from blame to people with certain
maladies. Syphilis and gonorrhea are easily understood historic
examples, but more recently, there is a certain reluctance to grant
full rights to those who suffer from emphysema caused by smoking,
hypertension due to obesity, or cirrhosis resulting from alcohol
abuse. AIDS is, of course, an extraordinarily complicated situation
with stigmatic implications for many of its victims. But even such a
blameless condition as epilepsy does not assure freedom from
blame for its victims.

Because of changes in the demographic makeup of the population, it is necessary to note that a further problem with chronic
illness conditions is that they are not distributed randomly among
the population but are prevalent among those of advanced age.
Normative expectations applied to older chronically ill people may
be ambiguous because of certain similarities between typical attributes of being ill and being old. Expectations concerning aging may,

like being sick, involve impairment of certain types of usual role performance, exemption from obligations, and other forms of permissiveness. Like the occupant of the sick role, the old person is not held responsible for his condition nor can he stop it by act of will. Again, such secondary gains as are associated with aging are mitigated by the lack of regard in which the state is often held. Ambiguities in the case of old, chronically ill people may have unwelcome consequences. There is evidence that older people with chronic illness tend to tolerate functional impairments unnecessarily (in that such impairments could be ameliorated with medical care) because they erroneously associate the symptoms with aging rather than illness. Professionals are not immune to this:

Physician: The reason you are having trouble with your knee is your age.

Patient: My other knee is the same age, and I'm not having trouble with it.

Illness Behavior

Understanding of the behavior of people when they are ill has been enhanced by viewing this behavior as a "sickness career."[6] The sickness career begins with a state of wellness. As already noted, being well or healthy will mean something different to different people. There have been a variety of studies of this phenomenon, and the general criteria by which people view themselves as well include:

1. A feeling of well-being.
2. An absence of symptoms.
3. An ability to perform normal personal and work functions.

Although these criteria will not be uniform from person to person, their meaning for any given individual will form a baseline of health against which to judge changes. When a change from a state of wellness is perceived, there will again be varying reactions. Most people—even those who feel well—are able to identify the presence at any time of some symptoms. Often they will view these symptoms as normal, although the symptoms may trigger a desire for further information.

Ultimately a decision must be made about the significance of symptoms. Twaddle and Hessler report that a variety of factors go into determining if a change in health status is significant.[6] These factors are:

1. Interference with normal activities and functions (for example, bowel habits, work ability, or leisure activity are affected).
2. Clarity of symptoms (sharp pains or symptoms visible to family or friends are likely to be judged important).
3. Tolerance threshold (some people can tolerate more pain, either because of personal characteristics, cultural factors, or the nature of their work).
4. Familiarity with symptoms (common symptoms that one has experienced previously and recovered from are likely to be viewed as less serious than those that have not been previously experienced).
5. Assumptions about cause (for example, in the case of chest pain, it may be viewed as anything from a heart attack to indigestion).
6. Assumptions about prognoses (if long-term incapacity or possible death is associated with the symptom, it is likely to be viewed as more serious).
7. Interpersonal influence (this item refers to effects of the lay referral system).
8. Other life crisis (in some cases, a symptom that might have been viewed as normal assumes greater proportions in the face of family or work crisis).

Symptom Control

Symptoms are at the heart of pharmaceutical care. Without them, patients usually will not seek assistance. They are the basis for self-medication (read the packages of OTC products). But the response to symptoms is by no means uniform or predictable.

According to Verbrugge and Ascione, people approach chronic and acute health problems in different ways.[7] For chronic ones, they devise strategies of care (determined partly by their roles, attitudes, and resources) over months and years and apply them during flare-ups. For acute problems, decisions about care are made in the short run and hinge mostly on symptoms. Analysis shows that actions

complement or substitute for each other. Self-care actions (nonprescription drug use and restricted activity) tend to co-occur, and so
do actions based on medical care (prescription drug use and medical
contact). The two domains substitute in one way (nonprescription
drug use greatly reduces chances of prescription drug use) and join
in another (restricted activity increases chances of medical contact).

Whatever sophisticated technical references there may be for his
symptoms, the person who has symptoms will be concerned primarily with whether he hurts, faints, trembles visibly, loses energy suddenly, runs short of breath, has had his mobility or speech impaired,
or is evidencing some kind of disfigurement. Aside from what these
may signify to him about his disease or his life span, such symptoms
can interfere with his life and his social relationships. How much
they interfere depends upon whether they are permanent or temporary, frequent or occasional, predictable or unpredictable, and publicly visible or invisible; upon their degree (as of pain), their meaning
to bystanders (as of disfigurement), the nature of the regimen called
for to control the symptoms, and upon the kinds of life-style and
social relations that the sufferer has hitherto sustained.

Even minor, occasional symptoms may lead to some changing of
habits. Thus, someone who begins to suffer from minor back pain is
likely to learn to avoid certain kinds of chairs and may even discover, to his dismay, that his favorite sitting position is precisely
what he must eliminate from his repertoire. Major symptoms, however, may call for the redesigning or reshaping of important aspects
of a life-style. A stroke patient writes: "Before you come downstairs, stop and think. Handkerchief, money, keys, book, and so
on—if you come downstairs without these, you will have to climb
upstairs, or send someone to get them." People with chronic diarrhea need to reshape their conventional habits like this person did:
"I never go to local movies. If I go . . . I select a large house . . .
where I have a greater choice of seats. . . . When I go on a bus . . .
I sit on an end seat or near the door."

Once a symptom is viewed as significant, a decision must be
made whether help is needed, and, if so, what kind of help is
needed. The enormity of sales of nonprescription drugs attests to
the fact that self-treatment is often the first treatment choice. If,
consistent with the sick role, the individual chooses to seek profes-

sional help, it is likely that the lay referral system (as well as prior experience) will again come into play, both in the initial decision to seek help and in the type of help chosen. Once help is sought, usually from a physician, the drug use process begins. Within this process, a wide range of patient behaviors remains, including the degree to which the patient will comply with professional recommendations. Research has shown that the degree of patient compliance varies greatly and is influenced, again, by a variety of personal, psychological, social, cultural, and economic factors.

The final step in the sickness career ranges from death to a chronically ill status to a return to wellness.

Health Behavior

Hardest of all health-related behavior to predict is that of the "healthy" individual. Included in such behavior are decisions to obtain preventive care or detection tests as well as activities designed to maintain a state of wellness, such as dental hygiene and good nutrition. The individual has no clear-cut symptoms to prompt such action, yet the time, effort, and money spent on this type of activity are potentially the most productive of those spent on any health care activity.

One of the most widely accepted explanations of people's health behavior is the Health Belief Model shown in Figure 2.1. Briefly, the model proposes that an individual is psychologically ready to take action relative to a given condition (e.g., cancer prevention). The degree of such readiness is influenced by the extent to which that individual feels susceptible to the condition and views the condition as one with serious personal consequences. The individual must also believe that a proposed action (e.g., a Pap smear) is both feasible and appropriate to use, that it would reduce susceptibility to the condition or to the seriousness of the condition, and that taking the action presents no serious psychological or other barriers. Finally, some sort of cue or stimulus is needed to trigger an action response.

The pharmacist may conceivably perform useful functions in any of these steps: in the first by health education efforts, in the second by providing the convenience factor, and in the third by providing cues. In any case, the pharmacist is likely to be involved because drugs are involved in almost any health problem scenario.

FIGURE 2.1. Health Belief Model for Predicting and Explaining Compliance Behavior

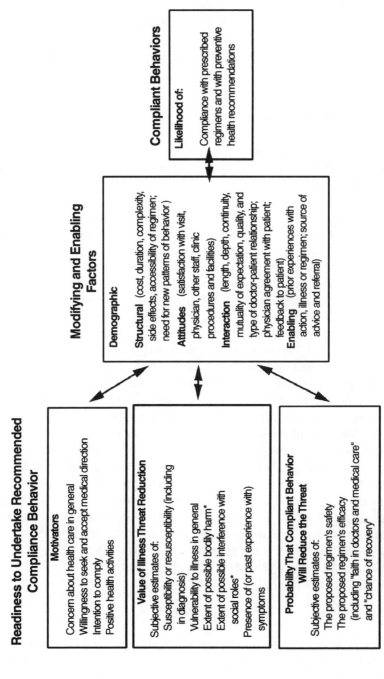

It is important to note that the Health Belief Model is especially applicable to medication compliance (see Chapters 13 and 14).

THE PHARMACIST AND PATIENT BEHAVIOR

It should be apparent from the foregoing discussion that variations in human behavior may be important determinants in whether and how an individual seeks and receives medical care. The stages in health, illness, and sick role behavior offer opportunities and obligations for pharmacy. Some of these are a traditional part of pharmacy practice; others have not been fully delineated. This in no way lessens their potential importance. A few examples may help to illustrate how the pharmacist may best serve the patient.

During the symptom experience stage, the potential patient is quite likely to enter the pharmacist's domain, to obtain a product for relief of one or more symptoms or to receive assistance in interpreting the symptoms, or both. The pharmacist has both the responsibility to offer only those products providing safe and effective symptomatic treatment and the responsibility to be knowledgeable enough to interpret symptoms sufficiently well so that the patient is neither unduly delayed in obtaining the services of a physician nor acts as a drain on our overburdened health care system without need.

As the patient assuming the sick role is expected to try to get better, he may choose self-medication as a means to that end. He should receive guidance in his choice if this is appropriate. The pharmacist may, in fact, be the first one approached in the medical care contact stage because he is readily available. Depending upon the nature of such contact, he may assist in finding access to the system, he may aid in locating the proper medical services, and, most important, he may provide the pharmaceutical services and products that the physician ultimately prescribes.

SUMMARY

Modern medicine understands disease as being specifically related to changes in specific tissues, caused by specific mechanisms, which affect the body in specific ways. The situation usually does not appear as such to the lay person, however, for when he is sick, he feels that something is wrong with him as a whole individual, and his sickness is likely to permeate all that he does and all the ways in which he perceives himself. In the strictest sense, disease is an *objective* phenomenon characterized by altered abnormal functioning of the body as a biological organism. Illness, on the other hand, is a *subjective* phenomenon in which individuals perceive themselves as sick.

Contrasted with disease, which is a sociobiological concept of pathology, illness is a psychosocial phenomenon that is apparent to the individual in terms of an altered state of his perception of himself. Because he feels sick, he may act in ways that are different from those that might normally be expected of him. These feelings of sickness may be related to symptoms of an actual disease, but it is also possible that persons may feel sick without any organic processes of disease being manifested. In this case, the sickness is either an imagined illness or it is a disease that cannot be scientifically validated by the physician. As health professionals, our problem is to discern (1) how the signs or symptoms become labeled or diagnosed as sickness in the first place, (2) how an individual gets to be labeled as sick, and (3) how social behavior is molded by the process of diagnosis and treatment.

How does health differ from illness and disease? If someone is ill, is he or she also sick? What does it mean to be sick? Health and illness may be thought of as endpoints along a continuum. An individual may find himself or herself at different points on the continuum at different times. Physical, mental, and social factors determine whether one is healthy or ill.

The distinction among the terms health, disease, illness, and sickness is facilitated by thinking about a person with diabetes. Is a diabetic ill? While it is true that a diabetic has a disease, he may not consider himself ill. Indeed, many diabetics think of themselves as healthy individuals. Disease is considered a physiological condition by medical sociologists. We think of hypertension, diabetes, asthma, renal impairment, and coronary heart disease (CHD) as diseases. Persons with these diseases may or may not feel ill.

Illness is described by medical sociologists (among others) as a personal perception. That is, you decide if you are ill. You may feel ill in the absence of disease. You may not feel ill in the presence of a severe disease such as bone cancer.

Sickness is a designation that arises from interaction with one's "significant others" (family, friends, peers, etc.). An individual may rise in the morning feeling rested and well, only to learn from those around her during the course of the day that she "looks sick." Significant others may find that we appear sick in the absence of perceived illness on our part.

Physicians are a particularly important source of legitimization of illness, that is, a person may feel ill, but the illness is not perceived or accepted by her significant others. For example, a person feels tired and seems to become exhausted easily. Coworkers may interpret behavior arising from this condition as laziness. Upon going to the physician, the person is found to have a cold. Whether this person actually has a disease (e.g., bacterial or viral infection) is somewhat irrelevant to his personal definition of illness. When the physician prescribes a medication, this person's illness is legitimized because he has been described as sick. Coworkers will likely also find the person sick *at this point* and perhaps instruct him to go home and get some rest.

In this chapter, we discussed the different ways in which a person may behave when he is ill and when he is not ill. The importance of the perception of the individual of his relative illness or wellness was expressed, and application of these behavior patterns in pharmacy practice was made. It should be apparent from the material presented in this chapter that the stages of illness behavior provide a multitude of opportunities for pharmacist activity, many of which have not been adequately exploited to date. The knowledgeable pharmacist who is willing to study the patient as an individual will be better able to fulfill his professional role.

REFERENCES

1. Bezold C, ed. Pharmaceuticals in the year 2000. Washington, DC: Institute for Alternative Futures, 1983.

2. Rosenstock IM. Why people use health services. In: Mainland D, ed. Health services research. New York: Milbank Memorial Fund, 1967:94-127.

3. Sigerist HE. On the sociology of medicine. New York: MD Publications, 1960.

4. Parsons T. Definitions of health and illness in the light of American values and social structure. In: Jaco EG, ed. Patients, physicians and illness. New York: Free Press of Glencoe, 1959:165-87.

5. Strauss AL. Chronic illness and the quality of life. St. Louis, MO: C.V. Mosby, 1975.

6. Twaddle AC, Hessler RM. A sociology of health. 2nd ed. New York: Macmillan, 1987.

7. Verbrugge LM, Ascione FJ. Exploring the iceberg. Med Care 1987;24: 539-69.

ADDITIONAL READINGS

For the Instructor

Although there are literally hundreds of relevant readings available, we recommend as a good review Chapter 25, "Models of Health Related Behavior," by Marshall Becker and Lois Maiman in David Mechanic's *Handbook of Health, Health Care, and the Professions* which–unaccountably–does not include a chapter on pharmacy (Free Press, 1983).

For the Student

1. Twaddle AC. Health decisions and sick role variations: an exploration. J Health Soc Behav 1969;10:105-15.

2. Roghmann KJ, Haggerty RJ. The diary as a research instrument in the study of health and illness behavior: experiences with a random sample of young families. Med Care 1972;10:143-63.

3. Demers RY, Altamore R, Mustin H, Kleinman A, Leonardo D. An explanation of the dimensions of illness behavior. J Fam Pract 1980;11:1085-92.

4. Conrad P. The meaning of medications: another look at compliance. Soc Sci Med 1985;20:29-37.

5. Green LW, Mullen PD, Stainbrok GL. Programs to reduce drug errors in the elderly: direct and indirect evidence from patient education. J Geriatr Drug Ther 1986;1(1):3-18.

6. Cousins N. Anatomy of an illness as perceived by the patient. New York: W. W. Norton Co., 1979.

7. Knauth P. A season in hell. New York: Harper and Row, 1975.

8. Elfenbein D, ed. Living with Prozac. San Francisco, CA: Harper, 1995.

SECTION B:
PATIENT RESPONSES TO SYMPTOMS

Chapter 3

The Meaning
of Signs and Symptoms

Dennis A. Frate
Juliet B. Frate

INTRODUCTION

All of the wide range of research associated with diseases, from epidemiological investigations into disease causation to clinical trials of disease therapy, is conducted to answer one of the two following questions.

1. How does a person get sick?
2. How does a person get well?

In this chapter we will not focus on the first question and develop etiological models of disease causation (e.g., infectious, chronic non-infectious, congenital, genetic, environmental, nutritional, or external).[1] Instead, we will focus on the question of how a person gets well. A discussion on this particular question could encompass such areas as drug development, clinical trials, quality of medical care, and doctor-patient relationships, to name a few. Such topical areas deal with the outcomes of clinical therapy. An examination of therapeutic outcomes is obviously highly relevant to any discussion on disease. In this chapter, however, the focus is not on the outcomes of medical treatment but rather on the process that leads to a patient receiving that treatment. Numerous questions can be raised when examining this process. Who seeks medical care? When do individuals seek medical care? Where do individuals seek medical care?

To examine these questions, we must begin with two basic premises. First, not everyone responds similarly to the signs and symptoms of a given illness. Zola's and Zborowski's classic works clearly demonstrate both the variation in symptom recognition and the variation in response to the universal phenomenon of pain.[2,3] Individuals differentially internalize and recognize symptoms of a given disease. In fact, one focus of health education is to create a population knowledgeable in disease recognition (e.g., the warning signs of cancer). Second, once a sign or symptom is recognized, not everyone perceives and uses the medical care system in the same fashion.[4] Self-treatment, physician-directed therapy, and no overt therapeutic response are but three of the numerous options open to an individual responding to disease symptoms. Different people seek different points of entry into the health care system in response to the same symptom. For example, some segments of the U.S. population rely on self-treatment for minor illnesses, while other segments of the population immediately enter the formal health care system for similar illnesses. A sore throat is a relatively definable condition. Some individuals experiencing a sore throat may self-treat with a home remedy or purchase an over-the-counter medication. Others may consult a pharmacist as the most efficacious therapeutic path. Still others may immediately seek treatment from their family physician.

We can interpret the signs and symptoms of an illness as merely biological phenomena. The physiological or nonbehavioral paradigm for signs and symptoms is reflected in its definition: an observable physical phenomenon and an organic or physiologic manifestation of disease.[5] However, a strong behavioral component exists in the recognition that a particular sign or symptom is disease related, the expression of that disease sign or symptom, and the health care response to that disease sign or symptom. Therefore, when examining the initial course of a disease, especially the seeking of therapy, we must recognize the importance of behavioral or nonbiological factors that influence care-seeking decisions. The behavioral aspects of care seeking and response to disease states have long been recognized by social scientists.[6,7] Numerous independent variables have repeatedly been shown to influence utilization of medical services, including age, gender, race, socioeconomic status, and type of cost reimbursements. For example, do females utilize the medical care system more than males? Do whites in this country visit a physician more frequently than African Americans? Do poor people tend to underutilize the medical care system?

Relatively elaborate theoretical models combining a wide range of independent variables have been developed to explain and predict the behaviors of individuals responding to a disease. Two of the more popular models used to explain health behavior are the Health Belief Model and the Health Locus of Control Model.[8,9] These theoretical models attempt, each in their own way, to examine the process of seeking therapy for a disease or the decision process used to maintain wellness. The goals of these models are to explain who seeks care and who does not, when they seek care and when they do not, and where they seek care and where they do not. While useful for purposes of discussion, such theoretical models tend to be heuristic in nature, i.e., merely a method of facilitating the description of a given complex phenomenon, in this case health behavior. A detailed discussion on such theoretical models is not the intent of this chapter, as sufficient work has already been done on the relative utility of these and other models defining the behaviors associated with response to diseases.[10,11] Nevertheless, a brief review is necessary to provide the reader with enough familiarity to recognize the importance of these heuristic models of health behavior for clinical practice.

Health Belief Model

The Health Belief Model grew out of psychological research conducted in the 1950s. It was first developed to help explain the preventive health practices of individuals, specifically the use of screenings for tuberculosis detection. The distinct beliefs expressed by an individual regarding susceptibility to a disease, perceived severity of the disease, and perceptions of the availability of both detection and therapy for the disease were all felt to influence the adoption of a particular health behavior, in this case disease detection.[8] Since its inception, the Health Belief Model has also been

applied to general health care seeking and compliance with therapeutic regimens, as well as other areas in health with distinct behavioral connections. The goal of such a model is not only to understand the health care action taken by an individual but also to
predict the likelihood of a particular action occurring.

For example, the topic of breast cancer screening has recently
been investigated to predict participation and nonparticipation.[12] In
this case, if a woman has a family history of breast cancer, she may
feel susceptible to the disease and thus have a mammogram performed. If a woman believes that undiagnosed breast cancer would
be fatal, she may respond to the severity of the disease and have a
mammogram done. Finally, a woman may opt for a mammogram if
she thinks that early detection of and therapy for breast cancer
would increase the likelihood of her survival. Studies examining the
health belief models of women would investigate which factors
predict those who would or would not participate in a breast cancer
screening activity. The results of such a study could then direct
health education efforts into particular behavioral areas such as
those mentioned above.

Studies focused on the role of health beliefs in predicting a particular health behavior or action are usually conducted on large
population samples. Such large-scale studies tend to be more generalizable in their findings than studies conducted on a few selected
individuals. The Health Belief Model is actually measured by a
survey instrument. Examples of the questions comprising this
model include:

1. Vulnerability (How likely are you to get the following illness symptoms?)

Illness Symptom	Not Likely						Very Likely
Fever	1	2	3	4	5	6	7
Sore throat	1	2	3	4	5	6	7
Blood in urine	1	2	3	4	5	6	7
Back pain	1	2	3	4	5	6	7
Chest pain	1	2	3	4	5	6	7

•
•

2. Need for Medical Attention (How likely is it that you will need to get medical attention for the following illness symptoms?)

Illness Symptom	Does Not Require Medical Attention				Requires Immediate Medical Attention
Fever	1	2	3	4	5
Sore throat	1	2	3	4	5
Blood in urine	1	2	3	4	5
Back pain	1	2	3	4	5
Chest pain	1	2	3	4	5
•					
•					

3. Consequences of No Medical Attention (If you do not get any medical attention for the following illness symptoms, what will happen?)

Illness Symptom	Symptoms Will Go Away		Symptoms Will Get Worse
Fever	1	2	3
Sore throat	1	2	3
Blood in urine	1	2	3
Back pain	1	2	3
Chest pain	1	2	3
•			
•			

4. Perceived Benefit of Medical Care (If you do get medical care for the following illness symptoms, how much improvement will there be?)

Illness Symptom	No Improvement				Much Improvement
Fever	1	2	3	4	5
Sore throat	1	2	3	4	5
Blood in urine	1	2	3	4	5
Back pain	1	2	3	4	5
Chest pain	1	2	3	4	5
•					
•					

5. Health Maintenance Index (How much protection against illness will you
    get if you . . . ?)

Activity	No Protection					Complete Protection
Exercise	1	2	3	4	5	6
Medical checkups	1	2	3	4	5	6
Diet	1	2	3	4	5	6
Sleep more	1	2	3	4	5	6
Take vitamins	1	2	3	4	5	6

•
•

Results from such surveys can help health professionals and health care organizations target specific groups of individuals whose belief systems exclude them from seeking medical care and/or from modifying specific high-risk behaviors. For example, if one particular group of individuals–white males–felt that chest pain did not require medical attention, resources could be shifted to target this population about the symptomatology of angina pectoris. This particular belief held by white males concerning chest pain may also vary by age, socioeconomic status, and/or region of the country. Consequently, the more we know about who has a particular belief, the more targeted our action can be.

Another example may help clarify the utility of this theoretical model. Twenty years ago, a large proportion of individuals in this country with hypertension or high blood pressure were undiagnosed.[13] Part of the reason for this phenomenon was the general belief held by specific segments of the population that the consequences of untreated hypertension did not warrant detection and therapy. A concerted effort was launched on all organizational levels of the health care system to alter this belief. Massive health education campaigns that focused on the possible end organ damage resulting from untreated hypertension were begun. Screening, detection, treatment, and control rates for hypertension changed rapidly as a result of these educational activities.

Health Locus of Control Model

The other theoretical model widely used to help explain health behavior is the Health Locus of Control Model. This model is centered on the premise that an individual's behaviors or actions are associated with that individual's views and the degree to which free will or relative control is involved in the individual's daily activities. Health behavior is one such activity. Simply put, a person's perception may be that he or she controls all aspects of personal behavior, including health; health, in this case, is perceived to be internally controlled. On the other hand, an individual may perceive that personal health is directed from the outside, by a physician for example, and is therefore beyond personal control; health, here, is perceived to be externally controlled.[14] Many factors may influence whether health is viewed as internally or externally controlled. Previous illness experience, religious beliefs (spirituality), educational level, and economic status are but a few of the factors that can relate to a person's health locus of control. As with the Health Belief Model, a person's health locus of control can influence whether that person seeks medical care, complies with treatment regimens, and/or alters high-risk life-styles. In addition, the Health Locus of Control Model is usually administered to large population samples and is measured by a survey instrument. The following are a few examples of the questions used to determine the health locus of control of an individual:

Statement	Strongly Disagree					Strongly Agree
1. If I get sick it's my own behavior which determines how soon I get well again.	1	2	3	4	5	6
2. No matter what I do, if I am going to get sick, I will get sick.	1	2	3	4	5	6
3. Having regular contact with my physician is the best way for me to avoid illness.	1	2	3	4	5	6
4. Most things that affect my health happen to me by accident.	1	2	3	4	5	6

Again, the results of such data can help explain patterns of health care utilization and also provide health professionals and health organizations with the information necessary to target those specific groups in the population that are under- or overutilizing the health care system. An example of this can be found in our research conducted in central Mississippi. The strong religious beliefs and sacred practices of the resident population indicated an external health locus of control. Data on the diagnostic and therapeutic status of hypertension in this population also indicated that an intervention to manage those with diagnosed high blood pressure would have a greater health impact than the implementation of detection activities in the community. As a result, hypertension control efforts were implemented in both African-American and white community churches.[15] The use of the church setting for health activities utilized the influence of religious beliefs to affect high blood pressure management activities in a positive manner.

In summary, why, when, and where individuals seek health care are very important questions that need to be addressed by all health care professionals, from practitioners to researchers to academics. The need for such information is clear to the behavioral scientist involved in health care research. This chapter, however, will demonstrate the utility of such data for the health practitioner as well. Such data will be beneficial for understanding the efficacious use of medical care services and for the reorganization of this delivery system as we enter the twenty-first century.[16] As will be shown in the following section, individuals respond differently to the signs and symptoms of a given illness. An understanding of why this occurs may not necessarily be important for daily practice in a clinical setting. The independent variables listed previously which relate to health behaviors (e.g., age, race, gender, and socioeconomic status) are ample for theoretical debate as to why differences occur in symptom response. What is important for clinical practice is the recognition of the existence of this phenomenon and of the need to understand its occurrence within the practice setting. No population is homogeneous in its response to illness. Sensitivity to this factor can lead to more appropriate and efficacious delivery of health care services.

EMPIRICAL DATA

Three different data sets will be presented in this section. Each of these data sets will broadly address questions centering on health care-seeking and patient response to symptoms. The objective of presenting these data is not to learn about the particular population studied; a population-specific interpretation is not the intent of this chapter. Rather, the objective is to document that the perceptions of symptoms and actual symptom responses by patients are not homogeneous. All of these data were collected by the Rural Health Research Program, Research Institute of Pharmaceutical Sciences, University of Mississippi, between 1981 and 1990 as part of larger research efforts focused on both basic health research and evaluations of chronic disease management and prevention efforts. Random sampling techniques were used for selection of study participants. Response rates varied but were never lower than 90%. These research activities were supported by two separate grants from the National Institutes of Health (grant numbers HL31370 and HL39200).

The data for these studies were collected on a biracial population in rural central Mississippi. Although a detailed characterization of this population is not essential for the purposes of this presentation, a brief description will make the data appear more real to the reader. The sampling area in central Mississippi covered over 3,000 square miles and was composed of five counties. According to the 1990 U.S. Census, over 83,000 individuals resided in the study area which can be characterized as biracial (approximately 50% were African American), rural (over 68% of the population resided in rural settings), undereducated (approximately 47% of persons 25 years of age and older had less than 12 years of education), and poor (approximately 36% of all residents lived below the poverty level).[17] Illustrating health status, this area had one of the highest rates of essential hypertension and one of the highest rates of obesity ever documented in a defined African-American population in this country.[18] The 1990 death rates for diseases of the heart (defined as ICD Codes 390-398, 402, 404, 410-429) were officially reported as 393.9 per 100,000 population.[19] Nationally, the mortality rate for diseases of the heart (defined comparatively) was 155.9 per 100,000 population.[20]

Alternative care has recently been receiving attention in both the scientific and popular literature.[21] Alternative care refers to treatment other than the more typical response of using the mainstream, primary care system and includes self-treatment.[22] The first data set illustrating health behaviors is from a large study that investigated health care seeking response ($n = 267$). These particular data explore whether individuals self-treat illness episodes. Broadly speaking, these data address the question of who seeks formal medical care and who does not. Such information has broad implications for health care practitioners in understanding the role of self-treatment for illness in any population.

Table 3.1 illustrates the use by study participants, both historically (ever) and for their last illness, of medical treatments other than those provided by a physician and including self-treatment. The term "illness" was intentionally left undefined here. Historically, nine out of ten individuals indicated that they had sought some type of nonphysician medical care, including self-treatment. This pattern was widespread, with no significant differences found by race (African American and white) or by gender. Self-treatment as the sole option of care for their last illness was used by three out

TABLE 3.1. Prevalence of Self-Care Without Seeing Physician, by Race and Gender (percentage "yes")

Self Care	African American	White	Male	Female	Total
Have you ever sought other medical treatment or taken care of yourself at home when you were sick instead of going to a medical doctor?	88.1	94.4	91.1	90.2	90.7
Did you take care of your last illness by yourself without seeking a medical doctor?[1]	16.4	48.1	33.1	25.9	29.4
[1]Significant difference by race ($p < 0.05$)					

of ten individuals. Here, a significantly larger proportion of whites than African Americans opted for self-treatment of their last illness.

Table 3.2 presents more detailed information on self-treatment. Focusing here on only home remedies, the data address whether study participants ever used a home remedy, whether it was effective, how they learned about it, how often they used such remedies, and whether they were currently using a home remedy. Again, overall use patterns were very high in that eight out of ten individu-

TABLE 3.2. Home Remedy Usage, Outcome, and Source of Information, by Race and Gender (percentage)

Home Remedy	African American	White	Male	Female	Total
Have you ever used a home remedy when sick? (yes)	84.3	84.3	85.5	83.2	84.4
Did you get well from your sickness after you used a home remedy? (yes)[1]	90.2	92.3	90.6	91.5	91.2
How did you learn about the last home remedy used?					
family or self	88.9	69.6	83.7	78.5	81.1
friend or neighbor	6.8	8.5	6.5	8.4	7.5
pharmacist	0.0	3.7	2.2	0.9	1.5
physician	0.0	9.8	0.0	7.5	4.0
television	2.6	4.9	4.3	2.8	3.5
other	1.8	3.6	3.3	1.8	2.5
How often do you use home remedies for illness?[1]					
often	14.3	22.0	14.2	20.3	17.3
sometimes	23.3	35.2	27.4	28.8	28.3
seldom	62.4	42.9	58.5	50.8	54.4
Are you presently using a home remedy? (yes)[1]	8.2	19.8	12.3	13.4	12.8

[1]Significant difference by race ($p < 0.05$)

als indicated historical use of a home remedy. This use pattern was similar by race and gender. Of those who used a home remedy, over nine out of ten individuals found it to be an effective therapy. This did vary by race in that a significantly greater proportion of whites than African Americans experienced a positive health outcome. The vast majority of these individuals, over 80%, learned about the home remedy from a relative or through their own experiential base. Proportionately, the lowest point of origin for information about a home remedy was a pharmacist, 1.5%. The sources of knowledge about the home remedy did not vary by race or gender. Of those who used home remedies, over 17% used them often, while over 50% seldom used them. In this case, a significantly larger proportion of African Americans than whites reported that they seldom used a home remedy. Almost 13% of the population were currently using a home remedy. Here, consistent with overall self-treatment, a significantly larger proportion of whites than African Americans were currently using a home remedy.

As shown in Table 3.3, almost the entire population surveyed–98.9%–had used an over-the-counter (OTC) medication to treat an illness at some point in their past. The distribution of use was similar for both races and genders. A vast majority–98.5%–of the individuals who used an OTC medication reported that it was effective in treating their illness. This high rate of the perceived effectiveness of the OTC medication is of special interest. The main sources of knowledge about the OTC medication were relatives or individual's own experiential base. Proportionately, the pharmacist was again the lowest point of origin for information about this particular type of self-treatment. This distribution was similar by race and gender. Frequency of OTC medication use was fairly evenly distributed in this population but did vary significantly by race. Finally, about one-fourth of all individuals were currently using an OTC medication. As with overall self-treatment and home remedy use, this distribution of current OTC medication use was significantly related to race in that over 40% of whites and only about 10% of African Americans were currently using an OTC medication.

In summary, a vast majority of respondents reported using non-physician-based medical care. Both the use of home remedies and

TABLE 3.3. OTC Usage, Outcome, and Source of Information, by Race and Gender (percentage)

OTC	African American	White	Male	Female	Total
Have you ever used over-the-counter or store-bought medicine (OTC) when you were sick? (yes)	98.7	99.1	98.4	99.3	98.9
Did you get well from your sickness after you used the OTCs? (yes)	97.5	100.0	99.2	97.9	98.5
How did you learn about the last OTC used?					
family or self	35.3	44.3	47.1	31.9	39.0
friend or neighbor	20.3	7.5	15.7	14.5	14.9
pharmacist	3.3	3.8	4.1	2.9	3.4
physician	16.3	15.1	13.2	18.1	15.7
television	21.6	21.7	16.5	26.1	21.8
other	3.3	7.5	3.3	6.5	4.9
How often do you use OTCs for illness?[1]					
often	27.6	42.5	36.9	30.7	33.3
sometimes	48.7	32.1	37.7	45.7	42.4
seldom	23.7	25.5	25.4	23.6	24.2
Are you presently taking an OTC medicine? (yes)[1]	10.2	43.9	23.0	24.6	23.7
[1]Significant difference by race ($p < 0.05$)					

the use of OTC medications were reported by a majority of the respondents. These findings indicate that the patient's view of the therapeutic system is not limited to physician care in that other options or alternatives are available and used.

Symptom- or disease-specific health-care-seeking response provides insight into how any population views the relative severity of the disease itself and the role of specific healers or health practitioners in this society for dealing with that particular disease. As part of a larger research activity conducted in central Mississippi ($n = 1,738$), a series of questions were asked about where in the health care system people would go when experiencing a particular symptom or disease. Symptoms/diseases ranged in relative severity from a backache to chest pain. Health-care-seeking responses ranged from "do nothing," or inaction, to "see a physician."

As shown in Table 3.4, six out of ten individuals would see a physician in response to a backache, over 25% would self-treat, and only 5.8% would consult another health professional such as a pharmacist. The distribution of treatment varied significantly by race, with a greater proportion of African Americans reporting that they would self-treat or see a physician and a greater proportion of whites reporting that they would do nothing or see another health provider. Response to a backache was not significantly distributed by gender.

The health-care-seeking response to a head cold showed a different distribution. In this case, self-treatment had the largest response in that seven out of ten individuals would elect to self-treat a head cold. The distribution of the health care response to a head cold was significantly associated with race. Here, a larger proportion of whites would not seek any health care, while a larger proportion of African Americans would see a physician in response to this disease. There were no significant differences by gender. The overall proportion of individuals who would seek help from another health professional for a head cold was small: 3.9%.

About two-thirds of the study population indicated that they would self-treat a sore throat. In the case of this symptom, an additional 26.3% would seek help from a physician, and only 4.4% would seek advice from another health professional. There were significant differences in health-care-seeking response to a sore throat by both race and gender. A greater proportion of African Americans would see a physician for sore throat, whereas a greater proportion of whites would do nothing, self-treat, or see another health care provider such as a pharmacist. In addition, a greater proportion of women would see a physician for sore throat, while a greater proportion of men would self-treat.

Generally speaking, chest pain presents a more severe clinical symptom. In response to this symptom, a vast majority of individu-

TABLE 3.4. Symptom Response, by Race and Gender (percentage)

Symptom/Response	African American	White	Male	Female	Total
Backache[1]					
do nothing	0.9	8.6	5.6	3.6	4.3
treat self	27.8	22.8	26.3	25.2	25.6
talk to relative/friend	1.0	2.7	1.6	1.8	1.8
see physician	66.9	56.8	60.6	63.6	62.5
see other health professional	3.3	9.0	5.9	5.8	5.8
Head cold[1]					
do nothing	0.8	4.2	2.7	2.1	2.3
treat self	69.7	69.3	70.5	68.9	69.5
talk to relative/friend	0.8	0.8	1.0	0.7	0.8
see physician	25.4	21.0	21.6	24.6	23.5
see other health professional	3.2	4.8	4.3	3.7	3.9
Sore throat[1,2]					
do nothing	0.6	2.3	1.9	1.0	1.3
treat self	66.2	68.0	69.9	65.4	67.0
talk to relative/friend	0.9	0.9	1.1	0.8	0.9
see physician	28.7	23.3	22.4	28.5	26.3
see other health professional	3.6	5.6	4.7	4.3	4.4
Chest pain[1]					
do nothing	0.2	1.5	1.0	0.6	0.8
treat self	5.2	2.4	3.7	4.1	4.0
talk to relative/friend	0.9	0.8	1.0	0.8	0.9
see physician	93.6	94.9	93.9	94.3	94.2
see other health professional	0.0	0.5	0.5	0.1	0.2
Skin rash[1]					
do nothing	0.6	3.7	2.0	1.9	2.0
treat self	18.8	29.5	24.6	22.8	23.4
talk to relative/friend	1.3	0.8	1.3	0.9	1.1
see physician	76.3	60.1	69.0	69.4	69.2
see other health professional	3.0	6.0	3.1	5.0	4.3
High temperature[1]					
do nothing	0.2	2.0	1.1	0.9	1.0
treat self	12.4	35.7	25.2	21.2	22.7
talk to relative/friend	0.5	0.7	1.0	0.4	0.6
see physician	84.8	60.1	70.3	76.0	73.9
see other health professional	2.0	1.6	2.4	1.5	1.8
Exhaustion[1]					
do nothing	3.7	9.1	7.0	5.5	6.0
treat self	34.3	53.1	42.3	42.7	42.6
talk to relative/friend	1.1	0.9	1.0	1.0	1.0
see physician	60.3	36.4	49.0	50.2	49.8
see other health professional	0.7	0.5	0.8	0.6	0.7

[1]Significant difference by race ($p < 0.05$)
[2]Significant difference by gender ($p < 0.05$)

als surveyed–over 90%–would seek out a physician for care. Specifically, almost 95% of whites and almost 94% of African Americans would see a physician for chest pain. Thus, a very small proportion of the population would self-treat this symptom. This distribution was significant by race, with a larger proportion of African Americans (5.2%) than whites (2.4%) who would self-treat chest pain. There was no significant difference in the distribution of health care-seeking response to chest pain by gender. Of particular interest here, is that only 0.2% of the individuals surveyed would seek care from another health care professional for chest pain.

Care-seeking response to skin rash indicated that almost 25% of those surveyed would self-treat that symptom, while over two-thirds (69.2%) would seek a physician's care. Approximately 4% would see another health care provider for that clinical problem. The therapeutic response for skin rash was significantly associated with race in that a much larger proportion of African Americans would seek a physician's care, while a much larger proportion of whites would self-treat. The distribution of health care responses to skin rash was not related to gender.

Almost three out of four individuals surveyed would seek physician-directed therapy for a high temperature or fever. However, almost 25% would self-treat. These two particular health care responses to high temperature account for over 96% of the individuals surveyed; only 1.8% would seek care from another health care professional, such as a pharmacist, for a fever. This response distribution was significantly associated with race, as a much smaller proportion of African Americans would self-treat a high temperature; correspondingly, a much larger proportion of this cohort would seek a physician's care.

Exhaustion as a clinical symptom was also listed in this study. Interestingly, 42.6% of all respondents would self-treat exhaustion, while a similar proportion, 49.8%, would seek physician care. Again, only a very small proportion would consult another health care provider if they experienced exhaustion. This distribution was significantly associated with the race of the respondents in that 34.3% of African Americans would opt for self-treatment and 60.3% would select a physician for their treatment path. These proportions were reversed for whites in that 53.1% would self-treat and 36.4% would see a physician. The distribution of care-seeking responses was not significantly associated with gender.

In summary, possibly based on perceived severity of and/or past experience with the symptom and/or related disease, anticipated response to the health care system varied widely. Self-treatment appears to be a primary response for individuals experiencing common, less severe symptoms such as a head cold. However, the separate symptoms of chest pain and high temperature warranted a response involving entrance into the more formal mainstream health care system, i.e., a physician's care. In all cases, care-seeking responses to a symptom and/or disease varied by race. For six out of the seven symptoms listed, a higher proportion of African Americans than whites would select entry into the formal health care system. Self-treatment as the preferred therapeutic path appears somewhat equally split by race. Gender was only a significant factor for one particular symptom, sore throat. Here, a greater proportion of males would self-treat, while a greater proportion of females would seek a physician's care. Seeking out another health care professional such as a pharmacist for medical care advice was not a major option or path selected by any group surveyed for any symptom and/or disease.

The final data set examines the health care-seeking response of a rural, biracial population to a hypothetical coronary heart disease event ($n = 313$). The intent of this particular inquiry was to determine which of the multiple pathways into the health care system would be taken by people who thought they were having a heart attack (coronary heart disease event). In this case, the question was not intended to measure symptom recognition. Rather, participants were told that they had just experienced a heart attack and were asked to state their anticipated health-care-seeking response. The various pathways reported included not only entry into the formal health care system (call a physician, go to a hospital, call an ambulance, and/or call 911) but also do nothing (wait), self-treat (take nitroglycerin), or contact a person without a medical background (call a family member).

As seen in Table 3.5, almost seven out of ten respondents would call a physician in response to a heart attack or coronary heart disease episode. A significantly larger proportion of whites than

TABLE 3.5. Perceived Response to CHD, by Race and Gender (percentage)

Response	African American	White	Male	Female	Total
Call physican[1]	50.9	89.7	71.4	66.7	68.9
Wait[1]	7.8	17.4	12.2	12.2	12.2
Go to hospital	7.2	7.6	4.8	9.8	7.4
Call ambulance	3.0	4.2	5.4	1.8	3.5
Call family[1]	47.6	33.1	36.5	44.9	40.9
Call 911	2.4	2.1	1.4	3.1	2.3
Take nitro	0.0	0.7	0.0	0.6	0.3
Do not know[1]	1.8	6.9	3.4	4.9	4.2
Other[1]	8.4	2.8	6.8	4.9	5.8
[1]Significant difference by race ($p < 0.05$)					

African Americans would call a physician, 89.7% and 50.9%, respectively. However, only a small proportion of the population surveyed reported other possible responses within the formal medical care system. For example, only 7.4% would go directly to a hospital, 3.5% would call an ambulance, and 2.3% would call 911. None of these choices was significantly different by race. In addition, the gender of the respondents was not statistically associated with any formal medical care response.

The next most frequently mentioned care-seeking response to a heart attack was contacting a family member, 40.9%. In this case, a significantly greater proportion of African Americans than whites gave this anticipated response, 47.6% and 33.1%, respectively. There was no difference by gender. Inaction or waiting as a response to a coronary heart disease event was mentioned by 12.2% of the population surveyed. A significantly greater proportion of whites than African Americans listed this response, 17.4% and 7.8%, respectively. Again, there was not a significant difference by gender. Over 4% of the respondents did not know what they would do in response to a heart attack. As with the decision to wait, a greater proportion of whites than African Americans did not know what they would do if they thought they were having a heart attack.

In summary, a larger proportion of whites than African Americans would call a physician in response to a heart attack. However, a larger proportion of whites also reported that they would wait and/or did know what they would do in this situation. Although multiple care-seeking responses were solicited and reported, the initial decision to wait could affect disease prognosis. In addition, not knowing what one would do in response to a heart attack could affect disease prognosis. On the other hand, a larger proportion of African Americans than whites would call a family member as one response to a coronary heart disease episode. This could also delay entry into the formal medical care system and thereby affect survival rates. An understanding of African-American culture provides some insight into this response. Social research on African Americans has clearly shown their familial orientation or response to crisis situations.[23] This familial response is evident not only with health problems but in the economic, social, and political spheres of life as well. Such a familial orientation is not as pronounced in the white community.

These different pathways to therapy for a heart attack provide insight into the development of health education campaigns targeted to defined populations. The issue of the familial response to a coronary heart disease episode, as well as other medical crises, could be addressed in a health education effort directed at the African-American community. The issue of waiting or not knowing what to do in response to a heart attack could also be addressed in a health education campaign directed at the white community. In addition, health care professionals should be made aware that an encounter with a patient may not necessarily involve the presentation of coronary heart disease symptoms, even though clinically present. Consequently, they should inquire about such symptoms for possible further clinical evaluation. As shown, the health-care-seeking response to a heart attack is not uniform both within and across populations.

DISCUSSION

A discussion section is generally devoted to reviewing the results covered in the previous section. Such a treatise is not necessary here, however, in that the intent of this chapter is not to understand the health behaviors of a particular population. That task will be left to future publications looking at culture-specific health-care-seeking responses to disease symptoms encountered by rural, Southern populations. As stated earlier, the focus of this chapter is to demonstrate empirically that responses to disease symptoms are not uniform in any given population. This is especially the case with the population described here.

There is little debate that although disease is a biological phenomenon, the response to a disease state and/or symptom by an individual has strong behavioral connections. However, when behavioral scientists investigate health behaviors, the focus is not on the individual patient but on a defined population. This approach is one reason that it is not always easy to demonstrate the utility of such data to the health practitioner. The practitioner treats the individual, not a population. Consequently, when the behavioral scientist presents probabilistic models of behavior–i.e., a given proportion of a defined population will respond in a particular way to a given symptom–the lack of total uniformity in response is disheartening to the practitioner. How, then, can the practitioner determine a particular patient's behavior based on the results of the Health Belief or Health Locus of Control theoretical models? The answer, of course, is that he or she cannot. Nevertheless, he or she can realize that any given patient is not necessarily responding merely to the biological impetus of a disease. Other factors are involved in the patient's recognition of disease, his or her decision to seek care, and the kind of care to seek. These factors were illustrated in the data sets presented.

For example, the data demonstrating that almost one-third of the study participants relied on self-treatment for their last illness indicate that prior instances of self-treatment should be questioned by the health practitioner. Self-treatment can involve both home remedies and/or the use of over-the-counter (OTC) medications. Some of these therapies could result in contraindications with prescribed medications. Therefore, for the health professional prescribing medications, knowledge of the patient's self-treatment practices may be an important addition to the medical history. Also, for the pharmacist filling a prescription or selling an OTC medication, inquiring about symptom presentation and questioning prior self-treatment may be a valuable guide. Such questioning could result in a change of the prescription or the OTC medication or a recommended change in the self-treatment practice. Consequently, the practitioner should never assume that any particular point in the health care system is the initial pathway taken.

The data indicating that different paths are selected by different patients for a given symptom or disease also provide other important information for the health practitioner. Why are certain pathways viewed as more efficacious in the therapeutic process? Why is self-treatment recognized as one such therapeutic option? Do the patient's perceptions of particular health care professionals need to be altered? In the population studied here, there appear to be two main health care responses: physician care and self-treatment. Other pathways play only a minor role. Specifically, compared to the physician, the pharmacist is viewed as only a minor player in the therapeutic arena. A concept of pharmacy care or a patient advocate role for pharmacists does not necessarily reflect the reality of current patient perceptions.

The data examining the health care seeking response to a specific disease (heart attack) again indicate a lack of uniformity in symptom recognition and perception of the health care system. A multitude of therapeutic pathways were selected despite the fact that the prognosis for heart attack is so dependent on response time and immediate entrance into the formal medical care system. These types of data have obvious use for the development and direction of health education activities. For example, awareness of specific self-treatment practices may indicate whether a patient is experiencing heart disease symptoms. In addition, patients at known risk for development of a coronary heart disease episode may be targeted on the more appropriate immediate response necessitated by the symptoms of a heart attack.

As seen in the empirical data presented, gender did not play a major role in predicting health care response. This finding could merely reflect the behavior of this population only, or it could indicate that gender plays a more complex role not definable by the basic statistical analysis conducted here. Race, on the other hand, was an identifiable behavioral variable. As shown, race was significantly involved in the use of self-treatment, selection of health care seeking response, and response to a coronary heart disease episode. Other studies have repeatedly shown that the variable of race is predictive of numerous other behaviors. For example, race as an independent variable could predict access to medical care services or be related to a cultural interpretation of disease causation and the therapeutic environment. At the very least, race as a distinct variable could be used by health professionals as an identifier indicating that different therapeutic paths may be selected for the same disease and/or symptom.

SUMMARY

An individual's response to a disease is tied to both the behavioral realm and the disease's biological manifestation. Health practitioners can better understand the patients they serve as well as the patients who exclude them from the health care system if they recognize that behavior plays a significant role in the therapeutic process. Disease therapy begins with illness recognition and proceeds to health care seeking. Acceptance of the role of behavior in both illness recognition and health care seeking can improve the practice of any health professional who deals with individual patients. On the aggregate level, such recognition can enable health care professionals or organizations to target more effectively those groups not efficaciously utilizing the health care system. Thus, understanding the role of behavior in health care seeking should have a direct impact on the quality of care delivered.

As stated earlier, individuals conducting health behavior research generally focus their efforts on predicting the particular action of a patient or explaining why a particular action occurred. The intent of this chapter, however, was not to develop or test any models predicting health behavior or explaining health care utilization. Rather, the focus of this chapter was to demonstrate, empirically, to the health professional the two premises previously stated. As the data indicated, individuals do respond differently to the signs and symptoms of disease. In addition, populations do not necessarily manifest homogeneous perceptions of the health care system; numerous pathways from inaction to physician care are taken when seeking clinical therapy.

Will we ever be able to achieve more complete uniformity in disease-specific health care seeking response and thus increase the effectiveness of health care utilization? As long as populations themselves vary, behavioral responses to disease will continue to reflect that heterogeneity. Systematic documentation of such population-specific health behaviors can guide the development of intervention efforts which, in turn, should have a positive impact on the utilization of health care services. Ultimately, such effective utilization should result in more positive health outcomes.

REFERENCES

1. Mausner JS, Kramer S. Epidemiology–an introductory text. Philadelphia, PA: WB Saunders, 1985.

2. Zola IK. Culture and symptoms: an analysis of patients presenting complaints. Am Sociol Rev 1966;31:615-30.

3. Zborowski M. People in pain. San Francisco: Jossey-Bass, 1969.

4. Twaddle AC. Health decision and sick role variations: an exploration. J Health Soc Behav 1969;10:105-14.

5. Dorland's pocket medical dictionary. 21st ed. Philadelphia, PA: WB Saunders, 1968.

6. Parsons T. The social system. New York: Free Press, 1951.

7. Twaddle AC. Sickness and the sickness career: some implications. In: Eisenberg L, Kleiman A, eds. The relevance of social science for medicine. Dordrecht, Holland: D. Reidel, 1980:111-33.

8. Rosenstock I. Historical origins of the Health Belief Model. Health Educ Monogr 1974;2:328-35.

9. Lau RR. Beliefs about control and health behavior. In: Gochman DS, ed. Health behavior: emerging research perspectives. New York: Plenum Press, 1988:43-63.

10. Fishbein M, Ajzen I. Belief, attitude, intention, and behavior: an introduction to theory and research. Reading, MA: Addison-Wesley, 1975.

11. Kirscht J. Preventive health behavior: a review of research and issues. Health Psychol 1983;2:277-301.

12. Bryant H, Mah Z. Breast cancer screening attitudes and behaviors of rural and urban women. Prev Med 1992;21:405-18.

13. Rowland M, Roberts J. Blood pressure levels and hypertension in persons aged 6-74 years: United States, 1976-1980. Advanced Data from Vital and Health Statistics, No. 84. DHHS Pub. No. (PHS) 1982:82-1250.

14. Wallston BS, Wallston KA. Locus of control and health: a review of the literature. Health Educ Monogr 1978;6:107-17.

15. Whitehead TL, Frate DA, Johnson SA. Control of high blood pressure from two community-based perspectives. Hum Organiz 1984;43:163-7.

16. White House Domestic Policy Council. Health security. The President's report to the American people. October, 1993.

17. Bureau of the Census. Census of population and housing, 1990: summary tape files 1A and 3A (Mississippi). Washington, DC: Department of Commerce, 1991.

18. Storer JH, Frate DA. Hunger, poverty, and malnutrition in rural Mississippi: developing culturally sensitive nutritional interventions. Hum Serv Rural Environment 1990;14(1):25-30.

19. Mississippi State Department of Health. Vital statistics, Mississippi, 1990. Jackson, MS: 1991.

20. National Center for Health Statistics. Health, United States, 1991. Hyattsville, MD: Public Health Service, 1992.

21. Eisenberg DM, Kessler RC, Foster C, Norlock FE, Calkins DR, Delbanco TL. Unconventional medicine in the United States: prevalence, costs, and patterns of use. N Engl J Med 1993;328:245-52.

22. Banahan BF, Frate DA. Use of home remedies and over-the-counter products among rural residents at risk for development of coronary heart disease. American Pharmaceutical Association, San Diego, CA, 1992.

23. Shimkin DM, Shimkin EM, Frate DA, eds. The extended family in black societies. The Hague, Netherlands: Mouton, 1978.

ADDITIONAL READINGS

1. Becker M, ed. The Health Belief Model and personal health behavior. Thorofare, NJ: Charles B. Slack, 1974.

2. Currer C, Stacey M, eds. Concepts of health, illness and disease: a comparative perspective. Providence, RI: Berg, 1993.

3. Eisenberg L, Kleinman A, eds. The relevance of social science for medicine. Vol. 1. Boston: D. Reidel, 1981.

4. Glanz K, Lewis FM, Rimer BK, eds. Health behavior and health education. San Francisco: Jossey-Bass, 1990.

5. Gochman DS. Health behavior: emerging research perspectives. New York: Plenum Press, 1988.

6. Johnson TM, Sargent CF, eds. Medical anthropology: contemporary theory and method. New York: Praeger, 1990.

7. Landy D, ed. Culture, disease, and healing: studies in medical anthropology. New York: Macmillan, 1977.

8. Mechanic D. Medical sociology. New York: Free Press, 1978.

9. Schwartz HD, ed. Dominant issues in medical sociology. New York: McGraw-Hill, 1994.

10. Shimkin DB, Golde P, eds. Clinical anthropology: a new approach to American health problems. New York: University Press of America, 1983.

Chapter 4

Acute vs. Chronic Problems

Peter D. Hurd

INTRODUCTION

How would you answer if you were asked to describe symptoms that you have recently experienced? Would you talk about some temporary pain or discomfort? Would you describe how that symptom changed something in your life? When students of college age answered this question, they often described symptoms in particular situations:

- After I had my wisdom teeth out, I would experience dizziness when I would get up from lying down.
- Occasionally, if I step the wrong way on the stairs, I will have a pain in one of my knees. I think it is from running track in high school.
- Lower back pain that occurs after I work all day. It also hurts when I have been sitting in one spot for a long time.

When senior Olympic athletes, who are over 55 years of age, are questioned about health problems, they often respond in terms of long-lasting symptoms or chronic disease:

- Injured shoulder while playing volleyball . . . has not been the same since, though I did work through the pain.
- Facial skin cancer removals every year; must avoid exposure to sun, which is impossible in long-distance running.
- Right knee joint stiffens up and is painful. Cannot walk far without pain.
- Herniated lumbar disk–resulted from auto accident.[1]

CHRONIC AND ACUTE SYMPTOMS

For most situations, acute symptoms are temporary changes that allow the individual to return to the way things had been. Acute symptoms are of brief duration and, consequently, appear rapidly. For most acute symptoms, no permanent change is required. An exception would be symptoms of short duration that lead to death or disability; these are generally not considered in this discussion or are considered as a combination of acute and chronic symptoms (e.g., physical trauma that leads to joint disability). Chronic symptoms imply long-lasting change, a situation that can require lots of adaptation because one's life is altered in some way. Chronic symptoms may be gradual in onset and could appear over an extended period of time, perhaps many months. They may be of life-long duration.

Both acute and chronic symptoms are experiences by an individual and indicate that something is not as expected. The individual is aware of these experiences and can communicate them to another, usually with concern about the underlying disease that is thought to be the cause. A health care provider combines those communicated symptoms with "signs" (e.g., blood pressure and heart rate) that are observed from examination of the patient to infer a causal disease. In medical usage, signs are not necessarily noticed by the patient and symptoms are communicated to the health care provider; both are combined as evidence of disease.

ATTRIBUTION ERRORS AND SYMPTOMS

Much as signs are noticed more by health providers than by the patient, traits and other personal characteristics are often perceived by others more clearly than by the individual. To explore this difference in personal perception, concepts from attribution theory will be applied to symptom experience and the perception of health: How can older individuals with all the signs and symptoms of chronic disease say they are healthy when they have chronic diseases?

A Health Paradox: Health in the Presence of Chronic Disease

Over 75% of the elderly have one or more chronic health problems (arthritis, hypertension, hearing impairment, heart disease, cataracts, and so on).[2] At the same time, about two-thirds of the elderly rate their health as good, very good, or excellent rather than as fair or poor.[3] Understanding this paradox–health in the presence of chronic disease–can help us understand how people react to chronic and acute symptoms.

A paradox is something that is seemingly contradictory. Conventional wisdom offers many of these paradoxes, for example: "absence makes the heart grow fonder" compared to "out of sight, out of mind." When a girlfriend and boyfriend are separated, do they forget about each other or strengthen the relationship by realizing how important that person is because of the absence? From a slightly different viewpoint, a paradoxical response in pharmacy means that the drug produced the opposite of the expected effect, as when a stimulant-type medication is used to calm a child with attention deficit disorder, which would include children who are overactive or easily distracted. One would expect this type of drug to be used for a condition such as narcolepsy, which is the uncontrollable desire for sleep; it is "seemingly contradictory" to use the drug for children with attention deficit disorder, even though this can be an effective treatment. Consider our health paradox: elderly individuals, despite the presence of chronic disease, rate their health as good or excellent.

The Meaning of Health

Part of the explanation for this paradox is found in the meaning of the word "health." An overly simplistic definition of health would be the absence of disease.[4] For acute problems, such as a cold, this definition is workable. An individual may consider himself to be "healthy as an ox" because he never gets sick, always comes to work, or has no diagnosed disease. Absence of disease, however, usually ignores large groups of "healthy" people. The World Health Organization (WHO) defines health as the state of complete physical, mental, and social well-being, not just the absence of disease or infirmity. This is an excellent political definition of health that highlights the importance of various dimensions in defining health beyond the absence of disease. The WHO definition, however, is too broad to help explain our health-in-the-presence-of-chronic-disease paradox.

Defining health as the ability to meet one's expectations to perform social roles, as the "optimum capacity of an individual for the effective performance of the roles and tasks for which he has been socialized," or as the ability to "function to the point that they can do what they want to do" allows an individual with hypertension who is active and happy to say that he or she is in good health.[5, 6] Because the individual can do what he or she wants to do, an attribution of health is a reasonable decision. Similarly, a person with arthritis can be active–and thus healthy–despite pain because the individual expects and accepts the limitations that come with this disease and with age.

One subtle difference between disease-based definitions of health and function-based definitions is the perspective. Absence of disease usually involves the judgment of a physician, following an examination, that the individual is healthy. This "health" is a characteristic attributed to the individual by another, in this case the doctor. Compare this to the decision that individuals make that they are healthy because they can do what they want to do. In this case, the individual is assessing a personal response to the changing situations encountered in life. This self-assessment is, potentially, quite different from the assessment made by another.

The Fundamental Attribution Error and Health

An attribute is a characteristic or property that an individual has: he is handsome, or she is industrious. The observer of another can also assign a characteristic to that person: I think he is honest, or I think she is trustworthy. The observer is making an attribution, a generalization about the causes of the person's behavior, based on something the observer has noticed.

If an observer is generalizing from one situation to all situations, erroneous attributions can be made. The person may always exhibit this trait, may exhibit this trait only in certain situations, or may seldom exhibit this behavior. In any event, a potential difference exists between the observer's attribution and the various situations that have not been observed.

People tend to attribute traits to others, in this case health, and to see their own behavior in terms of the various situations in which they operate on a day-to-day basis. They see themselves as actors and interpret their responses as more of a response to different situations, while they observe others and, acting as observers, tend to see stable characteristics in others.[7] Researchers have called this actor-observer tendency the "fundamental attribution error," and it helps us understand one of the differences between acute symptoms (situations) and chronic symptoms (traits).

Some of the most common symptoms are headache and backache. We experience these on, hopefully, an infrequent basis and seek situational explanations for the pain. The headache is because of the party last night, the lack of coffee in the morning, or the eyestrain from too much sun or poorly fitting glasses. The backache is the result of lifting a heavy load, the chair used at the desk, the amount of time standing up, or the work at home over the weekend. In each of these examples, we are responding to a symptom by seeking a situational explanation for the pain.

Situational attributions have many values to the actor. Changing one's self-perceptions is unnecessary because it is only the situation. Changing one's situation should allow a return to more desirable circumstances, assuring the individual control of symptoms. The pain and discomfort will not continue indefinitely.

For some, the symptoms of an aching back or a headache are more permanent. Occasionally, these symptoms trigger devastating changes in the individual. Having a "bad back" or migraine headaches becomes a characteristic of the individual. They avoid activities such as lifting heavy loads or active sports to prevent further injury. He or she avoids situations that tend to bring on headaches or reacts to early signs of a problem. The symptoms are indicative of a life-long characteristic and lead to fundamental changes in behavior across many situations.

The explanation of our paradox of good health with chronic disease is based on the fact that most people tend to make situational attributions about themselves. They are able to say they are healthy despite chronic disease because they can do the things that they expect to be able to do; they can respond as expected in the situations that are important to them. They make a judgment that is more a result of the situations than of the personal trait.

Practical Implications

This theoretical knowledge about how people tend to see each other is, itself, too absolute. The theory is part of a general store-house of knowledge about people that we use to interact with others on a day-to-day basis. What are the situations in which this approach can help us understand another's reactions to various symptoms?

Sensitivity to the Tendency to Attribute Traits

In the hospital setting, Mrs. Smith is referred to as the UTI (urinary tract infection) instead of as a patient.[8] In the retail pharmacy, she is a Medicaid customer. In the emergency room, the confused older man is referred to as a "gomer."[9] Our human nature encourages us to simplify complex situations and symptoms by assigning labels to people.

Patients, the actors to our observations, tend to see themselves in terms of their situation at the moment. Their pain is immediate and frightening, their confusion is a sense of embarrassment and humili-ation, and their need for care transcends their ability to pay. Their focus is more situational. The challenge is to see patients' symp-toms as they experience them and to react to patients in terms of the situation:

- When patients seem compelled to tell you about a particular situation, this is how they see their problem.
- When patients ask about a particular symptom in a given situa-tion, they are helping you understand how they are experienc-ing their problem.
- When you encourage patients to keep records of symptoms and the situations that accompany the symptoms, you are using an approach that will be in harmony with many people's views of their symptoms.
- Use experiential terms that will help patients understand the message (what will they see, hear, feel, or experience).
- Encourage patients to define a urinary tract infection in terms of a temporary infection rather than in terms of pain on urina-tion so that they will be more likely to continue treatment after the pain subsides.

• Avoid trait attributions based on symptoms or brief situational observation.

The compelling nature of the situation has been demonstrated in a number of psychological studies in the attribution literature. In one example, Nisbett and Schachter gave male subjects a series of electric shocks, measuring the highest intensity the subjects were willing to tolerate.[10] Half of the subjects then received a placebo pill but were told that the pill would cause feelings of arousal. The second group received a placebo pill with no explanation. When tested again, subjects who could attribute their arousal to the pill tolerated greater levels of electric shock than the control group. Aspects of the situation, then, have become very important to subjects in the choice of an explanation for their symptoms.

Focus on the Dynamic Disease Process

The tendency for observers to see relatively permanent characteristics when they evaluate another's symptoms implies that health care providers will see the individual in a more static or unchanging situation, while the patient is struggling with a very dynamic and changing process. Labeling an individual as hypertensive is very straightforward for the pharmacist and implies a rather regular visit to the pharmacy for the next refill of a prescription. For the patient, being diagnosed as hypertensive implies a very complicated process of change, over a long period of time, that includes diet, exercise, side effects, fear of sequelae, and medication on a regular basis. Rather than seeing the patient in a relatively static disease state, the health care provider must focus on a more dynamic process of change.

The symptoms that the patient experiences are a part of this changing process. Symptoms may be correctly interpreted as part of a chronic disease process, leading to less anxiety and better patient control. In one case, the patient realizes that his indigestion is causing his distress rather than his heart, which has been diagnosed as healthy. He feared that his chest pains were the prelude to a heart attack but found that his symptoms were the result of heartburn. Conversely, symptoms associated with a chronic disease may take on significant new meanings, as when the patient associates lethargy with a low potassium level that has been caused by the diuretic for hypertension.

Response to Situation Suggestions

Most of our experiences with over-the-counter (OTC) products are very situational. A symptom leads us to take an OTC tablet, which results in a change in the symptom. We stop taking medications when the symptom is gone.[11] OTC drug advertisements often focus on what is the proper situation to use this particular product.

Because people have such a focus on what is "immediate," getting them to change their behavior often works better if the message is focused on something that will happen right away. The antismoking literature and medical writings advocate emphasizing the immediate consequences of smoking—such as bad breath, burned clothes, or financial cost—rather than the long-term health consequences of tobacco use to help keep school-aged students from starting to smoke and to help smokers to quit.[12,13] The immediate effect of a diuretic (the need to urinate) is more compelling to a patient than the long-term effect of reduced blood pressure. Vitamins are often used because they are felt to be needed that day, even though they also have a long-term effect. Tragically, chronic drug abuse is focused on the immediate effects rather than on long-term consequences.

How can a long-range therapeutic goal be translated into more immediate symptoms and consequences? Day-to-day compliance records give people immediate feedback that they are taking their medication and "reward" them as they place another check mark on the chart. Regular blood pressure monitoring (and recording if desired) helps provide that immediate feedback. Providing patients with symptoms that can be monitored on a very immediate basis should help improve compliance.

Examining the program from the opposite perspective, making attributions of chronic illness as an explanation for symptoms can be less effective than focusing on a daily need that might result from that chronic illness. In the mental health area, patients who attributed a problem to a more physical aspect, "a physical, medical, or biological problem in contrast to a mental illness," reported more positive social relations and higher overall quality of life than those who made more of a personal trait attribution.[14] Using a medication example, patients who interpret their depression in terms of their body not making enough of some substance are making a more situational explanation, which hopefully is easier to accept than the interpretation of symptoms as "I am chronically depressed." Lithium is a "salt" that is taken to replace the daily need that goes unmet.

The medication may become a symbol of one's chronic illness and the symptoms that may come with that illness. For some epileptics and some diabetics, the regular schedule of medicines becomes a daily confirmation of disease. Consequently, skipping medications can serve as a confirmation that they are not really very sick. Taking the drug means "I am sick" instead of healthy rather than "taking the drug will make me healthy." In this case, the lack of unwanted symptoms, in a very situational context, implies better health:

- Try to associate everyday symptoms (or their avoidance) with the fact that the medicine is working.
- Look for immediate positive aspects of the medication, perhaps in terms of observable symptoms, to reinforce the medication-taking behavior.
- Avoid linking a drug's success only with long-term outcomes.

Unintended Drug Effects

A recently introduced medication presents another attribution dilemma for the pharmacist when it appears that the drug has an unintended effect. Is the cause of this effect the drug, the situation, or some combination of drug and patient? Should this event be reported in a postmarketing surveillance program? To determine this, the individual must make decisions about the specificity of the event (How many different factors can cause the event?), the consistency of the event (Does it happen regularly, regularly in a specific group, etc.?), and whether there is a dose-response relationship.[15] In some cases, a rechallenge will be made (giving the drug after a period of time to see if the same unintended drug effect is again produced). The tendency to see the drug as the cause of the event has to be carefully evaluated with various situational considerations.

SYMPTOMS AND CHANGE WITH CHRONIC DISEASE

The literature on how people deal with change provides a variety of insights to help the health care provider understand acute and chronic symptoms from the perspective of the patient. Of course, some patients will be highly motivated to accept change to reduce their pain or to increase their mobility, but long-lasting symptoms coupled with the patient's tendency to see life events from a situational viewpoint can lead to significant challenges for the health care provider. How does he or she assist the patient in accepting the transition from a current state of affairs to what is often a less desirable situation? In some cases, the patient will be asked to take medication that actually decreases the immediate quality of life. In other cases, the patient will be coping with the fact that the symptom implies a life-threatening illness such as cancer or heart disease.

Resistance to Change

People, by nature, are resistant to change at the same time that they are bored with a life without change (another paradox). Despite chronic symptoms, patients are often unwilling to embrace a solution that seems of obvious benefit to the health care provider. Of course, symptoms help overcome this resistance to change.

Dissatisfaction with symptoms is frequently what brings a patient to the physician or pharmacist.[16] Symptoms that are unusual, that are associated with perceived risk, and that interfere with desired roles are examples of symptoms that often lead to action on the part of the patient.[17] This dissatisfaction with the status quo may be enough to unstick the earlier behaviors and furnish the opportunity for the provider to replace old actions with new.

One way to conceptualize change is with Lewin's unfreeze-movement-refreeze change process (Figure 4.1).[18] The process of change includes an initial period where the individual must be "unstuck" from existing ideas and behaviors. Without this compelling need to do something different, change is unlikely or only transient. Once the individual is ready for change, the individual must be moved toward the desired behaviors, sometimes over a period of time. Changing to new behaviors, however, is not enough.

FIGURE 4.1. Lewin's Three-Step Change Model

Those new activities must be solidified, habituated, and reinforced so that they continue over time.

The nature of the symptoms is not the only way to generate a compelling need to change, but this sense of urgency will help the patient be more receptive to suggestions about specific changes. Again, links to particular situations can be effective. Knowing the patient, perhaps on a long-term basis, can help the health care provider identify the factors that could help heighten or intensify this need to change. Without this need to change, the patient will resist change or slide back into learned behaviors after an attempt to change.

Shared View (Vision) of What the Symptoms Mean

Occasionally, linking a patient's seemingly unconnected symptoms to a unifying explanation can help overcome the resistance to change. For example, the backache, sore knee, and blisters are all related to the fact that the runner is overcompensating for a problem with the arch of the foot. A shoe insert may help correct the problem. Bruising easily, frequent falls, and stomach distress may all be symptoms of a body that is unable to handle the amount of alcohol that is being ingested on a regular basis. Both of these individuals, the runner and the alcoholic, often have a great deal of emotion tied to their activity and can be quite resistant to change, even when the connection between symptoms is (carefully) mentioned.

Resistance to change and denial, even with an agreement on what symptoms mean, is common in the treatment of alcoholism.[19] One of the goals of alcoholism treatment is to get the patient to identify himself or herself and his or her symptoms as part of this physical addiction to alcohol. Another is to help the patient move from such situational attributions as "I drink to control stress," "I never drink alone," or "I only drink on weekends." The attribution is to a personal characteristic or disease that is treatable rather than to a negative personality trait or a totally situational explanation. Despite this more permanent trait or disease attribution, dealing with addiction is focused on the daily situations: taking things one day at a time.

A shared view of the future is an important component of moving the patient from a willingness to change to new behaviors that are more health-oriented. For chronic illness, this vision may combine a more trait-like acceptance of the disease, coupled with a more situational approach to the treatment of the disease. Treating an illness as a day-to-day activity has a great deal of merit from this situational point of view. Taking medications for chronic illness may become a habit, but the behavior is still a day-to-day requirement.

Knowledge of the Behaviors in Different Situations

Taking medications over an extended period of time will usually mean that the patient will experience a variety of changing situations. "Take with food" is a much easier instruction to follow for ten days than for ten years, where the patient experiences such situations as stomach flu with no desire to eat, attempts to lose weight, and fancy restaurants where taking pills might seem inappropriate. Pharmacists are aware of the different informational needs of patients with new-versus-refill prescriptions and serious-versus-less serious situations.[20] The USPDI provides patient information to help deal with such situations as: missed dose, side effects (both those that are serious and those not requiring a doctor's attention), and precautions (e.g., do not stop taking the medication without first checking with your doctor, check with your doctor before taking any of the above while on this medicine).[21]

People tend to learn things best when they have a clear need to know or use something. For patients, knowing where to find an answer or whom to ask can be as important as actually knowing the information in advance. One of the values of patient information in printed format is as a reference material to be consulted when needed.

Force-Field Analysis

To help the health care provider identify the dynamic state in which the patient is placed, a force-field analysis can be useful (Figure 4.2). On this day-to-day basis, forces change each day to both facilitate and inhibit the patient's responses to the symptoms of the illness. A force-field analysis is an attempt to depict these various forces visually.

For simplicity, forces are listed that encourage change (avoidance of pain, increased functional abilities, long life) and are contrasted with those that make change difficult (remembering the medication, taking the medication when not at home, side effects). In reality,

FIGURE 4.2. Force-Field Analysis

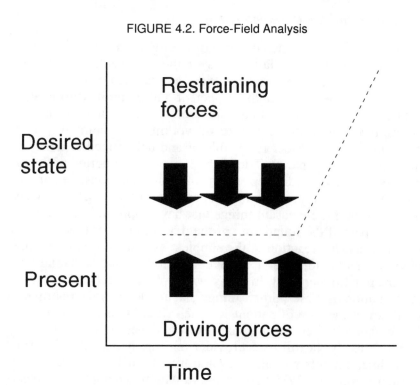

these forces are complex, and each may be both a force toward change (the desire for a happy life) and a force inhibiting change (the desire for a happy life *today*). Identifying these forces can be helpful to both patient and practitioner.

Those familiar with the Health Belief Model will see this as a benefits/barriers approach to behavior.[22] This component of the Health Belief Model has been helpful in various compliance models.[23,24] Force-field analysis also has some of the characteristics of the cost-benefit analysis of pharmacoeconomics.[25] Ideally, people will choose behaviors where the benefits exceed the costs.

Implications for care would include: determining what patients see as the inhibiting factors and enabling factors as they react to their symptoms, helping patients reduce the forces inhibiting desired behaviors, giving added significance to the factors promoting the desired changes in response to these symptoms.

Symptoms and the Development of Trust

One of the underlying concepts of the literature on change is the need for trust in the relationships or the organization. Trust is a combination of past experiences with the individual and the conviction that this person is both competent and committed to change. While other terms are often a part of change, such as personal ethics, the organizational culture, or working for a common vision, believing that the leader is capable of and determined to do something is central to change. Trust is built over a long period of time as promises are repeatedly kept and as multiple situations are continually handled with integrity and respect. From our situational viewpoint, all the little pleasant things that have happened in a relationship are part of the foundation of the development of trust.[26] Thus, the pharmacist's reaction to the simplest of situations and the most obvious of symptoms can lead, over a period of time, to a long-term patient-practitioner trust. In many ways, pharmaceutical care is the combination of these "little" things that are done by the pharmacist as a part of the care of patients.[27] Each situation is an opportunity, in the patient's eyes, for the pharmacist to develop that trust.

Much as the health care provider and the patient develop trust over time, patients will also develop trust in their own evaluations of their symptoms. While accurate evaluation of symptoms is desirable, symptoms that are familiar to the patient may be discounted or ignored when they represent some new health problem or a serious degeneration of an existing problem.[28] One of the challenges for a patient with chronic disease is to balance the influences of the situation, the potential changes in a chronic condition, and the decision to seek evaluation of this symptom.

EXAMPLES OF SYMPTOMS IN SITUATIONS

Temporomandibular Joint Dysfunction

Temporomandibular joint dysfunction can include painful and confusing symptoms, especially when the patient is unaware of the cause of the problem. Symptoms can include earache, facial pain, and increased pain after chewing. Garro examined the reports of patients with this problem.[29] When undiagnosed, this combination of symptoms was confusing, and patients often could not make sense out of it, sometimes thinking that they might be losing touch with reality. Patients reported experiencing their body as separate from the mind and in need of being fixed. Fear of pain was often reported, as well as an inability to control pain. The same person with the same goals prior to the pain began devoting lots of energy to the day-to-day activities of dealing with the pain, thus putting other goals on hold. Garro's report of these symptoms is highly focused on the situation and the concern of patients with that immediate situation and the need to cope with the moment.

Breast Cancer

In Facione's review of breast cancer research, symptoms other than a lump (symptoms such as pain, bleeding, or discharge, and other observable changes) were especially linked to patient delay in seeking diagnosis and treatment.[30] The provider delays, too, were greater when patients' symptoms were not consistent with the typical breast cancer picture. Women's denial leads to delays, as would be expected. Affective responses, such as anxiety, were a mixed bag, sometimes leading to delay and sometimes being associated with early treatment. The review found some suggestion that a prior history of benign breast disease (fibrocystic) was associated with increased delay. From an attributional viewpoint, the tendency to make situational attributions can lead to delay in diagnosis and treatment. The more confusing the symptoms, the more difficulty in making the correct attribution.

Multiple Sclerosis

The early symptoms of multiple sclerosis could include numbness and tingling sensations, coordination problems, vision disturbances, and fatigue, which are often interpreted in terms of the situation. The individual will see such issues as stress, work-related conditions, increasing age, or poor physical conditioning as the cause of the symptoms.[31] Because the disease is difficult to diagnose, incorrect attributions by others are likely, ineffective treatments will be prescribed, and attributions about psychological problems will be made before the disease is correctly diagnosed. The patient will struggle with incorrect situational explanations, inaccurate assessments from others, and the fact that a physical disease may be seen as a personality characteristic. In this case, the tendency to attribute traits to individuals can make the correct diagnosis of the disease more difficult.

SOME LIMITATIONS OF THIS APPROACH

Stress and Symptoms

The attributional approach tends to be a very cognitive and unemotional approach. The approach reflects the less emotional evaluation of the practitioner rather than that of the patient. Patient responses to symptoms include a variable level of stress and emotion. Students studying about disease will interpret their own benign symptoms as signs of serious illness, especially in the stressful times of a demanding medical education.[28] Mumford reviews a number of social, emotional, and cultural factors that encourage those experiencing symptoms to seek treatment, avoid it, or consider it irrelevant to the situation.[32]

Theory and Reality

Focusing on the situation rather than the individual is a recommendation based on differing perspectives. The approach will work better with some individuals than with others and, undoubtedly, will work better in some situations than others. The skilled practitioner will include this approach with other effective approaches for dealing with the patient.

The reality of a patient's chronic disease and accompanying symptoms should be dealt with realistically. The practitioner should be sure that the patient develops an understanding of the long-term implications of a chronic disease. Any misunderstanding of the significance of a disease because of a "situational" approach that has stressed immediate consequences should be avoided.

SUMMARY

Two different psychosocial approaches have been applied to the analysis of responses to symptoms: attribution theory and change management. The difference between actors and observers, the fundamental attribution error, served as an explanation for why individuals could experience the symptoms of chronic illness and still report their health as good or better than good. Essentially, individuals see more of their actions in terms of the situation, while those observing these individuals are more likely to make trait attributions about the individual. This tendency was then explored in terms of practical applications for patient care. Helping patients deal with chronic illness implies that their behaviors will need to change, and the change management literature highlighted some of the issues in managing a patient with chronic symptoms. Resistance to change is a common human response and can be dealt with by a variety of approaches, including a clear understanding of what the symptoms really mean and an appreciation of the various forces moving the individual both toward and away from change. The relationship between patient and provider that develops over time is based on a combination of many situations that have been effectively handled, leading to trust between these individuals. Keeping the patient's perspective of the immediate situation–the present problem–can help the provider accomplish the long-term goals of patient care.

REFERENCES

1. Fontane PE, Hurd PD. Self-perceptions of national senior Olympians. Behav Health Aging 1992;2(2):101-10.

2. National Center for Health Statistics. Current estimates from the National Health Interview Survey, 1990. Vital and Health Statistics. Washington, DC: Government Printing Office, 1991.

3. Mermelstein R, Miller B, Prohaska T, Benson V, Van Nostrand JF. Measures of health. In: Van Nostrand JF, Furner SE, Suzman R, eds. Health data on older Americans: United States, 1992. Vital and Health Statistics. DHHS Pub. No. (PHS) 93-1411. National Center for Health Statistics, 1992.

4. Atchley RC. Social forces and aging. 7th ed. Belmont, CA: Wadsworth, 1994.

5. Parsons T. Definitions of health and illness in the light of American values and social structure. In: Jaco EG, ed. Patients, physicians and illness. 2nd ed. New York: The Free Press, 1972.

6. Cockerham WC. Medical sociology. 5th ed. Englewood Cliffs, NJ: Prentice Hall, 1992.

7. Jones EE, Nisbett RE. The actor and the observer: divergent perceptions of the causes of behavior. In: Jones EE, Kanouse DK, et al., eds. Attribution: perceiving the causes of behavior. Morristown, NJ: General Learning Press, 1972:79-94.

8. Mizrahi T. Coping with patients: subcultural adjustments to the condition of work among internists-in-training. Rpt. in: Schwartz HD, ed. Dominant issues in medical sociology. 2nd ed. New York: Random House, 1987:241-51.

9. Leiderman DB, Grisso JA. The gomer phenomenon. J Health Soc Behav 1985;26:222-31.

10. Nisbett RE, Schachter S. Cognitive manipulation of pain. J Exper Soc Psychol 1966;2:227-36.

11. Self-medication in the '90's: practices and perceptions–highlights of a survey on consumer uses of nonprescription medicines. Washington, DC: Nonprescription Drug Manufacturers Association, 1992.

12. Green HL, Goldberg RJ, Ockene JK. Cigarette smoking: the physician's role in cessation and maintenance. J Gen Intern Med 1983;3:75.

13. Schroeder SA, McPhee SJ. General approach to the patient; health maintenance & disease prevention; principles of diagnostic test selection & use; & com-

mon symptoms. In: Tierney LM, McPhee SJ, et al., eds. Current medical diagnosis & treatment. Norwalk, CT: Appleton & Lange, 1993:1-21.

14. Mechanic D, McAlpine D, Rosenfield S, Davis D. Effects of illness attribution and depression on the quality of life among persons with serious mental illness. Soc Sci Med 1994;39:155-64.

15. Rogers AS. Unintended drug effects: identification and attribution. In: Hartzema AG, Porta MS, Tilson HH, eds. Pharmacoepidemiology: an introduction. 2nd ed. Cincinnati, OH: Harvey Whitney Books, 1991:64-74.

16. National Ambulatory Medical Care Survey: 1989 summary. Advance Data. DHHS Pub No. (PHS) 91-1250. July 1, 1991.

17. Mechanic D. Medical sociology: a selective view. New York: Free Press, 1968.

18. Lewin K. Field theory in social science. New York: Harper & Row, 1951.

19. Frances RJ, Franklin JE. Alcohol and other psychoactive substance use disorders. In: Hales RE, Yudofsky SC, Talbott JA, eds. Textbook of psychiatry. 2nd ed. Washington, DC: American Psychiatric Press, 1994.

20. Schommer JC, Wiederholt JB. Pharmacists' views of patient counseling. Am Pharm 1994;34(7):46-53.

21. United States Pharmacopeial Convention, Inc. USP dispensing information. Vol. II: Advice for the patient: drug information in lay language. 14th ed. Rockville, MD: U.S. Pharmacopeial Convention, 1993.

22. Becker MH, Maiman LA. Sociobehavioral determinants of compliance with health and medical care recommendations. Med Care 1975;13:10-24.

23. Hershey JC, Morton BG, Davis JB, Reichgott MJ. Patient compliance with antihypertensive medication. Am J Public Health 1980;70:1081-9.

24. Macrae FA, Hill DJ, St. John JB, et al. Predicting colon cancer screening behavior from health beliefs. Prev Med 1984;13:115-26.

25. Bootman JL, Townsend RJ, McGhan WF, eds. Principles of pharmacoeconomics. Cincinnati, OH: Harvey Whitney Books Company, 1991.

26. Covey SR. The 7 habits of highly effective people. New York: Simon and Schuster, 1989.

27. Hepler CD, Strand LM. Opportunities and responsibilities in pharmaceutical care. Am J Hosp Pharm 1990;47:533-43.

28. Mechanic D. Social psychological factors affecting the presentation of bodily complaints. N Engl J Med 1972;286:1132-9.

29. Garro L. Narrative representations of chronic illness experience: cultural models of illness, mind, and body in stories concerning the temporomandibular joint (TMJ). Soc Sci Med 1994;38:775-88.

30. Facione NC. Delay versus help seeking for breast cancer symptoms: a critical review of the literature on patient and provider delay. Soc Sci Med 1993;36:1521-34.

31. Stewart DC, Sullivan TJ. Illness behavior and the sick role in chronic disease: the case of multiple sclerosis. Soc Sci Med 1982;16:1397-1404.

32. Mumford E. Evaluating symptoms: seeking treatment. In: Medical sociology: patients, providers, and policies. New York: Random House, 1983:46-63.

ADDITIONAL READINGS

For the Instructor

1. Atchley RC. Social forces and aging. 7th ed. Belmont, CA: Wadsworth, 1994:100-106.

2. Becker MH, Maiman LA. Sociobehavioral determinants of compliance with health and medical care recommendations. Med Care 1975;13:10-24.

3. Bootman JL, Townsend RJ, McGhan WF, eds. Principles of pharmacoeconomics. Cincinnati, OH: Harvey Whitney Books Company, 1991.

4. Cockerham WC. Medical sociology. 5th ed. Englewood Cliffs, NJ: Prentice Hall, 1992.

5. Hepler CD, Strand LM. Opportunities and responsibilities in pharmaceutical care. Am J Hosp Pharm 1990;47:533-43.

6. Jones EE, Nisbett RE. The actor and the observer: divergent perceptions of the causes of behavior. In: Jones EE, Kanouse DK, et al., eds. Attribution: perceiving the causes of behavior. Morristown, NJ: General Learning Press, 1972: 79-94.

7. Leiderman DB, Grisso JA. The gomer phenomenon. J Health Soc Behav 1985;26:222-31.

8. Mumford E. Evaluating symptoms: seeking treatment. In: Medical sociology: patients, providers, and policies. New York: Random House, 1983:46-63.

9. Parsons T. Definitions of health and illness in the light of American values and social structure. In: Jaco EG, ed. Patients, physicians and illness. 2nd ed. New York: The Free Press, 1972:106.

10. Rogers AS. Unintended drug effects: identification and attribution. In: Hartzema AG, Porta MS, Tilson HH, eds. Pharmacoepidemiology: an introduction. 2nd ed. Cincinnati, OH: Harvey Whitney Books, 1991:64-74.

11. Self-medication in the '90's: practices and perceptions—highlights of a survey on consumer uses of nonprescription medicines. Washington, DC: Nonprescription Drug Manufacturers Association, 1992.

For the Student

1. Atchley RC. Social forces and aging. 7th ed. Belmont, CA: Wadsworth, 1994:100-106.

2. Becker MH, Maiman LA. Sociobehavioral determinants of compliance with health and medical care recommendations. Med Care 1975;13:10-24.

3. Bootman JL, Townsend RJ, McGhan WF, eds. Principles of pharmacoeconomics. Cincinnati, OH: Harvey Whitney Books Company, 1991.

4. Hepler CD, Strand LM. Opportunities and responsibilities in pharmaceutical care. Am J Hosp Pharm 1990;47:533-43.

5. Leiderman DB, Grisso JA. The gomer phenomenon. J Health Soc Behav 1985;26:222-31.

6. Rogers AS. Unintended drug effects: identification and attribution. In: Hartzema AG, Porta MS, Tilson HH, eds. Pharmacoepidemiology: an introduction. 2nd ed. Cincinnati, OH: Harvey Whitney Books, 1991:64-74.

SECTION C:
CHOOSING A SOURCE OF CARE

Chapter 5

Images of Pharmacists and Pharmacies

Sheryl L. Szeinbach

INTRODUCTION

Virtually everything we do from day to day tells others not only how we view ourselves but also how we want others to view us. Take a moment to think about your clothing choice today. Why did you select these particular clothes or shoes? What message are you trying to convey to yourself as well as to those around you? This self view or image is communicated to others through our apparel, the sports we watch and actively pursue, travels, food choices, and language. Usually, this image is more than just fitting in with the right crowd; it also identifies that part of you which is you. Your friends have their images, as do groups of individuals, small businesses, and large corporations. Of particular interest in this chapter is the image portrayed by pharmacists and community pharmacies. When you think about a pharmacist, what is the first image that forms in your mind?

BACKGROUND INFORMATION

Pharmacists are concerned with the same questions about image formation and professional image. Several studies on pharmacists' ability to provide medication information, communicate with patients, image, and educational training were performed in the 1960s.[1-4] Some of the major findings from these studies revealed that: (1) pharmacists are perceived as not fulfilling their potential capabilities, (2) pharmacists are placed closer to the concept of technician than to the concept of professional, (3) pharmacists are not being used as much as possible for their professional knowledge and expertise, and (4) pharmacists should educate patients about their educational background and willingness to provide their services to patients. Based on the findings of these studies, there was clearly a discrepancy between the ideal and actual images of pharmacists as perceived by different groups.

In the 1970s, one of the largest studies of pharmacy image was conducted and summarized in the Dichter Report. The objective of this motivational research study, entitled "Communicating the Value of Comprehensive Pharmaceutical Services to the Consumer," was to determine how the value of comprehensive pharmaceutical services can be most effectively communicated to the public so as to stimulate a demand for these services. Two central questions addressed in the study were: (1) Is the consumer aware that such services are available? and (2) Is the consumer aware of the value of these services?[5]

Results from this study revealed that most people interviewed by the psychologist of The Dichter Institute for Motivational Research did not regard the average pharmacist as the kind of professional most pharmacists envision themselves to be. The public tends to think of the pharmacist as being first and foremost a friend of the manufacturer, that is, a businessperson with primarily commercial motivations.[5] For this reason, pharmacists are viewed as having customers and not patients. In summary, there was a strong desire for the return of the personal pharmacist and the reestablishment of a professional relationship between the pharmacist and the patient.[6]

WHAT IS IMAGE?

Image formation relates to the way in which the pharmacy is defined in the shopper's mind, which consists partly of its functional qualities and partly of an aura of psychological attributes.[7]

The willingness to patronize and to remain loyal to one agency is related to several factors that contribute to the composition of the patient's attitudes, experiences, and expectations of service. In highly competitive environments such as the health care industry, patients may elect to shop according to price or service; in some situations, the location is determined by accessibility, availability, and attainability. Despite the market constraints of limited resources, decentralized services, and inadequate reimbursement policies, pharmacists must strive to develop an image that will be appealing to patrons. A careful analysis of the factors that influence image development will help pharmacists strategically position their pharmacies in this highly competitive health care market.

CHANGES AFFECTING PHARMACY

Retail pharmacy is experiencing changes in the provision of health-related services. When shopping for prescription medications, consumers have the option of patronizing hypermarkets, megamarkets, mail-order pharmacies, deep discount chains, large and small chain stores, independently owned pharmacies, or apothecaries.[8] Using the framework of Fisk, independent pharmacies (four or fewer stores), chain stores, and deep discount chain stores are distinguishable by store size, price, location, service, and assortment of merchandise.[9] Convenience and professional services are important motives for patronizing any pharmacy.[10,11] Pharmacists' ability to provide professional services, therefore, becomes important to retail outlets' continued success. Pharmacists' ability or willingness to provide such services may be related to their perceptions of retail pharmacy's professional image.

Unlike some industries, pharmacy is subject to many external controls. The distribution of prescription drug products is controlled and is subject to a number of federal and state regulations. As a health profession, pharmacy has formal training programs, a code of ethics, professional associations, and a fair degree of autonomy and self-regulation. Alternatively, pharmacists do not have complete control over the medications they dispense, their body of knowledge overlaps that of other health professionals, and there is a perceived business versus profession conflict.[12] This lack of control over medication distribution and use may well combine with the external forces affecting the business aspects of pharmacy to place stress on pharmacists and to threaten their self-perceived notions of pharmacy's image.

Implementation of the Omnibus Budget Reconciliation Act of 1990 (OBRA 90) will establish a new standard of pharmacy practice. Under OBRA 90 regulations, pharmacists will be required to maintain thorough patient medical records in addition to providing drug counseling and monitoring (i.e., drug utilization review). Compliance with OBRA 90 regulations will require pharmacists to interact more actively with patients and physicians regarding drug therapy. With these changes in health care delivery, pharmacists must examine their image and develop strategies for improvement.

CONCEPTUAL FRAMEWORK

As shown in Figure 5.1, image can be conceptualized from a number of perspectives. First, the corporate image of pharmacy is determined by external experiences, news media, and other sources. Second, when patients exchange information about their experiences with a particular provider, the organization will eventually develop a market image that is consistent with patients' valuation of products and services compared with those of competitors. Third, members demonstrate their commitment to professionalism through

FIGURE 5.1. Components of Pharmacy Image

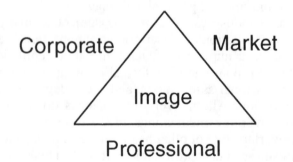

their behavior, attitudes, and beliefs about the organization and its role in meeting patients' expectations for service.[13]

CORPORATE IMAGE

Corporate image describes how patients view the company's goodwill toward society, employees, patients, and other individuals. Dependability, reliability, and responsibility are important factors in the distribution of pharmaceutical products and services. Regardless of the type of product dispensed, patients need to know they can get these products or services in a reasonable amount of time. If special directions are required for use, patients know they can rely on professionals to provide this information. In addition, if problems occur, patients need to know that problems will be resolved quickly.

Organizational culture is a function of company image and contributes to the reputation of the company as well as to the socialization process of its members. Organizational culture is defined as the pattern of basic assumptions that a given group has invented, discovered, or developed in learning to cope with its problems of external adaptation and internal integration and that have worked well enough to be considered valid and, therefore, to be taught to new members as the correct way to perceive, think, and feel in relation to those problems.[14] Culture is an integral function of an organization in that it holds together the different components of an organizational system and maintains its equilibrium. The various functions are manifest through the member's beliefs, definitions of the world, shared social knowledge, philosophies, values, assumptions, and attitudes. These concepts are vital to the establishment of corporate culture and ensure the ability of the corporation to adapt and to survive through an exchange process with the environment.

Image formation may also be linked to the process of organizational development, which relates to the enhancement of adaptive mechanisms within organizations. The goal is to make the organization more receptive to change, thus facilitating the realignment of the total organizational system into a more viable and satisfying configuration.[15] Key components of this system include the administration, production, technology, structure, culture, and organizational goals. Achieving congruency among these components will increase members' ability to plan strategically for the future as well as to create the image of a community leader.

MARKET IMAGE

The major concern for your products and services is market image. Market image is how patients and other health care providers rate the value of your products and services compared to that of competitors. If patients believe that they get high value from products and services relative to price, they will continue to purchase products and services from you.

Patients expect prices to be reasonable, consistent, and fair. Many companies rely on highly advanced computer systems, automated dispensing machines, and electronic claims processing to control costs and to ensure consistent pricing policies. Decisions to adopt these innovative technologies usually involve a complex analysis of the impact on organizational structure as well as personnel. Other considerations include the length of time expected to recapture the investment. Despite the enormous investment, however, information technologies will become the hallmark of organizational efficiency and effectiveness. Health care professionals will continue to rely on information technologies to maximize their patient time and minimize their administrative responsibilities.

Besides insurance information, home health care services, and professional consultations, referrals are another service that can be used to project the market image desired by a company. Patient referrals ensure continuity of care and demonstrate the agency's commitment to patient care. Moreover, patients will gain respect for members of the agency, and this will facilitate the development of patient rapport and trust.

PROFESSIONAL IMAGE

Professional image focuses on commitment to quality health care and the educational needs of the public concerning medication use.

A conceptual model of professional image developed by Hall suggests a correspondence between the structural side and the attitudinal side of professionalism.[16] Thus, the structural aspects would be linked to the professional's training, knowledge, code of ethics, service commitment, and occupational autonomy. The attitudinal side of professionalism is concerned with how practitioners view their own work and how it is incorporated into their internal cognitive framework for professionalism. In general, as determined from previous research, there is an inverse relationship between professionalism and bureaucratization. Moreover, studies performed in the 1960s and 1970s revealed the general need for pharmacists to improve their image and to expand patient services.

RECENT STUDIES ON PROFESSIONALISM AND IMAGE IN PHARMACY

As a result of the changing health care environment, movement toward self-care, and the need for pharmacists to become more involved in patient care, interest about pharmacy image has rekindled. Just as individuals' tastes for fashion, hobbies, and food change, pharmacists were wondering how their image has changed since the 1960s and 1970s studies were performed. Some questions addressed in more recent studies have included the issue of professional identification, pharmacy image, and pharmacists' attitudes about the pharmacy profession.

To some extent, pharmacists' attitudes about their practice will be related to professional identification. In 1984, a study was undertaken to determine if pharmacy patrons could recognize the pharmacist when approaching the pharmacy, identify licensure certificates and professional name tags, and distinguish the pharmacist from other members in the pharmacy.[17] Surveyors were instructed to place themselves in the role of a patient entering the pharmacy for the first time.

Results of the study revealed that professional attire was used to identify 69 (65.1%) of the pharmacists. Data were tabulated to determine whether ability to identify the pharmacist varied with pharmacy type. Chain pharmacies had the best score: in 88% the pharmacist could be identified. Regarding the ability to differentiate the pharmacist from other staff, it was determined that only 52 (62.7%) of the pharmacists were identifiable. In summary, professional identification is critical to pharmacists to guide those consumers who wish to seek them out for counsel. While individual pharmacists may differ in the ways they choose to display their professional identification, each should pause and think about ways to create a professional image.

Because pharmacists play a major role in monitoring medication therapy for patients receiving long-term care services, they are in a position of high visibility and can make a substantial contribution to the development of a company's professional image. To investigate the link between professionalism and pharmacy image, a nationwide study was conducted of 1,089 retail pharmacists who were randomly selected from the Hayes directory.[18,19] The questionnaire included 55 items relating to the structural and attitudinal aspects of image, such as practice environment, pricing practices, level of service, and pricing issues. Pharmacists were first asked to indicate on a seven-point scale whether the image of retail pharmacy has changed in the past five years. Following this question, pharmacists were asked to rate how each of the 55 items has influenced their opinion of the professional image of pharmacy in the past five years. Items were measured on a seven-point "strong negative impact" (–3), "no impact" (0), to "strong positive impact" (+3) scale.

When asked if the professional image of retail pharmacy has changed in the past five years, approximately three-fourths of the pharmacists reported a change in image. However, results from pharmacists were divided equally between a positive or negative change in image. As revealed by the *t* test (mean for chains = 4.76; mean for independents = 5.10; $t = 2.76$; $p = 0.006$), which indicates whether the means between the two groups were different, pharmacists working in independently owned pharmacies rated image significantly higher than pharmacists working in chain pharmacies.

Some of the items from the questionnaire that were linked to professionalism are shown in Table 5.1. Counseling services, public opinion, community involvement, code of ethics, advertising of pharmacy services, and pharmaceutical care were viewed as having a strong positive impact on the professional image of pharmacy. Alternatively, pharmacists had some concern for their work environment in

TABLE 5.1. Mean and Standard Deviation of Items Included for Assessment of Image by Retail Pharmacists

Item	Mean
Counseling services to patients	1.9
Public opinion poll results of pharmacists' trustworthiness	1.9
Community involvement by pharmacists	1.3
Pharmacy code of ethics	1.3
Advertising of pharmacy services by pharmacies	1.2
Pharmaceutical care	1.2
Specialization in pharmacy	1.1
Pharmacy technicians	0.4
Advertising by pharmacies	0.0
Pharmacist work environment	-0.2
Loss of autonomy in dispensing	-0.8
Advertising of prescription prices by pharmacies	-1.5
Remote mail-order dispensing services	-2.4

[a]Items were measured on a seven-point "strong negative" (–3) to "strong positive (+3) impact scale.

the retail setting and the possible loss of autonomy in dispensing. Other items that were rated negatively by retail pharmacists with respect to professional image included the advertising of prescription prices by pharmacies and the use of remote mail-order dispensing services. Pharmacy directors and administrators involved in providing long-term care services should meet and discuss how this information can be used to enhance the company's professional image.

In addition to the items identified by the study, professional courtesy is of paramount importance for projecting a positive company image. Members must be personable, knowledgeable, and willing to listen to patients' needs. Most of the elements that constitute professionalism can be incorporated into the pharmacy's policy where patient exposure to these elements will be high. Moreover, patients will most likely select a pharmacy that portrays a positive work environment and a high level of togetherness among its members.

Advertising can be used to enhance the professional aspects of the company. Besides standard choices for advertising, other avenues might include professional journals, community service centers, hospitals, nursing homes, long-term care facilities, and pharmacies. According to the study results, emphasis should be placed on professional services and other professional pharmacy services provided by the company rather than on prescription prices. Other promotions include radio and direct mailing. Many patients enjoy receiving personal recognition on birthdays, anniversaries, and other special occasions. Another opportunity to increase your company's image is to submit articles for publication in the local newspaper.

STRATEGIES FOR IMAGE IMPROVEMENT

Patient Education

Several strategies are available to help patients become familiar with the professional aspects of long-term care services. Patient education and training seminars can be performed to ensure that patients understand the use of their medication. These seminars can also be given to support staff at nursing homes and other long-term care facilities to ensure patient compliance and recovery. In addition, manuals can be developed and disseminated to inform administrators about the availability of educational services. Other efforts to enhance retail image include the portrayal of chain store pharmacists in advertisements as suppliers of medical information as well as information about medications.

Currently, the pharmaceutical industry is sponsoring advertising campaigns to counter the impression that it participates in price gouging. For example, expensive low-key advertisements in the lay press are consuming advertising dollars that would previously have been spent on splashy spreads launching new drugs in biomedical journals.[20] As a strategy to educate the public about how drugs are discovered as well as about their use, many of these direct-to-consumer advertisements do not even mention a product by name.

Community Service

Community involvement serves to enhance the agency's image. Involvement may include sponsorship of athletic events, civic organizations, and religious activities. Programs on the use of medication can be presented to these organizations by physicians, pharmacists, and nurses that target senior citizens. In addition, financial support of summer programs, charities and societies, and programs for the handicapped will also create a positive image for the company. Time and financial support will be appreciated by members in the community, and your visibility to others will attract patients to your business.

Integrated Care Networks

An interorganizational relationship is created when two or more organizations exchange resources of any kind (money, physical facilities and materials, patient, referrals, or technical staff services).[21] The interorganizational network can act as a unit and make decisions, perform actions, and pursue goals similar to an autonomous organization. Some of the goals of this relationship are to establish a distribution point to maintain the continuity of patient care, to promote areas of mutual interest, to obtain jointly and allocate a greater amount of resources than would be possible by each organization independently, and to resolve areas of conflict or competition. Therefore, the goal is to obtain resources that would normally be unavailable if the organization acted independently. This arrangement may necessitate the relinquishment of some autonomy; however, the corporation's image may improve through a relationship with another corporation whose image is already well-recognized and well-respected.

One goal of health care reform is to create a network of organized care delivery systems. These systems represent a network of organizations that provides or arranges to provide a coordinated continuum of services to a defined population and is willing to be held clinically and fiscally accountable for the outcomes and the health status of the population served.[22] Traditionally, acute inpatient care has been the hospital's focal or central point in this health care network. Although acute care services are vitally important, the provision of these services is no longer limited to hospitals. With the growing emphasis on primary care and preventive medicine, more patients are being channeled to outpatient settings. Consequently, hospital administrators are discovering that more hospital-based services are either contracted out as part of a decentralized network of care or offered through a managed care plan in conjunction with preferred provider organizations (PPOs), health maintenance organizations (HMOs), and community pharmacy services.

As part of this managed care network, pharmacists must seek professional growth through increased customer contact and greater efforts directed toward therapeutic outcomes. This increased professional role for pharmacists can serve as a basis for establishing a positive image in the marketplace. Pharmacists must also have greater personal contact with patients, increased availability of comprehensive pharmaceutical services, increased access to patient information, and more interaction with other health professionals. There is some concern that if pharmacists do not promote their professional qualities as a strategy for image enhancement, the public's perception of pharmacists will decline as chain pharmacies continue to grow in size and pharmacy practice becomes more corporate.[23-25]

CONCLUSION

Image is the total organization as perceived by patients. A positive image can be developed through planning, goals, communication, and commitment to patient care. Steps in launching a program to increase image include: (1) evaluating the strengths and weaknesses of the company's current image, (2) defining the image that the company wants to project, (3) determining the course of action that appeals to the largest possible number of patients, (4) creating patient-specific selling themes, and (5) coordinating communication channels to build the desired image.[26] These goals should be included in a mission statement that outlines major policies, encourages the work force, and explains the company's commitment to quality patient care.

REFERENCES

1. Knapp DA, Wolf HH, Knapp DE, Rudy TA. The pharmacist as a drug advisor. J Am Pharm Assoc 1969;NS9:502-5.

2. Knapp DA, Knapp DE, Engel JF. The public, the pharmacist and self-medication. J Am Pharm Assoc 1966;NS6:460-3.

3. Knapp DE, Knapp DA, Edwards JD. The pharmacist as perceived by physicians, patrons and other pharmacists. J Am Pharm Assoc 1969;NS9:80+.

4. Knapp DE, Knapp DA. Perceived credibility, expertise and effectiveness: pharmacist vs. physician. Soc Sci Med 1970;4:253-6.

5. Anon. What is the Dichter Institute saying about you? J Am Pharm Assoc 1973;NS13:638-41.

6. Anon. Communicating the value of comprehensive pharmaceutical services to the consumer. J Am Pharm Assoc 1973;13:23+.

7. Martineau P. The personality of a retail store. Harvard Business Rev 1958;36:47-55.

8. O'Reilly B. Rx for costs: drugs by mail. Fortune 1992;(Aug 24):116.

9. Carroll NV, Jowdy AW. Demographic and prescription patronage motive differences among segments in the community pharmacy market. J Pharm Market Manage 1987;1(4):19-33.

10. Gagnon JP. Factors affecting pharmacy patronage motives: a literature review. J Am Pharm Assoc 1977;NS17:556-9.

11. Wiederholt JB. Development of an instrument to measure evaluative criteria that patients use in selecting a pharmacy for obtaining prescription drugs. J Pharm Market Manage 1987;1(4):35-59.

12. Wolfgang AP. Challenging students to consider pharmacy's professional status. Am J Pharm Educ 1989;53:174-7.

13. Szeinbach SL. Image development for health care firms. Med Interface 1994;7(1):124-7.

14. Schein EH. Coming to a new awareness of organizational culture. Sloan Manage Rev 1984;26:3-16.

15. Smircich L. Concepts of culture and organizational analysis. Admin Sci Q 1983;28:339-58.

16. Hall RH. Professionalization and bureaucratization. Am Sociol Rev 1968;38:92-104.

17. Szeinbach SL, Fink JL. Identifying the pharmacist's professional status. Am Pharm 1984;NS24(9):60-3.

18. Szeinbach SL, Barnes JH, Summers KH, Banahan BF. The changing retail environment: its influence on professionalism in chain and independently owned pharmacies. J Appl Business Res 1994.

19. Summers KH, Szeinbach SL, Barnes JH. Pharmacists' perceptions of retail pharmacy's professional image: implications for pharmaceutical manufacturers. J Pharm Market Manage 1994;8(2):43-58.

20. Klein NC. Preserving pharmacy's public image. Am J Hosp Pharm 1994;51:2119.

21. Van de Ven A, Ferry D. The interorganizational field. In: Lawler EE III, Seashore SE, eds. Measuring and assessing organizations. New York: Wiley, 1980:296-346.

22. Shortell SM, Gillies RR, Anderson DA, Mitchell JB, Morgan KL. Creating organized delivery systems: the barriers and facilitators. Hosp Health Serv Admin 1993;38:447-66.

23. Anon. Public rates pharmacists number one in honesty, Gallup poll shows. Am J Hosp Pharm 1988;45:468.

24. Smith MC. Pharmaceutical marketing: strategy and cases. Binghamton, NY: Pharmaceutical Products Press, 1991.

25. Robinson B. Is the pharmacist's image changing for better or worse? Drug Top 1986;130(Jan 6):56-62.

26. Marken GA. Corporate image: we all have one, but few work to protect and project it. Public Relations Q 1990;35(1):21-9.

ADDITIONAL READINGS

1. Wolfgang AP. Challenging students to consider pharmacy's professional status. Am J Pharm Educ 1989;53:174-7.

2. Szeinbach SL. Image development for health care firms. Med Interface 1994;7(1):124-7.

3. Summers KH, Szeinbach SL, Barnes JH. Pharmacists' perceptions of retail pharmacy's professional image: implications for pharmaceutical manufacturers. J Pharm Market Manage 1994;8(2):43-58.

Chapter 6

Other Health Providers and the Pharmacist

Bernard Sorofman
Toni Tripp-Reimer
Richard Dehn
Mary Zwygart-Stauffacher

INTRODUCTION

With few exceptions, pharmacists practice their profession in concert with other health professionals. Working with both formal and informal health care teams, pharmacists center their role around the treatment the patient receives. Usually the pharmacist is involved in the selection, dispensing, and/or monitoring of the patient's pharmacotherapy. This chapter reviews the roles of members of the health care team and how they influence the pharmacist's activities.

It is commonly recognized that the patient enters the health care system at a point where diagnoses are performed. A patient may go to a physician (or nurse practitioner, physician assistant, or dentist) for a routine wellness checkup and discover that he or she requires pharmacotherapy–calcium supplementation for instance. Or, more likely, the patient is not well and is pursuing illness behavior, seeking treatment and a cure. It may be that a toothache, a pain in the side, swelling in the arm, or some other concern that is not "normal," has brought the patient to someone who will be able to determine the problem and how it can be corrected.

One major exception to this pattern exists. Patients may first enter the health care system by practicing self-care for their health/ illness concerns. This is a frequent event, especially in the ambulatory practice setting. Patients will determine that they are symptom-atic and will decide to act, often by visiting the nonprescription drug section of their local pharmacy. Pharmacists, acting in the course of their normal daily activities, advise patients on the use of nonpre-scription remedies for minor, self-identified aliments. During a patient's self-care interaction with a pharmacist, it is not unusual for the pharmacist to recognize that he or she has become the first health professional contacted in what will become a "team" activ-ity. Patients who require professional diagnostic advice about their conditions may be triaged by the pharmacist to a diagnostician–physician, dentist, nurse practitioner, physician assistant, or others.

CONTINUITY OF HEALTH CARE

Pharmacists do not practice in a vacuum. They are members not only of the general health care team but also of the pharmacy team that is composed of pharmacists from different practice settings. Communication is required between the members of the team to ensure optimal, seamless pharmaceutical care.

Seamless care is a buzzword of the 1990s. It stands for continuity of care, a critical aspect of quality health care delivery. As patients move from one health care setting to another–home to hospital, hospital to long-term care facility, hospital to hospital, hospital to home, and so on–the knowledge about the patient's care must be transferred from setting to setting (Figure 6.1). For the patient to receive optimal care, information must not be lost in the transfer. This becomes critical in the rapidly changing health care environ-ment where changing roles in the pharmacy arena may also disrupt communication. Dispensers of information and medications must also communicate as the patient moves from setting to setting. Pharmacists in hospital settings provide pharmaceutical care to inpatients. Community pharmacists generally provide care to those patients who are living in the home.

CHANGING HEALTH CARE ENVIRONMENT

The practice environment of the health care provider in the United States is a dynamic system. Challenges from the social

FIGURE 6.1. Continuity of Patient Care

Home	Community Pharmacist
Hospital	
Skilled Nursing Facility	Hospital Pharmacist
Intermediate Nursing Facility	
Residential Care Facility	Consulting Pharmacist
Hospice	
Locus of Patient Care	Patient Care Pharmacist

environment have created a different public view of the roles of physicians, nurses, and other caregivers. Roles are rapidly refined and redefined. Not only is this true for the profession of pharmacy, as presented in other chapters in this book, but for all the health care professions.

In the late 1970s, the phenomenon associated with "health maintenance," emerged and a new focus on patient care developed. Initially, the new organization of health care appeared to have a more holistic, positive view of the patient. Within the patient's daily life, health care was to be totally integrated. A patient would receive care when ill, but care would also prevent illness and encourage wellness. In the best interests of the patient, health teams were formed by groups of providers to deliver optimal care through the careful integration of services.

Health provider groups initially formed to provide total, integrated care to the patient. The model was built upon the patterns of health care practice found in the large medical clinics that emerged prior to the late 1970s and were created by the cooperative medical practices of groups of physicians, such as the Mayo Clinic of Rochester, Minnesota (begun in the 1930s). The Mayo Clinic evolved from a small practice of physicians who worked together informally to an integrated group medical practice that now provides health care to a large, geographically dispersed population throughout the midwestern United States and, because of the quality of its specialty practitioners, the world. On such a large scale health care became (or at least appeared) more efficient, and the specialist practices of high-profile group practices such as the Mayo Clinic were perceived as better quality.

In fact, what emerged was something quite different from what was originally imagined. As the health care groups organized, they worked carefully to provide more efficient care. Nonphysician personnel, important to the provision of complete, integrated health care, became a part of the organization. Through a process of decreasing costs and increasing revenues of the health maintenance organization, a better picture of who was the customer and who was the patient emerged. The health care environment changed to a program of "managed health care."

Simply, "[m]anaged care is a system that, in varying degrees, integrated the financing and delivery of medical care through contracts with selected physicians and hospitals that provide comprehensive health care services to enrolled members for a predetermined monthly premium."[1] This health care package was appealing to consumers, and as patients moved to managed care plans, so did physicians and other providers. In 1993, physicians received approximately 35% of their income by participating in nonfederal managed care systems. Nearly 75% of all physicians participate in some form of organized managed care.[1]

In managed care environments, the individual practitioner was no longer practicing with complete autonomy. Policies and procedures about medical practice were developed by the medical staff on behalf of the organization. Requirements such as the use of formulary drugs, generic over brand-name prescribing, limited quantity dispensing, and closed pharmacy networks began to evolve. No longer was the physician the autonomous provider or "the captain of the team," but a diagnostician and case coordinator.[2] The definition of "primary health care providers" expanded to include family medicine practitioners, internal medicine practitioners, obstetricians and gynecologists, and physician assistants in medicine and nurse-midwives and advanced practice nurses (including nurse practitioners) in nursing. As the practice of health care evolves in the United States, the image of the physician and of other health providers will continue to change.

An interesting way to view the changes was created by David Landy, a medical anthropologist.[3] In his observations of healers in a variety of cultures, Landy defined three basic provider reactions to culture change. Commonly, providers adapt to change and integrate the new practice ideas into their provision of care. As previously stated, many health providers in the 1990s are already part of some sort of managed health care organization. Pharmacy, in response to these new forces, has adopted the principles of "pharmaceutical care." Some providers stay the same and "attenuate" their role. They do not change their practices to fit the new social environment but continue to practice as they have in the past. The demand for practitioners in the "older" role diminishes, but does continue for a gradually reducing portion of the population. There will always be a few physicians who will practice solo in medicine, and there will always be a continuing role for "compounding" pharmacists. Finally, change creates entirely new roles for health practitioners. Nurse practitioners, physician assistants, and pharmacy benefits managers, for example, have all become new members of the health care team. The number of practitioners in new roles is strongly positively associated with favorable legal/regulatory environments.

PROFESSIONALS IN SOCIETY

To explain how health professionals behave in society, one may turn to an anthropological description of the curing role.[4] There are similarities in the roles of health providers from one culture to another. These similarities also exist between health providers within a single cultural setting. One may see that society has certain expectations and limits that it places on its healers. In general, these social expectations construct a set of characteristics and describe professional behaviors for each health profession.

Practice specialization is one characteristic of the health professional role. Each profession presents itself to the patient (consumer) as performing a unique function in the health arena. Classically, and certainly from the perspective of the patient, physicians are there to diagnose and determine treatment for illnesses, nurses to provide caring that facilitates the healing, and pharmacists to dispense and assist with medications. Also, one finds that within each profession subspecialization occurs. One becomes a hospital, clinical, or community pharmacist. A physician may be an internist or a surgeon, the two largest subcategories in medicine. Further, internists may specialize in cardiology, among many other options, and surgeons may belong to one of many specialties, such as transplantation surgery.

How one selects and is selected by a health profession varies between provider roles. In the United States, students select which health profession they are best suited for based on scholastic, financial, and family characteristics and apply to a program that will educate them within that role. Health professions are often thought of as being "inherited"–children of health professionals tend to adopt their parents' profession. The educational institutional then selects students from this applicant pool based on experience, letters of reference, and other less objective factors than academic performance.

The process of becoming a health professional varies slightly among the main health professions of concern in this chapter. Each one takes some form of formal classes (although, in the case of some nurses, not college course work), and each has some form of on-the-job training or apprenticeship. The educations of the main health professionals are both similar and different. One is taught the concept of health and illness, but the foci of the different programs are on those aspects of health that are the specialties of the specific profession.

The training of these health professionals leads to some form of basic certification for society to allow professional practice. The minimum certification is defined by licensure to practice, a power held by individual states. Each of the main health professionals described in this chapter must be licensed in some fashion. Beyond that, health professions seek additional professional certification. In pharmacy, certification may be formally sought in such areas as nuclear pharmacy, pharmacotherapy, and nutrition.

Image is also an important aspect of the health professional. Where one practices, how one dresses, and how one behaves is important in defining the health professional in society. The community pharmacist of the 1980s and early 1990s practiced dressed in white coats in sterile-looking, brightly lit pharmacies with counters that served as barriers. A good example of this is the hospital-based clinical pharmacist who has been historically seen as having a specific uniform–a long white coat; the community pharmacist's uniform is a short white coat.

Payment for services is a universal aspect of the healing role. However, how one gets paid does vary greatly. Currently, the system of payment for services is undergoing change and is described elsewhere in this book. Historically, however, one can describe the different forms of payment. Generally, a fee for service was paid to physicians, employee wages to nurses, and the purchase of a product to a pharmacist. In the changing environment, many individuals in all health professions are becoming employees.

There is an acceptance by patients of the "powers" inherent in each health professional role. Prestige, achieved by title and education, allows certain actions by health professionals to go unexplained. Patients agree to surgery, the cutting and/or manipulation of internal parts or their bodies, by individuals known only to them by their titles (roles). This reflects an ultimate acceptance by the public of the healers' skills and role in society.

THE ADAPTATION OF MEDICINE

For decades, the medical profession (allopathy) has dominated the decision-making processes of the health care system. Currently, there are over 600,000 practitioners of medicine and osteopathy in the United States (Table 6.1). For many professions, physician education and practice is the predominant model to emulate.

Prospective physicians are selected based on excellence in undergraduate college education and a desire to become a physician. Many social factors, such as a family history, gender, ethnic background, and a variety of other personal characteristics tend to determine physician selection.

The educational process is long. Generally, physicians receive two years of classroom lecturing and two years of experiential training before graduating with a medical degree. What follows are several years of advanced apprenticeship leading to specialization in one or more fields and certification at the end of the process.

For the most part, physicians practice in groups, a process driven

TABLE 6.1. Number of Employed Health Professionals, 1983 vs. 1992

	1983	1992
Dentists	126,000	162,000
Nurses	1,372,000	1,805,000
Pharmacists	158,000	198,000
Physicians	519,000	614,000
Physician Assistants	51,000	Not Available

Source: Ref. 5

by social and economic factors. Moving away from the demanding time and energy of a solo practice of medicine, physicians are forming professional practice organizations with other physicians to share resources, work loads, and other aspects of patient service. Initially, specialty practice groups, such as pediatricians, were the predominant form of group practice. Managed care initiatives have developed physician practice groups that provide a wide spectrum of care.

EMERGENCE OF PHYSICIAN ASSISTANTS

The physician assistant (PA) profession was formally established in 1965 with the opening of the first PA program at Duke University. By 1971, 17 PA programs were graduating approximately 430 PAs per year. By 1992, 55 accredited programs were graduating approximately 1,600 PAs per year. As of 1993, 24,800 PAs were practicing in the United States. As the number of educational programs increases it is expected that by the year 2000, 42,700 PAs will be in practice.

A PA practices as a dependent practitioner, which means that a PA works in conjunction with a physician who supervises the PA and is legally responsible for the PA's activities. In practice, a PA would be allowed to function at whatever level his or her supervising physician deemed appropriate.

A PA's practice has always been considered to be a legal extension of the supervising physician's practice. Physician assistant laws have followed this principle, and all 50 states now have laws that allow PA practice dependent upon physician supervision. In the 1970s, PAs were commonly under the legal jurisdiction of state medical boards; however, more recently, the trend has been for PAs to have licensing boards independent from physician licensing boards.

When the profession was created, wide variations in training, role definition, and legal status made it difficult for PAs to address the issue of prescribing seriously. Generally, prescribing was not addressed in early legislation, or it was specifically not allowed. In the early 1970s, many PAs practiced in rural and underserved settings, encouraging support for allowing PAs to prescribe because the realities of such practice settings necessitated it. In 1977, PAs had explicit legal prescribing authority in eight states, and by 1993, 36 states had legislation allowing prescribing by PAs.

In the 1970s, early PA prescribing legislation was often limited to PAs practicing in rural or underserved areas, and drug choices were often limited to a formulary from which addictive or potentially dangerous drugs were excluded. As PA practice has migrated away from a protocol-oriented philosophy toward one of individually negotiated supervised practice, prescribing legislation has taken on the philosophy that the PA's prescription is merely an extension of the supervising physician's prescribing privilege. Consequently, the most recent trends in PA prescribing legislation avoid the use of formularies and generally do not restrict PA prescribing to any specific geographical area or practice type. Most states have some restrictions concerning the prescribing of certain classes of controlled drugs, but again, the trend has been toward the removal of restrictions.

In the late 1980s, the pharmaceutical industry began to recognize the effects of liberalized prescribing laws. Sales and marketing activities were directed at PAs, thus confirming that prescriptions originating from PAs were becoming a noticeable part of the market share. Currently, most pharmaceutical companies use the same marketing strategies for PAs as for physicians.

In 1992, the U.S. Drug Enforcement Agency (DEA) began issuing registration numbers to PAs in states that permitted PAs to prescribe controlled substances. Prior to this action, prescriptions for controlled substances originating from PAs were filled under the supervising physician's DEA number. The DEA's action illustrates the general trend of increased prescribing rights for PAs.

While the issue of liberalizing PA prescribing laws has become less controversial, less consensus exists concerning drug dispensing by PAs. Laws regarding drug dispensing by PAs vary widely from state to state, although the issue is often discussed in debates relating to the liberalization of PA prescribing laws. Most states have viewed PA dispensing in the same fashion as other PA issues: PA dispensing is a logical extension of the supervising physician's practice, and thus PA dispensing is generally permitted under the authority of the supervising physician. Many states have limited dispensing privileges for PAs to specific settings where registered pharmacy services are not conveniently available, such as rural or underserved areas. Generally, where PAs have increasingly gained prescribing privileges, they have also gained dispensing privileges.

In the last 25 years, PA prescribing has progressed from no legal authority to prescribing laws in 36 states. Health care reform proposals generally call for the increased utilization of PAs, and health care policy researchers generally view restrictive prescribing laws as a barrier to PA utilization. It is expected that PA prescribing and dispensing laws will continue to be liberalized. It is likely in the future that a PA working under a physician's supervision will have the same prescribing and dispensing privileges as a physician.

An interesting question in pharmacy circles is whether the addition of a PA degree would be appropriate for advanced-practice pharmacists.[6] The advanced education would add diagnostic skills and certification in a time frame similar to advanced pharmacy doctorate training. This discussion is just beginning.

THE EVOLUTION OF NURSING

Nursing has been defined as the "diagnosis and treatment of human responses to actual or potential health problems."[7] In fulfilling their professional role, nurses may engage in a broad range of roles including: direct care provider, counselor, health educator, service coordinator, collaborator, consultant, and patient advocate.

Nurses represent the largest single group of health professionals in the United States. Nearly 1.8 million nurses are licensed to practice, about three times the number of physicians and nine times the number of pharmacists (Table 6.1). Nursing is an evolving profession with multiple levels of educational preparation and practice settings. Although the number of male nurses has increased dramatically in recent years, nearly 98% of all nurses are women.

Preparation for licensure as a registered nurse may be obtained by graduation from one of three types of nursing education programs: associate degree, diploma, and baccalaureate. Associate degrees are awarded after two years of nursing education in a community (junior) college. Diploma preparation means that the nurse has graduated from a three-year hospital nursing program. A baccalaureate degree is awarded through a four-year collegiate program. In some locations, nurses may receive preparation for basic nursing practice through master's or doctoral entry-level programs. However, overwhelmingly, the masters and doctoral degrees indicate advance practice/research preparation. In 1965, the American Nurses Association endorsed the baccalaureate degree as the basic preparation for professional nursing practice (noting that the two-and three-year programs provide a more technical level of preparation). However, the labor-force demands have been for nurses at the technical level. Approximately 1,500 programs in the United States provide entry-level nursing education; advanced preparation may be obtained at the 174 master's-level graduate collegiate programs and the 52 doctoral programs.

Following a basic educational program, nurses must pass a national examination to receive licensure in any state. Within each state, nursing practice is guided by the state practice statutes as well as the professional organization (American Nurses Association) standards of practice. Other professional practice standards (such as the Joint Commission for the Accreditation of Health Organizations) may also apply according to practice setting and clinical type.

Education for advanced clinical practice now occurs at the master's level. Baccalaureate-prepared nurses attend two-year programs in an advanced practice area, specialized clinical area (family, pediatric, geriatric practitioners and nurse-midwifery), or clinical specialty (critical care, oncology). Following completion of the master's program, nurses sit for certification through a designated national organization (e.g., American Nurses Association, National Association of Critical Care Nurses, American College of Nurse-Midwives). The scope and requirements of advanced nursing practice are subject to state laws, rules, and regulation.

Nursing is practiced in hospitals, long-term care facilities, community health agencies, clinics, occupational health, school offices, and hospices. Most nursing practice is institution based, rather than an independent, entrepreneurial profession such as pharmacy or medicine. As a result, most nurses have a dual or triple pattern of authority (administrative, physician, nursing supervisor). About 75% of all nurses are employed by hospitals or nursing homes.

In hospitals, nurses are responsible for the overall care and comfort of the patient. Nurses are also responsible for patient assessment and implementation of the therapeutic plan. While some of the therapeutic plan is developed by physicians or other health professionals (respiratory therapists, pharmacists), nurses also provide independent care ranging from preventive care (turning to avoid pressure sores) to educating patients and their families about home management of conditions after discharge.

Long-term care facilities are the second largest practice setting for nurses. Nursing practice in these facilities varies greatly depending on the institutional factors, but generally parallels hospital practice.

Community (public) health nursing is the third largest practice setting. In these agencies, nurses generally work with a team consisting of the nurse, social worker, nutritionist, dentist, physical therapist, physician, and pharmacist. Care offered through these agencies includes maternal-child care and counseling, control and monitoring of communicable diseases, care for persons with mental illness and drug/alcohol problems, and home health care. Nurses in the community are prepared to perform a broad spectrum of primary care interventions, including health assessment, risk appraisal, health education and counseling, as well as collaborative care as specified by physicians. Advanced practice nurses (registered nurse practitioners) also are prepared to diagnose and manage common acute and chronic illnesses.

An important responsibility of the nurse is the elicitation of the patient's medication history, even when this is obtained independently by other practitioners. Nurses obtain a medication history that includes current and recent prescription and nonprescription use, allergies and previous drug responses, and compliance.

In hospitals or long-term care facilities, nurses are responsible for the administration and monitoring of medications for patients. Specifically, nurses are responsible and accountable for ensuring that the right drug at the right dosage is given by the right method at the right time to the right patient and that the patient's questions are answered correctly. The nurse often prepares medications (independently, or collaboratively with pharmacists). For example, in the many settings where medications are not provided through a unit dose method, the nurse is responsible for checking the accuracy of the medication order, determining that the patient's clinical condition still warrants the administration of the medication, and making sure that the medication is the correct one. After administering the medication, the nurse evaluates the reaction to the drug and records the administration of the drug and any unusual response to it.

An important aspect of the nurse's role related to medications is to teach the patient about the medication regimen. This program includes appropriate administration, drug actions, and side effects, as well as the potential interactive effects of the prescribed medication with foods, alcohol, and over-the-counter remedies.

In most states, with advanced practice legislation, nurses may prescribe medications for patients in their practice specialty (geriatric nurse practitioner, pediatric nurse practitioner, nurse-midwife). Prescription privileges may be contested by other practitioners (physicians and pharmacists) but are authorized through state legislated nurse practice acts.

Nurses provide a broad array of care services to patients in a wide variety of care settings. The interventions that nurses employ may be either independent or collaborative and often overlap with the scope of practice of other health professionals, including pharmacists.

THE PRACTICE OF DENTISTRY

Dentistry is a health profession commonly associated with pharmacy. As one of the major health providers in a community, dentists frequently prescribe analgesics and antibiotics for their patients. Their education and practice aspects are similar to other primary healers. Usually, dentists achieve their degree in a multiyear post-baccalaureate program that includes classroom and experiential training.

OTHER HEALING PROFESSIONS

Many other professionals work with pharmacy practitioners on a regular basis. As stated earlier, the most common reason is related to patient pharmacotherapy. Both prescribers and patient advocates work closely with pharmacists. Other prescribers that pharmacists may have professional contact with include podiatrists, clinical psychologists, and optometrists. The recognition of these practitioners as prescribers varies from state to state. Their roles are generally limited to specified diagnoses and/or certain medications. Podiatrists, for instance, are limited to diagnoses concerning the feet.

There are other health professionals who act in a healing capacity with patients. These include medical social workers, health aides, and volunteers in patient advocate roles, such as members of a hospice program or from the Alliance for the Mentally Ill. Often in these interactions it is difficult to determine one's position regarding issues of patient confidentiality and client protection. The professional and lay patient advocates may pick up patient medications and receive patient-specific educational information to take back to the patient. This complicates the pharmacists' caregiving role.

PRESCRIBING AND DISPENSING

In contemporary pharmacy practice, it is the medical activities surrounding pharmacotherapy that usually, but not always, bring pharmacists and other health providers together. The common scenario: A diagnosis is made and the proper pharmacotherapy is determined; the patient receives a prescription that is taken to a pharmacy; the pharmacist provides information, the opportunity for counseling on medication issues and the drug product.

Who diagnoses the infirmities of the body? This, we know, is the primary role of the physician and is dominated by the medical profession. Other major or emerging diagnosticians include dentists, advanced practice nurses (e.g., nurse practitioners), and physician assistants. These professionals are also the major prescribers, using the prescription as the basis for summing up and carrying forward plans of action.[8] Because the physician is the most frequent prescriber, more is known about physician activities. This section focuses on our understanding of physician prescribing behaviors.

Prescribers learn about medications from many sources. Randomized clinical trials do influence drug use, but the information does not always reach the general prescribers. When the pharmacotherapeutic information does get to prescribers, they may act contrary to the indicated action. Promotion by pharmaceutical firms does provide information to prescribers.

A large number of social factors influence prescribing. In physician encounters, female patients tend to generate different discussions regarding health than male patients, and women get more prescriptions than men. Even the source of payment for the patient's care, such as Medicaid recipient versus private-pay patient, can indirectly determine whether a patient will get prescription or nonprescription drugs.

The environment in which the prescriber works influences prescription writing. Socially integrated physicians tend to learn from others, and practice settings tend to alter prescribing behaviors. A work environment that contributes to low job satisfaction is more likely to lead physicians to inappropriate prescribing and permitting ancillary staff to make more prescribing decisions. Geographic variation can and does occur for prescribing patterns or frequencies of selected surgical procedures. These patterns of health care appear to be based on clinical judgments and are not always explainable by health/illness patterns or the delivery of substandard or inappropriate care. Pharmacist practice patterns can and do influence prescribing.

Who dispenses medications? We know that the pharmacist is seen by both providers and patients as the primary dispenser of medications. Nearly all hospitals now have pharmacists as the primary dispensers of medications to patient care units. Nurses, physicians, and medication technicians (administratively pharmacy- or nursing-based) usually are the providers who actually give the med-icine to the patient. In hospital settings, pharmacists may administer medications as a part of their practice, but this is not common.

To physicians, it is important that pharmacists undertake activi-ties directly associated with dispensing–filling prescriptions, coun-seling patients on these prescriptions, and checking medication instructions. To a lesser extent, physicians feel pharmacists should counsel on nonprescription drugs. Physicians do not find it impor-tant that pharmacists advise patients on health matters (Table 6.2).

In ambulatory settings, the community pharmacist, working in the local pharmacy setting, dispenses medications to patients. Traditionally, the actions of pharmacists were not easily separated from the physical pharmacy environment; the pharmacist as busi-ness person was not easy to distinguish from the pharmacist as a health care provider. But it is not the pharmacist alone who dis-penses in the ambulatory setting. Physicians dispense medications in the normal course of their activities. It is estimated that approxi-mately 4 to 6% of all physicians dispense full therapeutic courses of medications to their patients. Dentists also provide medications in the ambulatory setting. Usually confined to antibiotics and pain

TABLE 6.2. Perceptions of Important Pharmacist Functions

	Percentage of Pharmacists Reporting As Important	Percentage of Physicians Reporting As Important
Advising Patients on Health Matters	57	6
Filling Prescriptions	46	64
Checking Physician's Instructions on Prescriptions	37	22
Counseling Patients on Prescriptions	20	37

Source: Ref. 9.

relievers, dentists may dispense full therapeutic regimens to their patients.

There are other opportunities beyond the notion of the prescription/drug product when the pharmacist participates as a part of the health care team. The pharmacist in the ambulatory setting is ideally established to be a health resource to the members of the community. In that capacity, the pharmacist may be called upon to recommend that patients contact clergy, social workers, psychologists, public health officials, etc. Although this broad notion of public health pharmacy may not easily be applied to the hospital and long-term care pharmacist practices, these opportunities exist no matter where a pharmacist practices.

PHARMACISTS AND THE HEALTH CARE TEAM

In general, the pharmacist's role on the health care team is to optimize pharmacotherapy. Professional activities related to the medication regimen of the patient are carefully documented. Patients receive care directed at their requirements for pharmaceutical interventions. Medications are provided by pharmacists in a timely and cost-effective fashion, and, in some settings, are administered by the pharmacist.

Several advantages are added to the health care team by the pharmacist. The pharmacist can save time for other members of the team on issues related to pharmacotherapy. The pharmacist is there to consult with and to educate practitioners and patients on the basic principles of medication use and patient-specific treatment. Patients can be monitored by the pharmacist to determine if the medications are being properly used, that no untoward effects of the drugs are causing harm to the patient, and that the regimen is improving the patient's health.

In performing the role of the medication specialist on the health care team, the pharmacist works with controversial issues regarding the domains of other health professionals. First, who has the authority to prescribe medications? In some pharmacy roles, prescribing is sought by pharmacists. Second, who has the final decision on the selected drug therapy? Pharmacist selection of manufacturer source (generic substitution, drug product selection) does, in some instances, change the prescriber's intended therapy. This is clearly seen when a pharmacist goes beyond drug product selection to therapeutic interchange, where the pharmacist changes the class of drugs selected by the prescriber (a cephalosporin may be changed to a penicillin). In the end, it is a question of who makes what decisions on the team.

To ensure optimal communication with physicians regarding patient therapy, there are several activities that pharmacists can undertake.[10] First, pharmacists can create practice patterns that provide information for patient care. Structured care plans, formal and systematic evaluations of the patient's conditions and therapy, should be developed for each patient. Skills in delivering care, such as monitoring, laboratory testing, and counseling should be expanded. Pharmacists can begin to document their patient care activities and, with documentation, begin to measure and record outcomes associated with the pharmacotherapeutic (and other) aspects of patient care. These can be stored in computerized retrieval systems and recalled when needed or transmitted as a matter of record and information to the patient's physician.

In working with the health providers on either formal or informal health care teams, the role of the pharmacist is clear.[11] The pharmacist must focus on the patient and the pharmacotherapy of their illness, suggesting alternatives for treatment and being available for treatment-related questions. During the course of treatment they must monitor the patient's progress. In interactions with the health care team, optimal care will be achieved by taking the initiative on medication issues and explaining the rationale behind pharmacist decisions. Above all, listening to each person involved in whatever role or setting will optimize not only patient health but the practice of pharmacy.

REFERENCES

1. Inglehart JK. Physicians and the growth of managed care. N Engl J Med 1994;31:1167-71.

2. Fuchs VR. Who shall live? New York: Basic Books, 1974.

3. Landy D. Role adaptation: traditional cures under the impact of western medicines. In: Landy D, ed. Culture, disease, and healing. New York: Macmillan, 1977:468-81.

4. Foster GM, Anderson BG. Medical anthropology. New York: Alfred A. Knopf, 1978.

5. Bureau of the Census. Statistical abstract of the United States, 1993. Washington, DC: Government Printing Office, 1993.

6. Dean JO Jr. Pharm.D. plus P.A.? AACP News 1994;25(11):4-5.

7. Anon. Nursing: a social policy statement. Kansas City, MO: American Nurses Association, 1980.

8. Smith M. The relationship between pharmacy and medicine. In: Mapes R, ed. Prescribing practice and drug usage. London: Croom Helm, 1980:157-200.

9. Anon. Schering Report XI. Pharmacists and physicians: professional allies or professional adversaries? New York: Schering/Key Laboratories, 1989.

10. Klotz RS. Pharmacist-physician link: keys to effective outcomes management. Am Pharm 1994;NS34(10):46-58.

11. Timmerman S. How to work with physicians. Am Pharm 1992;NS32(2): 39-40.

BIBLIOGRAPHY

1. Anon. Schering Report V. Pharmacists and physicians: attitudes and perceptions of two professions. New York: Schering/Key Laboratories, 1983.

2. Avorn J, Chen M, Hartley R. Scientific versus commercial sources of influence on the prescribing behavior of physicians. Am J Med 1982;73:4-8.

3. Ayanian JZ, Hauptman PJ, Guadagnoli E, et. al. Knowledge and practices of generalist and specialist physicians regarding drug therapy for acute myocardial infarction. N Engl J Med 1994;331:1136-42.

4. Batey M, Holland J. Prescribing practices among nurse practitioners in adult and family health. Am J Public Health 1985;75:258-62.

5. Beisecker A, Beisecker T. Patient information-seeking behaviors when communicating with doctors. Med Care 1990;28:19-28.

6. Bertakis K, Callahan E, Helms J, et. al. The effect of patient health status on physician practice style. Fam Med 1993;25:530-5.

7. Carter B, Helling D, Jones M, et. al. Evaluation of family physician prescribing: influence of the clinical pharmacist. Drug Intell Clin Pharm 1984;18: 817-21.

8. Chassin MR, Kosecoff J, Park RE, et. al. Does inappropriate use explain geographic variations in the use of health care services? JAMA 1987;258:2533-7.

9. Greco PJ, Eisenberg JM. Changing physicians' practices. N Engl J Med 1993;329:1271-4.

10. Lamas G, Pfeffer M, Hamm P, et. al. Do the results of randomized clinical trials of cardiovascular drugs influence medical practice? N Engl J Med 1992;327:241-54.

11. Leedy J, Schlager C. A unique alliance of medical and pharmaceutical skills. J Am Pharm Assoc 1976;NS16:460-2.

12. Lipton H, Bero L, Bird J, McPhee S. The impact of clinical pharmacists' consultations on physicians' geriatric drug prescribing. Med Care 1992;30: 646-58.

13. Mechanic D. Medical sociology. 2nd ed. New York: Free Press, 1978.

14. Mechanic D. Physicians. In: Freeman HE, Levien S, Reeder LG, eds. Handbook of medical sociology. 3rd ed. Englewood Cliffs, NJ: Prentice Hall, 1979:177-92.

15. Melville A. Job satisfaction in general practice: implications for prescribing. Soc Sci Med 1980;14A:495-9.

16. Meyer BM. Prescribing authority for nonphysicians. Am J Hosp Pharm 1994;51:308-11.

17. Miller DW, Poirier S, McKay JS, et al. Does drug insurance influence physician prescribing? J Res Pharm Econ 1993;5(2):19-35.

18. Mick SS, Moscovice I. Health care professionals. In: Williams SJ, Torrens PR, ed. Introduction to health services. 4th ed. Albany, NY: Delmar Publishers, 1994:269-96.

19. Segal HJ. The social risks and benefits of prescription writing. In: Mapes R, ed. Prescribing practice and drug usage. London: Croom Helm, 1980:19-32.

20. Sekscenski ES, Sansom S, Bazell C, et. al. State practice environments and the supply of physician assistants, nurse practitioners, and certified nurse-midwives. N Engl J Med 1994;331:1266-71.

21. Smith M. The prescription: everything you wanted to know but didn't think to ask. Am Pharm 1978;NS18(16):30-33.

22. U.S. Department of Health and Human Services. Secretary's Commission on Nursing: final report. Washington, DC: Government Printing Office, 1988.

23. Verbrugge L, Steiner R. Prescribing drugs to men and women. Health Psychol 1985;4(1):79-98.

24. Weeks WB, Wallace AE, Wallace MM, Welch HG. A comparison of the educational costs and incomes of physicians and other health professionals. N Engl J Med 1994;330:1280-6.

25. Wennberg JE, Freeman JL, Culp WJ. Are hospital services rationed in New Haven or over-utilized in Boston? Lancet 1987;(May 23):1185-9.

ADDITIONAL READINGS

For the Instructor

Dentists

1. Anon. Principles for dentist-pharmacist relationships. J Am Pharm Assoc 1975;NS15:588-9.

2. Anon. Principles for dentist-pharmacist relationships: guidelines developed in Connecticut. J Am Dent Assoc 1976;93:237-8.

3. Anon. New program promotes interaction between dentists and pharmacists. Hosp Form 1977;12:618.

4. Elliott P. Dentists and pharmacists–is it time to get your act together? Dental Econ 1979;July:49-54.

5. Young WO, Cohen LK. The nature and organization of dental practice. In: Freeman HE, Levien S, Reeder LG. Handbook of medical sociology. 3rd ed. Englewood Cliffs, NJ: Prentice Hall, 1979;193-208.

Nurses

6. Hall J. How to analyze nurse practitioner licensure laws. Nurse Pract 1993;18(8):31-4.

7. Lerner D, D'Agostino R, Musolino J, Malspeis S. Breaking with tradition: the new groups in professional nursing. Med Care 1994;32:67-80.

8. Mahoney DF. Nurse practitioners as prescribers: past research trends and future study needs. Nurse Pract 1992;17(1):44-51.

9. Ochs G. Mutual areas of cooperation between pharmacy and nursing. Am J Hosp Pharm 1960;17:475-8.

10. Reeder SJ, Mauksch H. Nursing: continuing change. In: Freeman HE, Levien S, Reeder LG. Handbook of medical sociology. 3rd ed. Englewood Cliffs, NJ: Prentice Hall, 1979:209-29.

Pharmacist-Physician Interactions

11. Albro W. How to communicate with physicians. Am Pharm 1993;NS33(4):59-61.

12. Bernstein L, Klett E, Jacoby K. Physicians' attitudes toward the use of clinical pharmaceutical services in private medical practice. Am J Hosp Pharm 1978;35:715-7.

13. Bootman JL, Hurd P, Gaines J. Physician and pharmacist attitudes toward medication use. Am J Hosp Pharm 1982;39:818-21.

14. Briggs G. Pharmacist-physician drug consultations in a community hospital. Am J Hosp Pharm 1974;31:247-53.

15. Brown D, Helling D, Jones M. Evaluation of clinical pharmacist consultations in a family practice office. Am J Hosp Pharm 1979;36:912-5.

16. Bryant S, Guernsey B, Pearce E, Hokanson J. Pharmacists' perceptions of mental health care, psychiatrists, and mentally ill patients. Am J Hosp Pharm 1985;42:1366-70.

17. Cowen D. Changing relationships between pharmacists and physicians. Am J Hosp Pharm 1992;49:2715-21.

18. Crackower S. Pharmacists were sabotaging my patient care. Am Pharm 1984;NS24(9):6-8.

19. Davis R, Crigler W, Martin H. Pharmacy and family practice. Drug Intell Clin Pharm 1977;11:616-21.

20. Ekwo E, Hendeles L, Weinberger M. Those who make decisions about management of children with asthma: pharmacist-physician interaction. Am J Hosp Pharm 1978;35:295-9.

21. Fisher D, Pathak Dev. Influence of attitudes, normative beliefs, and situational variables on physicians' use of pharmacists as drug information consultants. Am J Hosp Pharm 1980;37:483-91.

22. Grussing P, Goff D, Kraus D, Mueller C. Development and validation of an instrument to measure physicians' attitudes toward the clinical pharmacist's role. Drug Intell Clin Pharm 1984;18:635-40.

23. Haxby D, Weart CW, Goodman B. Family practice physicians' perceptions of the usefulness of drug therapy recommendations from clinical pharmacists. Am J Hosp Pharm 1988;45:824-7.

24. Lambert R, Wertheimer A, Dobbert D, Church T. The pharmacist's clinical role as seen by other health workers. Am J Hosp Pharm 1977;67:252-3.

25. McKay A, Jackson R. Attitudes of primary care physicians toward utilization of the pharmacist versus the physician's assistant in patient care. Am J Pharmacy 1976;148(Nov-Dec):157-67.

26. Morley A, Jepson MH, Edwards C, Stillman P. What do doctors think of pharmacists treating minor ailments? Pharm J 1983;231(Oct):387-8.

27. Phillips R, Lipman A. Physicians' perception of clinical pharmacy service. Hosp Form 1981;16:988-1000.

28. Ritchey F, Raney M. Effect of exposure on physicians' attitudes toward clinical pharmacists. Am J Hosp Pharm 1981;38:1459-63.

29. Segal R, Grines L, Pathak D. Opinions of pharmacy, medicine, and pharmaceutical industry leaders about hypothetical therapeutic-interchange legislation. Am J Hosp Pharm 1988;45:570-7.

30. Tootelian D, Blackburn W. How to build more productive relations with physicians. Calif Pharm 1985;(Nov.):22-35.

Physicians

31. Anderson L. Keeping tabs on prescription writers. Minn Med 1990;73:15-6.

32. Bush P. Prescribing is a social process. Pharm Int 1980;(Nov):VIII-X.

33. Friedson E. Profession of medicine. New York: Dodd, Mead, & Co., 1970.

34. Friedson E. Doctoring together: a study of professional social control. New York: Elsevier, 1978.

35. Fruen M, Cantwell J. Geographic distribution of physicians: past trends and future influences. Inquiry 1982;19:44-50.

36. Mechanic D. Practice orientations among general practitioners in England and Wales. Med Care 1970;8:15-25.

37. Penchansky R, Macnee C. Initiation of medical malpractice suits: a conceptualization and test. Med Care 1994;32:813-31.

38. Perri M, Kotzan J, Carroll N, Fincham J. Attitudes about physician dispensing among pharmacists, physicians, and patients. Am Pharm 1987;NS27(10): 57-62.

39. Robertson D, Groh M, Papadopoulos D. Family pharmacy and family medicine: a viable private practice alliance. J Fam Pract 1980;11:273-7.

40. Soumerai SB, McLaughlin TJ, Avorn J. Improving drug prescribing in primary care: a critical analysis of the experimental literature. Milbank Q 1989;67:268-317.

41. Spencer J, Edwards C. Pharmacy beyond the dispensary: general practitioners' views. Br Med J 1992;304:1670-2.

42. Waitzkin H. Doctor-patient communication. JAMA 1984;252:2441-6.

43. Weeks WB, Wallace AE, Wallace MM, Welch HG. A comparison of the educational costs and incomes of physicians and other health professionals. N Engl J Med 1994;330:1280-6.

44. Williams CB. Everybody wants to play doctor. Med Econ 1992;(Jan):37+.

Physician Assistants

45. Crandall L, Santulli W, Radelet M, et. al. Physician assistants in primary care. Med Care 1984;22:268-82.

46. Duryea W, Moyer J. Commentary on regulations for physician assistants. Pa Med 1990;93:32-4.

47. Fincham J. How pharmacists are rated as a source of drug information by physician assistants. Drug Intell Clin Pharm 1986;20:379-83.

48. Mittman D, Mirotznik J. PA prescribing behavior and attitudes: a profile. Physician Assistant 1984;(Mar):15+.

49. Repicky P, Mendenhall R, Neville R. The professional role of physicians assistant in adult ambulatory care practices. Eval Health Professions 1982;5: 283-301.

For the Student

1. Batey M, Holland J. Prescribing practices among nurse practitioners in adult and family health. Am J Public Health 1985;75:258-62.

2. Bush P. Prescribing is a social process. Pharm Int 1980;(Nov):VIII-X.

3. Cowen D. Changing relationships between pharmacists and physicians. Am J Hosp Pharm 1992;49:2715-21.

4. Duryea W, Moyer J. Commentary on Regulations for Physician Assistants. Pa Med 1990;93:32-4.

5. Klotz RS. Pharmacist-physician link: keys to effective outcomes management. Am Pharm 1994;NS34(10):46-58.

6. Lipton H, Bero L, Bird J, McPhee S. The impact of clinical pharmacists' consultations on physicians' geriatric drug prescribing. Med Care 1992;30: 646-58.

7. Mahoney DF. Nurse practitioners as prescribers: past research trends and future study needs. Nurse Pract 1992;17(1):44-51.

8. Meyer B. Prescribing authority for nonphysicians. Am J Hosp Pharm 1994;51:308-11.

9. Mick SS, Moscovice I. Health care professionals. In: Williams SJ, Torrens PR, eds. Introduction to health services. 4th ed. Albany, NY: Delmar Publishers, 1994:269-96.

10. Soumerai SB, McLaughlin TJ, Avorn J. Improving drug prescribing in primary care: a critical analysis of the experimental literature. Milbank Q 1989;67:268-317.

11. Tootelian D, Blackburn W. How to build more productive relations with physicians. Calif Pharm 1985;(Nov):22-35.

Chapter 7

Unorthodox Healing Systems

Albert I. Wertheimer

INTRODUCTION

Nothing is obvious. Something may appear to be obvious to all from one person's perspective, but the very next person may see that same item or solution not only as unobvious but as strange, even ridiculous. For better or worse, health care philosophy and treatment are included in the realm of different perceptions in different cultures, areas, age groups, religions, etc.

It should not be any surprise that if we were to ask ten residents in any street, "What do you do for a dry cough?" we could expect to receive seven, eight, or even nine different replies. In fact, there is even room within the domain of orthodox healing systems for wide diversities in approaches to symptoms and to disease states. We would expect a spectrum or continuum of triggers to which people respond across systems. However, some might go to the hospital emergency room upon the very first cough, while others might get around to self-medicating eventually. Still others might visit a primary care physician only after they have coughed up blood.

Around the world, the primary care provider could be an acupuncturist, a homeopath, a chiropractor, an herbalist, a witch doctor, a village elder, a nurse, an Oriental traditional healer, an osteopathic physician, a religious healer, or any of the practitioners of hundreds of different healing systems encountered around the world.

On this topic, another word is in order. Something may be considered unorthodox in one place or at one time by a certain group of followers, but it can be quite orthodox in other locales or situations. For example, 25 years ago, acupuncture was considered in the West as a type of quackery or as some Oriental placebo, but today, it has passed with high marks when it has been scrutinized by Western medical leaders. Acupuncture treatments can be found routinely in the West these days, and even insurance companies pay for it in many situations.

Payer points out in her extremely good, eye-opening book that even within the so-called orthodox health care system, medical cultures vary throughout the world.[1] Payer tells us how the French associate many pathologies with the liver in the way the Germans do with the heart and how high tech is mistrusted, avoided, or embraced immediately in various countries. Even with something as basic as medication dosage forms, where one would think little controversy could exist, there is an immense spectrum of orientations. Although it must be unfathomable to most Americans, some Europeans, and especially the French, use suppositories as a favored dosage form. Injections are very common and even expected in numerous Asian nations when one visits a physician. Well warned that a broad range of opinions, beliefs, and attitudes exists, it is time to explore what is orthodox and what has come to be called unorthodox.

Orthodox medicine is called allopathic medicine. Its practitioners are referred to as medical doctors (MD in the U.S. and MB in the U.K.). They study anatomy, physiology, pathology, pharmacology, and surgery. The germ theory is a key foundation here. Drugs and surgery are the principal tools to eradicate disease states or to provide symptomatic relief. Allopathic medical practitioners can be found in virtually every nation of the world, and this is rapidly becoming the dominant medical model as the Chinese pay greater attention to it.

Why then, one might ask, would one go to a different type of practitioner? There must be hundreds of answers to that question. Perhaps one of the major reasons is that allopathic medicine sometimes admits its limits and acknowledges that it cannot help further. We find this in disease states lacking effective cures (some viral infections, cancers, tumors), in other conditions where complete relief cannot be provided (some lower back pain problems), and in a host of inflammatory autoimmune diseases (rheumatoid arthritis, scleroderma, lupus, etc.).

Other reasons for consulting unorthodox healers include knowledge that someone else in the past, usually a relative, friend, or celebrity could not be helped by orthodox medicine when he or she had the same medical condition. Some persons use unorthodox healers because these healers have a toehold in a community or are more accessible, such as in a community where a chiropractic college and clinic exist. If one grew up in an area or within a subculture where an unorthodox system had made inroads, one can be expected to patronize such practitioners. If one was raised as a child in an area where one's parents used herbalists or traditional Chinese doctors, it is likely to expect that the next generation will continue such practices. This would be the same for persons who were exposed to homeopaths, naturopaths, osteopaths, holistic healers, etc.

The unorthodox healing systems can, for the sake of description, be categorized into one of several healing system types. The most common categories are:

1. Physical therapies
2. Hydrotherapy
3. Nutrition
4. Plant-based therapies
5. Wave and radiation
6. Mind and spirit healing
7. Self-exercise
8. Comprehensive systems

PHYSICAL THERAPIES

Physical Disciplines

In recent years, there has been a massive increase in interest and participation in a host of physical disciplines intended to help people remain healthy, look young, feel good, and have fun. Perhaps the greatest component here is the physical fitness arena. It is nearly impossible to avoid seeing joggers, and it is estimated that for each jogger there are at least two other people engaged in swimming, aerobics, bicycling, racquet sports, handball, team sports, and other non-Western martial arts.[2]

In working with the body, touch, one of the most natural human activities, is seen in a large number of formats. Some of these are designed to achieve greater relaxation abilities or states. Others attempt to improve body function and alignment through the spatial alignment or relationships among major body organs. Still other techniques enhance sensory awareness. Finally, the psychotherapeutic domain is applied to the relationship between body chemical levels and stress, tension, and fatigue. Of course, there is overlap when one creates artificial categories.

Massage

This is most often a rubbing of areas of the body with the hands or other apparatus to relieve muscle tension, stimulate circulation, and enhance joint flexibility. Esalen massage is a soft and gentle means to enhance relaxation by a smooth and rhythmical stroking of the skin. Swedish massage uses heavier pressure intended to tone muscles. Shiatsu is intended to relieve fatigue through the application of direct pressure on a sequence of points, based upon ancient Chinese medical principles. Here, the origins of physical disease are thought to be due to energy imbalances in different organ systems. The Japanese have used shiatsu for a wide variety of physical ailments.

Rolfing®

Rolfing is a system for improving body alignment. It strives for structural integration through a body that is ideally aligned and therefore functions with less muscular effort. Posture is important and is improved through stretching and lengthening tissues that have grown together from past trauma, poor posture, or inappropriate movement habits.[3]

Sensory Awareness

In sensory awareness, the work focuses on relaxation and aiming attention on immediate experience, on being a child again and absorbing and sensing the world in an open and playful manner. Sensory awareness offers some of the benefits found in more formal meditation.

The major work in this domain was done by Wilhelm Reich, the first Freudian analyst to recognize that the ego, id, and superego inhabit a body.[4] His psychotherapeutic techniques directed him to examine a patient's character structure as an interrelated whole rather than to focus solely on a problem or neurosis.

Although Reich's work has been out of favor, much of his work has been the basis for newer and more well-accepted concepts. He developed the theory of orgone energy (described later in the chapter) and of treatments to dissolve blocks of muscular tension.

Acupuncture

By penetrating the skin at specific spots, disease could be cured. This philosophy came to us centuries ago when Chinese warriors with arrow or spear wounds were cured of preexisting ailments. The *Nei Ching*, published 3,000 years ago, remains the basic source of information on acupuncture.

Early acupuncture needles were carved in stone. Next came bamboo and bone and then metal of various types. Stainless steel needles are used today. Early thinkers postulated that the stone, bone, or metal had curative properties, but scientists were discovering that specific points on the body controlled certain organs, such as the kidneys, heart, and stomach.

The contemporary thinking is that there exist in the body two flows of energy named yin and yang that together become the life force. Because everything is postulated to have a force and opposition force, the yin and yang were seen akin to hot and cold, moist and dry, etc. Yang is supposed to contract and stimulate and is the positive principle, whereas yin seems to sedate and is the negative principle. Health is an equilibrium within the body and within the universe.

Traditional healers "found" that the vital energy passes in the body along paths, or "meridians," such as the lymphatic system. Their texts tell us that there are 26 major circuits, each associated with a specific organ system. It is thought that there are nearly 1,000 acupuncture points. In disease, the flows are disrupted and an imbalance results. So, by piercing the skin at certain points, energy flow is stimulated or sedated. A traditional acupuncturist will evaluate the meridians by feeling the pulse of the radial artery, with very experienced practitioners being able to discern under- or overactivity, fullness, tension, and hundreds of other characteristics. Practitioners will know which meridians require some balancing by the insertion of needles to varying depth for a duration of a few seconds to several minutes.

There are reports of successes with ulcers, digestive problems, headaches, arthritis, rheumatism, psoriasis, hypertension, anxiety, and asthma.[3]

Reflexology

Reflexology is a treatment system using a unique massage and pressure application to the feet. This procedure has also been known as zone therapy. Not so dissimilar to acupuncture, it is a system which postulates that when an organ is malfunctioning the corresponding reflex in the foot will be tender under pressure. The therapist is to relieve tension, enabling a full blood supply to distressed areas. Practitioners can make diagnoses and monitor patient progress. In addition to the reflexes, there are ten zones dividing the body longitudinally.[5]

Moxibustion

Moxibustion is a derivative of acupuncture. The herb *Artemisia vulgaris* is shaped into a cone, placed on an acupuncture point, and lit. The cone is left to smolder until the patient can no longer tolerate the heat, at which time it is rapidly removed. This is done to stimulate the life energy. Sometimes acupuncture and moxibustion are used together.

Chiropractic

Chiropractic was developed by D. D. Palmer, who told us to look to the spine for the cause of disease because it contains and protects the spinal cord, through which vital forces flow and are spread to all parts of the body. Palmer devised a healing system hypothesis that misaligned spinal segments interfere with the flow of vital forces by impinging on nerves. Manual adjustments of the misaligned spinal segments are thought to help the status and optimal functioning of organs served by those nerves.

The system has caught the attention of the public, and today it represents the largest drugless healing philosophy in the world. It is thought that many patients seeking chiropractic care have lower back problems that have not been adequately corrected by traditional medical or surgical interventions. Other patients may wish to avoid surgery. It is quite difficult to define chiropractic accurately in the late twentieth century. Manipulations (now called adjustments) of the spine only are performed by some practitioners. Others use various combinations of holistic approaches, nutritional techniques, acupuncture, etc.

Palmer did not anticipate the ability to treat infections or the growth in the incidence of malignant and cardiovascular diseases where his procedures hold little value. Nevertheless, chiropractic is a common, popular treatment system. Perhaps accessibility and lower cost add to its appeal.

There could be a great deal of debate about whether chiropractic is an unorthodox healing system or just a system different from allopathy. Chiropractic has its own colleges, requires state licensure, and acknowledges conventional science.

Others

In addition to the meridian and manipulation therapies, there are a large number of other physical systems; they can be classified into the following subgroups:

- Mineral contact
- Air and light
- Heat
- Electrotherapy

There are nearly 100 different such techniques, but for the purposes of this book, we will discuss some of the most common of these and at least representative members of each group. A robust list of selected additional readings is offered for the serious student.

Mineral Contact

Gems have always fascinated mankind. The colors they reflect have been indexed to certain disease therapies. Each color has its own vibration rate. Kirlian photographs of body parts show radiation of all seven colors of the spectrum. These same colors are the basis of what psychic observers call the aura—the light deflected and refracted by the human body. Ancient yoga related these colors to the body and divided the color energies into seven concentrations called chakras. The seven chakras also relate to the principal endocrine glands: pineal, pituitary, thyroid, thymus, suprarenal, and ovarian or testicular. Violet, for example, relates to the top of the head. Gems are also relevant to numerology and the astrological signs.

Some healers use a gem in a room for vibratory rates and retinal impressions. Others want the gem worn around the neck of a patient. The objectives vary by gem. For example, zircon is for ear tumors, as well as those in the nose or throat.

Copper

Copper bracelets are a common sight due to the folklore of copper's value in treating rheumatoid arthritis. There is little or no evidence of efficacy. However, there are infrequent reports in the literature suggesting that copper from a bracelet could combine with a chemical in sweat, perhaps an amino acid, to penetrate the skin and the venous system to reach the inflammation site.

Mud and Clay

Mud and clay have been used as treatments for thousands of years. Clay has antibiotic and antiseptic properties, and its negative ions are supposed to absorb positively ionized toxins and even radioactivity. Mud baths are thought to be therapeutic, especially for some skin conditions, but there is little evidence of efficacy in properly controlled studies.

HYDROTHERAPY

There is a convenient division for the various hydrotherapy approaches. We can separate them into external and internal activities.

Externals

Baths

Hot baths cause arteries to dilate and result in perspiration. They are employed to lose weight, to clean out the body, or to allow optimal absorption of drug substances through open pores. Salts and herbs are used for rheumatic problems, among others.

Cold baths are used for up to a few minutes and are supposed to have an invigorating or tonic effect.

A sitz bath is usually a bath in which the abdomen is submerged. Such a bath is recommended for genital, urinary, or female reproductive disorders. Drugs may be added to the water.

Douches

Douches fall into this therapy category. A douche is a spray or stream of water aimed at all or parts of the body. Douches are similar in effect to baths but provide the additional stimulation of the spray hitting the skin. Naturally, there is a wide variety of alternating hot and cold schemes.

Others

Other hydrotherapy systems include packs and compresses. An ice pack is a cold compress. These have a wide range of legitimate and unproved uses. It does appear that hot and cold packs have some beneficial effect in treating inflammatory problems.

Internals

Perhaps the most is known about mineral water therapies, enemas, and colonic irrigation, but the reader will be spared a review of the more arcane of these.[6]

Colonic Irrigation

Colonic irrigation, while of recognized value for constipation patients, is usually employed as a component of nutritional and naturopathic spa therapies. Like the enema, it cleanses the lower bowel, or colon.

Water at body temperature is instilled into the rectum (its path of departure as well) through a two-chambered tube. Repeated washings for about one-half hour each can loosen impacted fecal material. Many of colonic irrigation's claimed benefits are controversial, and some effects are even negative, such as the washing out of helpful bacteria.

Enemas

Many diseases in ancient (and not-so-ancient) times were believed to be caused by toxins or other "dirty" substances that must have entered the body. The enema is one means of inducing the elimination of unclean substances. Liquids, usually warm or soapy water, are injected into the rectum to remove unexpelled feces as the water leaves. Enemas should be relied upon for short periods of time and/or for special purposes. Long-term use disturbs the bacterial equilibrium.

Inhalation

The use of vapors for respiratory complaints is thousands of years old. Breathing the air in mines, in caves, on mountain tops, and elsewhere has been commonplace therapy. Drugs can be introduced in this fashion, but the most common procedure is the use of steam for sinus and other upper-respiratory difficulties. Modern aerosols and nasal spray products continue to employ this dosage route.

There has been a great deal written about this topic, and we can expect that inhalations will become much more important in the future, with protein, peptide, and other biotechnology products being delivered to the bloodstream through the lungs instead of being destroyed by the oral/gastric route of administration.[7]

NUTRITION

There must be thousands of schools of thought about nutrition and health. Many are quite similar but differ based upon the indigenous plants and other natural products in a region or climate. If a society obtains its nutritional requirements from a particular animal or plant family, there will be numerous healing systems that, in essence, prevent the digestion of those common nutrients.

In the twentieth century, there has been interest in natural, nonprocessed foods with the idea that commercial processing removes some beneficial principals or adds substances in the name of freshness or preservation that may be harmful to consumers.

Macrobiotics

Macrobiotics was popularized in Japan. With an appropriate diet, one may feel full of life. Foods are divided into yin or yang, and one's diet must reflect individual needs and situations at any place and/or point in time. In fact, a large number of factors may affect diet. Water, air, soil, and sunlight are taken in by the vegetable kingdom, and to eat is to partake in this entire environment. By balancing yin and yang foods, one has an opportunity for a balanced diet. Examples of such foods include:

Yin	Yang
Fruits and leaves	Dry
Hot, aromatic foods	Growing below ground
Containing water	Salty, sour foods
Growing in hot climate	Growing in cold climate

When all is said and done, macrobiotics is a common-sense approach to nutrition for healthy or ill persons.[8]

Food Supplement Therapy

Food supplement therapy is a broad and rather imprecise nomenclature for a wide variety of therapies that employ the medical qualities of some natural foods. The most common of these are mentioned in our lore, and the names are familiar to all. As with the other therapies, a serious student or reader should refer to the references found at the end of this chapter; only a barely superficial treatment is to be found here.

Apple Cider Vinegar

Apple cider vinegar has been popularized by Dr. D. C. Jarvis of Vermont. This is prepared by fermenting apple juice and is used for arthritic and related inflammatory conditions. It is to be consumed with honey and warm water.[9]

Molasses

Molasses is considered a general tonic, and it has a laxative effect as well. It is a by-product of the sugar refining process and contains protein and minerals of interest to its advocates. It was popular as a spring tonic.

Ginseng

Ginseng root has a reputation as a tonic and aphrodisiac. It has been used for thousands of years, primarily in the Orient. The root is available for tea preparation or sold in forms for chewing, mixing with foods in meals, and even as tablets. It is popular as a refresher, to counter stress, and for preventative purposes.

Honey

Honey–containing pollens, minerals, and trace elements–has been thought to be useful for allergies, as a nonspecific tonic, and as a source of energy.

Brewer's Yeast

Brewer's yeast contains B vitamins: thiamin, riboflavin, pyridoxin, and niacin. It also is a source of protein. It is commonly sold in health food stores and pharmacies in tablet form as a dietary supplement to treat a large number of medical problems associated with B vitamin needs.

Others

Other nutritional philosophies include eating only raw foods, vegetarian products, high-protein diets, high-fiber diets; megadose vitamin or mineral supplementation; and other diets restrict items instead of adding them, such as variations of periodic fasting.

PLANT-BASED THERAPIES

Perhaps the most widely used healing system in the world is the variety of diverse therapies that may be collectively called herbal medicine. Herbs may be obtained from roots, plants, trees, flowers, etc. Herbs assist in the recovery process by restoring the balance of humors and fluids the homeostasis or equilibrium of functions performed by the various organs.

Plant remedies were recorded by the Chinese and by the Egyptians more than 3,000 years ago. Little has changed beyond, perhaps, the purification and technical sophistication of the extraction processes. It is usually the case that the active ingredients are prepared as teas and other infusions, lotions, inhalants, and most other historical dosage forms.

Tinctures

Tinctures (water and alcohol solutions) were popular early dosage and storage forms. After positive identification of the plant, the extract is prepared according to the traditions of a culture. Herbal products are said to have efficacy for virtually all known pathologies. In fact, various compendia provide compilations of plant remedies having effectiveness for most of the commonly encountered conditions. Some examples include:

- Anxiety
- Appetite improvement

- Asthma
- Boils
- Bronchitis
- Colds and flu
- Constipation
- Cramps
- Depression
- Diarrhea
- Eczema
- Gout
- Insomnia
- Menopausal symptoms
- Migraine
- Nausea and vomiting
- Rheumatism
- Ulcerations
- Warts
- Wounds[10]

Herbal practitioners study at formal educational institutions in the Orient and by apprenticeship in the West. Interest is increasing as greater numbers of patients fear the side effects of synthesized organic chemicals and desire a back-to-nature or green-wave orientation. In Western Europe and the United States, this movement is, in large part, moving hand-in-hand with another movement attempting to demedicalize common health problems and to return control to consumers.

Aroma Therapy

Aroma therapy is mentioned in ancient writing, but its mention as a healing system independent of other concepts is more recent. René Gattefossé, a French cosmetic chemist, is credited with repopularizing this system. His discovery of the skin treatment properties of essential oils began the field. Others in France, England, and the United States continued his work, discovering additional essential oils having clinical effectiveness. When massaged into the skin in vegetable oil vehicles, absorption of essential oils is excellent, leading some to speculate about systemic treatment possibilities. Moreover, the volatile nature of the oils leads some to believe that respiratory absorption takes place.

Less is known about several other plant-based therapies that are mentioned in professional journals and in the lay press from time to time. These include Bach flower remedies, vita flour, and exultation of flowers.

WAVE AND RADIATION

In the nineteenth and twentieth centuries, there has been much fascination with electricity, electromagnetic radiation, ionizing radiation, magnetism, and other wave forms (such as vibrations) for their possible health benefits. Many cults, schools, philosophies, and healing systems predated the rigorous testing of treatments today via randomized clinical trials using double blinding (neither practitioner nor patient knows whether he or she has actual drug or placebo). Nevertheless, let us look at some of the most well-known wave and radiation systems.

Orgone Therapy

Wilhelm Reich developed the concept of orgone therapy in the 1930s. Orgone therapy employs an accumulator cabinet to capture and direct orgone energy. Atmospheric energy is captured and used against a number of diseases by changing the electrical charge of tissue. It therefore strengthens the immune system. A patient breathes in accumulated orgone energy through a tube connected to an accumulator cabinet.

The U.S. Food and Drug Administration forced a halt to this work because there is no evidence that orgone energy even exists.

Pyramid Therapy

Pyramid therapy was popularized in the 1930s and further developed in Europe in the 1950s. In essence, it has been postulated that there is a relationship between the shape of the space enclosed by a pyramid and the biological and physical processes that occur within that space. Supposedly, cuts, wounds, and bruises heal more rapidly under a pyramid. Patients have been placed in the center of a pyramid-shaped chamber for treatment of inflammation, burns, digestive problems, and migraine headaches, to name a few examples.

Others

Countless variations of these and other such systems exist. There have been endeavors to put arthritic patients in abandoned uranium mines; to use magnets of differing shapes, strengths, and sizes; and also to use electromagnets. The use of electricity in low voltages directly in contact with patients or in higher voltages with patients nearby has been touted. Heat therapy from ultraviolet lamps, sounds above and below our ability to hear, infrared bulbs, and pulses in water aimed at patients situated in tanks have been used to treat patients. Radio waves, visible light separated by prisms into components of the spectrum, heat, cold, and subliminal messaging have all been attempted and described.

Surely, new energy forms and pulse sources will be the foundation for new curiosities and experiments in this realm in the future.[11]

MIND AND SPIRIT HEALING

Philosophers have divided man into components for thousands of years. Even medical care today employs different specialists for the mind versus the body. The former uses psychiatrists, psychologists, social workers, and other types of counselors. Some healing systems recognize the relationship between mental and organic conditions, and during the twentieth century, great progress has been made in understanding the nature of stress, fatigue, anxiety, fear, and other factors on physical health status. In the very late twentieth century, we are beginning to accept a psychiatric or mental component in asthma, allergies, and hypertension, among a growing list of somatic pathologies.

Linking of body and mind in holistic practices has not met with overwhelming success in much of the West, probably because the concept has been embraced by unorthodox healers and is being promoted by what have come to be called fringe or even cult systems. For better or worse, the germ theory, coupled with the work of Pasteur, Jenner, Virchow, and others in describing somatic illnesses along with preventive and curative efforts, has diminished the psychosomatic focus.

Erlich and the German chemists added credence to a physical sector healing system. Yet, we are shortchanging the interface when we ignore stress-induced diseases. Pelletier makes the most convincing argument when he cites a large number of studies linking mental health with physical health.[12] Pathogenic personalities were hypothesized in the 1940s, and with the findings of Friedman and Rosenman about the Type A personality, we cannot ignore such a link.[13] If, in fact, there is a component of psychosomatic origin in nearly all disease states, then it makes perfect sense to help patients use their minds to restore their health. It appears that one's mental state is involved in one's reaction or response to illness.

Biofeedback

It should be no surprise that many, many healing systems have developed that are intended to help the mind resist illness, prevent illness, or treat illness. One of the most popular such techniques is biofeedback. It offers one the ability to monitor the external world and to control some components of it voluntarily. Persons trained in this technique can modify their heart rates, for example. The key to success is to monitor, continuously, sensory signals representing physiological functions. One can learn to breathe differently or to modify or control a pulse or brain wave. Biofeedback has had some success with tension and migraine headaches, hypertension, insomnia, stress-related problems, and phobias.

Hypnosis

Hypnosis is described as a physical state of relaxation that contributes to an altered state of consciousness. Suggestions are accepted, acted upon, and forgotten or repressed memories might be brought to the surface. Dr. James Braid developed the theory and techniques of medical hypnotism after studying mesmerists' work and learning about animal magnetism. Despite scientific publications and demonstrations, however, the medical community remained skeptical until the twentieth century, and then for the next 50 years, hypnosis was practiced only by a small number of psychiatrists and surgeons.

Unfortunately, even though hypnosis has proven its worth in some behavior modification uses, as an alternative to general anesthesia, and in treatment of chronic pain, there are only a handful of practitioners experienced in its use. Calls for hypnosis as a required component of general medical education have been ignored.

Spiritual Healing

Spiritual healing and a host of related systems require direction of energy, usually through prayer, to will a recovery. Often a healer helps a patient. The healer's hands are placed on an afflicted area of the patient's body. Some systems use the focusing of the patient's energy, as directed and guided by the spiritual healer, to chase away the devil or the evil spirits responsible for the illness.[14]

Psychic Surgery

Psychic surgery uses body energy to change the physical body. Patients journey to the Philippines, where the patient, in street clothes, is placed on a table, exposing only the site of the operation. The patient feels no pain, and anesthetics are seen as unnecessary. Often, there are no scars. The patient walks away when the procedure is completed. Supposedly, tumors are removed by such healers, who explain that they receive guidance and direction from spiritual sources.

Noland studied this and found it to be a fraud when compared to orthodox surgery. Whether belief is critical or there really is some basis to this treatment remains to be discerned.

Autosuggestion was a popular treatment system in France early in the twentieth century. Emile Coué advocated it, and in effect, it can be seen as a type of hypnosis.

Meditation

Many persons think of meditation and transcendental meditation as hippie or Indian cult activities, but in actuality, various types of meditation are practiced throughout the world. Meditation begins with relaxation, but extends far beyond that phase. Persons familiar with meditation claim that one can modify, measurably, physiological effects such as oxygen consumption and pulse rate.

Transcendental Meditation (TM)

Transcendental meditation was originated by Maharishi Mahesh Yogi. In a meditation state, an individual should have a clear mind and be open to new ideas and higher concepts. While a mantra is assigned to followers, it is merely a vehicle or aid, many feel, for the neophyte. TM advocates claim decreased stress-related medical problems such as hypertension, anxiety, and insomnia.

Psychodrama

Psychodrama was developed as a therapeutic group method by Dr. J. Moreno, a colleague of Freud. He searched for a means to create an emotional catharsis for an individual to shed light on his or her realities and fantasies. He postulated that people have problems because they have to adjust to constantly changing multiple roles (parent, spouse, worker, leader, etc.).[15]

Psychodrama has several key components required for its effectiveness. These include a protagonist—the individual who is the counselor or therapist, auxiliaries—the others in a role-playing situation, a stage where one is not afraid to say what his character wants to say, and an audience—the others in the room. Mirroring is what the audience does to provide feedback to the protagonist on what his/her soliloquy meant to them.

It appears that the major contribution of this therapy system is to give an individual the perspective of what others think about his/her stance or orientation in an issue or conflict. Like many of the other therapies described in this chapter, it should be attempted only by those skilled and experienced in its proper use.

Primal Therapies

Primal therapies describes a group of systems or techniques that help patients go back to their childhood in search of the origin of neurosis or selected other emotional imbalances. Hypnosis was used in the early phases of primal therapy development. Most readers will have heard of primal scream systems popular in the 1960s and 1970s.[15]

Another therapy that developed popularity and then decreased in importance and/or popularity is cocounseling, where two persons rotate the roles of patient and counselor. This provides the opportunity for individuals to let out (catharsis) guilt and other suppressed feelings of anger, denial, and previous trauma, to name a few uses.

Encounter Groups

Encounter groups were very popular 30 years ago and permitted people to challenge societal norms and behavioral limits. Based upon the work of Maslow and Laing, encounter groups offer participants an opportunity to experience joy, to recognize their feelings and reactions, and then to relate to other group members with honesty.

Sensitivity Training

Sensitivity training is quite similar to encounter groups in many ways. About a dozen participants in a T (training) Group discover the impact of their actions and opinions on the others in the group. It became popular in the 1950s, and employers sent their workers to sessions to learn to deal with each other in a more positive manner.

Transactional Analysis

Transactional analysis was described by Eric Berne in the 1950s. It is a therapeutic system that describes itself as giving participants a better understanding of their modes of current functioning without having to go back to early childhood or birth. Today, it is used less for pathologies and more for helping "normal" people improve their relationships in marriages or the workplace, for example.

The serious student of these mind-based systems may desire to explore systems not mentioned in this chapter. Others include music therapy, color therapy, enlightenment intensive groups, Gestalt, psychomuscular release therapy, bioenergy, somatography, arica, autogenic training, and psychosynthesis.

SELF-EXERCISE

Yoga

Yoga is undoubtedly the major entity in this category. Simply stated, the goal of yoga is to create and sustain healthy minds in healthy bodies. It is a philosophy that has been mentioned for over 5,000 years.

In essence, yoga is practiced in daily sessions of exercises done in various positions with a counter position for each. Deep relaxation is learned, along with breathing exercises. Hatha Yoga is probably the most widely practiced form in the West, although that may not be the case in India, where yoga originated.

The breathing control and meditation of yoga are claimed to help with nervous and anxiety-related problems.

Others

Dance therapy, sport therapy, T'ai Chi, and eurythmy are other exercise-based systems. T'ai Chi is practiced by many millions in China, but it has not achieved widespread adoption by non-Chinese in the West.

COMPREHENSIVE SYSTEMS

Ayurvedic Medicine

Ayurvedic medicine is the major body of Hindu Indian medicine. Its holistic approach is slowly gaining acceptance and popularity in the West. It is not possible to describe the complete and complex system of Ayurvedic medicine and do it justice here. Some of its innovations have been copied in the West, along with its knowledge of preventive medicine and immunology.

In its most basic essence, Ayurvedic medicine believes in a unified person with interacting psychological and physical aspects. This rests on a theory of Tridosha: the three elements of water, air, and fire. A patient is examined with reference to a balance of the three elements.

Oriental Medicine

Oriental medicine is based on the two fundamental and opposing forces in nature–yin and yang. By balancing foods, the environment, and the psyche, health is obtained. It is believed that all objects contain some minute energy, called chin in Chinese. This energy produces vibrations, and we arrange foods, medications, etc., to balance an excess or depletion to be in harmony with the environment.

This system is taught today, thousands of years after its original presentation. Proper colleges train its practitioners in medicine and pharmacy. There are seven principles of the order of the universe and another 12 unifying principle theorems that guide its use. Then there are the five elements, which become the basis for specialized and advanced Oriental medical practices such as acupuncture.

We will be seeing more of this system now that greater commerce and interaction with the Orient is taking place. Furthermore, it should be pointed out that while Western medicine and surgery have been available in China for about 100 years, a great proportion (perhaps a majority) of the 1.2 billion Chinese continue to use this system, presumably with some degree of satisfaction.

Homeopathy

Homeopathy is the discovery of a Dr. Hahnemann who established a medical practice in 1790. He noted that the side effects of quinine on a healthy person were identical to the symptoms it was to cure in ill patients. From this, Hahnemann developed his Law of Similars. His writings indicate that the value and effectiveness of a drug is inversely proportional to its volume or amount present. He wrote his formulae to dilute a drug with ten parts of an inactive solvent or vehicle. This step is repeated, usually six times, so that a drug is present one part per million. Some dilutions leave no measurable trace of the active ingredient.[16]

Medications in the official *Homeopathic Pharmacopeia* are used by its followers to create the symptom from which one wants treatment or relief. During the nineteenth century and into the twentieth century, homeopathic medical schools existed in England, Germany, and the United States. Today, properly educated medical doctors may elect to use homeopathic remedies, which are still available. It is estimated that only a tiny percentage of physicians use these drugs, but there is a great resurgence of self-medication using homeopathic products purchased in health food stores and in pharmacies.

SUMMARY

In the interval between the writing of this chapter and its publication, several more therapeutic systems will probably have been described and promoted. We can expect new systems as long as man finds dead ends, treatment failures, and the word "incurable" in orthodox medicine. Our desire for self-preservation will keep us searching for relief from or a cure for a problem. While most therapeutic systems appear to be legitimate and well intentioned, the consumer should always have a healthy level of skepticism. If a therapeutic system were miraculous in its accomplishments, we could expect word of such successes to spread like wildfire.[17]

It is important that the pharmacist practicing in the community at least have access to reference sources or information about the major unorthodox healing systems. Finally, it would probably be inappropriate for a pharmacist to answer a patient's request for information or guidance with a definitive negative reply. Faith is an important component of healing and self-limiting diseases. The placebo effect and surprise recoveries and remissions are all possible and do occur from time to time.[18]

REFERENCES

1. Payer L. Medicine and culture. New York: H. Holt and Company, 1988.

2. Cantwell JD. Running. JAMA 1978;239:1409-10.

3. Kruger H. Other healers, other cures. Indianapolis: Bobbs-Merrill, 1974.

4. Boadella D. Wilhelm Reich: the evolution of his work. London: Vision, 1974.

5. Bayly DE. In: The visual encyclopedia of unconventional medicine. New York: Crown, 1979.

6. Carroll D. The complete book of natural medicines. New York: Summit Books, 1980.

7. Hastings A, Fadiman J, Gordon J, eds. Health for the whole person. Boulder, CO: Westview, 1980.

8. Salmon JW. Alternative medicines: popular and policy perspectives. New York: Tavistock, 1984.

9. Law D. A guide to alternative medicine. Garden City, NY: Dolphin Books, 1976.

10. Walker B. Encyclopedia of meta-physical medicine. London: Routledge and Kegan Paul, 1978.

11. Schaller WE, Carroll CR. Health, quackery and the consumer. Philadelphia, PA: Saunders, 1976.

12. Pelletier KR. The mind in health and disease. In: Hastings AC, Fadiman J, Gordon JS, eds. Health for the whole person. Boulder, CO: Westview, 1980.

13. Friedman M, Rosenman R. Type A behavior and your heart. New York: Knopf, 1974.

14. Grant PH. Holistic therapy. Secaucus, NJ: Citadel Press, 1978.

15. Bannerman RH, Burton J, Wenchieh C. Traditional medicine and health care coverage. Geneva: World Health Organization, 1983.

16. Hafner AW, Carson JG, Zwicky JF. Guide to the American Medical Association Historical Health Fraud and Alternative Medicine Collection. Chicago: AMA, 1992.

17. Miller L. Alternatives go on the record. USA Today 1995 Jan 12:6D.

18. Watson WH. Black folk medicine. New Brunswick, NJ: Transaction Books, 1988.

ADDITIONAL READINGS

Because there has been little scientific inquiry into the unorthodox healing systems discussed in this chapter, it is impossible to identify definitive, scientific articles or books on the systems. The additional readings listed below are, therefore, intended for those interested in further research on alternative therapies.

For the Instructor

1. *Journal of Alternative and Complementary Medicine: Research on Paradigm, Practice and Policy.* Peer-reviewed quarterly. Editor: Dr. Micozzi of the National Museum of Health and Medicine.

2. *Alternative and Complementary Therapies.* Six issues per year (1-800-654-3237).

3. *Alternative Therapies in Health and Medicine.* Six issues per year. Peer reviewed (1-800-345-8112).

4. *Alternative Health Practitioner: The Journal of Complementary and Nature Care.* Springer Publishing. Three issues per year. Peer reviewed.

5. *Advances: Journal of Mind-Body, Health.* Fetzer Institute, Kalamazoo. Quarterly.

6. Jarvis DC. Folk medicine. New York: Fawcett, 1973.

7. Law DC. Concise herbal encyclopedia. New York: St. Martins, 1973.

8. Iyengar BK. Light on yoga. New York: Schocken Books, 1972.

9. Kleinman A. Patients and healers in the context of culture. Berkeley: University of California Press, 1980.

10. Leslie C, ed. Asian medical systems. Berkeley: University of California Press, 1976.

For the Student

1. Acuna HR. Cross cultural communication: its contributions to health in the Americas. Bull Pan Am Health Organiz 1979;13:111-6.

2. Bently UW. Let herbs do it. Boston: Houghton-Mifflin, 1973.

3. Harding TW. Traditional healing methods for mental disorders. WHO Chron 1977;31:436-40.

4. Inglis B. Fringe Medicare. New York: Putnam, 1969.

5. Mason AS. Hypnotism for medical and dental practitioners. London: Secker and Warburg Publishers, 1960.

6. Stanway A. Alternative medicine: guide to natural therapies. London: Macdonald and Jones, 1980.

7. Veith I. Use of acupuncture in modern health care. WHO Chron 1980;34:294-301.

8. Veith I, transl. The yellow emperor's classic of internal medicine. Berkeley, CA: University of California Press, 1967.

9. Vithoulkas G. The science of homeopathy. New York: Grove Press, 1980.

10. Wallace RK, Benson H. The physiology of meditation. Sci Am 1972;226:84.

SECTION D:
CHOOSING A THERAPEUTIC AGENT

Chapter 8

Determinants of Prescribing Behavior

Dennis W. Raisch

INTRODUCTION:
WHY ARE WE INTERESTED?

Drug prescribing comprises one of the principal forms of treatment for illness. Each year, more potent and potentially harmful drugs become available. However, there is substantial evidence that prescription drugs are often used inappropriately.[1,2] As many as 80% of prescriptions for antiulcer drugs have been shown to contain prescribing errors.[3] Estimated costs associated with microbial resistance due to overuse of antibiotics are significant.[4]

Prescribing errors can be classified in several ways. They can be described as errors of commission or omission.[5] Errors of commission involve those that are incorrectly written. Errors of omission are those in which the prescriber fails to specify a required element in the prescription (e.g., missing strength or dosage). Errors can also be classified by type of error made in the use of the drug, in indication, in dosage, or in duration.[6] For example, an indication error occurs if the drug is being used for the wrong indication or the drug of choice for the patient is not prescribed. A dosage error occurs when the dosage is too high or too low, and a duration error occurs when the drug is prescribed for an inadequate or excessive period of time.

These errors can cause many problems for patients, ranging from decreased quality of life and prolonged illness to unnecessary costs. If a patient receives two drugs that have a similar effect—for example, two nonsteroidal anti-inflammatory drugs—there will not only be unnecessary costs but also the potential for additive or potentiation of side effects, such as ulcerogenic effects. Other documented health outcomes associated with inappropriate prescribing include additional drug prescribing to treat adverse effects, hospitalizations, outpatient visits, and prolonged illness.[7,8]

In community pharmacies, errors that require pharmacists to contact prescribers occur in 2 to 3% of prescriptions.[5,9,10] In settings where pharmacists have access to more patient information (hospital, clinics), errors have been documented as high as 21% of prescriptions.[11] Thus pharmacists have an opportunity, indeed a responsibility, to detect prescribing errors and significantly affect patient care by addressing those problems. To affect patient care, however, pharmacists must be successful in their endeavors to influence prescribers to change and improve therapy. To be successful in influencing prescribing, pharmacists should not only be able to identify errors but also to understand what factors affect prescribing. This chapter is intended to provide you with (1) an understanding of the causes of prescribing errors, (2) an overview of methods used to influence prescribing, and (3) some tools for improving efforts at influencing prescribing.

PRESCRIBING MODELS/THEORIES: HOW ARE PRESCRIBING DECISIONS MADE?

Prescribing models provide insight into the drug decision process, helping us understand why practitioners prescribe in certain ways and how prescribing errors occur. There are three types of models. The first type focuses on demographic and practice variables.[12-16] The second describes psychosocial issues related to physician/patient interactions.[17-19] The third type examines cognitive theories behind prescribing decisions.[20-24]

Demographic and Practice Variables Associated with Prescribing

Relationships between prescribing and practice site and personal characteristics of practitioners have been studied.[12-16,25,26] The practitioner's educational experiences, specialty, age, and relationships with colleagues influence prescribing decisions. Specialty physicians tend to be well versed in the use of drugs for illnesses within their specialty but may be less knowledgeable about drugs outside of their specialty. Patterns of adoption of new drug therapies are different for specialists than for general practitioners.[25] Specialists treat patients more aggressively than generalists.[27] Knowledge of these factors may help target educational interventions to specific prescribers. For example, nonpsychiatrists who prescribe antidepressants may need additional training in the use of these agents. In addition, patient demographics such as gender, age, and ethnicity all affect prescribing decisions.[28,29]

Organizational and practice factors have been shown to be related to prescribing. For example, physicians in group practice prescribe differently from those in solo practice.[12] Practice constraints placed upon prescribers in some settings, such as health maintenance organizations and government hospitals, also affect prescribing decisions. Obviously, it is not possible to alter demographic variables or organizational factors, but awareness of them can help us to understand prescribing errors and to design more effective and efficient programs for improving prescribing by helping us select those practitioners who will benefit most from our efforts.

Psychosocial Dynamics Related to Prescribing

Psychosocial factors have been ascribed to drug prescribing, and symbolic meanings of the prescription have been identified.[17] One symbolic meaning is the expression of power and authority: the practitioner has the power and authority to provide drugs and to heal.[17,18] The prescription also is a means of expressing concern for the patient.[17,18] Other psychosocial dynamics of prescribing involve the physician/patient interaction. For example, prescribing is a method of forestalling lengthy discussions and ending the patient visit. It also serves as a verification to patients that they are, in fact, ill.[19] It is important to consider the patient's influence on the prescribing process when attempting to change prescribing. Patients may request or even demand a specific drug, and this can have a profound effect on the physician's decision.[30] Figure 8.1 shows how patients interact in prescribing decisions.

Cognitive Models of Prescribing

How the prescriber weighs and evaluates information regarding drug products is the focus of cognitive models. The practitioner's decision-making process is of critical importance in these models.[20-24,31] Factors such as the physician's determination of the seriousness of the disease and the perceived risk-to-benefit ratio of a particular therapy strongly influence the decision process. For example, if a disease is mild and the drug can potentially cause serious side effects, the practitioner is unlikely to prescribe the drug. On the other hand, a highly toxic cancer chemotherapy agent will be used consistently due to the seriousness of the disease process. In these models, the prescriber's valuation of drug attributes (dosage frequency, cost, length of action, etc.) and outcomes associated with drug use (side effects, cure rates, etc.) are vitally important.

The Drug-Choice Model is a cognitive prescribing model related to Vroom's expectancy theory, a management theory used to explain work motivation. In the Drug-Choice Model, a practitioner's drug choice is based upon beliefs about the likelihood that certain outcomes will occur when the drug is used and the value the practitioner places on the outcomes.[23] The outcomes identified as important to practitioners have been divided into six groups: (1) control of the disease, (2) patient compliance, (3) side effects, (4) cost, (5) satisfying the patient's demand, and (6) criticism from colleagues.[24] In a retrospective study, responses to a survey using the Drug-Choice Model were shown to correlate with previous prescribing decisions.[24]

The following is an example considering just two of the six outcomes. If a drug has a high cure rate but is associated with frequent side effects, the prescriber's choice will be dependent on the assessment that these outcomes will occur together with the relative importance of each outcome. Curing a runny nose with a drug that causes drowsiness may or may not be judged more desir-

FIGURE 8.1. The Patient's Influence in Prescribing Decisions

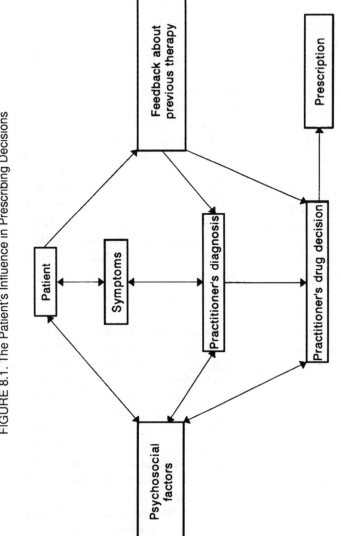

able than curing it with another drug that causes muscle aches. The decision depends on how the prescriber assesses these two outcomes and his or her perception of the likelihood that each will occur.

Cognitive models of prescribing focus on which prescribing outcomes and drug attributes are important when practitioners make prescribing decisions. The process by which these assessments are constructed should be considered when developing and evaluating methods of influencing prescribing.

A practical implication from cognitive prescribing models is that efforts to influence prescribing decisions should include a discussion of important outcomes and information regarding the likelihood of the outcomes relative to other drugs available in the therapeutic class. Also, these models have shown that cost of therapy is not as important to physicians as other aspects of drug outcomes such as cure rates and side effects. Therefore, discussions of costs are irrelevant unless more important outcomes (efficacy, side effects, etc.) have been addressed.[31] Figure 8.2 shows how cognitive processes affect prescribing decisions.

Sources of Information

Understanding that the practitioner's attitudes and beliefs about drugs are a basis for prescribing decisions, we must now consider the sources of drug information. First, scientific training received in health education directs us to rely upon the scientific literature. Ideally, every relevant new study would be reviewed and evaluated on its merits and used to form the basis of prescribing decisions. However, the growth of scientific literature is incredible, with a predicted growth rate as much as 40% per year.[32] Even specialists have difficulty keeping up with literature relevant to their fields. Pharmacists have recognized this problem, and subspecialties have been formed within the profession.[33]

Another source of information is continuing education programs, including professional meetings. In these programs, specialists share their research and present the latest literature regarding their specialty area. Health professional licensure is usually dependent on participation in continuing education. However, there is great variation in the quality of programs and degree of involvement of participants. Also, the level of sponsorship by pharmaceutical man-

FIGURE 8.2. Cognitive Processes in Prescribing Decisions

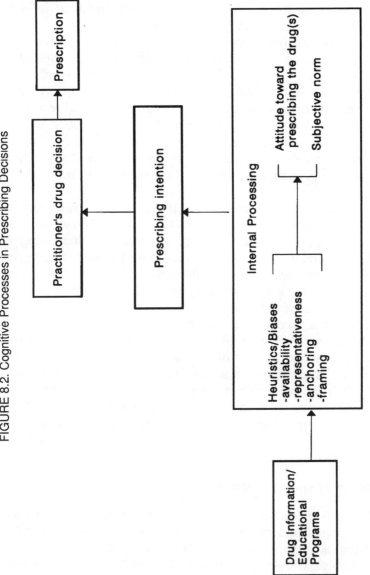

ufacturers indicates a need for careful assessment of information gained through continuing education programs.

Professional colleagues comprise an important source of information.[26] Interactions with respected colleagues can strongly influence the way information is acted upon, as well as affecting the information itself. Professional colleagues include pharmacists, as they are often consulted regarding drug therapy, and recommendations made by them are well accepted.[10,34] The theory of reasoned action includes subjective norm as a major variable.[35] The subjective norm involves the idea that opinions of respected authorities affect the decision being made. This theory has been applied to prescribing.[31,36]

Pharmaceutical advertising and pharmaceutical sales representatives are another source of information. The pharmaceutical industry spends a large portion of its gross revenue on marketing. Much of this is direct-to-prescriber advertising through pharmaceutical sales representative visits and mailings. Although prescribers tend to downplay the effects of these methods, it has been shown that they do affect prescribing.[37]

Another source of information is the patient. Patients provide feedback about previous therapy and/or make requests for specific drug prescribing. In light of the psychosocial processes previously discussed, the patient can significantly affect prescribing. Not only will this information affect current prescribing, but future prescribing for other patients may also be affected by feedback from patients. Realizing these effects, pharmaceutical manufacturers are increasingly aiming advertisements directly to patients through television and magazines.

Theories of Human Inference

Theories of human inference help explain how drug information is interpreted and evaluated. Inferences concerning the effectiveness and adverse reactions of drug therapy affect attitudes toward prescribing. The same inferences that help prescribers categorize the large amounts of drug information, obtain meaning, and draw conclusions about drug therapy can cause errors.

When people make inferences, they generally employ one or more of a variety of judgmental heuristics. Judgmental heuristics can be defined as cognitive strategies, rules of thumb, or shortcuts used to classify and interpret new information and thereby form opinions or make decisions about it. We use these heuristics constantly without even realizing it. Usually they are accurate and help us accomplish our daily activities. Sometimes, however, they lead us into judgmental errors. Four major types of judgmental heuristics are: representativeness, availability, framing, and anchoring.[38]

The representativeness heuristic involves a similarity judgment between events or objects. It tells us how well a certain event "fits" with other events. When an event occurs, explanation of it can involve assigning an antecedent to it. Errors in judgment relating to representativeness occur when an antecedent is inappropriately connected to an event and/or when an event or object is incorrectly evaluated as similar to (representative of) another event. Relationships between events or objects are, therefore, sometimes overstated or misstated.[38]

For example, if a patient improves after a drug is given, the prescriber may decide the patient was cured (or symptoms alleviated) by the drug. However, the improvement may have occurred as a result of any of a number of other factors, such as normal fluctuations in the illness, improved diet, nonprescription drug therapy, or decreased stress. The error in judgment occurs when these other factors are not considered. When a prescriber states that "personal experience" has shown that a drug is effective for a medical condition, even though there is little scientific or theoretical basis for it, the representativeness heuristic may be at work.[39] The prescriber may have considered "personal experience" in a few patients to be representative of a form of clinical trial, even though none of the basic tenets of scientific method was included (randomization, placebo control, double blind).

The availability heuristic helps us judge frequency, probability, and causality. According to this heuristic, new information is assessed according to information that is more readily available from memory. Specifically, relationships between events and causality are evaluated based upon what is more readily retrievable from memory.[38] If we see the same problem repeatedly, it affects how we evaluate the next problem that is similar. A practitioner, having seen the same set of symptoms in a series of patients, may make an inference that a new patient with similar symptoms has the same illness. If proper tests are not used to verify the diagnosis, an error may occur. This can result in inappropriate prescribing.

Availability is affected by vividness. Events and objects that are more interesting, emotion provoking, or concrete may have an inordinate effect on our decisions.[38] Vivid pictures, stories, and actual patient cases are more available, thus affecting our judgment. It is easier to recall these images than statistics, summaries, or abstractions. Often we see pictures of patients or quotations in drug advertisements. Prescribing can be affected by these images through vividness. Another example is when an individual patient's serious adverse reaction to a drug causes a practitioner to use it less frequently in new patients than indicated.

Framing is another type of judgmental heuristic. The concept of framing involves how decision alternatives are presented. People make judgments based upon the desirability of potential outcomes. The phrasing of alternatives affects perceptions of desirability. An individual's propensity to take risks affects the effects of framing.[40] Framing affects prescribers by causing them to avoid particularly undesirable outcomes, even when the probability is low and the potential benefits outweigh the risks. This might occur when a drug causes cancer in a small percentage of patients. Since cancer is a very undesirable outcome, practitioners may avoid using the drug even though it is clearly beneficial. A real-life example of framing bias occurred when saccharin was nearly taken off the market due to its association with cancer in rats exposed to massive doses.

Another heuristic principle is anchoring. We often make initial assessments and decisions, then make adjustments until we find the "correct" answer.[41] However, sometimes our adjustments are inadequate, even in the presence of conflicting information. Thus the initial judgment inordinately influences us. This resistance to change from the initial decision is called anchoring. Anchoring can affect prescribing when a practitioner uses a drug and fails to discontinue it after diagnostic tests indicate it is not needed.

Theories of human inference are relevant to medical decision making.[42-52] Table 8.1 summarizes selected studies in which these theories have been applied to clinical decision making. Errors in medical judgments have been related to these theories. The relation-

TABLE 8.1. Theories of Human Inference and Effects on Medical Practice

Heuristic Principle	Effects/Implications	References*
Representativeness	- Failure to incorporate prevalence rates into diagnoses - Reliance on clinical experience - Errors in assessment of mortality rates - Diagnosis based upon subset of available information	39, 45, 46, 48
Availability	- Patient assessment of susceptibility to illness - Risk aversion for medical outcomes that are rare but highly undesirable	51, 52
Anchoring	- Difficulty altering original diagnosis - Early information affects diagnosis more than late information - Inadequate consideration of supplemental information - Pre-existing beliefs have inordinate effects on treatment choices	45, 48-50, 52
Framing	- Patient choices depend on how statistical information is presented - Treatment choices are affected by how statistical information is presented - Treatment choices are affected by undesirability of toxicities	47, 50, 52

*Often several theories are addressed in a single study.

ships between judgmental heuristics and cognitive processes involved in drug decisions are shown in Figure 8.2.

In summary, prescribing models and theories of human inference help us appreciate and understand the complexity of drug prescribing. These concepts show us what factors influence prescribing and give us insight into how mistakes can be made.

METHODS USED TO INFLUENCE PRESCRIBING

An understanding of how prescribing decisions are made is important for developing a clear picture of drug use. It is also the foundation for designing programs to correct errors that have a significant impact on patient care. Applying this information to practice means incorporating these principles into interventions designed to correct errors and prevent their recurrence. Methods used by pharmacists to influence prescribing generally fall into two categories: administrative and educational. Sometimes a combination of both methods is employed.

ADMINISTRATIVE PROGRAMS USED TO AFFECT PRESCRIBING: LIMITING PRESCRIBER CHOICES

Administrative programs are those in which policies are established by an organization to control prescribing. These policies limit drug choices and prescribing activities directly. Administrative programs can be divided into the following types: prescribing restrictions, financial incentives, required specialty consultations, and medical management protocols. The relationships between administrative programs and prescribing are shown in Figure 8.3.

A formulary is one of the most commonly used prescribing restrictions. Drug availability is limited by establishing a list of drugs that can be prescribed within the health care setting. High-cost and/or newly marketed products are kept off the formulary unless they are therapeutically necessary. The Pharmacy and Therapeutics Committee (P&T Committee) consists of prescribers, pharmacists, and administrators who evaluate and approve requests for additions to the formulary. Formularies have been an effective way of controlling prescribing and drug expenditures in hospitals and health maintenance organizations, especially for particular therapeutic categories.[53,54] However, in Medicaid programs, the results are less universally positive due to enforcement difficulties and effects of limiting access to available treatments. In some studies of Medicaid formularies, decreased drug expenditures have been offset by increased hospital, ambulatory care, or other costs.[55,56]

Another type of prescribing restriction is required consultation. Under these programs, specific drugs are restricted to prescribing only by a subset of physicians, usually specialists (e.g., psychiatrists, infectious disease department). For a patient to receive the drug, a specialty consultation is needed.[57] These programs are

FIGURE 8.3. Relationships Between Administrative Programs and the Prescribing Process

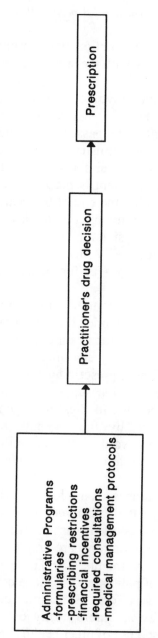

sometimes used in conjunction with formularies; a drug may be added to the formulary with the stipulation that only certain specialists can prescribe it.

Financial incentives involve tying prescriber reimbursement to drug expenditures. The idea behind these incentives is that excessive costs or savings are passed on to the prescriber. To avoid ethical dilemmas, these are usually administered through a risk pool rather than through direct reimbursement. This means that a portion of prescriber reimbursement is placed in a separate fund to cover excessive drug costs. At the end of the year, the prescribers receive a share of the balance remaining in the risk pool. Although financial incentives to affect physician behavior seem logical, success with them has been limited.[58,59] A negative type of financial incentive is not providing reimbursement for nonapproved or nonefficacious therapies.[60]

Another type of administrative program is the use of medical management protocols. Under these programs, prescribing is preprogrammed. A protocol for treatment is established by achieving a consensus among experts regarding appropriate treatments. If a diagnosis is established, prescribing occurs according to the protocol. Computerized decision support systems are helpful in establishing and implementing medical management protocols. Practice guidelines developed by the Agency for Health Care Policy Research are a step toward increasing uniformity in medical treatment for specific conditions.[61] These guidelines can also be used as an educational tool.

Table 8.2 summarizes examples of research of administrative programs. Although the effectiveness of administrative programs has been well demonstrated, certain difficulties can occur. First, when prescribing restrictions are implemented, prescribers may resort to substitute treatments.[62-64] These substitutes may result in decreased quality of patient care if the substitute is less effective or causes more adverse effects. Substitutes can also nullify potential cost savings if they are more costly or cause medical problems resulting in increased use of other health care resources. It is unclear whether the use of substitutes consistently offsets the cost savings of administrative programs, but it is clear that use of substitutes

TABLE 8.2. Administrative Programs Used to Influence Prescribing

Type of Program	General Conclusions	References
Prescribing restrictions, including formularies	- Decreased drug costs - Increased use of substitutes - Variable impact on other medical costs	54, 57, 62-64, 68-73,
Financial incentives to control prescribing costs	- Decreased drug costs - Improved prescribing practices - Potential conflict of interest and ethical problems - May increase use of substitutes offsetting cost savings	58-60, 74, 75
Required specialty consultations for use of particular drugs	- Decreased drug use - Revert back to original prescribing problems when enforcement discontinued	65, 76
Medical management protocols including prescribing protocols	- Variable results from no change to improved quality of care	66, 67, 77

should be considered in evaluating the outcomes of administrative programs.

Second, prescribing will usually revert to previous patterns when administrative programs are not well enforced.[65,66] Also, administrative programs can cause decreased cooperation between practitioners and administrators because practitioners may see the program as interfering with medical practice.[65] Another difficulty is that there are medical-legal liability concerns when restrictions are placed on medical practice.[67]

CHANGING PRESCRIBING KNOWLEDGE AND DRUG ASSESSMENT: HOW CAN PHARMACISTS EDUCATE PRESCRIBERS TO BETTER UTILIZE DRUGS?

Providing drug information and employing educational programs to change prescribing has the advantage of changing the prescriber's knowledge and assessment of drug use. Thus even after the program has ended, its effects may persist. The relationship between drug information/educational programs and the prescribing process is shown in Figure 8.4. Selected research of educational programs is summarized in Table 8.3. Education regarding prescribing is provided in several ways. First, printed educational material containing general prescribing information and recommendations can be distributed. This is accomplished through direct mailings, newsletters, or distribution of minutes of P&T Committee meetings. Printed material has the appeal of ease of getting the information out to prescribers in a timely manner, but a problem regularly cited with this method is that printed material may not be read. Competition for the attention of health care practitioners is great, several direct mailings from manufacturers may be received daily, so printed material may be discarded prior to being read. Another problem is that the information may not seem directly applicable to the practitioner's patients. Thus the message may be interpreted as, "Yes, but this doesn't really apply to my patients." Well-designed, controlled studies of this type of printed material have shown little, if any, effect when used alone.[78-82,94,95] Two studies that compared printed material with one-to-one educational meetings revealed the ineffectiveness of printed material when used alone.[78,79]

Another type of printed information is group or individual feedback about past prescribing. In these materials, results of drug use evaluations (DUEs) are given, followed by recommendations for improvements. Printed feedback can also consist of individualized reports with comparisons to peer prescribers. These reports show prescribers how their prescribing habits compare with others, implying that changing toward the median (or below) should occur. A problem with this type of feedback is that where one falls on the range of drug prescribing may not be directly related to quality of prescribing. Other factors, such as patient mix in terms of age and disease, will affect these patterns. Also, just because one is at or below the median does not mean that prescribing habits are optimal. Effectiveness has been demonstrated in some well-designed studies of printed individual feedback.[87-92] However, in one of the positive studies, only a marginal effect after several months was noted, while in another, only initial changes were produced.[88,89]

FIGURE 8.4. Relationships Between Drug Information/Educational Programs and the Prescribing Process

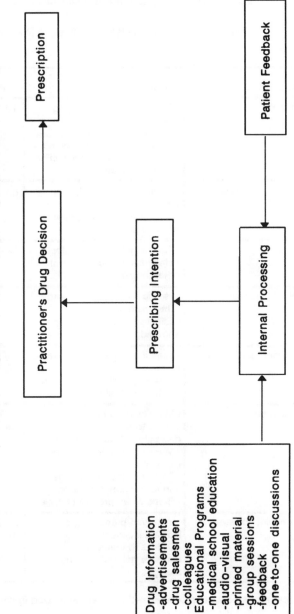

TABLE 8.3. Educational Programs Used to Influence Prescribing

Type of Program	General Conclusions	References*
Printed material	- Not effective at changing prescribing - Some changes in responses on hypothetical cases found - Can be useful in conjunction with other methods	78-85
Printed individual feedback	- Moderate improvement in prescribing - Can take several months to cause changes	83, 86-92
Survey feedback	- Generally little if any effects - Helpful in conjunction with other methods	82, 93-95
Audio-visual educational programs	- Helpful in conjunction with other methods - Effective at changing knowledge	93, 96
Group educational programs	- Little if any effects when used alone - May augment other methods - Better in small groups	83, 92, 94, 95, 97, 98
One-to-one education	- More likely to be effective - Need to repeat, effects dissipate without follow-up - More effective than printed material - Tested in several settings and for several drug classes - Barriers can prevent changes	6, 78, 79, 82, 99-102
Clinical pharmacist activities	- Improved prescribing on individual patients - Difficult to change or measure general prescribing changes	34, 84, 103-105

* Studies appear repeatedly when several methods were employed in conjunction or in comparison.

Another type of individualized feedback is the "Dear Doctor" letter. In these letters, a specific patient and prescribing problem are discussed and/or potential prescribing problems identified. Sometimes compliance with prescribing criteria is presented. A response may be requested. These letters have the potential of directly affecting a patient's care; however, it is unclear if or how care of other patients might be affected by these letters. It is critical that information and recommendations contained in "Dear Doctor" letters are accurate because the credibility of the intervention program and those involved with it is at stake. These letters can also invoke defensive reactions and may affect administration-practitioner relationships. Physicians' attitudes toward "Dear Doctor" letters are less positive than they are toward other types of drug use evaluation interventions.[106] Pharmacists may be included in this type of intervention to increase their participation in influencing prescribing.[107]

Group educational programs are sometimes held to influence prescribing. These can be conducted as staff conferences and as continuing education programs. The advantage of group programs is that a large number of prescribers can be provided information at one time, and interaction between them may be stimulated. A disadvantage is that understanding, attention, and relevance to practice may vary considerably among practitioners. Success with group lectures has been limited and variable.[94,95]

One-to-one educational interactions have been shown to be more uniformly successful at influencing prescribing. These can be accomplished through several means: scheduled meetings, unscheduled informal meetings, and telephone discussions. Scheduled meetings have been used successfully in several studies. During these meetings, pharmacists or another health practitioners present specific, predetermined prescribing information in a discussion with the prescriber. Depending on how the program is designed, individualized feedback from prescribing records may be provided and/or a specific patient case discussed, as well as general prescribing guidelines provided. Often a handout or summary chart is provided to the prescriber. Research has shown that pharmacists can perform one-to-one interactions successfully.[78,84,108-111] This activity represents an important role for pharmacists in health care.

Unscheduled informal meetings consist of such activities as

attending rounds in hospitals or meeting a prescriber on the hospital nursing unit, ambulatory clinic, or other setting. These activities are a part of clinical pharmacists' activities, and effects may be difficult to measure separately from other activities such as patient counseling. These meetings are usually centered around problems identified in the treatment of a specific patient, but general drug information is often provided in these discussions. Clinical pharmacists' activities and resultant changes in prescribing have been researched.[34,103,104] Improved quality of care and cost savings have been documented.

Telephone discussions are a common type of one-to-one interaction between pharmacists and prescribers. Again, the discussions are usually centered around a specific prescribing problem, with general drug information being less of a focus. These interactions can be initiated by the pharmacist, patient, or prescriber. Pharmacists successfully influence or change prescribing in a large majority (80% or more) of interactions.[9,10,34,112] In addition, pharmacists can provide information about patients (e.g., compliance assessment) that is otherwise unavailable to prescribers.[113] Performing these interactions is an integral part of pharmaceutical care. A personal interest in the patient's well-being is necessary to identify potential drug therapy problems and to take the risks necessary in confronting the prescriber. This is, of course, a cornerstone of pharmaceutical care.

Communication techniques are critical for success of one-to-one interactions. Avoiding a defensive reaction from the prescriber takes skill and practice. Being prepared for several types of responses will help to make sure the point comes across. Having the literature references available may be necessary to substantiate one's position. Comprehensive professional knowledge, ability to interpret literature, and confidence in that knowledge and ability will help make these kinds of interactions more successful and less stressful.

Application of Theories of Human Inference

Theories of human inference can be used to evaluate and design educational methods for influencing practitioner prescribing. The representativeness heuristic helps explain why feedback from group audits has been ineffective.[82,94] Group audit results may be perceived by the practitioner to relate only indirectly to a specific clinical situation and therefore are not as likely to affect prescribing decisions.[52] It has been recommended that when feedback on prescribing habits is given, it should be given frequently, should include examples of specific patients, and should be provided through a recognized authority.[114] Frequent feedback increases availability. Examples of clinical events can provide vividness because interesting, concrete events are more readily recalled and incorporated into judgments than generalities.[51] The need for establishing relevance to medical literature or an expert is related to availability because this increases the impact and makes the information more retrievable from memory.

Theories of human inference also help explain the effectiveness of one-to-one educational discussions. Specific occurrences are incorporated into judgments more readily than statistical data summaries.[38] One-to-one interactions are, by definition, more vivid than group meetings or printed information and, therefore, more readily remembered (available). Incorporating discussions of the practitioner's patients can help assure that the information is representative of the practitioner's practice. This procedure has been shown to be effective.[78]

Framing of educational information is also important.[47,50] Researchers have found that physicians respond to quality of care issues more readily than to cost containment issues.[21,22,24] This suggests that focusing information on quality of care as well as cost can increase the likelihood of prescribing changes.

Anchoring may decrease the effectiveness of educational programs. Once a practitioner has made a decision to use a drug in a patient, it can be difficult to incorporate new information and modify therapy.[45,48-50,52] The use of repeated educational interventions can address this problem in the long run by eventually influencing initial decisions.

DESIGNING PROGRAMS TO IMPROVE PRESCRIBING: A MODEL OF METHODS FOR INFLUENCING PRESCRIBING

The model displayed in Figure 8.5 is a compilation of factors associated with prescribing and literature related to methods of

FIGURE 8.5. A Model of Methods for Influencing Prescribing

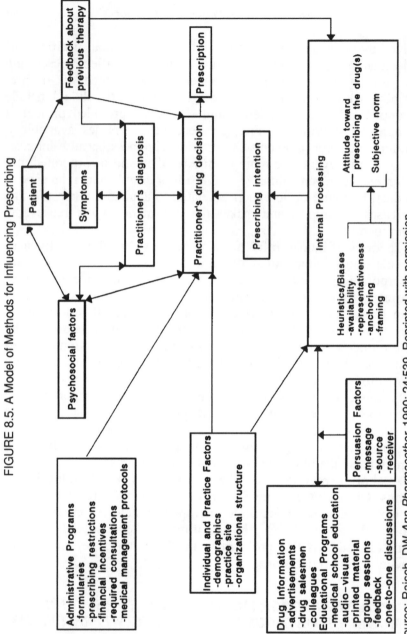

Source: Raisch, DW *Ann Pharmacother,* 1990; 24:539. Reprinted with permission.

influencing prescribing.[115] Methods of influencing prescribing are shown as two types: administrative and educational. Administrative methods restrict prescribing or otherwise influence prescribing without necessarily affecting prescribing intention. Educational methods cause changes in prescribing through changing the thought process that results in the practitioner's drug decision.

Administrative methods of influencing prescribing are often supplemented with educational methods. In the case of a formulary change, pharmacy newsletters or memos describing the rationale for the change are distributed. When a nonformulary drug is prescribed, one-to-one interactions occur when practitioners are contacted by pharmacists. These can be educational provided the opportunity is used correctly: to educate and not just inform regarding availability or control. In these situations, the P&T Committee and pharmacists represent educational influence from colleagues, while the formulary itself represents a direct administrative restriction in choice of alternatives. When the educational aspect is lacking (for example, a new physician on staff), the formulary itself represents only an administrative influence on prescribing.

Internal processing is the thought process involved in evaluating information. Internal processing is affected by the heuristic principles of human inference, which can result in biases. If information is incorrectly classified or evaluated, inaccurate assessments may result in errors in judgment and cause inappropriate prescribing. Biases can affect belief and attitude formation about prescribing specific drugs and potential outcomes associated with their use.

Desirability of potential outcomes and probabilities of achieving those outcomes are evaluated for each potential therapy. The various outcomes are also weighted according to which outcome is most important to the prescriber. The results of these thought processes are prescribing intentions for specific drugs. This portion of the model is derived from cognitive models of prescribing.[21-23]

The term "prescribing intention" is derived from the behavioral intention model, which signifies the difference between an intention and the actual prescribing decision.[35] Once an intention to prescribe a drug has been formed, the practitioner's actual drug decision can still be altered. For example, a practitioner may be convinced by a journal article to prescribe a particular antibiotic but cannot do so because the antibiotic is not on the formulary. So, administrative methods for influencing prescribing and organizational policies that limit availability of drugs can affect the prescribing decision contrary to intention.

Patients, too, influence prescribing. First, the symptoms presented by the patient affect the practitioner's diagnosis and drug selection. Second, psychosocial factors involving interactions between the patient and practitioner, such as patient attitudes and behaviors, affect the practitioner's diagnosis and eventual drug decision.[17, 18] Patient requests for drugs and their assessment of effectiveness of previous treatments directly affect prescribing. Feedback about previous therapy also enters into the internal processing portion of the model. As patients report effectiveness (or lack of it) of drug therapy, changes in prescribing intention will be established. Thus, patients not only directly influence prescribing decisions but also indirectly influence prescribing by providing another source of drug information.

An important difference between administrative and educational methods of influencing prescribing is that if administrative programs are lessened, ceased, or just not enforced, they will no longer affect prescribing. This can occur if a pharmacist does not follow through with formulary restrictions and dispenses a nonformulary drug without contacting the prescriber. Adequate follow-through is critical to the success of administrative prescribing controls. In contrast, educational methods may continue to influence prescribing after the programs have ceased due to changes in attitudes, beliefs, and prescribing intentions. To bring about long-term change, education must be provided repetitively. Research indicates that a single educational meeting can only bring short-term changes in prescribing.[80,95,114,116] This may be due to the constant onslaught of information provided to prescribers from a multitude of commercial and noncommercial sources and/or other factors such as patient requests. Thus, problems with administrative programs can be addressed by using educational programs concurrently. Changing physicians' beliefs about outcomes associated with drug therapy are more likely to affect future behavior.[117]

Using Persuasion to Improve Efforts to Influence Prescribing

An important element of performing educational interventions to influence prescribing is persuasion. Certainly there is great variation in how persuasive different techniques of presenting information are. A simple persuasion model includes the source, the message, and the receiver.[118] The effectiveness of educational programs is mediated by these variables. The source is the presenter of the information and/or what references (authorities) the presenter is using in the message. The message involves how information is presented and what the information is. The receiver is the person who is getting the message, in this case the prescriber. The relationship between persuasion variables and educational interventions is shown in Figure 8.5.

The source must be credible for the message to be persuasive. Taking steps to enhance credibility can improve effectiveness of educational interventions.[119] Such steps include providing background information such as the title and experience of the source, citing medical literature during the message, and specifying medical committees or physicians who were involved in developing the intervention. The presenter must be confident, speak authoritatively, and conduct himself or herself in a professional manner. If meeting face-to-face, the presenter must be dressed appropriately. Preparation is key to achieving these requirements.

Using effective messages is also important in implementing persuasive educational interventions. Develop presentations that are oral rather than just written. Use visual aids such as charts and summaries to augment the message. Attempt to phrase information in a manner that avoids defensive responses. For example, rather than telling the prescriber that he or she has made an error, provide information related to the case that may direct the prescriber to the problem being addressed. Another tactic is to ask for additional relevant information about the patient, then relate that information to the prescribing problem. In addition to decreasing defensiveness, this additional information may clarify the patient's situation, changing the intervention or making it unnecessary. It can prevent the embarrassment of making an inappropriate recommendation.

The characteristics of the receiver (prescriber) also affect the success of the intervention. Because the receiver's level of involvement with the issue and confidence in his or her opinions can alter the likelihood of persuasion, more explicit, quantitative proof may be needed for specialists than for general practice physicians.[120] After interacting with a prescriber a few times, it may be possible to identify methods that are more successful for that person.

QUALITY OF CARE ISSUES

Programs are implemented to influence prescribing for two general reasons: to decrease costs or to improve the quality of care. Optimally, both of these issues are addressed simultaneously. However, this is not always the case. Restriction of the use of propoxyphene in a Medicaid program was found to be associated with higher drug costs for the program. However, quality of care may have been improved by the restrictions because there was evidence that prescribers substituted nonsteroidal anti-inflammatory drugs for propoxyphene.[62] In another study, lower drug costs resulting from restricting the availability of antiulcer agents were related to decreased quality of care.[69] These findings direct us to be wary of programs that simply aim at lowering drug costs without evaluating the overall picture of patient outcomes related to changes in prescribing.

Measurements of improvements in prescribing usually focus upon costs of drugs and/or assessment of therapy according to specific criteria. Evaluating therapy according to criteria provides a process measure rather than an outcome measure of quality of care. A direct relationship between improved health outcomes and prescribing practices consistent with preestablished criteria is then assumed. Lack of measurement of health outcomes, however, obscures this relationship. It is important to identify and measure patient outcomes to get a clear picture of the effects of the program.

The patient's opinion comprises another perspective of quality of care resulting from programs to influence prescribing. Quality of care is usually viewed from the perspectives of the prescribers and/or health care organization. It is important to realize, however, that restricting access to drug products may be assessed as undesirable by the patient, while it represents improved therapy (or at least lower-cost therapy) from the health care organization's perspective. In some cases, patients may not receive needed drug therapy due to restricted access and/or financial limitations. Decreased access can result in long-term health problems and increased costs, even though short-term costs are decreased. Optimal quality of care measures should include patient evaluations as well as prescriber and organizational perspectives.

IMPLICATIONS FOR PRACTICE: PUTTING IT ALL TOGETHER

How can this information be used to improve our interactions with prescribers? Here are ten basic principles applicable to everyday practice as well as to specialized pharmacist activities such as providing educational programs to improve prescribing:

1. Patient demand can affect prescribing and make changes difficult. Alternative therapies must be discussed, as well as methods for dealing with patient requests for drug treatment.
2. Administrative programs must be adequately enforced. Without enforcement, these programs will falter because attitudes toward prescribing and intention to prescribe nonsanctioned (nonformulary or restricted) drugs are not altered. Pharmacists sometimes are reluctant to follow through with formulary requirements because of lack of confidence, not wanting to do so, or thinking they do not have time to contact prescribers.
3. When administrative methods (e.g., formularies) are used to change prescribing, an educational component should also be included. This can help affect change in prescribing intention, promoting sustainable prescribing changes.
4. Heuristic principles can help explain errors in prescribing, and applying these principles can help design educational programs. For example, educational drug information should be framed in terms of quality rather than cost.
5. One-to-one educational interactions are more effective than

printed material or group meetings. Whenever possible, discuss prescribing problems directly with prescribers.

6. Educational programs must have follow-up. Changes in prescribing are incorporated into everyday practice over time.

7. Include specific patient examples, preferably the prescriber's own patients. This assures relevance to the prescriber's situation.

8. Incorporate concepts related to persuasion. Citing literature or recognized authorities lends credibility. Be prepared for arguments contrary to the message being provided. Strive to develop confidence in the ability to change prescribing.

9. Provide the same information through several media.[119] In addition to discussing the problem, give printed materials such as written summaries, charts, or articles that support a given position.

10. Include individualized feedback that relates directly to a practitioner's prescribing patterns if at all possible. This type of feedback is more effective than results of group audits.

Do not give up. Every successful improvement in prescribing affects a patient's health and well being.

CONCLUSIONS

Evaluating drug prescribing and improving drug use is a cornerstone of pharmaceutical care. However, to provide pharmaceutical care, pharmacists must be successful at getting prescribers to change their drug prescribing decisions when errors are detected. Knowing what factors influence prescribing helps us understand why errors occur and general factors that result in prescribing decisions. Being familiar with research regarding methods for influencing prescribing will help us design programs that are most likely to be effective and to help conserve resources devoted to improving prescribing. As drug products become more powerful, more toxic, and more costly, it will be critical for pharmacists to be active participants in the prescribing process and to promote appropriate drug use.

REFERENCES

1. Wilcox SM, et al. Inappropriate drug prescribing for the community-dwelling elderly. JAMA 1994;272:292-6.

2. Beers M, Avorn J, Soumerai SB. Psychoactive medication use in intermediate-care facility residents. JAMA 1988;260:3016-20.

3. Mead RA, McGhan WF. Use of histamine$_2$-receptor blocking agents and sucralfate in a health maintenance organization. Drug Intell Clin Pharm 1988;22:466-9.

4. Phelps CE. Bug/drug resistance. Med Care 1989;27:194-203.

5. Rupp MT, Schondelmeyer SW, Wilson GT, et al. Documenting prescribing errors and pharmacist interventions in community pharmacy practice. Am Pharm 1988;NS28:574-80.

6. Raisch DW, Bootman JL, Larson LN, McGhan WF. Altering prescribing of anti-ulcer therapies in an HMO through one-to-one educational meetings. Am J Hosp Pharm 1990;47:1766-73.

7. Col N, Fanale JE, Kronholm P. The role of medication noncompliance and adverse drug reactions in hospitalizations of the elderly. Arch Intern Med 1990;150:841-5.

8. Hallas J, Harvald B, Worm J, et al. Drug related hospital admissions: results from an intervention program. Eur J Clin Pharmacol 1993;45:199-203.

9. Hawkey CJ, Hodgson S, Norman A, et al. Effect of reactive pharmacy intervention on quality of hospital prescribing. Br Med J 1990;300:986-90.

10. Raisch DW. Relationships among prescription payment methods and interactions between community pharmacists and prescribers. Ann Pharmacother 1992;26:902-6.

11. Shaughnessy AF, Nickel RO. Prescription-writing patterns and errors in a family medicine residency program. J Fam Pract 1989;29:290-5.

12. Becker MH, Stolley PD, Lasagna L, McEvila JD, Sloane LM. Correlates of physicians' prescribing behavior. Inquiry 1972;9(3):30-42.

13. Greenland P, Mushlin AI, Griner PF. Discrepancies between knowledge and use of diagnostic studies in asymptomatic patients. J Med Educ 1979;54:863-9.

14. Renaud M, Beauchemin J, LaLonde C, Poirier H, Berthiaume S. Practice settings and prescribing profiles: the simulation of tension headaches to general practitioners working in different practice settings in the Montreal area. Am J Public Health 1980;70:1068-73.

15. Cherkin DC, Rosenblatt RA, Hart LG, Schneeweiss R, LoGerfo J. The use of medical resources by residency-trained family physicians and general internists. Med Care 1987;25:455-69.

16. Hemminki E. Review of literature on the factors affecting drug prescribing. Soc Sci Med 1975;9:111-5.

17. Pellegrino ED. Prescribing and drug ingestion symbols and substances. Drug Intell Clin Pharm 1976;10:624-30.

18. Hall D. Prescribing as social exchange. In: Mapes RE, ed. Prescribing practice and drug usage. London: Croom Helm, 1980:39-57.

19. Smith MC. The relationship between pharmacy and medicine. In: Mapes RE, ed. Prescribing practice and drug usage. London: Croom Helm, 1980:157-200.

20. Knapp DE, Oeltjen PD. Benefit to risks in physician drug selection. Am J Public Health 1972;62:1136.

21. Harrell GD, Bennett PD. An evaluation of the expectancy value model of attitude measurement for physician prescribing behavior. J Market Res 1974;11: 269-78.

22. Lilja J. How physicians prescribe their drugs. Soc Sci Med 1976;10: 363-5.

23. Segal R, Hepler CD. Prescribers' beliefs and values as predictors of drug choices. Am J Hosp Pharm 1982;39:1891-7.

24. Segal R, Helper CD. Drug choice as a problem-solving process. Med Care 1985;23:967-76.

25. Ferguson JA. Drugs prescribing habits related to characteristics of medical practice. J Soc Admin Pharm 1990;7:34-47.

26. Peay MY, Peay ER. Patterns of preference for information sources in the adoption of new drugs by specialists. Soc Sci Med 1990;31:467-76.

27. Engel W, Freund DA, Stein JS, et al. The treatment of patients with asthma by specialists and generalists. Med Care 1989;27:306-14.

28. Hohmann, AA. Gender bias in psychotropic drug prescribing in primary care. Med Care 1989;27:478-90.

29. Hooper EM, Comstock LM, Goodwin JM, Goodwin JS. Patient characteristics that influence physician behavior. Med Care 1982;20:630-8.

30. Sleath BL. Patient gender and psychotropic prescribing during the physician-patient interaction [Dissertation]. Madison, WI: University of Wisconsin-Madison, 1993.

31. Denig P, Haaijer-Ruskamp FM, Zijsling DH. How physicians choose drugs. Soc Sci Med 1988;27:1381-6.

32. Naisbitt J. Megatrends: ten new directions transforming our lives. New York: Warner Books, Inc., 1982.

33. Directions for specialization in pharmacy practice, Part 1. Am J Hosp Pharm 1991;48:469-500.

34. Stansilav SW, Barker K, Crismon ML, et al. Effect of a clinical psychopharmacy consultation service on patient outcomes. Am J Hosp Pharm 1994;51: 778-1.

35. Ajzen I, Fishbein M. Understanding attitudes and predicting social behavior. Englewood Cliffs, NJ: Prentice-Hall, Inc., 1980.

36. Chinburapa V, Larson LN, Bootman JL, McGhan WF, Nicholson G. Prescribing intention and the relative importance of drug attributes: a comparative study of HMO and fee-for-service physicians. J Pharm Market Manage 1987;2(2): 89-105.

37. Avorn J, Chen M, Hartley R. Scientific versus commercial sources of influence on the prescribing behavior on physicians. Am J Med 1983;308: 1457-63.

38. Nisbett R, Ross L. Human inference: strategies and shortcomings of social judgment. Englewood Cliffs, NJ: Prentice-Hall, Inc., 1980.

39. Raisch DW, Barreuther AD, Osborne RC. Evaluation of a non-FDA approved use of cimetidine—treatment of pruritus resulting from epidural morphine analgesia. DICP Ann Pharmacother 1991;25:716-8.

40. Kahneman D, Tversky A. Choices, values, and frames. Am Psychol 1984;39:341-50.

41. Tversky A, Kahneman D. Judgment under uncertainty: heuristics and biases. Science 1974;185:1124-31.

42. Cebul RD, Beck LH, eds. Teaching clinical decision making: a handbook for instructors. New York: Praeger, 1985.

43. Schwartz S, Griffin T. Medical thinking, the psychology of medical judgment and decision making. New York: Springer-Verlag, 1986.

44. Weinstein MC, Fineberg HV. Decision analysis. Philadelphia, PA: WB Saunders Company, 1980.

45. Wallsten TS. Physician and medical student bias in evaluating diagnostic information. Med Decis Making 1981;1(2):145-64.

46. Detmer DE, Fryback DG, Gassner K. Heuristics and biases in medical decision-making. J Med Educ 1978;53:682-3.

47. Meyerowitz BE, Chaiken S. The effect of message framing on breast self-examination attitudes, intentions, and behavior. J Pers Soc Psychol 1987;52: 500-10.

48. Dubeau CE, Voytovich AE, Rippey RM. Premature conclusions in the diagnosis of iron-deficiency anemia. Med Decis Making 1986;6:169-73.

49. Friedlander ML, Stockman SJ. Anchoring and publicity effects in clinical judgment. J Clin Psychol 1983;39:637-43.

50. McNeil BJ, Pauker SG, Sox HC Jr, Tversky A. On the elicitation of preferences for alternative therapies. N Engl J Med 1982;306:1259-62.

51. Rook KS. Encouraging preventative behavior for distant and proximal health threats: effects of vivid versus abstract information. J Gerontol 1986;41: 526-34.

52. Elstein AS, Holzman GB, Ravitch MM. Comparison of physicians' decisions regarding estrogen replacement therapy for menopausal women and decisions derived from a decision analytic model. Am J Med 1986;80:246-58.

53. Hazlet TK, Teh-Wei Hu. Association between formulary strategies and hospital drug expenditures. Am J Hosp Pharm 1992;49:2207-10.

54. Sloan FA, Gordon GS, Cocks DL. Hospital drug formularies and use of hospital services. Med Care 1993;31:851-67.

55. Jang R. Medicaid formularies: a critical review of the literature. J Pharm Market Manage 1988;2(3):39-61.

56. Kozma CM, Reeder CE, Lingle EW. Expanding Medicaid drug formulary coverage: effects on utilization of related services. Med Care 1990;28: 963-77.

57. Abramowitz PW. Controlling financial variables-changing prescribing patterns. Am J Hosp Pharm 1984;41:503-15.

58. Hunt DD. Effects of incentives on economic behavior and productivity of psychiatric residents. J Psychol Educ 1980;4(1):4-13.

59. Martin AR, Wolf MA, Thibodeau LA, Dzau V, Braunwald E. A trial of two strategies to modify the test-ordering behavior of medical residents. N Engl J Med 1980;303:1330-6.

60. Soumerai SB, Ross-Degnan D, Gortmaker S, et al. Withdrawing payment for nonscientific drug therapy. JAMA 1990;263:831-9.

61. Depression Guideline Panel. Depression in primary care. Vol. 2: Treatment of major depression. AHCPR Publication No. 93-0551. Washington, DC: Government Printing Office, April 1993.

62. Kreling DH, Knocke DJ, Hammel RW. The effects of an internal analgesic formulary restriction on Medicaid drug expenditures in Wisconsin. Med Care 1989;27:34-44.

63. Shenfield GM, Jones AN, Patterson JW. Effect of restrictions on prescribing patterns for dextropropoxyphene. Br Med J 1980;281:651-3.

64. West SK, Brandon BM, Stevens AM, et al. Drug utilization review in an HMO. I. Introduction and examples of methodology. Med Care 1977;15:505-14.

65. Craig WA, Uman SJ, Shaw WR, Ramgopal V, Eagan LL, Leopold ET. Hospital use of antimicrobial drugs: survey of 19 hospitals and results of antimicrobial control program. Ann Intern Med 1978;89:793-5.

66. McDonald CJ. Protocol-based computer reminders, the quality of care and the non-perfectibility of man. N Engl J Med 1976; 295:1351-5.

67. Wachtel T, Moulton AW, Pezzullo J, Hamolsky M. Inpatient management protocols to reduce health care costs. Med Decis Making 1986;6(2):101-9.

68. Zullich SG, Grasela Jr TH, Fiedler-Kelly JB, et al. Impact of triplicate prescription program on psychotropic prescribing patterns in long-term care facilities. Ann Pharmacother 1992;26:539-46.

69. Bloom BS, Jacobs J. Cost effects of restricting cost-effective therapy. Med Care 1985;23:872-80.

70. Green ER, Chrymko MM, Rozek SL, Kitrenos JG. Clinical considerations and costs associated with formulary conversion from tobramycin to gentamicin. Am J Hosp Pharm 1989;46:714-9.

71. Huber SL, Patry RA, Hudson HD. Influencing drug use through prescribing restrictions. Am J Hosp Pharm 1982;39:1898-901.

72. Britton HL, Schwinghammer TL, Romano MJ. Cost containment through restriction of cephalosporins. Am J Hosp Pharm 1981;38:1897-900.

73. DeTorres OH, White RE. Effect of aminoglycoside-use restrictions on drug cost. Am J Hosp Pharm 1984;41:1137-9.

74. Brook RH, Williams KN. Effect of a medical care review on the use of injections. Ann Intern Med 1976;85:509-15.

75. Ellenor GL, Frisk PA. Pharmacist impact on drug use in an institution for the mentally retarded. Am J Hosp Pharm 1977;34:604-8.

76. Paris M, McNamara J, Schwartz M. Monitoring ambulatory care: impact of surveillance program on clinical practice patterns in New York City. Am J Public Health 1980;70:783-8.

77. Komaroff AL, Flatley M, Browne C, Sherman H, Fineberg SE, Knopp RH. Quality, efficiency, and cost of a physician-assistant protocol system for management of diabetes and hypertension. Diabetes 1976;25:297-306.

78. Avorn J and Soumerai SB. Improving drug-therapy decisions through educational outreach. N Engl J Med 1983;308:1457-63.

79. Schaffner W, Ray WA, Federspiel CF, Miller WO. Improving antibiotic prescribing in office practice. JAMA 1983;250:1728-32.

80. Evans CE, Haynes RB, Birkett NJ, et al. Does a mailed continuing education program improve physician performance? JAMA 1986;255:501-4.

81. Cummings KM, Frisof KB, Long JM, Hrynkiewich BA. The effects of price information on physicians' test-ordering behavior. Med Care 1982;20: 293-301.

82. Everett GD, deBlois S, Chang P, Holets T. Effect of cost education, cost audits, and faculty chart review. Arch Intern Med 1983;143:942-4.

83. Gurwitz JH, Noonan JP, Soumerai SB. Reducing the use of H_2-receptor antagonists in the long-term-care setting. J Am Geriatr Soc 1992;40:359-64.

84. De Santis G, Harvey KJ, Howard D, et al. Improving the quality of antibiotic prescription patterns in general practice: the role of educational intervention. Med J Aust 1994;160:502-5.

85. Hershey CO, Goldberg HI, Cohen DI. The effect of computerized feedback coupled with a newsletter upon outpatient prescribing charges. Med Care 1988;26:88-93.

86. Frazier LM, Brown JT, Divine GW. Can physician education lower the cost of prescription drugs? Ann Intern Med 1991;115:116-21.

87. Gehlbach SH, Wilkinson WE, Hammond WE, et al. Improving drug prescribing in a primary care practice. Med Care 1984;22:193-201.

88. Hershey CO, Porter DK, Breslau D, Cohen DI. Influence of simple computerized feedback on prescription charges in an ambulatory clinic. Med Care 1986;24:472-81.

89. Cohen DI, Jones P, Littenberg B, Neuhauser D. Does cost information availability reduce physician test usage? Med Care 1982;20:286-92.

90. Eisenberg JM, Williams SV, Garner L, Viale R, Smits H. Computer-based audit to detect and correct overutilization of laboratory tests. Med Care 1977;15:915-21.

91. Wones RG. Failure of low-cost audits with feedback to reduce laboratory test utilization. Med Care 1987;25:78-82.

92. Berwick DM, Coltin KL. Feedback reduces test use in a health maintenance organization. JAMA 1986;255:1450-4.

93. Schroeder NH, Caffey EM, Lorel TW. Antipsychotic drugs: can education change prescribing practices? J Clin Psychiatry 1979;40:186-9.

94. Schroeder SA, Meyers LP, McPhee SJ, et al. The failure of physician education as a cost containment strategy. JAMA 1984;252:225-30.

95. Eisenberg JM. An educational program to modify laboratory use by house staff. J Med Educ 1977;52:578-81.

96. Wang VL, Terry P, Flynn B, Williamson JW, Green LW, Faden R. Evaluation of continuing medical education for chronic obstructive pulmonary diseases. J Med Educ 1979;54:803-11.

97. Stross JK, Bole GG. Evaluation of a continuing education in rheumatoid arthritis. Arthritis Rheum 1980;23:846-9.

98. Gutierrez G, Guiscafre H, Bronfman M, et al. Changing physician prescribing patterns: evaluation of an educational strategy for acute diarrhea in Mexico City. Med Care 1994;32:436-46.

99. Ray WA, Blazer DG, Schaffner W, Federspiel CF. Reducing antipsychotic drug prescribing for nursing home patients: a controlled trial of the effect of an educational visit. Am J Public Health 1987;77:1448-50.

100. Klein LE, Charache P, Johannes RS. Effect of physician tutorials on prescribing patterns of graduate physicians. J Med Educ 1981;56:504-11.

101. McConnell TS, Cushing AH, Bankhurst AD, Healy JL, McIlvenna PA, Skipper BJ. Physician behavior modification using claims data: tetracycline for upper respiratory infection. West J Med 1982;137:448-50.

102. Inui TS, Yourtee EL, Williamson JW. Improved outcomes in hypertension after physician tutorials, a controlled trial. Ann Intern Med 1976;84:646-51.

103. Herfindal ET, Bernstein LR, Kishi DT. Effect of clinical pharmacy services on prescribing on an orthopedic unit. Am J Hosp Pharm 1983;40:1945-51.

104. Stergachis A, Fors M, Wagner EH, Sims DD, Penna P. Effect of clinical pharmacists on drug prescribing in a primary-care clinic. Am J Hosp Pharm 1987;44:525-9.

105. Haig GM, Kiser LA. Effect of pharmacist participation on a medical team on costs, charges, and length of stay. Am J Hosp Pharm 1991;48:1457-62.

106. Pierson JF, Alexander MR, Kirking DM, et al. Physicians' attitudes toward drug-use evaluation interventions. Am J Hosp Pharm 1990;47:388-90.

107. Brown CM, Lipowski EE. Pharmacists' reactions to the Wisconsin Medicaid drug-use review program. Am J Hosp Pharm 1993;50:1898-902.

108. Yeo GT, de Burgh SP, Letton T, et al. Educational visiting and hypnosedative prescribing in general practice. Fam Pract 1994;11:57-61.

109. Sutters C, Keat A, Lant A. Improving prescribing of non-steroidal anti-inflammatory drugs in hospital: an educational approach. Br J Rheumatol 1993;32:618-22.

110. Sumpton JE, Frewen TC, Rieder MJ. The effect of physician education on knowledge of drug therapeutics and costs. Ann Pharmacother 1992;26:692-7.

111. McPhee JA, Wilgosh CP, Roy PD, et al. Effect of pharmacy-conducted education on prescribing of postoperative narcotics. Am J Hosp Pharm 1991;48:1484-7.

112. Mueller BA, Abel SR. Impact of college of pharmacy-based educational services within the hospital. DICP Ann Pharmacother 1990;24:422-5.

113. Poulsen RL. Some current factors influencing the prescribing and use of psychiatric drugs. Public Health Rep 1992;107(1):47-53.

114. Eisenberg JM. Doctors' decisions and the cost of medical care. Ann Arbor, MI: Health Administration Press, 1986.

115. Raisch DW. A model of methods for influencing prescribing: part II. DICP Ann Pharmacother 1990;24:537-42.

116. Rhyne RL, Gehlbach SH. Effects of an educational feedback strategy on physician utilization of thyroid function panels. J Fam Pract 1979;8:1003-7.

117. Hepler CD, Clyne KE, Donta ST. Rationales expressed by empiric antibiotic prescribers. Am J Hosp Pharm 1982;39:1647-55.

118. Shelby AN. The theoretical bases of persuasion: a critical introduction. J Business Communication 1986;23(1):5-29.

119. Soumerai SB, Avorn J. Principles of educational outreach ("academic detailing") to improve clinical decision making. JAMA 1990;263:549-56.

120. Jaccard J. Toward theories of persuasion and belief change. J Pers and Soc Psychol 1981;40:260-9.

ADDITIONAL READINGS

For the Instructor

1. Eisenberg JM. Doctors' decisions and the cost of medical care. Ann Arbor, MI: Health Administration Press, 1986.

2. Haig GM, Kiser LA. Effect of pharmacist participation on a medical team on costs, charges, and length of stay. Am J Hosp Pharm 1991;48:1457-62.

3. Hazlet TK, Teh-Wei Hu. Association between formulary strategies and hospital drug expenditures. Am J Hosp Pharm 1992;49:2207-10.

4. Rupp MT, Schondelmeyer SW, Wilson GT, et al. Documenting prescribing errors and pharmacist interventions in community pharmacy practice. Am Pharm 1988;NS28:574-80.

5. Soumerai SB, Avorn J. Principles of educational outreach ("academic detailing") to improve clinical decision making. JAMA 1990;263:549-56.

For the Student

1. Avorn J, Soumerai SB. Improving drug-therapy decisions through educational outreach. N Engl J Med 1983;308:1457-63.

2. Bloom BS, Jacobs J. Cost effects of restricting cost-effective therapy. Med Care 1985;23:872-80.

3. Jang R. Medicaid formularies: a critical review of the literature. J Pharm Market Manage 1988;2(3):39-61.

4. Kozma CM, Reeder CE, Lingle EW. Expanding Medicaid drug formulary coverage: effects on utilization of related services. Med Care 1990;28: 963-77.

5. Kreling DH, Knocke DJ, Hammel RW. The effects of an internal analgesic formulary restriction on Medicaid drug expenditures in Wisconsin. Med Care 1989;27:34-44.

6. Nisbett R, Ross L. Human inference: strategies and shortcomings of social judgment. Englewood Cliffs, NJ: Prentice-Hall, Inc., 1980.

7. Schaffner W, Ray WA, Federspiel CF, Miller WO. Improving antibiotic prescribing in office practice. JAMA 1983;250:1728-32.

8. Segal R, Helper CD. Drug choice as a problem-solving process. Med Care 1985;23:967-76.

9. Sloan FA, Gordon GS, Cocks DL. Hospital drug formularies and use of hospital services. Med Care 1993;31:851-67.

Chapter 9

Pharmacists' Performance in Drug Product Selection and Therapeutic Interchange

Nancy J. W. Lewis
Steven Erickson

INTRODUCTION

The profession of pharmacy is evolving into a health profession intricately involved in patient care. Recognition of the critical role pharmacists have in drug therapy decisions is most apparent in the areas of drug product selection and therapeutic interchange. Both of these activities require pharmacists to apply product and patient-specific information to the selection of medications, therapeutic monitoring, and patient counseling. This chapter describes the role of the pharmacist in these professional activities.

DRUG PRODUCT SELECTION

Definition

Drug product selection is the act of dispensing a different brand or an unbranded medication that has the same active chemical ingredients as the prescribed medication.[1] Drug product selection can only be performed if the medication is available in more than one brand-name version or in a generic version (i.e., the medication is a multisource medication). The goal of drug product selection is to provide the patient with a medication that will provide the same therapeutic response as the prescribed medication but at a reduced cost.[2] Achieving this goal requires the pharmacist to consider numerous factors, including the bioequivalence of the product selected to the product prescribed, the effect of substitution on patient health status and compliance, and drug product costs.

Authority to Perform Drug Product Selection

Gaining authority for drug product substitution is a prized achievement for the pharmacy profession. During the 1950s, most states passed laws requiring pharmacists to dispense the exact drug product written by the prescriber.[3,4] These laws denied pharmacists recognition for their knowledge of drug properties and gave control of drug product selection solely to physicians.

In the 1970s, concerns about the rising costs of medications led policymakers to seek ways to increase the use of generic medications. At the same time, pharmacists saw an opportunity to gain recognition for their professional knowledge, and lobbied to have authority to perform drug product selection. Together, these efforts resulted in states overturning their antisubstitution laws.[5]

State Laws Regulating Drug Product Selection

The right to perform drug product selection is governed by each state, just as the practice of pharmacy is regulated under state law. The state laws allow pharmacists to practice drug product selection for products on a specific list (a positive formulary) or to practice substitution for all drugs except those on a specific list (a negative formulary).[3] Some states mandate that pharmacists dispense a generic product whenever one is available unless the prescriber specifically indicates that the brand-name product must be dispensed. By 1993, 13 states mandated generic substitution, while the remainder permitted pharmacists to practice drug product selection as they deemed appropriate.[6] Most states require patient consent or that the patient be informed that generic substitution is occurring. Many states also require that the cost savings from the dispensing of a generic product be shared with the patient.[6,7]

Prevalence of Drug Product Selection

Initially, pharmacists infrequently performed drug product selection. In the late 1970s and mid-1980s, substitution rates ranged from 6.6% to 9.6%.[1,8] In the early 1980s, state Medicaid programs reported low substitution rates, with no program reporting rates greater than 20% for multisource prescriptions.[1]

Several major events served to increase the dispensing of generic products. First, the availability of generic pharmaceuticals was increased due to the Drug Price Competition and Patent Restoration Act of 1984. Second, concerns about health care costs led third-party payers to provide financial incentives to pharmacists and patients to use generic medications. A third event, the generic drug scandal of the 1980s, negatively affected the image of generic medications, but its effects on utilization were minor. Each of these events is described below.

The Drug Price Competition and Patent Restoration Act of 1984

The availability of generic pharmaceuticals was significantly enhanced by the passage of the Drug Price Competition and Patent Restoration Act of 1984 (also called the Waxman-Hatch Act of 1984). Since 1962, the Food and Drug Administration (FDA) had required a generic pharmaceutical manufacturer to demonstrate that its product was effective, safe, and of comparable bioavailability to the innovator product.[2] The Waxman-Hatch Act of 1984 allowed manufacturers to get marketing approval for generic products through an Abbreviated New Drug Application (ANDA) mechanism that eliminated the need for duplicative efficacy and safety testing.[9] Under this act, the generic product must be shown to be bioequivalent to the brand-name product and must meet the same FDA standards for manufacturing and quality as the pioneer drug.[9] The act thereby eliminated a significant regulatory barrier for the generic industry and stimulated the development of new generic products.[10]

Payer Incentives to Increase Generic Medication Use

As prescription drug prices increased, the use of generic medications became a primary tool for cost containment.[11] Federal and state governments assumed a major role in promoting the use of generic products via the establishment of maximum allowable cost (MAC) reimbursement for multisource prescription drugs within the Medicare and Medicaid programs.[11] Under this payment scheme, upper limits for reimbursement of selected multisource products are instituted. The payment limit is set at the lowest price at which the generic drug is widely and consistently available.[12] Many private-sector payers also adopted a MAC reimbursement method as a means of promoting generic medication use.

Because pharmacists are reimbursed at an amount corresponding to the cost of a generic product, this payment method encourages pharmacists to dispense generic medications when possible. The system allows for the reimbursement of brand-name medication costs only if the physician indicates that the brand-name product is medically necessary. This type of reimbursement system is credited with increasing generic substitution for multisource prescriptions from 19% in 1987 to 42% in 1992.[11]

The use of generic medications was also stimulated through the design of pharmacy benefit plans that required patients to share the cost of receiving a brand-name product if a lower cost generic substitute was available. In 1992, 70% of HMOs required a patient to pay the difference between the price of the brand-name product and the price of the generic product if they chose to receive the brand-name product. Twenty-seven percent of HMOs required higher copayments for brand-name products than for generic pre-scriptions.[13]

The Generic Drug Scandal

In 1988, the increase in generic medication use was dampened by the generic drug scandal. Certain employees at the FDA Division of Generic Drugs were found to have accepted money and other bribes from generic firms in exchange for information and assistance in the ANDA approval process. In addition, certain manufacturers were found to have submitted false or fraudulent data to the FDA to expedite market approval. Manufacturing practices at several generic manufacturers were also found to be deficient.[4,11]

These events resulted in the recall of certain generic products, a slowing of ANDA approvals, and the loss of some generic manufacturers from the market. The FDA reorganized its generic drug review division and imposed tough sanctions on generic manufacturers convicted of fraudulent practices related to the submission of an ANDA.[4]

The scandal raised doubts among health professionals and consumers about the quality of generic pharmaceuticals.[8,14] Pharmacists were particularly concerned because they were responsible for the dispensing of generic medications. National and state pharmacy associations worked to ensure that pharmacists were informed of specific generic drug recalls and provided guidance to pharmacists on how to assure both patients and prescribers of the overall quality and cost-effectiveness of generic products.[15]

Pharmacist Response to These Events

Generic substitution rates increased significantly in the late 1980s and 1990s. In 1991, the generic product dollar market had increased to about 19% of the total prescription drug market.[8] Within health maintenance organizations (HMOs), 39% of prescriptions were filled with generic medications in 1992.[9] By 1994, one-third of all prescriptions were filled with generics.[3]

One survey found that pharmacists were most willing to dispense a generic product if it is requested by the patient.[16] Thus, increased drug product selection may reflect an increased demand for generics. This finding may reflect pharmacists' responses to patient concerns about the out-of-pocket costs associated with brand-name medications. When asked for the most important reason for not substituting a generic for a brand-name product, pharmacists mentioned FDA bioequivalence ratings most often. To a lesser extent, pharmacists also considered factors associated with the clinical condition of the patient or the degree of risk associated with the drug class.[16]

Although financial incentives to pharmacists and patients have increased the use of generic products, pharmacists have not always welcomed payer influence in drug product selection. One survey revealed that the majority of pharmacists felt that a higher copayment on brand products interfered with their professional prerogative and negatively affected their relationship with patients.[17] Some pharmacists stated that higher copayments would lead them to try to get physicians to require that the brand-name product be dispensed.

Pharmacists have also been concerned about the reimbursement limits set by MAC reimbursement programs. Some pharmacists have found that their purchase price for certain generic pharmaceuticals is above the MAC payment limit.[12] Thus, dispensing those generic products results in a financial loss.

Considerations in Drug Product Selection

The pharmacist's role in drug product selection is multifaceted. It includes the evaluation of drug product bioequivalence, the application of this information to an individual patient's therapy, and the provision of information about the quality of generics to prescribers and patients. While pressures to contain expenditures may lead pharmacists to give primary consideration to cost when performing drug product selection, a more comprehensive evaluation is necessary to provide quality patient care.[18]

In 1990, the American Pharmaceutical Association (APhA) published guidelines to assist pharmacists in performing product selection.[19] The guidelines address issues related to both dispensing and purchasing decisions. The major factors to be considered in these processes are listed in Table 9.1.

Determining Drug Product Bioequivalence

Drug product selection requires an assessment of the bioequivalence of the prescribed and substituted product. Product bioinequivalence is a primary reason for pharmacists not substituting a generic for a brand-name product.[16] Being able to determine the bioequivalence of two products requires certain references as well as the ability to interpret pharmacokinetic data findings. Drug product selection decisions should not be based on one patient's experience with a specific product.[2] An individual's response to a medication is as likely to be the result of a physiological, behavioral, or emotional condition as the manufacturing source of the medication.

Pharmacists can use four sources of information to determine if drug product selection can be appropriately performed: state regulations, the FDA's list of therapeutic equivalence evaluations, the published literature, and manufacturer information.

TABLE 9.1. Factors to Consider When Dispensing or Purchasing Generic Pharmaceuticals

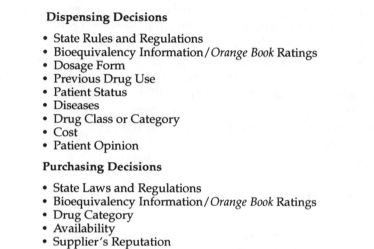

Dispensing Decisions

- State Rules and Regulations
- Bioequivalency Information/*Orange Book* Ratings
- Dosage Form
- Previous Drug Use
- Patient Status
- Diseases
- Drug Class or Category
- Cost
- Patient Opinion

Purchasing Decisions

- State Laws and Regulations
- Bioequivalency Information/*Orange Book* Ratings
- Drug Category
- Availability
- Supplier's Reputation
- Cost

Source: APhA Bioequivalency Working Group Guidelines

State Regulations

Pharmacists should follow state formulary guidelines when performing drug product selection.[19] A state's determination of bioequivalence may or may not be the same as the FDA's. It is the responsibility of the pharmacist to be aware of the formulary status of each generic medication.

FDA Bioequivalence Ratings

The FDA determines bioequivalency based on results of single-dose studies of the generic and brand-name product conducted in normal healthy volunteers aged 21 to 35 years.[2,20] There must be 90% confidence that the average values for the area under the curve (AUC), the maximum drug concentration (Cmax), and the time to maximum concentration (Tmax) of the generic product are within 80% to 120% of the brand product. In addition, results from disintegration and dissolution tests must be similar.[2] The FDA assumes that bioequivalence can be translated into therapeutic equivalence because there is no scientific basis to believe that products that have similar pharmacokinetic profiles differ in their clinical effects.[13]

Several concerns have been raised about the appropriateness of the FDA's bioequivalence testing procedure. The process does not include an evaluation of potential differences in bioequivalence due to gender, age, or the presence of disease.[21,22] Some have suggested that multiple-dose testing should be required, while others have questioned the evaluation process used for certain drug product types, such as dermatologicals and inhalers.[23-25]

The FDA Orange Book

The FDA bioequivalence ratings are listed in the *Approved Drug Products with Therapeutic Equivalence Evaluations*, commonly called the *Orange Book*. The *Orange Book* is published by the FDA and provides a list of all multisource drug products grouped as pharmaceutical equivalents (i.e., drug products with the same active ingredients in the same concentration, dosage from, and route of administration).[26] The FDA ratings are listed in Table 9.2.

The *Orange Book* has some limitations as a source of information. Bioequivalence information is not given for products marketed before 1938, for products marketed between 1938 and 1962 which were approved for safety but are still under review for efficacy, or for pharmaceutical equivalents that are marketed after the filing of a new drug application (NDA).[3] Because the information is listed by the NDA or ANDA applicant and not the manufacturer, it may be difficult to find the FDA ratings for some products.

The *Orange Book* provides information comparing a specific generic product to the innovator product; no information is provided comparing one generic product to another. One study reported that various generic formulations of phenytoin were bioequivalent to the brand-name product, but that significant pharmacokinetic variations were found among the generic formulations.[27] These results emphasize the need for pharmacists to compare the

TABLE 9.2. The Bioequivalence Rating Used in the *Orange Book*

AA	Bioequivalent products in conventional dosage forms
AB	Products therapeutically equivalent to other AB products under that heading
AN	Solutions and powders for aerosolization
AO	Injectable oil solutions
AP	Injectable aqueous solutions
AT	Topical products considered therapeutically equivalent
BC	Extended-release tablets, capsules, and injectables
BD	Active ingredients and dosage forms with documented bioequivalence problems
BE	Delayed-release oral dosage forms
BP	Active ingredients and dosage forms with potential bioequivalence problems
BR	Suppositories or enemas that deliver drugs for systemic absorption
BS	Products having drug standard deficiencies
BT	Topical products with bioequivalence uncertainties
BX	Insufficient data

pharmacokinetic profiles of generic products before substituting one generic product for another.

Published Literature

Published studies comparing the therapeutic effects of generic and brand medications are relatively uncommon. Some studies have found that generic medications do produce similar therapeutic effects resembling those of the innovator, while in other cases, differences in therapeutic responses have been reported.[21,22,25,28,29] While pharmacists need to assess every generic substitution situation, Table 9.3 lists those medications that the pharmacist should be particularly diligent in evaluating before performing drug product selection. These medications represent the products least likely to have bioequivalent generic substitutes.

For one medication, carbamazepine, some comparative studies

TABLE 9.3. Medications Posing Special Problems in Drug Product Selection

---

- Narrow therapeutic range medications (e.g., digoxin, phenytoin, warfarin)
- Delayed-release or extended-release product formulations
- Dermatologicals
- Inhaled medications
- Suppositories or enemas

The bioequivalence of the generic versions associated with these types of medications should always be verified before the prescription is dispensed.

report no therapeutic difference between the brand-name product and a generic product, while case reports of differences between the brand-name product and the generic product have also been published.[30-33] A possible explanation for these divergent results is that the FDA has found only some generic versions of carbamazepine to be bioequivalent to the innovator.

Potential problems with drug product selection have been most widely discussed for medications with narrow therapeutic ranges.[16,20,34,35] For these medications, it is critical that serum blood concentrations remain within a small range to achieve efficacy without the occurrence of medication side effects. Although the FDA has reviewed generic versions of these medications and found them to be bioequivalent, there are a few documented cases of therapeutic mishaps due to changes in manufacturing source.[36] For example, a change in the brand of warfarin sodium used at an anticoagulation clinic was shown to decrease coagulation control.[37]

The literature has also documented the dangers associated with performing drug product selection for products available in both an immediate-release formulation and a controlled-release formulation.[38] One may expect that differences in onset and duration of action of these formulations may affect patient response. In one instance, problems reportedly due to the use of a generic medication were actually the result of a patient being switched from a sustained-release product to an immediate-release product.[39]

Much of the difficulty in evaluating the appropriateness of published reports lies in the anecdotal nature of reported problems with generic substitution. Although many have voiced their opinions about generic substitution, few double-blind control studies have documented the extent to which changes in drug manufacturer source alter patient medication responses.

Manufacturer Information

Pharmacists may need to rely on the manufacturer's pharmacokinetic information to make a professional judgment about the bioequivalence of two pharmaceutically equivalent products. The information must be reviewed to determine if significant differences exist among the Cmax, Tmax, and AUC values of the products. In some instances, manufacturers may be able to provide data from clinical trials that compared generic and brand-name products.

Individualizing Drug Product Selection Decisions

Even if two products are bioequivalent, the pharmacist must assess the individual patient and his or her therapy before performing drug product selection. Table 9.4 lists questions that the pharmacist should consider before performing drug product selection. A review of the patient's past drug use is necessary to determine whether the patient has previously received a brand-name or a generic product. Patients having poor health status, having multiple diseases, or receiving a drug with a narrow therapeutic range should be maintained on a medication from the same manufacturing source.[40] If treatment is initiated with a generic medication, continuing this same product is preferred to switching to a brand-name product.[19] Pharmacists should also consider product costs and the patient's past preferences for generic products.

Drug product selection requires pharmacists to be actively involved in patient monitoring. The patient should be asked about his or her response to the medication and the occurrence of any side effects associated with the medication. Although similar side effects are likely to occur with any manufacturer's version of a medication, differences in inactive ingredients may cause different side effects or allergic reactions.[9] When a patient reports side effects from a generic medication, the pharmacist must discern whether the event is due to the active ingredient or to a change in product manufacturing source.

TABLE 9.4. Patient Assessment Consideration in Drug Product Selection

1. Is a generic product available that has been shown to be equivalent by the FDA, published study results, or some other reliable source to the medication prescribed?

2. Has the patient taken the medication previously? If so, was a generic medication or brand-name product dispensed?

3. Does the patient need to receive a product from the same manufacturing source every time the prescription is dispensed? (e.g., Is the patient receiving a narrow therapeutic range medication?) If yes, can consistency in manufacturing source be guaranteed if a generic product is dispensed?

4. Does the patient have a preference for a generic or a brand-name product?

5. If the patient is taking multiple medications, is the medication being dispensed sufficiently different in appearance so that it will not be confused with another medication?

6. Will patient compliance be affected positively or negatively?

7. Is the generic product available in the dosage form needed by the patient? (e.g., a liquid)

Factors that may influence medication adherence should also be considered. Because generic medications often look different from brand-name products, patients must be informed if a different manufacturer's product is being dispensed. If not informed, a patient may worry that the wrong medication was dispensed. It is important to stress that the generic product is bioequivalent to the brand-name product so that the patient has faith in its effectiveness.[40] Differences in color or shape and different flavors may make a product more acceptable or less acceptable to a patient.[10] If a patient is on multiple medications, ensuring that medications look different may avoid patient confusion regarding the prescribed dosage regimen for each medication. Compliance may improve with the use of a generic medication if noncompliance had been the result of high medication prices.

Informing Prescribers and Patients

The provision of pharmaceutical care encompasses meeting the information needs of both prescribers and patients. Prescribers and patients may be unaware of the availability of a generic product or may have an unfounded belief that the generic version is less suitable than the brand-name product.

Pharmacists can play a major role in increasing the cost-effectiveness of therapy by promoting the prescribing of generic pharmaceuticals. Physician education by pharmacists regarding the availability and appropriateness of generic products can increase generic prescribing.[41] One program aimed at lowering drug costs found that pharmacist interventions led to a 6% increase in the dispensing of generic medications.[42]

When performing drug product selection, pharmacists should be prepared to counsel patients about generic medications. Patients often have misconceptions about generic medications as revealed in a recent survey that found only 65% of people aged 65 and older thought generic drugs were equal in quality to brand-name drugs.[43] Pharmacists may need to explain what a generic medication is and the process involved in gaining FDA approval for marketing generic medications. Because patients may be concerned that a generic medication may be less effective or cause more side effects than a brand-name product, pharmacists must reassure them of the drugs' therapeutic equivalency. Informing patients about how they can monitor their own therapy may give them more confidence in taking a generic medication.

THERAPEUTIC INTERCHANGE

Definition

Therapeutic interchange is the act of dispensing a therapeutic alternative for a prescribed drug product under prescriber prior authorization.[44,45] A therapeutic alternative has a different chemical structure than the prescribed drug but belongs to the same pharmacological and/or therapeutic class and usually can be expected to have similar therapeutic effects and adverse reactions when administered to patients in therapeutically equivalent doses.

Therapeutic alternatives are usually selected from drug categories that have multiple products with similar efficacy and side effect profiles and are associated with significant expenditures. The chosen alternative is usually the drug with the best therapeutic and cost profile. Table 9.5 lists the most common drug categories in therapeutic interchange programs.

Therapeutic substitution, a term that has been used synonymously with therapeutic interchange, is the act of dispensing a therapeutic alternate for a prescribed drug product without prior authorization of the prescriber. In most states, this practice is illegal under pharmacy practice law. Today, therapeutic interchange is the concept most accepted by pharmacy and medical organizations.

Drug categories likely to be involved in therapeutic interchange include those with multiple products that have similar efficacy and side effect profiles and usually are associated with significant expenditures. By choosing to promote one medication over others (i.e., making the product the therapeutic alternative), an organization can bargain for lower prices from the drug manufacturer. Thus, therapeutic interchange lowers the costs of care without affecting the quality of care.

TABLE 9.5. Drug Categories Most Often Involved in Therapeutic Interchange

- Cephalosporins (oral and injectable)
- Oral vitamins
- Dermatologicals
- Nonsteroidal anti-inflammatory agents
- H_2 antagonists (oral and injectable)
- Antacids
- Angiotensin converting enzyme inhibitors
- Calcium channel blockers
- HMG-CoA reductase inhibitors
- Oral contraceptives
- Nitroglycerin patches
- Potassium salts
- Laxatives
- Beta blockers

Source: Refs. 67-70.

Support for Therapeutic Interchange

Professional Association Support

Therapeutic interchange has the support of health care providers and payers. Pharmacists view therapeutic interchange as a means of using their expertise to promote cost-effective care. Payers support therapeutic interchange because it can assist in controlling prescription benefit costs without lowering the quality of care provided.

A number of health professional associations support therapeutic interchange by pharmacists. The American Pharmaceutical Association, the American Society of Hospital Pharmacists (ASHP), the American College of Clinical Pharmacy, and the American Medical Association all support therapeutic interchange by a pharmacist if it is authorized in accordance with previously established and approved written guidelines or protocols within a formulary system.[46,47] These organizations support the practice of therapeutic interchange in all health care settings, including outpatient and long-term care settings, where a Pharmacy and Therapeutics Committee (or a similar committee) can oversee the process. The Coalition for Consumer Access to Pharmaceutical Care, an organization composed of most major pharmacy organizations, states that preapproved product interchange should be included as part of the pharmaceutical care benefit in health care reform.[48]

In 1993, the National Association of Boards of Pharmacy (NABP) passed resolutions stating that state boards of pharmacy should review and seek revision of statutes or rules to provide a legislative basis for therapeutic interchange. Furthermore, those statutes or rules should ensure that patients receive the most appropriate medication therapy. NABP also recommended that state boards work cooperatively with state drug use review boards to establish protocols that would allow therapeutic interchange by pharmacists in all practice settings.[49]

Not all associations endorse the concept of therapeutic interchange. The Pharmaceutical Research and Manufacturers Association (PhRMA) is against therapeutic interchange outside of the institutional setting. It believes that therapeutic interchange may adversely affect a patient's health, may be questionable legally, and is inefficient and intrusive. The PhRMA does recognize that the process is

acceptable in institutional settings where direct physician involvement is mandatory.[47] In contrast, the Generic Pharmaceutical Industry
Association supports therapeutic interchange by pharmacists.

Health Payer Support

Health payers, through pharmacy benefit plans, promote therapeutic interchange as a means of containing prescription drug costs.
By choosing to promote one medication over others (i.e., making a
product the therapeutic alternative), an organization can bargain for
lower prices from the drug manufacturer in return for increased
sales of the therapeutic alternative. Thus, therapeutic interchange
can lower the costs of care without affecting the quality of care.

Pharmacists may be provided an incentive to comply with the
formulary via the receipt of a higher dispensing fee if the chosen
therapeutic alternative is dispensed. Some pharmacy benefit plans
may penalize the pharmacist if therapeutic interchange is not performed by disallowing reimbursement for the dispensing of a nonformulary item. When provided with financial incentives to perform therapeutic interchange, pharmacists must be certain that
patient care concerns override their own financial interests.

Several researchers explored the cost savings associated with
therapeutic interchange. Using data from the 1987 ASHP National
Survey of Pharmaceutical Services, the American Hospital Association, and the Health Care Financing Administration, Hazlet and
Hu suggested that therapeutic interchange is associated with lower
pharmacy drug expenditures.[50] A report by Sloan and colleagues
showed a financial benefit to the pharmacy department when formularies and therapeutic interchange were used. However, they also
found that hospital expenditures were increased for patients with
cardiovascular disease and that costs were unchanged for patients
with infectious diseases.[51]

Other smaller studies have confirmed cost savings with therapeutic interchange programs. Intravenous antibiotics and intravenous
histamine-antagonists were the most frequently studied drug
classes.[52-59] Several studies found that allowing only one therapeutic alternate on formulary controlled costs related to antiulcer drugs
while maintaining high-quality medical care.[60,61]

State Laws Regulating Therapeutic Interchange

It is important that pharmacists understand the concept of therapeutic interchange and when it can be legally practiced. Therapeutic interchange should only be performed if such practices are legal under pertinent state laws and if prior authorization has been obtained from the prescriber.[62]

In 1980, a group of researchers surveyed boards of pharmacy and other regulatory agencies responsible for the enforcement of laws and regulations of pharmacy practice to determine their views of therapeutic substitution in hospitals.[63] Fifty-one jurisdictions were surveyed, and 34 responded. Fifteen agencies said therapeutic interchange was illegal, 17 said it was in accordance with policies established by Pharmacy and Therapeutics Committees and not subject to state or federal regulation, and two gave qualified responses. Few state laws, at that time, addressed the practice of therapeutic interchange, and thus, the opinions of regulatory agencies varied.

Some states have developed legislation or laws that place limits on therapeutic interchange. Many allow it only in the hospital setting and require specific guidelines or policies to govern the process. In 1981, the State of Washington adopted regulations permitting pharmacists to initiate and to modify drug therapy under physician authorization and under a protocol on file with the state board of pharmacy. California, in 1984, authorized pharmacists under physician authority to initiate drug therapy for patients in institutions. Wisconsin passed legislation allowing therapeutic interchange only in hospitals under carefully controlled conditions with physicians approving every substitution.[44]

Additional states have attempted to pass legislation regarding therapeutic substitution. In 1985, the Iowa legislature passed a bill giving pharmacists the right of therapeutic interchange if authorized by a prescriber, however, this bill was vetoed by the governor. In Illinois, a last-minute legislative amendment halted the passage of a bill that would have allowed therapeutic interchange in HMOs.[44,64,65]

Legal concerns often revolve around the appropriate mechanisms for physician consent.[66] Physician consent in the hospital or managed care program, in the absence of legal or statutory regulation, is allowed as legal authority if the physician has given prior consent by agreeing to adhere to the bylaws of the institution. Some institutions have physicians sign documents stating they are fully aware of formulary policies. Other institutions may require the practitioner to check a box or sign on a certain line if therapeutic interchange is permitted. Another method of obtaining consent is the use of pre-printed order sheets with statements such as "therapeutic interchange is permitted unless otherwise specified."[66]

An institution that chooses an agent to represent a class of drugs may be at increased liability. If the institution adheres to certain procedures in choosing a therapeutic alternate, the likelihood of liability may be reduced.[67] First, if the institution chooses therapeutic alternatives based on known standards of care, it can be inferred that its actions are reasonable. Second, the selection process should follow established procedures and be based on the analysis of relevant scientific literature.

Prevalence of Therapeutic Interchange Practices

As the number of medications within various therapeutic classes increased, health care institutions began to implement therapeutic interchange as a means of improving patient care and controlling increasing drug costs. Therapeutic interchange is often a component of the drug formulary process, serving to increase compliance with Pharmacy and Therapeutics Committee decisions. Specific guidelines for the therapeutic interchange process are established.[68]

Therapeutic Interchange in Hospitals

Therapeutic interchange is commonly practiced in hospitals. One survey found that approximately 40 to 50% of hospitals allowed the stocking of single products to represent a given therapeutic category. Thirty percent of hospitals allowed automatic dispensing of a therapeutically equivalent drug product without contacting the physician for permission.[69,70]

Therapeutic interchange was most frequently practiced in hospitals having drug formularies, drug utilization review programs, medical school affiliations, and documentation of significant savings due to therapeutic interchange. Therapeutic interchange was also more prevalent in federal hospitals and those located in states with laws allowing therapeutic interchange. Hospitals gave the following reasons for not engaging in therapeutic interchange: lack of acceptance by physicians, interference with physicians' right to select the drug, unnecessary risk of civil liability, violation of laws, and expected benefits do not justify the cost.

Therapeutic Interchange in Managed Care Organizations

Many managed care organizations have embraced therapeutic interchange as a means of promoting appropriate therapy and containing costs. In 1992, 23% of managed care organizations allowed therapeutic interchange, with staff-model HMOs having the highest prevalence (46%) and independent practice associations (IPAs) having the lowest (17%).[13]

A survey of HMOs conducted by Doering and colleagues found that approximately 30% of respondents allowed therapeutic interchange, with the practice being more prevalent in staff- or group-model HMOs and less prevalent in IPAs.[71] Policies allowing therapeutic interchange were found more frequently in HMOs with in-house pharmacies than in those using contracted pharmacy services. Physician knowledge of patient-specific therapeutic interchange was not required by HMO policy in 30% of the reporting HMOs that allowed the practice to occur.

The Role of the Pharmacist in Therapeutic Interchange

Pharmacists are involved in therapeutic interchange programs at both the policy and the practice level. Through their involvement in Pharmacy and Therapeutics Committees, pharmacists participate in the selection of therapeutic alternatives and the development of therapeutic interchange guidelines. As practitioners, pharmacists implement therapeutic interchange guidelines. This process includes assessing patients individually to determine the health and economic impact of performing therapeutic interchange, informing patients about the dispensing and use of therapeutic alternatives, and monitoring patient responses.

Establishing Therapeutic Interchange Guidelines

In the 1980s, the concept of therapeutic interchange came under attack by the medical profession.[72,73] Many of the arguments against therapeutic interchange were based on concerns that interchange might occur without physician knowledge or authorization. Recent emphasis on the implementation of therapeutic interchange guidelines has decreased organized medicine's concerns about this practice.[74]

The process used to select therapeutic alternatives is important to ensure quality patient care, provider and patient acceptance, and positive economic outcomes. Factors to be considered in determining that two or more products are therapeutic alternatives are listed in Table 9.6. The development of a therapeutic interchange policy should include a thorough evaluation of the products, solicitation of prescriber support, Pharmacy and Therapeutics Committee review and approval of therapeutic interchange guidelines, and provider notification of the therapeutic interchange policy.[68]

The Pharmacy and Therapeutics Committee is an important aspect of the therapeutic interchange process. It is the cooperative

---

TABLE 9.6. Considerations When Selecting Therapeutic Alternatives

---

- Drug efficacy
- Drug side effect profile
- Availability of various dosage forms
- Suitability for use in various populations (pediatrics, elderly)
- FDA-approved indications for use
- Literature-supported indications for use
- Total therapy costs (i.e., costs related to drug product, monitoring, drug failures, and side effects)
- Current prescribing patterns
- Impact on patient compliance
- Impact on patient quality of life
- Impact on patient satisfaction
- Product-related educational services offered by pharmaceutical manufacturer

---

work of physicians, pharmacists, and other health care professionals on the committee that determines the therapeutic categories in which therapeutic interchange will occur, the agents that will be the therapeutic alternatives, the procedure for therapeutic interchange, and the monitoring of outcomes associated with this process.

Establishing therapeutic interchange policies through a committee process facilitates the attainment of physician authorization for the process. Often, physicians admitting patients to hospitals with therapeutic interchange programs must agree to the program before they are allowed to admit patients. When a pharmacist selects and dispenses a therapeutic alternate, such action must occur within the framework of the Pharmacy and Therapeutics Committee guidelines.[75]

To assess the effectiveness of a therapeutic interchange program, there must be a quality assurance program that measures patient and financial outcomes. If patient satisfaction and health-related quality of life data are available for the population, the patient perspective may also prove important in deciding whether to continue with the therapeutic interchange policy as designed.

Performing Therapeutic Interchange

Achieving optimum patient outcomes should be the pharmacist's goal in practicing therapeutic interchange. When evaluating a patient's therapy to determine whether a therapeutic alternative is appropriate, pharmacists must assess both patient and economic factors.[76] Table 9.7 lists the questions that pharmacists should consider before performing therapeutic interchange.

Certain products may require particular caution before a therapeutic interchange is performed. Variance in the pharmacokinetic profiles of sustained-release products requires that therapeutic interchange be performed cautiously, if at all, and that patient responses are closely monitored.[77] Care must also be taken when interchanging oral contraceptives because packaging can differ from one product to the next, leading to patient confusion and noncompliance.[78] In addition, these products also have a fairly narrow therapeutic index. Blood levels outside of the normal range can lead to side effects such as breakthrough bleeding and nausea.

Therapeutic interchange requires particular care in the elderly. Older people, as a group, exhibit a greater variation in pharmacoki-

TABLE 9.7. Patient Assessment Considerations in Therapeutic Interchange

---

1. For this patient, is the therapeutic alternative likely to have the same efficacy and side effect profile as the prescribed drug, given the patient's age, race, gender, and comorbidities?

2. Will the therapeutic alternative result in any interactions with the patient's current drug regimen?

3. Will therapeutic interchange result in a change in therapy for the patient? If so, will the patient be unduly confused by the regimen change?

4. Does the patient's age preclude therapeutic interchange due to differences in the pharmacokinetic/pharmacodynamic profiles of the prescribed drug and the therapeutic alternative?

5. Will patient compliance be affected positively or negatively?

6. Is the therapeutic alternative available in the dosage form needed by the patient (e.g., a liquid)?

---

netic and physiologic processes than younger populations; thus, therapy changes require careful monitoring. For some elderly, particularly those in nursing homes, the use of medications that have a liquid formulation may be important.

Providing Provider and Patient Information

Effective communication between pharmacists and other health care professionals is very important if a therapeutic interchange program is to be successful. While therapeutic interchange policies implemented by institutions or pharmacy insurance plans are usually communicated to the appropriate prescribers, additional prescriber notification by the pharmacist may be prudent.[38] Such notification may promote acceptance of the process and help ensure that the physician has an accurate record of the patient's therapy, thus preventing confusion as the prescriber reviews response to the therapy.

It is also important that pharmacists who practice therapeutic interchange communicate the practice to patients. For patients beginning therapy, such communication will alleviate concerns when a medication with a different name than the prescribed medication is dispensed. For continuations in therapy, counseling should focus on explaining the authorization for the interchange and the similarities of the therapeutic alternative to the prescribed medication. Counseling patients offers an opportunity to discuss any concerns they may have regarding the interchange process. Failure to communicate the fact that therapeutic interchange is occurring could lead to patient mistrust and possible legal action.

CONCLUSIONS

As the professional role of the pharmacist expands, drug product selection and therapeutic interchange will continue to serve as foundations for the provision of pharmaceutical care. As health care payers seek to contain costs by promoting generic substitution and therapeutic interchange, pharmacists must assume responsibility for evaluating the scientific soundness of these policies. As health practitioners, pharmacists must assess patient responses to drug therapy, communicate with other professionals and patients regarding the intricacies of drug therapy, and above all, advocate high-quality patient care.

REFERENCES

1. Culkin TT, Mendell S. "Generic substitution" in New Jersey, 1979-87. Am Pharm 1989;NS29:25-30.

2. Rheinstein PH. Therapeutic inequivalence. Drug Safety 1990;5:114-9.

3. Bentley JP, Summers KH. Drug product selection and the *Orange Book*. Drug Top 1994;(Suppl):41S-8S.

4. Silver RB, Abrutyn E, eds. Generic drug industry handbook. New York: Lehman Brothers, 1994.

5. Vivian J, Slaughter R. Legal and clinical aspects of drug product selection for pharmacists. Drug Top 1993;4:4-11.

6. Colligen BH, ed. Pharmaceutical benefits under state medical assistance programs. Reston, VA: National Pharmaceutical Council, Inc., 1993.

7. Bloom BS, Sierz DJ, Pauly MV. Cost and price of comparable branded and generic pharmaceuticals. JAMA 1986;256:2523-30.

8. Lamerton J, ed. The U.S. generic drug industry. New York: The NatWest Investment Banking Group, 1992.

9. Pelsor FR. Drug product selection: a discussion of generics from the FDA perspective. Mich Pharm 1986;24:28-30.

10. Nightingale SL, Morrison JC. Generic drugs and the prescribing physician. JAMA 1987;258:1200-4.

11. Opportunities for the community pharmacist in managed care. Washington, DC: American Pharmaceutical Association, 1994.

12. Torielli G, Gagnon JP, Lingle EW. MAC study confirms pharmacy losses. Am Pharm 1982;NS22:31-4.

13. Marion Merrell Dow. Managed care digest, HMO edition. Kansas City, MO: Marion Merrell Dow, 1993.

14. Anon. AAFP approves policy against mandatory generic substitution. Am Fam Physician 1989;40:297+.

15. Anon. Ask your pharmacist—what you should know about generic drug products. Pharm Today 1990;29:4.

16. Smith M, Monk M, Banahan B. Factors influencing substitution practices. Am Druggist 1991;88-96.

17. Colaizzi JL, Barone JA. Physician and pharmacist attitudes toward a generic incentive program. N J Med 1986;83:153-6.

18. McGregor T. Generics: friend or foe? Am Pharm 1987;NS27:18-21.

19. APhA Bioequivalency Working Group. Guidelines for pharmacists performing product selection. Am Pharm 1990;NS30:40-1.

20. Huntington, CG. Generic drugs reexamined. Physician Assistant 1990;14:13+.

21. Carter BL, Corsoma LM, Williams CO, Schabold K. Once-daily propranolol for hypertension: a comparison of regular-release, long-acting, and generic formulations. Pharmacother 1989;9:17-22.

22. Carter BL, Noyes MA, Demmler RW. Differences in serum concentrations of and responses to generic verapamil in the elderly. Pharmacother 1992;13:359-68.

23. Williams RL. Bioequivalence of topical corticosteroids: a regulatory perspective. Int J Dermatol 1992;31:2-5.

24. Blake KV, Harman E, Hendeles L. Evaluation of a generic albuterol metered-dose inhaler: importance of priming the MDI. Ann Allergy 1992;68:169-74.

25. Jackson DB, Thompson C, McCormack JR, Guin JD. Bioequivalence (bioavailability) of generic topical corticosteroids. J Am Acad Dermatol 1989; 20(5 Pt 1):791-6.

26. Hare D, Foster T. The *Orange Book*: the Food and Drug Administration's advice on therapeutic equivalence. Am Pharm 1990;NS30:35-7.

27. Soryal I, Richens A. Bioavailability and dissolution of proprietary and generic formulations of phenytoin. J Neurol Neurosurg Psychiatry 1992;55:688-91.

28. Sharoky M, Perkal M, Tabatznik B, Cane RC, Costello K, Goodwin P. Comparative efficacy and bioequivalence of a brand-name and a generic triamteriene-hydrochlorothiazide combination product. Clin Pharm 1989;8:496-500.

29. Teresi ME, Riggs CE, Webster PM, Adams MJ, Noonan PK, O'Donnell JP. Bioequivalence of two methotrexate formulations in psoriatic and cancer patients. Ann Pharmacother 1993;27:1434-8.

30. Hartley R, Aleksandrowicz J, Ng PC, McLain B, Bowmer CJ, Forsythe WI. Breakthrough seizures with generic carbamazepine: a consequence of poorer bioavailability? Br J Clin Pract 1990;44:270-3.

31. Oles KS, Penry JL, Smith LD, Anderson RL, Dean JC, Riela AR. Therapeutic bioequivalency study of brand name versus generic carbamazepine. Neurology 1992;42:1147-53.

32. Jumao A, Bella I, Craig B, Lowe J, Dasheiff RM. Comparison of steady-state blood levels of two carbamazepine formulations. Epilepsia 1989;30:67-70.

33. Welty TE, Pickering PR, Hale BC, Arazi R. Loss of seizure control associated with generic substitution of carbamazepine. Ann Pharmacother 1992;26:775-7.

34. Report of the Therapeutics and Technology Assessment Subcommittee of the American Academy of Neurology. Assessment: generic substitution for anti-epileptic medication. Neurology 1990;40:1641-51.

35. Colaizzi JL, Lowenthal DT. Critical therapeutic categories: a contraindication to generic substitution? Clin Ther 1986;8:370-9.

36. Anon. Narrow therapeutic range drugs audited by FDA. FDC Rep 1990;(Sept 17).

37. Richton-Hewett S, Foster E, Apstein CS. Medical and economic consequences of a blinded oral anticoagulant brand change at a municipal hospital. Arch Intern Med 1988;148:806-8.

38. Arnold RJ, Kaniecki DJ. Selection of oral controlled-release drugs: a critical decision for the physician. South Med J 1993;86:208-14.

39. May JR. Setting the record straight on generic procainamide. Consultant Pharm 1991;6:685.

40. Thompson LF. After the scandals: new generic counseling. Am Pharm 1990;NS30:31-3.

41. Erramouspe J. Impact of education by clinical pharmacists on physician ambulatory care prescribing of generic versus brand name drugs. DICP 1989;23:770-3.

42. Knowlton CH, Knapp DA. Community pharmacists help HMO cut drug costs. Am Pharm 1994;NS34:36-42.

43. Rosendahl I. Consumers on generics: some clues on their views from AARP. Drug Top 1994;(Suppl):54S-6S.

44. Segal R, Grines LL, Pathak DS. Opinions of pharmacy, medicine, and pharmaceutical industry leaders about hypothetical therapeutic-interchange legislation. Am J Hosp Pharm 1988;45:570-7.

45. APhA and AMA define terms in interprofessional dialog. Am Pharm 1984;NS24(Feb):12.

46. Anon. AMA policy on drug formularies and therapeutic interchange in inpatient and ambulatory patient care settings. Am J Hosp Pharm 1994;51:1808-10.

47. Guidelines for therapeutic interchange. ACCP position statement. Pharmacother 1993;13:252-6.

48. Coalition for Consumer Access to Pharmaceutical Care. An outpatient pharmacy services benefit in a reformed health care system. Am J Hosp Pharm 1993;50:1464-6.

49. Anon. Technician status, therapeutic interchange are subjects of National Association of Boards of Pharmacy. Am J Hosp Pharm 1993;50:1538.

50. Hazlet TK, Hu TW. Association between formulary strategies and hospital drug expenditures. Am J Hosp Pharm 1992;49:2207-10.

51. Sloan FA, Gordon GS, Cocks DL. Hospital drug formularies and use of hospital services. Med Care 1993;31:851-67.

52. Martin LA, Watkins JB, Greene SA, Haddow AD, Gerecht WB, Powell CW. Procedural compliance and clinical outcome associated with therapeutic interchange of extended-spectrum penicillins. Am J Hosp Pharm 1990;47:1551-4.

53. Smith KS, Briceland LL, Nightingale CH, Quintiliani R. Formulary conversion of cefoxitin usage to cefotetan: experience at a large teaching hospital. DICP 1989;23:1024-30.

54. Oh T, Franko TG, Rumsey KR, Dominguez C. Automatic interchange and selective culture and sensitivity reporting of cefazolin sodium. Am J Hosp Pharm 1988;45:1278.

55. Guastella C. Cost savings realized from interchanging ceftrizoxime for cefoxitin. Am J Hosp Pharm 1988;45:2376-7.

56. McCloskey WW, Johnson PN, Jeffrey LP. Cephalosporin-use restrictions in teaching hospitals. Am J Hosp Pharm 1984;41:2359-62.

57. Guernsey BG, Berina LF, Lazarus MC, Hokanson JA, Prohaska C, Allen JA, Doutre WH, Mader JT. Cost containment through therapeutic substitution of aminoglycosides. Hosp Pharm 1985;20:82-93.

58. Brown GR, Clarke AM. Therapeutic interchange of cefazolin with metronidazole for cefoxitin. Am J Hosp Pharm 1992;49:1946-50.

59. Lawrenz CA, Cole P, Theodorou A, Cook RL, Bermann L. Therapeutic interchange of ampicillin-sulbactam for cefoxitin. Am J Hosp Pharm 1991;48:2150-4.

60. Oh T, Franko TG. Implementing therapeutic interchange of intravenous famotidine for cimetidine and ranitidine. Am J Hosp Pharm 1990;47:1547-51.

61. Rich DS. Experience with a two-tiered therapeutic interchange policy. AJHP 1989;46:1792-1798.

62. Simonsmeier LM. Legal issues emerging from changes in pharmacy practice. Am J Pharm Educ 1987;51:87-90.

63. Doering PL, McCormick WC, Klapp DL, Russell WL. State regulatory positions concerning therapeutic substitutions in hospitals. Am J Hosp Pharm 1981;38:1900-3.

64. McLeod DC. Therapeutic drug interchange: the battle heats up! DICP 1988;22:716-8.

65. Anon. Wisconsin restricts therapeutic interchange. Am J Hosp Pharm 1988;45:741.

66. Doering PL, McCormick WC, Klapp DL, Russell WL. Therapeutic substitution and the hospital formulary system. Am J Hosp Pharm 1981;38:1949-51.

67. Liang FZ. Hospital risk in selecting therapeutic alternates. Am J Hosp Pharm 1988;45:1052.

68. Abramowitz PW. Controlling financial variables–changing prescribing patterns. Am J Hosp Pharm 1984;41:503-15.

69. Doering PL, Klapp DL, McCormick WC, Russell WL. Therapeutic substitution practices in short-term hospitals. Am J Hosp Pharm 1982;39:1028-32.

70. Stoler MH. National survey of hospital pharmaceutical services–1985. Am J Hosp Pharm 1985;42:2667-78.

71. Doering D, Russell W, McCormick W, Klapp D. Therapeutic substitution in the health maintenance organization environment. DICP 1988;22:125-30.

72. Schwartz LL. The debate over substitution policy: its evolution and scientific basis. Am J Med 1985;79:38-44.

73. Nelson EB. Drug substitution and rational therapeutics: old problems and new challenges. Postgrad Med 1989;86:247-50.

74. American College of Physicians. Therapeutic substitution and formulary systems. Ann Intern Med 1990;113:160-3.

75. Penna RP. Pharmacy: a profession in transition or a transitory profession. Am J Hosp Pharm 1987;44:2053-9.

76. Smith ME. The cost of noncompliance and the capacity of improved compliance to reduce health care expenditures. In: Improving medication compliance: proceedings of a symposium. Washington, DC: National Pharmaceutical Council, 1984:35-43.

77. Hendeles L. Slow-release theophylline: do not substitute. Am Pharm 1989;NS29:22.

78. Ansbacher R. Interchangeability of low-dose oral contraceptives: are current bioequivalent testing measures adequate to ensure therapeutic equivalency? Contraception 1991;43:139-47.

ADDITIONAL READINGS

1. Allnut RF. PMA's concerns about therapeutic substitutions. Am Pharm 1990;NS30:39+.

2. Brust M, Hawkins CF, Grayson D. Physicians' attitudes toward generic drug substitution by pharmacists. Tex Med 1990;86:45-9.

3. Chinburapa V, Larson LN. The availability of new drugs in health maintenance organizations. J Res Pharm Econ 1991;3(1):91-110.

4. Considerations in the selection of multisource pharmaceuticals: a monograph and slide series by Scientific Therapeutics Information supported by Schien Pharmaceuticals, Inc.

5. Cowen DL. Changing relationship between pharmacists and physicians. Am J Hosp Pharm 1992;49:2715-21.

6. Feldman JA, DeTullio PL. Medication noncompliance: an issue to consider in the drug selection process. Hosp Formul 1994;29:204-11.

7. Gibofsky A. Legal implications of therapeutic substitution. Drug Ther 1989;19:15+.

8. Gore MJ. Cost, safety, and efficacy: defining the pharmacist's role in drug product selection. Consultant Pharm 1991;6:771-89.

9. Heenan J. Prescription drug benefits in a managed care plan: balancing quality and costs. Med Interface 1994;(Jan):84-92.

10. Knoben JE, Scott GR, Tonelli RJ. An overview of the FDA publication *Approved Drug Products with Therapeutic Equivalence Evaluations.* Am J Hosp Pharm 1990;47:2696-700.

11. Knowlton CH, Knapp DA. Give patients their Rx rights! Am Pharm 1990;NS30:38+.

12. Kralewski JE, Pitt L, Dowd B. The effects of competition on prescription-drug–product substitution. N Engl J Med 1983;309:213-6.

13. Lamy PP, Palumbo FB. Therapeutic substitution: putting the patient first. Pharm Times 1990;56:41-5.

14. LeBlanc TR. Legal pitfalls of therapeutic substitution. Am Druggist 1991;205:20+.

15. Levy RA. Therapeutic inequivalence among pharmaceutical alternates. US Pharm 1984;9:44+.

16. Liang FZ, Greenberg RB, Hogan GF. Legal issues associated with formulary product-selection when there are two or more recognized drug therapies. Am J Hosp Pharm 1988;45:2372-5.

17. Meyer MC, Straugh AB. Biopharmaceutical factors in seizure control and drug toxicity. Am J Hosp Pharm 1993;50(12 Suppl 5):S17-22.

18. Nightingale CH, Gousse GC, On A, Quintiliani R. Antibiotic use and formulary considerations. J Pharm Pract 1991;4:153-7.

19. O'Connor TW, Pisano D. Estimated cost savings through the use of antacid treatment as an alternative to H2-histamine antagonists. Adv Ther 1991;8:305-16.

20. Oddis JA. Future practice roles in pharmacy. Am J Hosp Pharm 1988;45:1306-10.

21. Patchin GM, Carmichael JM. Measuring drug compliance using predetermined drug-use criteria. P and T 1994;(Mar):225-34.

22. Perry PA. Therapeutic substitution and better purchasing significantly reduce hospital pharmacy costs. Hosp Mater Manage 1990;15:10-11.

23. Pulmeri PA, Crane VS. Legal and medical issues in therapeutic interchange: implications for pharmacists, physicians, and P and T committees. Hosp Form 1992;27:1040-50.

24. Shepherd MD, Salzman RD. The formulary decision-making process in a health maintenance organization setting. Pharmacoeconomics 1994;5:29-38.

25. Shulkin DJ, Giardino AP, Freenock TF, et al. Generic versus brand name drug prescribing by resident physicians in Pennsylvania. Am J Hosp Pharm 1992;49:625-6.

26. Shulman SR, Dicerbo PA, Ulcickas ME, Lasagna L. Survey of therapeutic substitution programs in ten Boston area hospitals. Drug Info J 1992;26:41-52.

27. Thompson LF. After the scandals: new generic counseling. Am Pharm 1990;NS30:31-3.

28. Zoeller JL. Does therapeutic interchange hurt patients? Am Druggist 1991;203:66+.

Chapter 10

Interprofessional Relations in Drug Therapy Decisions

Jennifer A. Gowan
Louis Roller
Alistair Lloyd

INTRODUCTION

Pharmacists' participation in decisions about drug therapy has been evolving since the development of clinical pharmacy practice in hospital wards and the widespread adoption of computer-based patient medication histories in community practice. Their involvement in decisions about drug therapy can occur:

- When their knowledge of drugs and medicines is used in the production of therapeutic guidelines for the prescribing professions, in the development of formularies within institutions, and by involvement in drug information services, when these are used to influence prescribing;
- When their knowledge is used for individual patients, when decisions are being made to initiate new therapy, as well as when that therapy is reviewed during the dispensing process.

Initiation of Drug Therapy in Patients–Generally

Pharmacists are valued members of multidisciplinary teams involved in the production of therapeutic guidelines for prescribers in a wide range of clinical categories–such as those developed by the Victorian Drug Usage Advisory Committee in Australia–as well as in the development of formularies of approved medicines for individual hospitals, residential care institutions, and even for medical practice groups. Input into the development of treatment protocols is also being extended to pharmacists being involved directly with the education of prescribers about the therapeutic guidelines, as well as monitoring adherence to them.[1]

Initiation of Drug Therapy in Individual Patients

Medical and other prescribing practitioners have a responsibility to diagnose their patients' condition and to initiate appropriate treatment, including drug therapy. In some hospitals and residential care settings, pharmacists (particularly on wards) may be consulted by the prescriber on aspects of the choice of that therapy. In community practice, there are occasional reports of individual pharmacists who develop sufficient personal rapport with their local prescribers to be asked to provide advice about drug therapy for individual patients. However, the ultimate responsibility for the appropriateness of that therapy must still rest with the prescriber, limited only by the pharmacist's duty of care to influence that decision if the pharmacist believes it is warranted for the safety or well-being of the patient.

Review of Drug Therapy Once Prescribed

Pharmacists are involved in drug therapy decisions about individual patients after that therapy has been initiated, particularly:

- During the process of dispensing, when actual or potential drug problems may be detected and negotiations about their resolution instituted with the prescriber
- Subsequent to the medicine being supplied when a pharmacist becomes aware of the need to review therapy, for example, by noticing an adverse drug reaction (ADR) or other drug-related problem while the patient is taking the therapy. The practice of pharmaceutical care will no doubt enlarge this review considerably.

FORMAL AND INFORMAL NETWORKS

The Pharmacist-Prescriber Interaction

Hepler has postulated the concept of the pharmacist becoming a cotherapist for pharmaceutical care.[2] Whether this is achieved or the more conventional "consulting colleague" model remains, the norm will depend on how effective and accepted pharmacists can become in this field. Whatever the future, getting there involves continuing struggle for the pharmaceutical profession to establish a basis of mutual respect and recognition with the medical and other prescribing professions.

The process of interaction between medical practitioners and pharmacists should commence at the undergraduate level, where the mutual interdependence of the two professions can be fostered.[3,4] This should involve the various organizations that represent pharmacists and prescribers working closely together at all levels to ensure that both professions understand, respect, and accept how their members may work together to achieve the desired patient outcomes.

An example of such a cooperative effort is the recently published joint statement on interprofessional communication by the Royal Australian College of General Practitioners and the Pharmaceutical Society of Australia.[5] This statement proposes an efficient means for general practitioners (GPs) and pharmacists to communicate with each other by using formal referral forms and by extending the range of prescription writing protocols. Of even more interest, it encourages general practitioners to include on their prescriptions an indication of the purpose for which the medicine has been prescribed so that pharmacists may perform their role appropriately.

The concept of pharmacists and medical practitioners being encouraged to work and participate together in continuing professional education at regional and local levels is also being promoted actively in many countries.[3,6-8]

Interprofessional Consultation and Networking

Pharmacist involvement in drug therapy decisions occurs in hospital pharmacy with the acceptance of pharmacists on wards, but it is uncommon in community pharmacy settings.[9] Few studies have found that community pharmacists influence prescribing, except in identifying inappropriate selection of drugs for a specific patient or clarifying the prescriber's intention.

Cooperation between prescribers and community pharmacists is encouraged in some settings more than others.[10,11] Differences in these relationships associated with practice settings may be due to prescription volume, patient mix, extent of participation in third-party reimbursement schemes, and financial incentives.[11] There is a greater recognition of the pharmacists' role in health maintenance organizations, where the cost of drugs is of more direct concern to prescribers.[12]

This has also been seen in the United Kingdom, where general practitioners' attitudes toward pharmacists working in the health center differ markedly from the attitudes of GPs who work in relative isolation from pharmacists. GPs in health centers were more likely to report using the pharmacist as a source of information and were more supportive of the pharmacist's role in giving advice about medicines than those who worked in health centers without a pharmacy.[13]

Community pharmacists are reticent to venture outside their pharmacies to meet physically with other local health professionals. The establishment of good rapport would be facilitated if pharmacists were prepared to meet their local prescribing health professionals (e.g., specialist physicians, medical practitioners, dentists, optometrists, and podiatrists).

Methods of strengthening informal networks emphasizing the benefits and creating a communication climate between community pharmacist and prescriber could include:

- Leaving a professional business card at the surgery at the time of initial meeting.
- Cross-referral by:

    a. Use of referral forms between pharmacist and doctor, for example, (1) a patient who has sought advice at the pharmacy about a symptom where a prescribed medicine would be more advantageous compared to an over-the-counter (OTC) medicine or (2) the issue of a card for a suspected adverse drug reaction.[5] Blenkinsopp, Jepson, and Drury

showed that almost three-fourths of patients who were issued cards reported to their doctors. These cards were well received by patients, doctors, and pharmacists.[14]
b. Offer for the doctor to refer to the pharmacist for the demonstration of an asthma nebulizer, inhaler technique, or use of a blood glucose monitoring machine.

• Patient information leaflets supplied by the pharmacist to reinforce prescriber's and pharmacist's verbal instructions, copies to be supplied regularly to the doctor.
• Regular newsletter advising about new products, regulations, adverse drug reactions, reports from continuing professional education activities, invitations to combined education lectures or seminars, etc.
• Proactive reporting of general issues, e.g., alerting a prescriber to decreased compliance problems generally instead of only reacting to inappropriate prescribing to an individual.
• Offer of up-to-date drug information to assist the prescriber in drug therapy decisions. For example, pharmacists who use computerized drug information programs (e.g., Micro-Medics® containing Drug-Dex®, Posindex®, Martindale®, etc.) would benefit by inviting prescribers to explore the scope of these programs, followed up by offering to assist prescribers to access data for specific patients. A library and journal service at the pharmacy will also encourage prescribers to discuss difficult or unusual prescribing problems.
• Availability of computerized patient medication histories in all pharmacies facilitates pharmacists' detection of drug abuse by patients (e.g., doctor shopping) in addition to more routine monitoring of individual histories.
• Establishment of community pharmacist and prescriber liaison groups. Initiatives to contain prescribing costs have provided a focus for such groups in the United Kingdom.[15, 16] These groups could also develop regional or practice-based formularies, adapt and make recommendations for national prescribing formularies, establish drug monitoring protocols, lobby health authorities and local government organizations, and participate in collaborative research.

IMPROVEMENT OF THERAPEUTIC OUTCOMES

Improved communication between health professionals offers the possibility of improved therapeutic outcomes for patients. Some specific benefits are listed below.

Reduction in Prescribing Errors

A number of studies have shown that interventions by community and hospital pharmacists have reduced prescription errors.[17,18]

Reduction in Number of Medications and Adverse Drug Reactions

Lamour and colleagues report that the risk of adverse drug reactions increases rapidly with the number of drugs used.[19] The use of multiple prescribed medicines (often called polypharmacy) was predicted by recent hospitalization, increasing age, female sex, and increasing depression.[20] Visits by pharmacists to housebound people have brought drug-related problems to the attention of prescribers, thus facilitating medication reviews.[21,22]

Reduction in Errors in Administration of Drugs

Intervention by a pharmacist has increased efficiency in monitoring drug therapy, thereby reducing administration errors and dosage modifications and increasing serum drug and biochemical tests.[18] The use of computers by pharmacists also increases medication surveillance and detection of unwanted drug interactions.[23]

Improvement in Compliance

Medication counseling using written information is effective in reinforcing verbal information.[24,25] Noncompliance with dosage regimens following discharge has been shown to be responsible for up to 16% of hospital readmissions.[26] Underuse of medication is also an area for concern, particularly with hypertensive and asthmatic patients.[27]

Rational Prescribing

The report of the Committee of Inquiry into Pharmacy in the United Kingdom recommended developing closer relationships between GPs and pharmacists so GPs could have the same benefits as hospital doctors in access to pharmaceutical expertise. The inquiry noted that there was "one [drug] company representative for every six GPs" and that the representatives' role was to "promote the use of their companies' products rather than to supply an impartial scientific advice service to doctors."[6]

Prescribing practice may be influenced more impartially by the use of academic detailers, educational campaigns, and prescribing guidelines.

Academic Detailers. The use of academic detailers to change prescribing practice has been proposed recently. Academic detailers are usually pharmacists who provide prescribers with unbiased drug information to promote better prescribing.[28] The pilot studies typically seek to influence the prescribing of a few specially targeted drugs, while prescribers typically order about 150-200 drugs each.[29] Pharmacists in the Netherlands are providing drug information to physicians through local Prescribing and Therapeutic Committees.[30]

Educational Campaigns. Educational campaigns have been used in hospital settings with significant changes in prescribing habits for antibiotics.[31] This initiative has also been applied successfully to community practice.[1]

Drug Usage Guidelines. Drugs and Therapeutics Committees of hospitals—consisting of pharmacists, prescribers, and administrators—determine prescribing protocols, taking into account the efficacy and cost-effectiveness of the competing drug treatments.

In 1978 in Australia, the Victorian Drug Usage Advisory Committee identified a need for uniform prescribing guidelines for both hospital and community settings. The series includes guidelines for analgesic, antibiotic, cardiovascular, gastrointestinal, psychotropic, and respiratory drugs. The guidelines are reviewed and reprinted biannually, with recommendations being obtained from community medical and pharmacy practitioners; they are endorsed nationally by the Australian Medical Association. The impact of the guidelines on prescribing has also been significant.[32] This is an excellent example of pharmacist-prescriber interaction in decisions regarding drug therapy.

Adequate Knowledge of a Patient's Drug Regimen

Inadequate knowledge of a patient's total drug regimen by the prescriber has the potential for inappropriate prescribing. Many patients visit various doctors and specialists for a multiplicity of conditions. Each prescriber may be unaware of the medication prescribed by the other. Encouraging patients to attend one pharmacy only where complete drug history records are maintained would help prevent this problem. Patient education on this point could be implemented, as many consumers of prescription medication are unaware that this service exists. Additionally, the ultimate introduction of "smart" cards or linked computer systems on a state or even national basis would increase the completeness and accuracy of patient records. These could also be linked to the prescriber's computer. These issues, of course, bring up all sorts of potential ethical and confidentiality problems.

Each pharmacist must review a patient's drug history prior to dispensing any medication. A review permits the identification of possible inappropriate prescribing, where the pharmacist may identify a drug cause for a presenting symptom and then contact the prescriber to suggest modification of the prescribed regimen to remove the cause. Examples are:

- Prochlorperazine for nausea when the patient is taking digoxin and verapamil.
- Nitrazepam for sleep when the patient is taking a monoamine oxidase inhibitor late in the afternoon.
- Allopurinol for gout when the patient is taking a thiazide diuretic.
- Temazepam for sleep when the patient is found to be taking metoprolol at night.

Detection of Drug Misuse

Improved communication between health professions encourages the reporting of excessive medication use, for example, when patients use multiple prescribers to obtain supplies of drugs of addiction or benzodiazepines.

Identification of At-Risk Populations

Mass screening (e.g., blood pressure testing, cholesterol testing, blood glucose testing) can be used by pharmacists and other health professionals to detect populations at particular risk who may require referral and further investigation.

Pharmaceutical Care

Proactive monitoring of patients during the whole of the drug therapy by the pharmacist will cause consultation with the prescriber if the medication is or has the potential of producing drug-related problems, as well as encourage better compliance (e.g., blood pressure monitoring, weight reduction, usage monitoring).[33] Cross-referral should occur between pharmacist and prescriber.

Research and Development

Pharmacy practice research needs to be collaborative and multidisciplinary to achieve improved therapeutic outcomes for patients. Areas of research include compliance, patient education, doctor shopping, unfilled prescriptions–original and refills, reporting of side effects, adverse drug reactions, etc.

Health Promotion

Local health professionals need to cooperate to encourage people to take responsibility for their own health. Guidelines can be developed for consistent health promotion, screening procedures, and local community health programs, all involving pharmacists and all other health professionals.

To achieve the optimum use of medicines, it is important to maximize appropriate selection of medication, to individualize therapy, and to provide suitable patient information and support. Prescribers and pharmacists have different but complementary expertise. Recognition and strengthening of their professional links, both formally and informally, is essential to maximize their ability to work synergistically to achieve the desired outcomes and improve the patient's quality of life.

BARRIERS TO INTERPROFESSIONAL RELATIONS

There are a number of barriers preventing pharmacists from communicating effectively with other health professionals. Attitude, time, and knowledge are major concerns.

Attitude

The attitude and expectations of pharmacists are built on prior experiences. The health hierarchy has caused many pharmacists to believe that the prescriber is unapproachable; therefore, these pharmacists do not approach any encounter with confidence. Many pharmacists simply do not like contacting prescribers unless it is absolutely necessary. Part of the problem appears to be the perceptions that general practitioners and community pharmacists have of each other.

A conference on interprofessional relations in the U.K. reported that "[p]harmacists saw doctors as diagnosticians while doctors saw themselves as healers and doctors saw pharmacists as retailers (or laboratory workers) or suppliers of medication, while pharmacists saw themselves as drug use experts."[34]

Pharmacists must educate prescribers about their (pharmacists') expertise and then gain acceptance as members of the health care team by their performance.

Time and Money

Lack of time is an excuse for any activity an individual chooses to avoid. Nevertheless, a desire to achieve a goal is usually sufficient for finding the time necessary to do so. Lack of remuneration or incentives for providing extended clinical activities to improve patient care is also often seen as a real barrier by some community pharmacists and is something that needs careful negotiation with fund providers who must be cognizant of the value-added component of pharmacists' input, which increases cost efficiency as well as patient care.[35]

Skill and Knowledge

Many pharmacists do not communicate on a professional basis with their colleagues or counsel their patients due to a lack of skill and knowledge.[35] A major barrier is the inability of the pharmacist to obtain relevant information due to barriers in the existing system as well as anticipated attitudes by prescribers. Pharmacists may also lack knowledge of the patient's clinical details such as changes in medication, complete patient medication history, and potential medication problems due to failing to keep up to date. Pharmacists may be unable to apply previously gained knowledge to a clinical situation, or fail to communicate effectively with patients to identify compliance problems.

Major concerns in the GP's and pharmacist's role in consumer drug education were identified by the RISK Study in general practice.[36] GPs were concerned that pharmacists might sometimes give patients advice that conflicted with their own. Pharmacists themselves are concerned about offering conflicting advice. Pharmacists said that they were constrained when they did not know the patient's clinical details and were not informed about changes in medication, particularly when directions were omitted from prescriptions or unusual doses were prescribed.[35]

In an attempt to clarify and resolve these issues, the RISK Study facilitated workshops with "natural working pairs" of local GPs and pharmacists with consumers. As a result of these meetings, improved guidelines for writing and dispensing prescriptions, providing drug information to patients, handling unusual situations, and adapting the prescription to serve as a practical aid for improving communication between GP and pharmacist were developed. These are summarized later in this chapter.[36]

A joint national conference of community pharmacists and medical practitioners in Britain identified examples of misconceptions hindering closer contact. General practitioners were concerned that confidentiality might be put at risk and were unaware of the importance of pharmacy ethics. Another common belief was that commercial considerations would influence a pharmacist's advice on the need for and choice of drug treatment.[34] There is also resistance by the medical profession when pharmacists attempt to acquire new roles because the medical profession feels its territory is being threatened. Pharmacists' clinical role in hospitals has generally been accepted by doctors and nurses.[37,38] Ultimately, it is the doctors' reactions to clinical pharmacy that determine whether they perceive pharmacists as part of a medical care team and accept a drug therapy recommendation. A survey of doctors' opinions in West Australia showed that 65% considered that medication recommendations by pharmacists were clinically relevant "most of the time."[38]

New roles that open for pharmacists tend to be those that have been neglected by doctors. An English study suggests general acceptance by the medical profession of drug-related activities by pharmacists such as counseling patients on the side effects of drugs, reporting adverse drug reactions, encouraging compliance with prescribed regimens, managing minor health conditions, and participating in primary health care teams.[39-41] Sixty-one percent of GPs saw a role for community pharmacists in advising GPs about cost-effective prescribing.[40,42]

Pharmacists' view of themselves as being capable of offering adequate advice about nonpharmacological treatment and informing on the effects and adverse effects of lipid-lowering drugs was not shared by Swedish physicians.[43] This study shows that there is need for better communication, trust, and understanding between Swedish pharmacists and physicians.

Use of electronic networks would assist in interaction among all health professionals, particularly for hospital discharge patients. Orders for postdischarge medication could be sent by modem from the hospital to the local pharmacy and to the patient's usual prescriber. Alternatively, the increased use of patient health information in the form of computerized cards containing full patient records would assist in reducing multiple prescribing and also in recording OTC medicines. This latter aspect would probably not be as applicable in the U.S. or most Asian countries, as OTC medicines are available on a self-select basis in those countries.

COMMUNICATION SKILLS TO IMPROVE INTERPROFESSIONAL RELATIONS

Pharmacists have become increasingly concerned with how well they communicate with patients, and a number of studies show the need for improvement in communication.[44,45] Evidence of deficiencies in communication skills is also apparent between general practitioners and patients and between general practitioners and specialists.[46,47]

Pharmacy courses now aim to teach students patient communication techniques, i.e., the skills of establishing rapport, maintaining ongoing relationships, empathy, interviewing techniques including body language, and providing information and ensuring the patient understands that information. These same principles must be applied to communication with other health professionals. The establishment of rapport is helped by face-to-face contact, but this may not always be possible. Initial contact is often by telephone, which may be followed by a written report. Community pharmacists' written communication skills have had low priority in the past, but this must now change to improve pharmacists' professional standing.

For successful communication, a professional, confident approach must always be used. Quintrell states: "As a pharmacist, you have the right to your professional existence and your professional opinion: and the right to have that existence and opinion respected."[48] A truly professional interchange between prescriber and pharmacist is based on mutual respect. How can a pharmacist gain respect when criticizing prescribing habits? Firstly, the pharmacist must prepare and obtain information to substantiate the case. The pharmacist must be able to talk the language of the prescribers and understand their needs, focusing on the benefits to the patient and to the prescriber–not to the pharmacist.[44] Quintrell suggests that there are four stages in the interchange: statement, reply, negotiation, and resolution. The first stage is a clear, confident statement about the situation, without apologies or indecision. During the second stage, there must be careful and respectful listening to the needs and wishes of the other person. The third stage involves negotiation with offers of help; it is followed by the fourth stage in which the problem is resolved with the maximum possible satisfaction, focusing on the benefits to the patient.[48]

Quintrell's four stages of interchange is illustrated in the following example. You are working on a hospital Drug Usage Committee and wish to change drug prescribing protocols. You would need to cite reputable research studies that show that the new drug is as effective, or more effective, than the currently used drug and then point out the benefits of the recommendation for change, e.g., reduced side effects, better method of administration, reduced toxicity, and less necessity for monitoring. Cost is the final issue that must be addressed, but not before optimum patient care issues have been addressed. Cost should also not be based solely on drug cost but on total cost of therapy, including nursing care and length of hospital stay. The use of hydrocolloid dressings illustrates this point: the initial cost of the actual dressing is more expensive, but the reduced time spent changing dressings by nurses offers reduced overall costs plus improved quality of wound healing.[49]

Pharmacists who are successful negotiators will be constantly asking themselves prior to and during the interactions: What are the needs of the other party? Success will depend on good timing, a suitable negotiation environment, assertive behavior, the ability to listen, and adequate preparation of the case. An awareness of power interchanges and the resistance to change is important. The responses made by the pharmacist will also influence the outcome. A summary of aggressive, assertive, and nonassertive ways of acting is listed in Table 10.1.

Let's look at a typical situation of pharmacist intervention. A pharmacist has received a prescription for lithium carbonate and enalapril for Mrs. Anyone. You recognize that the interaction between the drugs is potentially harmful. You decide to call the pre-

TABLE 10.1. Summary of Assertive, Aggressive, Nonassertive Behavior (Adapted from Quintrell)[48]

	Assertive	**Aggressive**	**Nonassertive**
Personally you	- enhance own self-respect	- deny others self-respect	- decrease your own self-respect
	- decide for yourself	- decide for others	- allow others to decide for you
	- may achieve your goal	- achieve your goal at other's expense	- fail to achieve your goal
	- are left feeling good about yourself	- may feel good, but at other's expense and loss of future positive outcomes	- are left hurt, anxious, and afraid to try again
	- express your needs and listen to others	- express your need but ignore others	- are inhibited and do not express your needs
Verbally, your **speech** is	- open, honest, clear and direct	- arrogant, dictatorial, superior, accusing, blaming	- apologetic, vague, hedging, concealed, imprecise
your **voice** is	- firm, warm, honest, assured	- loud, demanding, obnoxious, offensive	- weak, indecisive, hesitant
your **body** is	- balanced, relaxed, and open	- tense and over-bearing	- tense and closed, with eyes averted

scriber. It could go like this (based on a scenario proposed by Quintrell):[48]

Pharmacist: *Hello, this is Everyman's Pharmacy. May I speak to Doctor Dogood, please?*

Receptionist: *Can I help you?*

Pharmacist: *No. It's something I will need to discuss with the doctor.* (Good! You haven't let yourself be put off.)

Receptionist: *He's consulting at the moment. Could you call back?*

Pharmacist: *Certainly. I do have Mrs. Anyone waiting in the phar-*

macy. How soon could I call him? (You are being polite but firm. You understand the doctor's needs and respect them, but also let the receptionist know that you and Mrs. Anyone have needs also.)

Receptionist: *I will get him to call as soon as he finishes with his current patient.*

Pharmacist: *Thanks. I'll let Mrs. Anyone know that she has to wait a little while. My number is. . . .*

(A SHORT TIME LATER)

Doctor: *Hello. You wanted to talk to me.*

Pharmacist: *Yes. Thank you for calling back, Dr. Dogood. I have your prescription for Mrs. Anyone for lithium and enalapril. There is a potentially serious interaction between these two. The sodium loss associated with ACE inhibitors may cause a rise in lithium toxicity.* (You state your concern and explanation clearly and briefly.)

Doctor: *Well, I've just repeated what they had her on at the hospital.* (He is on the defensive and calls on higher authorities to justify himself.)

Pharmacist: *Well, I can see we have a problem. Would you like me to call the hospital and check? Or would you prefer to do that?* (You are firm. Your professional judgment sees a problem, but you respect his feelings and are quite prepared to help find a solution. Note that you give him two options, both of which are acceptable to you.)

Doctor: *I suppose we could call the hospital. Is there any alternative we could try?* (Now he is entering into the spirit of negotiation, but not only that, he is relying on you to come up with a viable alternative.)

Pharmacist: *What about a calcium channel blocking agent—say felodipine?* (Now you are confirming that you know what you are doing as well as letting him down gently.)

Doctor: *That sounds like a good idea. Yes. Give her felodipine instead*. (See, he has acquiesced to your suggestions and saved face.)

Pharmacist: *That's fine. Do you want me to tell her anything about the change?* (You have achieved your goal, so you can afford to be pleasant and helpful.)

Doctor: *Yes. Just tell her I've decided that this medication will be a better one for her.* (He holds his own self-respect and you have retained yours, and you have reached a satisfactory compromise and safeguarded the therapeutic well-being of the patient.)

Pharmacist: *Fine. I'll send the prescription over for alteration. Thanks for calling back. Goodbye.* (A satisfactory outcome; what's more, you both feel good about the interaction.)

However, it could have gone quite differently:

Pharmacist: *Hello, Doctor. I'm sorry to bother you, but I thought I should call and check Mrs. Anyone's prescription.* [You've started by putting yourself in a "one-down" position. You begin with an apology (why?) and haven't stated the problem.]

Doctor: *What's the problem?*

Pharmacist: *You've prescribed lithium and enalapril for Mrs. Anyone. I see in the USPDI that there's an interaction.* (You do not say, "There *is* an interaction," stating your professional judgment. Instead, you call on other authority right from the beginning.)

Doctor: *Well, I've just repeated what they put her on at the hospital.*

Pharmacist: *Oh. What would you like me to do, then?* (You have sold out, handing all power to the prescriber.)

Doctor: *Look, I don't think it will matter. She's only got to take the enalapril daily.*

Pharmacist: *Well, as long as you think it's alright.* (You have abdicated responsibility for your patient and recanted on your professional judgment.)

Doctor: *Just tell her to stick to the dose and see me again next month.*

Pharmacist: *All right. Sorry to have bothered you.* (You have done it again! Apologized for your opinion. The prescriber has left with his authority and self-respect intact and yours is in tatters. Further, that particular prescriber will not have a very high opinion of your professional ability. This view can in turn have a halo effect of negative attitudes of that prescriber to all pharmacists.)

PATIENT, PRESCRIBER, PHARMACIST

Hepler outlines a model in which the prescriber and the pharmacist have direct responsibilities to the patient and to each other, but both share responsibility for the quality of the patient's entire drug therapy.[2] An adaptation of this model is outlined in Figure 10.1. The prescriber has greater social authority than the pharmacist to prescribe medication, but this does not decrease in any way the pharmacist's responsibility to advise the patient and the prescriber—whether or not either has requested the advice.

FIGURE 10.1. Patient/Prescriber/Pharmacist Interaction

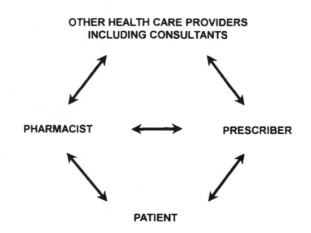

Informed Consent

It is vital that in any drug therapy decision the uniqueness of the patient is considered by the prescriber and the pharmacist. There must be interchange among all parties before decisions are made. Patients are entitled to make their own decisions about treatment and therefore must be given adequate information in a form that is appropriate for them to make such a decision. Although the prescriber or pharmacist should give advice, the patient should be free to accept or reject it. Guidelines for informed consent in treatment provide standards for practitioners.[50] Verbal advice, supplemented by written information, must be given to consumers to allow them to make informed decisions regarding drug therapy. It is vital that prescribers and pharmacists work cooperatively to tailor drug therapy to the patient's needs, e.g., dexterity, sight, cultural and religious beliefs, work schedules.

Once the selection of a drug has been made, effective drug therapy does not occur if any link in the chain from prescriber to pharmacist to patient is broken. The following is suggested as a mechanism to ensure that a team approach is used throughout the duration of therapy for optimal drug usage.

Recommended Good Practice for Communication Between Doctor and Pharmacist

Doctors are not aware of the full role of the pharmacist. This may be due to a lack of education of doctors about the pharmacist's role or to poor communication. Only 38% of hospital doctors identified that pharmacists have expertise in medication counseling.[51] To facilitate interchange between doctors and pharmacists and allow for further expansion and acceptance of the pharmacist's role in drug therapy decisions, the following ideas are suggested.

Guidelines for Pharmacist Ascertaining Patients' Conditions Without Breach of Ethics

The following options are suggested for pharmacists as methods of initiating counseling if details of the reason for therapy have not already been supplied by the prescriber:[5]

- Pharmacist asks the patient what the doctor said about the reasons for prescribing medication.
- Pharmacist seeks relevant information from the patient.
- Pharmacist requests patient's permission to ask doctor about patient's condition and medication.
- Pharmacist refers the patient back to prescriber for discussion.

In the majority of cases, the pharmacist does not require full diagnostic and clinical information about the patient to counsel the patient effectively about medication, although some information on the purpose for prescribing is required. If the doctor were to provide the full diagnosis to the pharmacist, the patient's written consent would be required. The situation differs in the hospital setting or residential care facility, where the pharmacist has access to the full medical record. The above situation of requesting information from the patient would not always be appropriate, for example, with an immunocompromised patient being treated for *Pneumocystis carinii* pneumonia with large doses of sulfamethoxazole/trimethoprim. In this case, personal communication between the doctor and the patient's regular pharmacist would assist, or if this did not occur, awareness and sensitivity by the pharmacist would be required to avoid a breach of confidentiality. Other examples might be the use of cyclosporine for other than the rejection after an organ transplant or the use of a medication for an unapproved condition.

Use of Prescription for Information and Instructions[5]

Prescribing information should be clearly detailed on the actual prescription to improve compliance by the patient with reinforcement by the pharmacist. The problems of poor communication from doctor to pharmacist by illegible writing and vague instructions have been described.[52]

If directions are omitted from the prescription, the pharmacist should take responsibility for ensuring the patient knows and understands how a medication should be taken or used by consulting the patient, using standard directions, and/or consulting the prescriber. Both professions discourage the use of "take as directed" as labeling.

The prescription can be used to notify the pharmacist if a new treatment has been initiated or if there is a change in dose.

If complex medication regimens are used, both pharmacist and patient should be notified in writing, e.g., decreasing doses of prednisolone, initiation of treatment with an ACE inhibitor following treatment with a diuretic and beta blocker.

The uncommon use of medications could be highlighted to assist pharmacist counseling by an explanation on the prescription, e.g., amitriptyline "for trigeminal pain," bromocriptine "for stopping breast feeding," carbamazepine "for lower jaw pain," nortriptyline "to prevent nocturnal enuresis."

The prescription should be used to indicate special needs of a patient, e.g., the patient is vision impaired or has confusion with medication indicating need for a dose administration container.

For discharge medications, the patient's local doctor and pharmacist should be supplied promptly with information about the patient's discharge medications following hospitalization.

Reinforcement of Information

Prescribers should be in a position to assume (by prior agreement) that every patient will receive from the pharmacist written information (Consumer Product Information–CPI) on first time of dispensing and on request and verbal counseling emphasizing the three or four most important points about the medication and that pharmacists will then check the understanding of the patient or caregiver as to the optimum use of the medication. Patients will have different counseling and information needs depending on whether they are presenting initial or repeat prescriptions.

Communication to Clarify Prescription Problems

An official mechanism of referral to the prescriber is required for patient queries not resolvable by the pharmacist, such as side effects. The decision as to whether there should be urgent referral via telephone or there should be a written request for total medication review prior to or at the next standard appointment should be resolved by the patient and the pharmacist. Other medication issues such as multiple sources of prescribing, interactions, and dosage adjustments will need to be referred to the prescriber by the most appropriate means based on the judgment of the pharmacist. To allow resolution of problems, it is vital that prescribers be aware of the benefits and necessity of being readily accessible to pharmacists by telephone. Their receptionists should be educated about the importance of and need for this accessibility.

Pharmaceutical Care Plan

Under modern pharmacy practice standards, pharmacists are expected to produce pharmaceutical care plans for their individual patients so that any actual and potential drug-related problems can be identified and managed. Pharmacists have willingly provided care intervention during dispensing processes up to the time they supply the prescribed drugs. In institutional settings, care intervention by pharmacists during the course of drug therapy has also become much more common, but in community (retail) practice, the idea is only beginning. A challenge that the profession is now facing is how to provide pharmaceutical care to *all* clients.

Pharmaceutical care planning has been well described.[53] The original plan involves nine steps, which can be summarized as:

> Establishing an understanding and acceptance by the patient that the pharmacist will provide pharmaceutical care to achieve the desired outcome with the drug therapy, identifying and ranking the actual and potential drug-related problems of the patient, considering alternatives and selecting the most appropriate course of action; designing and implementing monitoring plans to observe the results of the plan;

> and

> Following up with the patient, their carer and their other professional health providers when necessary, to modify the drug therapy as a result of that observation.

The importance of the recording of all steps of the planning process to ensure continual care and accountability is stressed.[53]

Local Area Services

All health providers should regularly be provided with a list of pharmacies, doctors, paramedicals, and support groups in the area to assist with referral and consultation. Guidelines should be provided for joint projects in local health promotion.

INTERVENTION

Despite the desire for pharmacists to be *proactive* when dealing with other health professionals, the reality is that, in the main, they are *reactive*. Generally speaking, pharmacists react to the receipt of a prescription or a patient providing information. In this section, we shall look at the extent and effectiveness of pharmacist intervention in what Main calls the "Fragile Therapeutic Chain," which illustrates the continuum of the initial appearance of symptoms, the help-seeking process leading to visits to a diagnostician, and involvement with a pharmacist leading to the desired therapeutic outcomes.[52,54] At each link along the chain, there is possibility of error and potential risk for the patient. The integrated relationship among the patient-physician-pharmacist is of great importance here (Figure 10.1).

Figures for the incidence of actual or potential drug errors in community medicine have been shown to average about 1.2% of all prescription orders written.[23,55]

Interventions by pharmacists in institutional settings are now being widely recorded, and as part of an international move to ensure "quality assurance" of these services, outcome measures are also being developed.[17,56,57] There is a constant theme underlying the need for intervention both in hospital and community settings; that constant is human error, which is a major "hazard of hospitalization," as it is one of the major causes of accidents in modern society.[58-63] Hospital studies have demonstrated that medication errors represent one of the leading causes of iatrogenic complications.[64] Anecdotal evidence suggests that community pharmacists ameliorate such errors on an almost daily basis without giving thought to the value they add to the health care system in the provision of such services. Documentation of these interventions and their evaluation is essential if resources are to be made available to extend pharmacists' roles and the economic benefits of pharmacists' interventions are to be demonstrated.

Some research and case studies have demonstrated the value of pharmacy interventions and clinical services in terms of reduced risk to the patient; enhanced patient knowledge; better compliance; and reduced cost of drugs, hospitalization, and other expensive health care. Some of these studies explore the potential for expanded clinical roles for pharmacists; all demonstrate the quality added to patient health care by the pharmacist. McGhan and colleagues illustrated the effectiveness of pharmacists in a comparative study with physicians on the quality of prescribing for ambulatory hypertensive patients.[65] This study showed that pharmacists can competently and effectively reduce provider workload by undertaking the management of certain chronic patients. This was also reported by McKenney, who previously demonstrated that clinical services provided by pharmacists to patients with essential hypertension, in comparison to a control group, elicited significantly better knowledge about their hypertension and its treatment, significantly better compliance, and significantly greater numbers keeping their blood pressure within normal ranges during the study period. It was also noted in this study that patient acceptance of this role for the pharmacist was high, reflected in the 92% compliance with appointments for consultations.[66]

Other studies in the U.S. have proved the effectiveness of pharmacy-based interventions in improving medication compliance.[66-68] One study of hypertensive patients attending community pharmacies in Australia demonstrated exceptional compliance with medication taking and prescription refill (approximately 90%), with 86% reporting their blood pressure to be under control.[69]

Davis described a retrospective community-based study in which pharmacists managed acute pathology. Results of 50 patients receiving diagnostic throat cultures and treatment from medical practitioners, physician assistants, and nurses were compared with 58 patients treated and managed by pharmacists. Data generated by this study showed that use of pharmacists in this manner released medical practitioners for more beneficial activities, reduced time between culturing and intervention, and reduced therapeutic costs in comparison with the other groups. Tentatively supported was a finding that there were also fewer return sick visits with the pharmacy-treated group.[67] Fortner studied the cost-effectiveness of an ambulatory clinical pharmacy service in U.S. health care centers and demonstrated considerable savings compared to another health center with comparable panel size, demography and patient encounters per provider, and similar consultant support.[70]

Cunningham-Burley and Maclean, a team of medical sociologists from Glasgow, have also investigated and found credibility in a greater clinical role for pharmacists. They identified a significant role for pharmacists in providing primary health care for children with minor complaints, stating that:

> Chemists do provide a valuable service to many mothers with young children: the flexibility and informal nature of this contact are important features. The chemist enables mothers to cope with a symptomatic child who does not need the attention of a doctor. They also provide some external source of advice, thus reinforcing a mother's feeling that the child does not need a doctor, with the safeguard that the chemist will suggest such referral if necessary.[71]

Christensen, in a study of outpatient problem interventions by pharmacists, documented the utility of pharmacists detecting and preventing prescribing errors.[72] A detection rate of two to four potential drug therapy problems per 100 dispensed prescriptions was reported. While the significance of the problems was not determined, the interventions usually resulted in some form of action, with many leading to explicit changes in the patients' medication regimen. This study led to the implementation of a revised method of measuring pharmacists' activities and included not only prescription volume by pharmacists, but also problems detected, review efforts undertaken, other drug information handled, and other time on professional activities.

A major area of interest appearing in the review literature is that of economic evaluation of pharmacists' interventions and the lack of measures of economic evaluation for community pharmacy practice. In a study of 89 community pharmacies in five states in the U.S., Rupp estimated that intervention activities by pharmacists avoided approximately $123 in medical costs in problematical prescriptions, or $2.32 for each new prescription.[73] Similarly, Dobie and Rascati, in a study of two rural Texas community pharmacies in 1994, calculated that appropriate intervention resulted in savings of about $3.50 in medical care costs per prescription processed.[74]

An interesting Australian study by DeSantis and colleagues demonstrated a marked improvement in antibiotic prescription patterns in general practice achieved through an educational mailing campaign with a follow-up by a visit by a pharmacist to discuss campaign messages.[1] The study had the objective of assessing the quality of antibiotic prescribing by Victorian general practitioners and the effectiveness of educational intervention techniques in improving prescribing. The design utilized a randomized, controlled, parallel group trial in rural and metropolitan Victoria; 182 general practitioners (78 control, 104 intervention) began and 103 (41 control, 62 intervention) completed the study. Participants recorded their antibiotic prescribing for tonsillitis. The intervention group received an educational mailing campaign. A pharmacist visited each prescriber to discuss campaign messages. The percentages of prescriptions of antibiotics for tonsillitis complying with those recommended in the *Antibiotic Guidelines* were assessed.[75] The results showed that in the intervention group, prescriptions consistent with recommendations in the guidelines increased from 60.5% before the campaign to 87.7% afterward. Improvement also occurred in the control group: from 52.9% to 71.7% of prescriptions. The improvement within the intervention group was significantly greater than that within the control group.

What Does the Pharmacist Want from the Prescriber?

The pharmacist is required to ensure that the prescription complies with legal requirements. The prescriber's intentions must also be clear, i.e., does the doctor really want Seldane® for arthritis? Feldene® for hay fever? Terbutaline (a sympathomimetic) for urticaria? Terfenadine (a nonsedating H_1-antihistamine) for a fungal infection? Terbinafine (an antifungal agent) for hay fever?

This leads to a consideration of brand-name (or trade-name) versus generic prescribing. Many prescribers in general practice tend to order by brand names. Provided the prescriber *knows* what the drug actually is and has not gone from diagnosis to brand name, rather than from diagnosis, to which drug therapy, to generic, to trade or brand name, there should not be any problem.

For example, in the late 1970s, when clonidine was in vogue for the treatment of hypertension and for migraine, many combination prescriptions were written as follows:

Rx   Catapres 150mcg tabs   Repeat × 2
100 Sig 1 tds for blood pressure

Rx   Tabs Dixarit   Repeat × 2
100 Sig 1 bd for 7/7, then 1 tds, for migraine
Dr. XXY

These trade products both contain the same ingredient—clonidine. The histories of most of the individuals who had these combinations prescribed were that they had been stabilized on Catapres® (clonidine 150mcg) for hypertension and had at some stage complained of "migraine." Boehringer Ingelheim had recently introduced Dixarit® (clonidine 25mcg) for the treatment of migraine and was promoting it heavily by trade name. The addition of more clonidine to the patient's regimen would not help the patient's migraine and could certainly increase the toxicity of clonidine. This is a case of brand prescribing without really thinking. A similar example might be a physician prescribing a beta blocker for the prophylaxis of migraine in addition to an existing beta blocker (same or different) for the treatment of hypertension.

Another problem occurred with the introduction of Augmentin® in the 1980s (i.e., amoxicillin/potassium clavulanate). Upon receipt of prescriptions for penicillin or cephalosporins, pharmacists ask patients about prior use of penicillin, with a view of determining whether patients might be allergic to penicillin. There are now many documented cases of patients being asked such a question with respect to Augmentin and replying, "But I'm allergic to penicillin. That's why the doctor prescribed this new antibiotic!"

What Does the Prescriber Really Want on This Prescription?

Pharmacy schools strongly suggest that pharmacists-to-be should not contact doctors for trivial matters. However, it is expected that pharmacists will contact prescribers when a real problem relating to the therapeutic need of the patient arises and that a discussion and resolution of the problem will occur in a professional manner. Some of the general areas of concern that a pharmacist would be expected to contact a prescriber about are:

- Inappropriate doses/strengths/routes of administration
- Drug-drug interactions
- Contraindications
- Inappropriate medication

Inappropriate Doses

Overdoses occur most often in children and particularly with narcotic analgesics, aspirin, acetaminophen (paracetamol), antihistamines, trimethoprim/sulfamethoxazole, digoxin, iron salts, theophylline, metoclopramide, and atropine.[76] *Extra* vigilance and care should be exercised when prescribing or dispensing these substances for children. Other circumstances associated with overdoses in children are:

- Where there are adult-strength mixtures, where a normal 5mL dose may not be appropriate for a child.
- Where a liquid preparation is designed to be given by dropper where the normal dose volume will be small.

Overdoses may arise in two ways: too frequent administration or too large an individual dose. In too frequent administration, beware of drugs with long half-lives, sustained release preparations, and situations with complicated directions. In too large an individual dose, beware of very potent drugs, drugs with a narrow therapeutic index (individualization of dose may be necessary), situations where more than one tablet or capsule is ordered, situations where preparations are available in different strengths, drugs where the maintenance dose is higher than the initial dose, drugs given parenterally, and new drugs.

In regard to strengths of topical preparations, beware of the potential for increased absorption in children, potentially caustic compounds, instances where large amounts are ordered or large areas of application are involved, instances where the preparation will be applied to mucous surfaces, instances where normal barrier properties of the skin will be disrupted, and instances where non-standard (nonformulary) preparations are involved.

It should be noted that inappropriate doses may also be *under-doses*. Underdoses should be viewed as being potentially as serious as overdoses. It is up to the pharmacist to ensure that the patient receives the appropriate dose at all times. Some adult patients, for one reason or another, may be unable to take oral solid dose forms, and a liquid preparation is required. Many of these preparations (e.g., antibiotics) exist only as pediatric formulations. Sometimes the prescriber may inadvertently order a pediatric dose of the medicine.

Drug-Drug Interactions

Pharmacists are a source of knowledge and information on clinical pharmacology, therapeutics, and drug interactions. Enterprising pharmacists can increase their professional reputation by the use of sophisticated drug information programs as well as the texts required by law. Price and Goldwire have illustrated this point by stating: "Pharmaceutical care is placing an increasing demand on community pharmacists to provide concise, accurate drug information."[77]

A drug interaction may be defined as a measurable modification in magnitude or duration of the action of one drug by prior or concomitant administration of another substance, including prescription and nonprescription drugs, cigarette smoke, food, or alcohol.[78] The principle is that the prescriber should know the risks of and set clearly defined therapeutic goals and guidelines for therapeutic or toxic effects. The pharmacist's role is to alert the prescriber to a potential problem or, if suggesting an alternative medicine, to offer an alternative without potential risk. Drug interactions per se are no threat to a patient; a physician's or pharmacist's ignorance of an interaction is dangerous. Often, adjusting doses may be more appropriate than discontinuing needed therapy.

For example, how important is the interaction of an asthmatic patient with theophylline plasma levels of 12mcg/mL who is given a 250mg qid dose of erythromycin? Compare this with, say, an asthmatic patient with theophylline levels of 19mcg/mL of plasma who is prescribed 400mg bd of norfloxacin (therapeutic level of theophylline is 10-20mcg/mL). Note the importance of the role of the prescriber and the pharmacist here in questioning and counseling the patient *and* talking to each other.

Summary of Important Features of Drug Interactions. Pharmacists should be aware of the following:

- The clinical effects of one drug can be *increased* or *decreased* by the concurrent administration of another drug.
- Many drug interactions can be predicted and thus avoided by a knowledge of the pharmacological and pharmacokinetic effects of the drugs dispensed.
- Drug interactions with a *pharmacological* basis involve actions at the same receptors or physiological systems or modification of drug response as a result of changes in fluid or electrolyte balance.
- Drug interactions of a *pharmacokinetic* nature are caused by the effect of one drug on the absorption, distribution, metabolism, or excretion of another drug.
- Interactions will not necessarily occur in all patients receiving a combination of drugs known to have potential interaction capabilities.
- Pharmacokinetic interactions demonstrated with one drug combination should not be extrapolated to other combinations utilizing closely related drugs.
- Rational prescribing dictates that prescribers should always attempt to minimize the number of drugs any individual patient is prescribed.
- The most common interactions are those involving mutual potentiation of central nervous system depression by hypnotics, tranquilizers, analgesics, antidepressants, alcohol, antihistamines, antiepileptics, and other centrally acting drugs.
- A common source of "interactions" is the ignorant use of trade or brand names, i.e., where the physician is not aware that the brand she or he prescribed contains the same (or similar) generic as another product the patient is already taking (see earlier).
- Another source of "potentiation interaction" is where a patient has visited more than one prescriber without making the prescribers aware of this situation.

• It is thus important to think, "Does it matter under the circumstances?" and, "Can we reduce any unwanted effects by an alternative medication or altering dosages?" and, "Do we need to contact the prescriber?"

Contraindications/Side Effects

Before pharmacists contact prescribers for a consultation regarding a patient's medication, they should consider the following:

• The medical condition of the patient.
• The medication history of the patient.
• The duration of treatment.
• The indication for which the drugs are being taken.
• Any factors such as pregnancy, breast-feeding, smoking, financial status, or if patient is a child or an older person.
• The significance of the problem.
• The possible solution to the problem.

An example of a drug interaction situation may be when a prescriber orders an ACE inhibitor for hypertension with, say, spironolactone in a postmenopausal woman for hirsutism. The combination can give rise to an increased risk of hyperkalemia. The pharmacist should be prepared to suggest an alternative drug to the prescriber, such as switching to a beta blocker or a calcium channel blocking agent, or conversely switching to an antiandrogen for the patient's hirsutism.

Another example of a drug interaction that can easily be resolved by appropriate communication with the prescriber would be the combination moclobemide (an MAOI-type A antidepressant) with the H_2-antagonist cimetidine. Suitable alternatives could easily be ranitidine, famotidine, or nizatadine, none of which would cause increases in plasma levels of moclobemide.

A common problem seen in pharmacy is when a prescriber has ordered drugs to treat (unknowingly) a drug-induced adverse reaction or an overdose. A too-frequent example of such a situation is when a patient has been stabilized on a cardiac glycoside but complains of palpitations (arrhythmias), so the doctor prescribes quinidine. Sometime later, the patient presents with flu-like symptoms, nausea, and color vision disturbances. The remedy may be to prescribe an antibiotic (still being prescribed for viral infection by some doctors) and/or an antinauseant). The real problem, of course, is that the quinidine competes for the excretion of the cardiac glycoside, and gradually, the serum levels of the cardiac glycoside rise to the extent of eliciting adverse reactions which in time could be life threatening. The pharmacist needs to take prompt and decisive action in such a situation.

A common problem is that of patients visiting different physicians, which–if physicians are not informed of the situation–can lead to dire consequences. For example, an asthmatic patient being treated quite appropriately by her local doctor for asthma (inhaled corticosteroids, inhaled bronchodilators) is switched by her ophthalmologist from pilocarpine eye drops to timolol drops for glaucoma. Because of lack of communication by the patient (in this case) with her two prescribers, she has placed herself at risk of developing an acute asthma attack because of the beta blocker drops.[79] The pharmacist, with access to the patient's full medical history and by careful questioning of the patient, can make an appropriate approach to the prescriber(s) to avert a dangerous situation.

Inappropriate Choice or Use of a Medication

Examples of inappropriate choice of medication are:

- Use of antibiotics for viral infections.
- Use of barbiturates as sedative/hypnotics.
- Use of a broad-spectrum antibiotic for streptococcal throat infection.
- Use of the wrong drug by inadvertent use of a similar sounding trade name.
- Nonadjustment of dosage in older or hepatic or renally impaired persons.
- Use of a drug contraindicated by a concomitant condition.
- Polypharmacy where drugs are added without a review of the total medication regimen.
- Use of drugs that interfere with the patient's quality of life, e.g., diuretics causing incontinence in a disabled arthritic person; beta blockers for a 40-year-old sports enthusiast; tri-

cyclics in older persons, increasing the risk of incontinence, blurred vision, and falls.

• Represcribing the same drug in a different formulation, leading to nonbioequivalence. For example, the inadvertent change to nifedipine capsules of one strength from the same strength tablets can lead to potentially serious increases in plasma levels of nifedipine.

Pharmacists have a profound knowledge of the clinical pharmacology and pharmacokinetics of drugs and can act as resource for prescribers at all levels of practice. Both pharmacists and physicians are professionals with legal, professional, and ethical expectations and obligations. Each can contribute to the other's practice to the mutual benefit of the patient while at the same time making both roles much more satisfying.

CONCLUSION

Because of their in-depth understanding of and knowledge about drugs, today's pharmacists have the potential to improve the drug therapy and quality of life of patients. Unfortunately, few health care systems or other health practitioners fully recognize or accept this fact, with the result that much of pharmacy practice today–particularly in community practice–remains in a system that regards the pharmacist merely as the supplier of medicinal products.

Fortunately, new concepts of health care delivery are emphasizing that systems must quickly and significantly improve the outcomes of drug therapy and install procedures and use all available expertise to ensure that the drugs used by the system–already a major part of the health budget–are used in the most cost-effective manner to improve patients' quality of life.

Pharmacists are in a position to make a considerable contribution to achieving this goal, as they are already involved in the drug supply aspect of the system. They have the basic scientific knowledge and the applied knowledge and skill to do so. What is lacking is a system that facilitates and makes their expertise available. Most health care systems still only remunerate pharmacists for their supply of drugs alone and provide no remuneration or incentive to pharmacists to apply the knowledge and skill they possess. In fact, many systems often provide disincentives for doing so!

Prescribers, in general, do not recognize or accept the very significant contribution that pharmacists can make to rational and cost-effective drug use and the achievement of desired drug therapy outcomes, if only pharmacists were accepted as a respected part of the drug therapy management team.

The challenge, then, for pharmacists–individually and collectively–in their personal practices and in their relationships with the prescribing and caring professions, with employers, with other health care providers (e.g., hospital managements), with health funding agencies (e.g., government, the health insurance industry), and with the health bureaucracy, is to be clever, assertive, and effective in demonstrating and actually delivering real value in their participation as full members of the drug therapy team. What pharmacists should be doing in the new area of pharmaceutical care is now being increasingly well described in the pharmaceutical literature. What they can now achieve individually and as a profession is in the hands of individual practitioners and their representative organizations. If collectively they do not accept the challenge confronting them or are unsuccessful in doing so, pharmacy has the potential of being sidelined as a health care profession and being relegated to performing a nonprofessional supply role. If, however, pharmacists achieve respected interprofessional relationships with the other health care stakeholders at all levels, the profession will remain a vital component of the health care team.

REFERENCES

1. DeSantis G, Harvey KJ, Howard D, Mashford ML, Moulds RFW. Improving the quality of antibiotic prescription patterns in general practice. Med J Aust 1994;160:502-5.

2. Hepler CD. Pharmaceutical care and specialty practice. Pharmacotherapy 1993;13(2 Pt 2):64S-9S.

3. Commonwealth of Australia. Prescribed health. A report on the prescription and supply of drugs. Part 3–Pharmacy and medicinal supply (Jenkins Report). Canberra: Australian Government Publishing Service, November 1992.

4. Anon. Doctor-pharmacist trainee exchange scheme commended. Pharm J 1988;240:117.

5. Ruth D. General practitioners' and pharmacists' interprofessional communication. Aust Pharm 1994;13:431-4.

6. Committee of Inquiry into Pharmacy. Pharmacy: report of the Committee of Inquiry. London: Nuffield Foundation, 1986.

7. Saltman DC. Primary health care: the role for pharmacists and GPs. Aust Pharm 1994;13:215-7.

8. Stewart BJ, Drury M, Greenfield S. Continuing education for pharmacists and general practitioners. Br J Gen Pract 1991;1(344):126-7.

9. Koda-Kimble MA. The United States experience. Aust J Hosp Pharm 1991;24:16-20.

10. Macklin J. Issues in pharmaceutical drug use in Australia. National Health Strategy Issues Paper No. 4, 1992.

11. Raisch DW. Interactions between community pharmacists and prescribers: differences between practice settings. J Soc Admin Pharm 1993;10:42-51.

12. Mitchell JL. Building cooperation with physicians: an interview with Charles Fortner. Am Pharm 1990;NS30(2):24-6.

13. Harding G, Taylor KMG. Professional relationships between general practitioners and pharmacists in health centers. Br J Gen Pract 1990;40:464-6.

14. Blenkinsopp A, Jepson MH, Drury M. Using a notification card to improve communication between community pharmacists and general practitioners. Br J Gen Pract 1991;41:116-8.

15. Blenkinsopp A, Booth TG, Kennedy EJ, King H, Purvis JR, Taylor PA. Community pharmacist and general practitioner liaison groups: (2) getting started. Pharm J 1993;251:477-8.

16. Drummond MF. The rise of pharmacoeconomics: implications for pharmacists. Aust J Hosp Pharm 1994;24:28-44.

17. Edgar TA, Lee DS, Cousins DD. Experience with a national medication error reporting program. Am J Hosp Pharm 1994;51:1335-8.

18. Leversha A. An analysis of clinical pharmacy interventions and the role of clinical pharmacy at a regional hospital in Australia. Aust J Hosp Pharm 1991;21:222-8.

19. Lamour I, Dolphin R, Baxter H, Morrison S, Hooke DH, McGrath BP. A prospective study of hospital admissions due to drug reactions. Aust J Hosp Pharm 1991;21(2):90-5.

20. Simons LA, Tett S, Simons J, et al. Multiple medication use in the elderly of prescription and non-prescription drugs in an Australian community setting. Med J Aust 1992;157:242-6.

21. Fairbrother J, Mottram DR, Williamson PM. Doctor-pharmacist interface, a preliminary evaluation of domiciliary visits by a community pharmacist. J Soc Admin Pharm 1993;10:85-91.

22. Colgan J. Liaison pharmacy—the Orange experience. Aust Pharm 1994;13:315.

23. Paes AHP. Drug monitoring in community pharmacy. J Soc Admin Pharm 1992;9:29-34.

24. Baker DM. A study contrasting different modalities of medication discharge counselling. Hosp Pharm 1984;19:545-54.

25. Cromdos S, Allen B. Patient comprehension of discharge medication: do ward pharmacists make a difference? Aust J Hosp Pharm 1992;22:202-10.

26. Blackbourne J. Readmission to Freemantle Hospital. Who comes back—when and why? Freemantle Hosp Drug Bull 1991;15:9-16.

27. National Heart Foundation. Risk prevalence study. Canberra: National Heart Foundation, 1989.

28. McNeece J. The Drug and Therapeutics Information Service. Aust J Hosp Pharm 1994;24:28-31.

29. Macklin J. Issues in pharmaceutical drug use in Australia. Issues Paper No. 4. Canberra: National Health Strategy, 1992.

30. Paes AHP. Pharmacists and general practitioners in consultation? J Soc Admin Pharm 1990;7:99.

31. Landgren FT, Harvey KJ, Mashfold ML, et al. Changing antibiotic prescribing by educational marketing. Med J Aust 1988;149:595-9.

32. Harvey K, Stewart R, Hemming M, Moulds R. Use of antibiotic agents in a large teaching hospital: the impact of antibiotic guidelines. Med J Aust 1983;2: 217-21.

33. Strand LM, Cipolle RJ, Morely PC, Ramsey R, Lamsam GD. Drug-related problems: their structure and function. Ann Pharmacother 1990;24:1093-7.

34. Anon. Working together; benefits and barrier. National Conference for Community Pharmacists and General Practitioners. Pharm J 1993;251:126-7.

35. Ortiz M, Walker WL, Thomas R. Development of a measure to assess community pharmacists' orientation toward patient counselling. J Soc Admin Pharm 1992;9:2-10.

36. Murphy B, Ruth D, Murray Hodge M. The use of qualitative research in the development of the "Heartwise" program for general practitioners. Med J Aust 1993;158:626-8.

37. Mesler MA. Boundary encroachment and task delegation: clinical pharmacists on the medical team. Sociol Health Illness 1991;13:310-31.

38. Clifford RM, Jessop JB, Lake JM. Evaluation of clinical pharmacy services: a survey of doctor's opinions. Aust J Hosp Pharm 1993;23:11-6.

39. Gerrett D, Willcocks AJ. Community pharmacists' mandate from general medical practitioners for a drug counselling role; a retrospective and prospective study. Pharm J 1991;247:R38.

40. Spencer JA, Edwards C. Pharmacy beyond the dispensary: general practitioners views. Br Med J 1992;304:1670-2.

41. Sutters CA, Nathan A. The community pharmacist's role: GPs' and pharmacists' attitudes towards collaboration. J Soc Admin Pharm 1993;7:70-83.

42. Woodward J. GPs and community pharmacists: a study of attitudes. Pharm J 1992;248:99-101.

43. Troein M, Rastam L, Selander S. Physicians' lack of confidence in pharmacists' competence as patient informants. J Soc Admin Pharm 1992;9:114-22.

44. Albro W. How to communicate with physicians. Am Pharm 1993;NS33: 59-61.

45. Butenya GB, Lauwo JAK. Communication about dispensed drugs between prescriber, the dispenser and the patient; a study from Port Moresby General Hospital. Aust J Hosp Pharm 1993;23:182-5.

46. Del Mar CB. Communicating well in general practice. Med J Aust 1994;160:367-70.

47. Graham PH. Improving communication with specialists. The case of an oncology clinic. Med J Aust 1994;160:625-7.

48. Quintrell N. Communication skills. A manual for pharmacists. Canberra: Pharmaceutical Society of Australia, 1982.

49. Sussman G, Gruen R, Chang S, MacLellan D. Proceedings of the Third European Conference on Advances in Wound Management. Harrogate, U.K., October 19-22, 1993. London: Macmillan.

50. ASHP. Board of Directors report on the Council on Educational Affairs. Am J Hosp Pharm 1994;51:907-27.

51. Clifford RM, Jessop JB, Lake JM. Evaluation of clinical pharmacy services; a survey of doctor's opinions. Aust J Hosp Pharm 1993;23:11-6.

52. Main A. Elderly patients and their drugs. Pharm J 1988;240:537-9.

53. Strand LM, Cipolle RJ, Morley PC. Pharmaceutical care: an introduction. Current Concepts. Kalamazoo, MI: Upjohn, 1992.

54. Mechanic D. Medical sociology: a selective view. New York: The Free Press, 1968.

55. Cargin C, Weaver J, Roller L. Prescribing errors in Victorian community pharmacies. Aust J Pharm 1990;71:606.

56. Western Australia Clinical Pharmacists Group. Recording clinical pharmacist intervention: is there a better way? Aust J Hosp Pharm 1991;21:158-62.

57. Cooper JW. Clinical outcomes research in pharmacy practice. Am Pharm 1993;NS33:S7-S8.

58. Schimmel EM. The hazards of hospitalization. Ann Intern Med 1964;60:100-10.

59. Steel K, Gertman PM, Crescenzi C, Anderson J. Iatrogenic illness on a general medical service at a university hospital. N Engl J Med 1981;304:638-42.

60. Dubois RW, Brook RH. Preventable deaths: who, how often, why? Ann Intern Med 1988;109:582-9.

61. Brennan TA, Leape LL, Laire NM. Incidence of adverse events and negligence in hospitalized patients: results of the Harvard Medical Practice Study I. N Engl J Med 1991;324:370-6.

62. Leape LL, Brennan TA, Laire NM. Incidence of adverse events and negligence in hospitalized patients: results of the Harvard Medical Practice Study II. N Engl J Med 1991;324:377-84.

63. Norman DA. Post-Freudian slips. Psychol Today 1980;10(4):41-50.

64. Raju TNK, Kecskes S, Thornton JP, Perry M, Feldman S. Medication errors in neonatal and paediatric intensive-care units. Lancet 1989;2:374-6.

65. McGhan WF. A comparison of pharmacists and physicians on the quality of prescribing for ambulatory hypertensive patients. Med Care 1983;21:435-44.

66. McKenney JM, Slining JM, Henderson R, Devins D, Barr M. The effect of clinical pharmacy services on patients with essential hypertension. Circulation 1973;34:1104-11.

67. Davis S. Evaluation of pharmacist management of streptococcal throat infections in a health maintenance organization. Am J Hosp Pharm 1978;35: 561-6.

68. McKenney JM. Clinical pharmacy and compliance. In: Haynes RB, Taylor DW, Sackett DL, eds. Compliance in health care. Baltimore: Johns Hopkins University Press, 1979:260-77.

69. Gilbert A, Owen N, Innes JM. High levels of medication compliance and blood pressure control among hypertensive patients attending community pharmacies. J Soc Admin Pharm 1990;7:78-83.

70. Fortner CR. Report on an ambulatory clinical program and its cost effectiveness. Group Health J 1985;(Fall):3-6.

71. Cunningham-Burley S, Maclean U. The role of the chemist in primary health care for children with minor complaints. Soc Sci Med 1987;24:371-7.

72. Christensen DB. Documenting outpatient problem intervention activities of pharmacists in an HMO. Med Care 1981;19:104-17.

73. Rupp MT. Value of community pharmacists' intervention to correct prescribing errors. Ann Pharmacother 1992;26:1580-5.

74. Dobie RL, Rascati KL. Documenting the value of pharmacist intervention. Am Pharm 1994;NS34:50-4.

75. Antibiotic Guidelines. 7th ed. Melbourne: Victorian Medical Postgraduate Foundation Inc., 1992.

76. Adverse Drug Reactions Advisory Committee. Metoclopamide—choose the dose carefully. Aust Adverse Drug React Bull 1990;15:387.

77. Price KO, Goldwin MA. Drug information resources. Am Pharm 1994; NS34(7):30-9.

78. Roller L, Gowan JA. The principles of drug interactions: Part I, How they happen. N Z Pharm 1994;14:328-33.

79. Bauer K, Brunner-Ferber F, Distlerath LM, et al. Assessment of systemic effects of different ophthalmic beta blockers in healthy volunteers. Clin Pharmacol Ther 1991;49:658-64.

ADDITIONAL READINGS

For the Instructor

1. Bailey RA, Ashcraft NA. Pharmacist-physician drug fair for educating physicians in cost-effective prescribing. Am J Hosp Pharm 1993;50:2088-9.

2. Bjornson DC, Hiner WO Jr, Potyk RP, Nelson BA, et al. Effect of pharmacists on health care outcomes in hospitalized patients. Am J Hosp Pharm 1993;50:1875-84.

3. Blom AT, Paes AH, Bakker A, Koopman CJ, van-der-Meer C. Pharmacist-physician co-operation at a regional level. Pharm World Sci 1994;16(1):13-7.

4. Cooper JW. Clinical outcomes research in pharmacy practice. Am Pharm 1993;NS33(12 Supp):S7-S13.

5. Deleo JM, Pucino F, Calis KA, Crawford KW, Dorworth T, Gallelli JF. Am J Hosp Pharm 1993;50:2348-52.

6. Dukes MNG, Swarts B. Responsibility for drug-induced injury. Amsterdam: Elsevier, 1988.

7. Emmerton L, Benrimoj SI. Product selection by pharmacists and medical practitioners: a review of methods and models. J Soc Admin Pharm 1994;11:46-53.

8. Fitzgerald M. Pharmacist-doctor alliance in asthma surveillance. IPU-Rev 1989;14:303-6.

9. Gardner SF, Stanek EJ, Munger MA. Medical versus pharmaceutical continuing education: are both appropriate for the pharmacist. DICP 1991;25:336-8.

10. Guglielmo BJ, Schweigert BF, Kishi DT. Pharmacist-managed therapy in California hospitals. Am J Hosp Pharm 1989;46:1366-9.

11. Hargie OD, Morrow NC. The effectiveness of microtraining in developing pharmacists' communication skills: a study of personality and attitudes. Med Teach 1989;11(2):195-203.

12. Hawkey CJ, Hodgson S, Norman A, Daneshmend TK, Garner ST. Effect of reactive pharmacy intervention on quality of hospital prescribing. Br Med J 1990;300:986-90.

13. Kennedy EJ, Purvis JR, Blenkinsopp A. The nature and extent of interprofessional contacts between pharmacists and general practitioners. Pharm J 1993;251:R39.

14. Kitching JB. Communications and the community pharmacist. Pharm J 1986;237:451-2.

15. Lau NR. Medication error analysis and intervention by a sole pharmacist. Aust J Hosp Pharm 1993;23:243-6.

16. Liddell M. Rational prescribing and professional standards. Med J Aust 1994;160:564-7.

17. Marklund B, Karlsson G, Bengtsson C. The advisory services of the pharmacies as an activity in its own and as part of a collaboration with the primary health care services. J Soc Admin Pharm 1990;7:111-6.

18. Neville RG, Robertson F, Livingstone S, Crombie IK. A classification of prescription errors. J R Coll Gen Pract 1989;39(320):110-2.

19. O'Shea A. Pharmacist-doctor relationship in practice. IUP-Rev 1987;12:193-4.

20. Philia NJ, Einarson TR, Dunn LA, Ilersich AL, Pillon L. Development of a community-based drug use review (DUR) model. AACP Annual Meeting 1993;94:VII-4.

21. Plumridge RJ. Intervention strategies aimed at modifying prescribing behaviour. Aust J Hosp Pharm 1984;14:93-8.

22. Raisch DW. Barriers to providing cognitive services. Am Pharm 1993;NS33(12):54-8.

23. Roddick E, Maclean R, McKean C, Virden C, Sykes D. Communication with general practitioners. Pharm J 1993;251:816-9.

24. Roller L. Illness/wellness: a psychosocial viewpoint. Soc Pharmacol 1989;3:315-35.

25. Rupp MT. Value of community pharmacists' intervention to correct prescribing errors. Ann Pharmacother 1992;26:1580-5.

26. Rupp MT, Schondelmeyer SW, Wilson GT, Krause JE. Documenting prescribing errors and pharmacist interventions in community pharmacy practice. Am Pharm 1988;NS28:574-80.

27. Sherman C. When the doctor is wrong. Am Druggist 1991;203(56):61-3.

28. Sisca TS. Pharmacist-manned drug therapy helps meet requirements for drug-use evaluation. Am J Hosp Pharm 1992;49:81-3.

29. Smith FJ. The extended role of the community pharmacist: implications for the primary health care team. J Soc Admin Pharm 1990;7:101-9.

30. Sumpton JE, Frewan TC, Rieder MJ. The effect of physician education on knowledge of drug therapeutics and costs. Ann Pharmacother 1992;26:692-7.

31. Sutters C, Keat A, Lant A. Improving prescribing of non-steroidal anti-inflammatory drugs in hospital; an educational approach. Br J Rheumatol 1993;32:618-22.

32. Sutters CA, Nathan A. Can the pharmaceutical industry promote collaboration between pharmacists and general practitioners. Pharm J 1993;251:546-9.

33. Sylvestri MF. Managed care. US Pharm 1993;18(8):39-54.

34. Tuan S. Clinical pharmacy services to a nursing home–a unique situation. Aust J Hosp Pharm 1994;24:22-5.

35. Western Australia Clinical Pharmacists Group. Recording clinical pharmacist interventions: is there a better way? Aust J Hosp Pharm 1991;21:158-62.

36. Wilson M, Robinson EJ, Blenkinsopp A, Ranton R. Customers' recall of information given in community pharmacies. Int J Pharm Pract 1992;1:152-9.

For the Student

1. Albro W. How to communicate with physicians. Am Pharm 1993;NS33:59-61.

2. Balon ADJ. Counselling. Pharm J 1986;237:452-4.

3. Blenkinsopp A, Booth TG, Kennedy EJ, King H, Purvis JR, Taylor PA. Community pharmacist and general practitioner liaison groups: (1) the Airedale/Bradford experience. Pharm J 1993;251:430-1.

4. Hepler CD, Strand LM. Opportunities and responsibilities in pharmaceutical care. Am J Hosp Pharm 1990;47:533-43.

5. National Conference for Community Pharmacist and General Practitioners. Working together; benefits and barriers. Pharm J 1993;251:126-7.

6. Quintrell N. Communication skills. A manual for pharmacists. Canberra: Pharmaceutical Society of Australia, 1982.

Chapter 11

Consumer Behavior Regarding the Choice of Prescription and Nonprescription Medications

Michael Montagne
Lisa Ruby Basara

INTRODUCTION
TO CONSUMER MEDICATION CHOICE

After reading any magazine or newspaper or watching television for a few hours, one can easily gauge the level of interest in health and medical treatment options in the United States. This growing attention is a result of many factors, including the consumerism movement that surfaced in the late 1960s, the generic drug scandal of the early 1980s, the growth of managed care organizations that limit patient choice, the concern about spiraling health care costs, and the popularity of consumer-oriented medical and health publications. In reaction to the same social changes, the amount of control that prescribers now wield in medication selection has declined, resulting in greater freedom for patients to identify and select medical therapy options.

Traditionally, physicians were responsible for the evaluation and determination of appropriate medical therapy. Pharmacists might have suggested appropriate pharmaceutical alternatives, but physicians possessed final decision-making authority. Apart from their physical symptoms, patients rarely influenced prescription drug selection directly. Today, in addition to more active pharmacists, many *patients* are unwilling to accept physicians' decisions to use particular medications. A growing proportion of patients want to understand both the health-related and financial repercussions of medication use and to choose whether to take certain medications based on complete information.

Active participation by patients in medical care is valuable. Knowledgeable patients can improve the quality of health care decisions by participating with physicians, nurses, and pharmacists. Particularly, such patients describe symptoms, communicate health status changes, and inform physicians and pharmacists of medical therapy preferences. Educated patients are also aware of self-care options available in the form of nonprescription or over-the-counter (OTC) medications and alternative medicine (e.g., chiropractors, homeopathic medicines, etc.) Depending on their health status, some patients even track the development of new drugs and experiment with unapproved therapies.

Of course, some patients, while desiring more information, also have preconceived notions about how and why medications work, the speed at which they alleviate symptoms or cure a condition, and their appropriateness for particular diseases. For example, some people believe that prescription medications are an instantaneous answer to whatever ailments are diagnosed. Similarly, the pharmaceutical industry, while charged with unfair pricing and marketing policies, is expected to produce a medication for each disease identified or popularized in the media (i.e., Alzheimer's disease, AIDS, chronic fatigue syndrome) immediately and regardless of the cost.

Because of the spectrum of attitudes and beliefs about prescription medications that patients have, as well as the varying levels of sophistication with which they gather and interpret information, health care professionals need to understand the nature of patients' attitudes, their origin, and strategies that are most effective when communicating drug information to patients. To help you understand the nature of consumer attitudes toward prescription medications and the processes by which consumers evaluate and choose among treatment options, these issues are explored in this chapter. Before investigating prescription medications, however, it is important to understand consumer decision making in a general context.

CONSUMER BEHAVIOR

Think about the last time you made a major purchase (e.g., car, stereo, etc.). What factors entered into your decision? How much time did you spend evaluating alternatives, talking with other people, and reading up on the products? Who influenced your decision? Are you happy with your choice?

As you can tell from your own experiences and those of others, the ways in which people make decisions are highly complex and variable. Decision making is influenced by (among other factors) past experience and knowledge, the amount of risk associated with the decision, the quality and availability of information, and the opinions of others. These factors are important in all purchase decisions, including the purchase of goods and services. Even when consumers are choosing among health care treatment options, such as medications, these factors play a role.

It is important for those who manufacture, market, and distribute goods and services to understand why consumers make the purchase decisions that they do, whether the goods are groceries or headache remedies. In the case of medications (both OTC and prescription products), pharmacists need to understand the attitudes, concerns, and health requirements of pharmacy patrons. Knowing the factors that influence consumer decisions helps pharmacists to identify, target, and satisfy patrons, as well as ensure that selected medication therapies are best for patients.

The next section of this chapter reviews the application of consumer behavior models to both traditional (consumer-oriented) and health-care-specific (patient-oriented) purchase situations. The factors that influence consumer nonprescription and prescription medication choices are then highlighted through discussion, examples, and case studies throughout the rest of the chapter.

GENERAL CONSUMER BEHAVIOR MODELS

A model is systematically designed to reproduce a phenomenon by specifying important elements and their relationships. Usually, a model is developed with the intention of explaining and predicting how a phenomenon has occurred or will occur. Several models have been proposed and refined throughout the past 30 years to explain and predict consumer buying behavior. However, because of their lasting value and acceptance, the models of two groups of researchers warrant detailed consideration.

Howard Buyer Behavior Model

Marketing educators John Howard and Jagdish Sheth published one of the most well-known consumer buying behavior models in their book *The Theory of Buyer Behavior.*[1] According to the Howard model, the primary elements that drive consumer purchase of a product or service include purchase intention, brand attitude, and brand comprehension. Brand attitude is affected by the amount and nature of information available, consumers' exposure to this information, their motive to pay attention to this information, and the criteria that the consumer uses to judge the product or service of interest. While the model is useful from a conceptual standpoint, research conducted to test the relationships it hypothesizes has not been very successful. Because of these limitations, the Howard model was revised by Farley and Lehman.[2] The revised model, which includes more concrete variables and relatively uncomplicated relationships, is shown in Figure 11.1.

EKB Buyer Behavior Model

While the Farley revision of the Howard model is not as complex as the original, it still has flaws. To address these, three researchers at Columbia University developed a new way to look at consumer buying behavior. The Engel-Kollat-Blackwell (EKB) model of consumer behavior, as shown in Figure 11.2, reorganizes and redefines several variables to clarify relationships. The key elements of the decision-making process include (1) recognition of a problem, (2) information search, (3) alternative evaluation, (4) product choice, and (5) outcomes.

The first step, realizing that a problem exists, is influenced by one's previous experiences, the amount and types of information input (mass, personal, general information), and one's ability to process information (exposure, attention, and reception of information), as shown on the left side of the EKB model. The second step, search, is directly influenced by the nature of the recognized problem and one's own beliefs and attitudes about available products. Indirectly, personality, life-style, family, culture, and the circumstances of the situation also influence the nature and extent of information search activities.

FIGURE 11.1. Howard and Sheth Model of Consumer Purchase Behavior

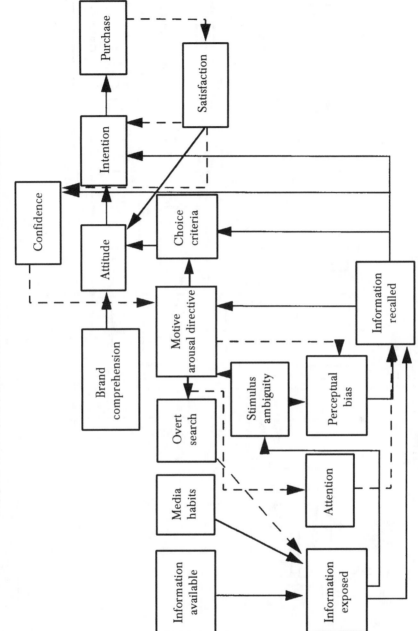

FIGURE 11.2. Engel-Kollat-Blackwell Model of Consumer Purchase Behavior

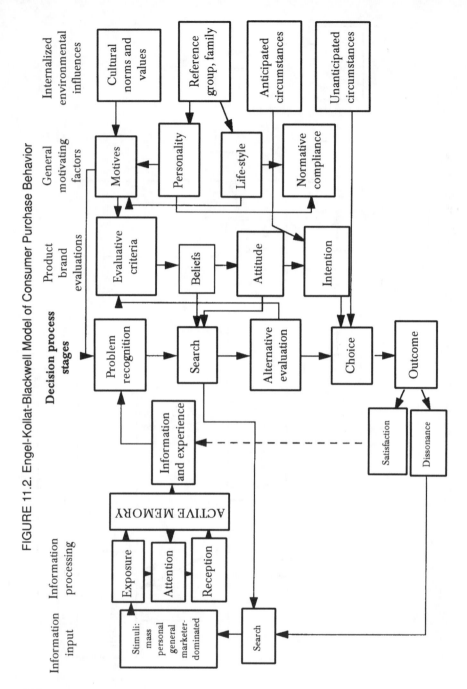

After a search has been conducted, alternative products are evaluated using criteria that have been established using both information collected and personal experiences and influences. Next, a product choice is made after considering the alternatives in the context of the purchase situation. Finally, the purchase decision results in either satisfaction or dissatisfaction (dissonance), which in turn, affects future buying decisions.

If the EKB model seems complicated, review it by thinking about a purchase decision that you have recently made. Depending on the situation, of course, certain variables are more important than others. If your car has broken down and you have no alternative way to get to school, you are likely to conduct a much different information search than if you are leisurely considering the purchase of a new car. Similarly, assume you are interested in buying a personal computer. When making this decision, information collected through review of magazines, talking with salespeople in computer stores, and the experiences of friends or colleagues with computers should be more valuable than information collected through mass media or from your family. Furthermore, while your personality and lifestyle might influence the evaluative criteria that you use, the anticipated circumstances of product use (e.g., for word processing, data processing, games) should be more influential in determining the criteria that are used to decide among different options.

High vs. Low Involvement in Purchase Decisions

A product decision like the one described for personal computers is termed "high involvement." High-involvement decisions are risky (both financially and socially), time- and resource-consuming, and long-term. Home appliances, cars, clothing, education, furniture, and vacations are other examples of high-involvement products and services. Most people are not immediately familiar with all of the attributes, benefits, and risks associated with these goods. Additionally, none of these products or services can be easily exchanged if you are not satisfied with them. It is fundamental to success for manufacturers of such products to provide understandable and accessible information, including desired product features, and a guarantee of satisfaction.[3]

On the other hand, low-involvement decisions are very easy to make and entail little risk. Groceries, small household items (e.g., dish detergent, toilet paper), and commodities (e.g., gasoline) are purchased with little evaluation or information search activities. For manufacturers of these types of goods, price and convenience serve as primary marketing appeals.[3]

To determine the extent to which prescription medicine purchases are high or low involvement, Everett conducted random, geographically stratified telephone interviews with 238 Denver residents. Prescription drug attributes that consumers rated as important included possible side effects (97%), physician recommendation (90%), strength (73%), prior use (72%), price (58%), and the availability of a generic version (51%). Overall, consumers focused on rational product attributes and the recommendations of physicians and pharmacists, demonstrating that the medication selection process is *not* perceived as a trivial, low-involvement issue.[4]

Transformational (Positive) vs. Informational (Negative) Purchase Motivation

In addition to low- or high-involvement, purchase decisions can be characterized as either transformational (positive) or informational (negative).[3,5] Products that are purchased to minimize or prevent negative situations (e.g., cleaning agents, auto repairs) are often purchased relatively quickly and on the basis of perceived benefits and convenience. For these products, impartial information about product attributes is more important to purchasers than emotional repercussions of product use.

Transformational, or positive, purchases are made when one yearns to buy a certain item because it enhances or generates a positive situation or state of mind (e.g., new clothes, fine wine, vacation). Consumers often consider several options and product offerings when positively motivated and enjoy the shopping experience (browsing, sampling) as a part of the purchase decision. Because of the higher level of emotional involvement in such products, subjective appeals that influence emotion and address personal goals (e.g., social status, reputation, superiority) are more effective in generating interest and product purchase.

To understand the extent to which motivation plays a role in purchase decisions, Widrick and Fram conducted a study of 387 households in one region of the United States.[5] Consumers were asked their feelings about the purchase of 14 different products and services, including birth control products, vitamins, and OTC medications. Specifically, they were asked to imagine (or remember) shopping for each item and to tell how much they would like (liked) or would dislike (disliked) the experience. They were also asked to indicate whether their general feelings about *use* of each product or service were positive or negative. Results showed that birth control products represented the third greatest *negative* purchase situation (following auto repairs and extermination services) and that OTC medications and vitamins were perceived as neutral purchases. This study, as well as the one conducted by Everett, implies that because prescription and OTC drugs are considered high-involvement negative-to-neutral products by most patients, the need for unbiased and accurate information is substantial.[4]

PRESCRIPTION MEDICATIONS AND CONSUMER GOODS ARE DIFFERENT

While reviewing the Howard and EKB models of consumer buying behavior, think about how different types of consumers (e.g., young, old, educated, uneducated, experienced, inexperienced) would approach the decision to buy an OTC allergy medicine or the decision to seek treatment for an ulcer. What are the circumstances of their problem? What types of information would they want? Where would this information be obtained? How would they use the information?

When considering these scenarios, it is relatively obvious that the decision to use a medication, especially a prescription medication, is more difficult than the decision to purchase a new consumer good, such as a television. While all of us have watched television, most people are not familiar with prescription antiulcer medications. Consumers know (or can easily find out) what characteristics of a television are preferred. However, they have only a vague idea about what constitutes a good antiulcer product.

Additionally, when consumers are looking for a new television, time or emotion is not usually a limiting factor. When they are sick, however, immediate symptom relief is essential. Finally, the risk of television mispurchase is relatively low (and is primarily financial in nature). While a financial risk is involved for prescription medications, the risk that one's health might worsen (and that one's life might change) if the condition is treated incorrectly is a relevant concern. Thus, a consumer product decision is relatively rational or logical, but the decision to use a prescription medication is often irrational.

Impact of Decreasing Physician Autonomy

One of the reasons that consumers can risk irrational decision making when it comes to prescription drugs is the presence of their physicians. Appropriate or not, most people trust that their physicians (and pharmacists) will make logical product choices on the basis of experience with other patients, medical wisdom, and superior diagnostic skills. Furthermore, because comparative information for prescription medicines is difficult for consumers to access and understand (there are no *Consumer Reports* editions for antiulcer drugs), it is easier for patients to rely on the accuracy of information obtained from and presumed intelligence of their doctors.

However, the traditional paternal relationship between patients and physicians is less common in today's health care environment. The family physician has been replaced by any one of a group of doctors to whom a patient might be assigned upon entering a physician's office. Essentially, to see the same physician over time and build trust in his or her abilities has become a luxury only available to those who can afford fee-for-service care. As a result, more than ever before, patients must rely on themselves to ensure the high quality of their health care. Resourceful patients are now making health care decisions in the same way that they make high-involvement product purchase decisions, rather than blindly trusting their physicians and making low-involvement health care decisions. Prescription drug importation, investigational drug use, and positive reactions to direct-to-consumer advertising (DTCA) and other information-containing media are just some of the manifestations of increased consumer involvement in prescription drug selection. These patient activities and their impact are discussed later in this chapter.

Summary

While consumer product purchase decisions are relatively well understood and heavily studied, the process by which patients make decisions about their health care, particularly in the selection of prescription and nonprescription drugs, is less clear. Changes in the health care system and physicians' decision-making autonomy have compelled patients to reconsider traditional methods of pharmaceutical selection and purchase. Pharmacists, as the most trusted and accessible health care professionals, must understand this shift in patient behavior, including the rationale and repercussions on medication use. Both appropriate and inappropriate ways in which prescription and OTC medications are used are described in the following sections to help you obtain this understanding.

STUDYING MEDICATION USE FROM A SOCIAL AND BEHAVIORAL PERSPECTIVE

Consumer Medication Use in a Social Context

The ways in which consumers use medications are not solely a function of biomedical or clinical knowledge and experience. All medication use occurs in a *social* context.[6] Consumers' decisions to seek care, specific medication choices (whether prescribed or self-selected), and product use are influenced not only by the physical symptoms they are experiencing, but also by what they know, who they interact with regarding their problem, and the situation in which their problem occurs.

Social knowledge refers to collective understanding that is based mostly on available information and past experiences. Social knowledge consequently influences behaviors and the nature of experiences. What consumers know about medications from reading, listening to media and promotional campaigns, receiving descriptions of others' experiences, and recalling their own previous experiences will affect their actual medication use. Social knowledge about the effects of a certain medication will also influence patients' experiences with that medication.[7,8]

There are some dominant "pharmacomythologies"–societal conceptions about medications that are illusory–that have achieved the status of and exist as fundamental principles in peoples' minds.[9, 10] For example, many people believe that a specific medication produces only one effect: its "main" effect, which is positive. If other effects are experienced, they are considered "side" or negative effects. Medications, in fact, produce a whole array of physiological and psychological changes. Additionally, many people believe that a medication produces the same main effect every time it is taken and in each person who takes it. In essence, drug effects are *caused* by the medication that is ingested, and by extension, drug effects are a property of and reside in chemical compounds and are not a function of some change in living organisms.

Consequently, there is a general belief that most prescription medications and many nonprescription medications cure disease. However, the vast majority of medications do not cure; they only alleviate symptoms, prevent exacerbation of acute conditions, or limit the extent of damage from chronic conditions, usually only over a restricted period of time. Most drug effects occur only as long as the medication is present physically in the organism. Because medications affect biological systems, and Western scientific cultures (and many other cultures through Western influences) believe that illness has a biological basis only, medications are considered the only effective medical treatment.

Methods for Studying Consumer Medication Use

The development of theories and methods to assess medication use has progressed in recent years with a resurgence of interest in drug epidemiology, or as it is now called, pharmacoepidemiology.[11] In essence, epidemiology provides a theoretical basis for examining the source of supply and flow of medications throughout a population. From this perspective, medication use occurs within a relationship that involves the consumer, his or her environment, and the medication that is being used.

The basic idea of pharmacoepidemiology is to determine the frequency and distribution of medication use practices in a population. From this information, the nature and extent of specific types of medication use can be determined in a given population, and potential or existing problems can be identified. The focus of phar-macoepidemiological research is: (1) what drug is being used, (2) how it is being used, including how much, where, when, and by whom, and (3) why it is being used, including the functions that pharmaceuticals serve in society.

Specifically, the World Health Organization (WHO) conducts pharmacoepidemiologic efforts to assure the quality, safety, and effi-cacy of medications and their use in specific patient populations.[12] WHO performs pharmacoepidemiologic studies specifically to describe current patterns of medication use; determine changes in drug use over time; measure the impact of information, education, promotional activities, media accounts, and price on medication use; detect inappropriate medication use and associated problems; esti-mate medication needs in terms of disease patterns and outbreaks; and plan the selection, supply, and distribution of medications.

There are many different sources of data and information on medication use (Table 11.1). Each source has its own unique advan-tages and limitations. Many studies only report frequencies of use without any basis in the population from which they were derived. Epidemiologic measures are provided as rates, with a numerator divided by a denominator (much like a percentage). Reporting that 100 patients are using a new medication (the numerator), without giving a sense of whether this is occurring in a population of 1,000 or 100,000, only provides marginally useful information.

Consumer Medication Use

Greater consumer control over medication use has emerged as a result of the pervasive "consumerism" movement in American society.[13] This expanding consumer consciousness and the desire to know more about what is going on and to have a say in it are very evident in health care. Patients want more information about their state of health. They want personal control over their own health care. They want to be "in charge" of staying healthy or getting better when they become ill.

The perception of current or potential symptom states by the patient, along with that person's social knowledge, creates and defines a health problem situation (Figure 11.3). A specific action is then taken in response to this problem. The patient may not fully

TABLE 11.1. Sources of Data on Medication Use

- Institutional Record Systems and Databases
  - hospital-based medical audits (inpatient)

- System-Wide Databases
  - institutionally based reviews (outpatient)
  - health insurance groups and third-party payers
  - commercial vendors of marketing studies and sales data

- National Databases
  - government-sponsored studies
  - essential drug lists and inventory data
  - pharmacoepidemiologic surveillance systems

- Local and Regional Studies
  - private or nonprofit surveys
  - industry-based (adverse drug reaction reporting)

- Field Studies
  - drug marketers, retailers, distributors
  - drug-taking behaviors of individuals and small groups

- Experimental Data
  - clinical trial results

recognize the symptoms as an illness and may ignore them, or the patient may choose to do nothing about the symptoms, deciding to live with them and wait for them to go away. In about two-thirds of the situations, the patient decides to act and chooses a particular approach to resolve the health problem.

When patients take action, they make contact with the health care system only about one-quarter of the time. Otherwise, they self-diagnose and self-treat, using some type of medication, either a prescription or OTC pharmaceutical, an herbal medicine, a home remedy, or some combination of these. In most illness situations, patients self-medicate with a nonprescription product.

The actual process of choosing a particular medication is influenced by social control over medication availability and acces-

FIGURE 11.3. Actions Taken in Response to Health Problem Situations

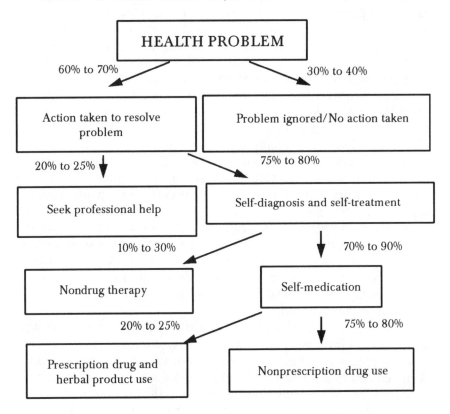

sibility and by sources of information used in decision making, among other factors. The consumer's ability to obtain and use certain products is controlled socially by laws and regulations. "Prescription" and "nonprescription" or "OTC" are constructs or legal terms, just as "approved" and "unapproved" with regard to indications of use in therapy and "illicit" with regard to substances that are used for presumably nonmedical reasons. Differences between prescription and nonprescription products are determined on the basis of safety and habit-forming potential. Safety is a major issue for giving OTC status to a medication, because, once a product becomes OTC, fewer mechanisms are in effect to monitor its use.[14]

Lack of *availability* of a particular drug means that the drug itself cannot be provided by pharmaceutical manufacturers to legitimate dispensers or that it cannot be provided by those dispensers to the general public. Situations in which a drug is unavailable are: the product has not been developed by an American company; it has not been approved by the Food and Drug Administration (FDA); it has been removed from the market; or its use has been limited to specific conditions or certain types of patients. In other cases, the product is usually available, but some problem in its production or distribution has made it temporarily unavailable.

Lack of *accessibility* to a drug means that the drug is available but consumers are not able to obtain it. Four main reasons for problems of accessibility are: lack of prescription or professional approval to use a specific product; geographic distance from the dispenser or supplier of the drug; the cost of the drug product and whether the patient is covered to some extent by insurance; and patient's fear, distrust, or dissatisfaction with the traditional health care delivery system. Consumers who decide to use a drug product that is not accessible usually go around the system, often becoming involved in illegal activities.

Consumer medication choice is also influenced by those who provide input into decision making and by sources of drug information and products. At one extreme, the consumer's treatment plan is designed and monitored by a health professional, who controls most decisions. In many situations, such as self-medication, the consumer may make important decisions but still seeks advice and guidance from a health professional. Consumers may consult alternative healers for different treatment approaches. At the other extreme, consumers seek input only from their lay referral network, or they rely solely on their own knowledge and beliefs in making decisions.

General Model of Consumer Choice in Medication Use

The structure of consumer medication behaviors consists of basic decision-making and problem-solving skills (Figure 11.4). Such behaviors usually begin with an actual or potential health problem. The person perceives the appearance of a symptom or set of symp-

FIGURE 11.4. Decision-Making Process of Consumer Medication Use

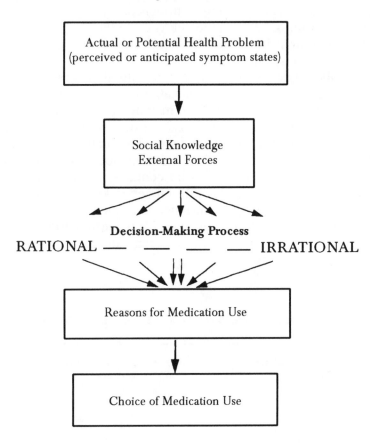

toms or anticipates that he or she is at risk for becoming afflicted. As a result, the person chooses specific actions to achieve or maintain health.

It should not be assumed that consumer decisions are made in a rational manner. We might think or wish that all decisions in life are made on a rational basis, but this is rarely the case. Decisions or choices made by different individuals, and even by the same person in different circumstances, can vary greatly. Each patient has the capability of making health care decisions, along a continuum, that can be very rational or very irrational in nature. The degree of rationality in decision making is influenced by social knowledge of health and illness, external forces (i.e., other people, the media, promotions), and the nature of the specific health problem being experiencing or avoided.

A number of behavioral, social, and cultural factors can influence consumer choice and use of medications (Table 11.2). One important factor in medication decision making is the consumer's social networks. Social networks are those sets of contacts or relationships with others through which the individual maintains a social identity and receives ideas, information, services, social support, and the opportunity to develop new social contacts. In health care, these social networks are called lay referral networks and consist of relatives and friends whom the consumer consults to gain guidance and understanding in decision making. Ideas, information, attitudes, behaviors, and even actual drug products are disseminated within these lay referral networks.

TABLE 11.2. Social and Behavioral Factors in Consumer Medication Use

• Consumer mind-set
  – mood
  – personality
  – attitudes
  – health beliefs
  – suggestibility
  – previous experiences
  – expectations
  – motivations for use
  – social/cultural background
  – physical state prior to use
	• Environment of use
	  – physical
	  – social
	
	• Rituals of use
	
	• Promotional campaigns
	
	• Mass media reports

• Symptom awareness
  and perception
	• Social networks
	  – lay referral system
	  – health care professionals

• Sources of drug information/ products	• Modeling and social learning
• Availability of drug products	• Accessibility of drug products

PRESCRIPTION MEDICATION USE

Patient and Societal Demand for Prescription Medications

Patient demand for specific prescription medications is a growing phenomenon and is related to the growth of general consumer consciousness. However, patient demand in a social context has also been influenced by changes in pharmaceutical marketing (i.e., direct-to-consumer advertising), in drug development and approval processes (e.g., orphan drugs), in the amount and nature of drug information directly accessible to consumers, and in the ways in which new pharmaceuticals and medication use problems are reported in the media.

Only a generation ago, it would have been unthinkable for patients to ask, let alone demand, their physician to prescribe a specific medication for their symptoms or conditions. This lack of patient influence could be attributed to both the nature of the patient-doctor relationship, in which patients acted passively while the paternalistic physician commanded them to engage in treatments, and to the minimal information about diseases and medications to which patients had access. This situation has changed greatly as consumers have become more active in their health care and as they have identified or developed different ways of accessing new drug information.

Direct-to-Consumer Advertising

Vignette 1: Advertising Drug Products Directly to Consumers

While reading through his most recent issue of *Time* magazine, Mark was surprised to see an advertisement for Moodzac, a new prescription antidepressant. The ad says that depression is a common and underdiagnosed condition that affects millions of Americans and recommends that people with particular symptoms contact their physicians for more information about the new product, which has fewer adverse effects and better efficacy than its competitors. Mark, a graduating law student, has been unhappy for several weeks because he broke up with his girlfriend and has not been able to find a job in his city of preference. Although he first tried to drown his sorrows in alcohol at the local bar, the next day's hangovers were too severe. Since then, Mark has been unable to sleep and feels listless during the day. While he does not usually believe in taking medicines to treat everyday stress, Mark believes that his situation may warrant a trial of this new medicine. He decides to see his physician, taking the *Time* advertisement with him so that he does not forget what drug to request.

Direct-to-consumer advertising (DTCA) and direct-to-patient (DTP) communication are broad terms used to encompass television and print advertising, public relations activities, direct-mail campaigns, patient support or compliance programs, and other consumer-directed initiatives conducted by pharmaceutical companies.[15] Since the 1980s, these types of programs have increased in prevalence and sophistication to reflect consumers' information needs and influence in selecting prescription medications.

Currently, DTCA campaigns appear in virtually all magazines and newspapers and on television (Table 11.3). Anecdotally, it is interesting to observe not only changes in advertising copy (i.e., increased product detail, such as brand name, and more quality-of-life appeals) but also the heightened amount of eye-catching artwork and photographs in DTC ads. The types of advertised medications have changed from products that prevent disease (e.g., vaccines and Pneumovax®), manage social conditions (e.g., hair loss and Rogaine®), or are brand new to the medical community (e.g., Cardizem CD®, nicotine patches) to products that compete in large, well-established markets (e.g., NSAIDs), treat severe conditions (e.g., epilepsy, herpes, migraine), and are alternatives to surgical techniques (i.e., Proscar®, Hytrin®).

The debate regarding the impact of DTCA and the need for government regulation began in the early 1980s and continues today.[16-20] For example, it has been argued that DTCA generates inappropriate consumer expectations, negatively influences the doctor-patient relationship because patients will be demanding medications that they do not need, and adds to the already high costs of prescription medicines. Others argue that patients have a

TABLE 11.3. Prescription Medications Advertised to Consumers in Print, January to October 1994

PRODUCT	MANUFACTURER	INDICATION	ADVERTISE-MENT TYPE
Depo-Provera (medroxyprogesterone)	Upjohn	contraception	P-S
Diabinese (chlorpropamide)	Pfizer	type II diabetes	INFORM
Estraderm (estradiol)	Ciba Geigy	ERT	P-S
Felbatol (felbamate)	Wallace	epilepsy	P-S
Hismanal (astemizole)	Janssen	allergy	P-S
Hytrin (terazosin)	Abbott	BPH	P-S
Imitrex (sumatriptan)	Glaxo/Cerenex	migrane	INFORM
Mevacor (lovastatin)	Merck	antilipemic	P-S
Nizoral (ketoconazole)	Janssen	dandruff	P-S
Norplant (levonorgestrel)	Wyeth-Ayerst	contraception	P-S
Premarin (conjugated estrogens)	Wyeth-Ayerst	ERT	P-S
Proscar (finasteride)	Merck	BPH	P-S
Relafen (nabumetone)	SmithKline Beecham	arthritis	P-S
Rogaine (minoxidil)	Upjohn	hair loss	P-S
Zantac (ranitidine)	Glaxo	heartburn	INFORM
Zantac (ranitidine)	Glaxo	ulcer	P-S

P-S   = product-specific message (brand name +/− indication)
INFORM = informational / help-seeking message
ERT   = estrogen replacement therapy
BPH   = benign prostatic hyperplasia

right to comprehensive information about the medications that they use; increasing awareness of new medicines helps consumers identify symptoms and seek treatment; the doctor-patient relationship is enhanced by additional communication; and because advertising helps increase competition, drug prices are lower.[21] To date, however, no comprehensive studies have been conducted to demonstrate either positive or negative consequences of DTCA.

Orphan Drugs and Consumer Activism

Consumer demand for medication can go beyond an individual patient requesting a prescription from his doctor. Together, consumers who demand particular medications or health care services can substantially influence society and the government, even to the point of generating legislation. Patients with rare diseases (i.e., diseases that affect less than 200,000 people in the U.S.) demonstrated this influence in the early 1980s when they helped design and convinced Congress to pass the 1983 Orphan Drug Act.[22]

Because they are used by a very small number of patients, drugs that treat rare diseases (orphan drugs) have little economic value and, without the 1983 legislation, were often not manufactured or distributed by traditionally profit-motivated companies. As a result of the Orphan Drug Act (ODA), drug companies were given regulatory breaks and tax incentives to facilitate the production and distribution of these desperately needed medications to patients with rare diseases. Imagine the frustration of patients who knew that a pharmaceutical existed to treat their disease, but, prior to the ODA, were unable to access it due to lack of approval or lack of availability!

Patients with rare diseases and their families pushed the ODA through letter-writing campaigns, lobbying, and demonstrations in Washington. Their consumer group, the National Organization for Rare Disorders (NORD), was formed specifically to address lack of access to and availability of lifesaving medicines. Today, the organization provides a national database of rare-disease patients, comprehensive disease and drug information, support, and guidance regarding medication use and involvement in trials of investigational orphan drugs. Consumer groups such as NORD, similar organizations for people with AIDS, the American Association of

Retired Persons, and the Consumer Union illustrate the substantial power that people with a common goal can wield in society.

Importing Pharmaceuticals

Vignette 2. Importing Prescription Drugs

Shirley has severe seasonal allergies, but none of the OTC products seems to work well, and all of them make her sleepy. She asked her pharmacist for a product that would help alleviate her symptoms. He informed her of the availability of non-sedating antihistamine medications, which are effective and have fewer, less severe side effects. However, these products require a prescription. Shirley visits her physician to ask about these new medicines and receives a prescription. While she is quite happy because the medicine works well and does not produce the unpleasant side effects, the product is expensive, and her medical insurance does not cover prescription drugs. While reading a health magazine, Shirley sees an advertisement for a Canadian drug supply company that mails medicines to people in other countries. She calls and learns that her antihistamine is available OTC in Canada for one-third of the price of the U.S. version. She orders a six-month supply.

Entrepreneurs have recognized consumers' demand for certain prescription medications that are not yet available in the U.S.[22] They have responded by setting up consumer buying groups through foreign companies, foreign clinics, and off-shore mail-order pharmaceutical supply houses. These organizations take different approaches to issues of consumer rights and legality. At one end of the spectrum, some companies are engaged in outright illegal activities. Not only are they providing drug products illegally, but also the quality of those drug products is highly suspect.

Somewhere in the middle are companies that are facilitating the distribution process for certain patients who are allowed to import drugs into the U.S. If the patients are following FDA guidelines, what they are doing is correct and legal, but the extent to which "foreign" companies can help them is not clear. The products they provide are usually manufactured by reputable companies, and they are of high quality. Price lists are available simply by writing or making a telephone call. There are no mechanisms, however, for ensuring quality control.

Finally, at the other end of the spectrum are organizations that provide information about pharmaceutical products, but they themselves are not involved in the actual buying and selling of drugs. Examples of this approach are various AIDS patient support groups and the National Organization for Rare Disorders (NORD), both of which help patients find information, investigational drugs, and other patients with the condition or disease.

NONPRESCRIPTION MEDICATION USE

Before the initiation of laws and regulations to control drugs, especially in the late nineteenth century during the height of the patent medicine era, consumers' self-directed use of therapeutic substances was extensive.[23,24] In fact, self-medication really was the first type of medical drug use, practiced by ancient people before designated healers became a part of structured societies. Substances considered today as dangerous addictive drugs, such as opiates, cocaine, and cannabis, were popular and effective ingredients in patent medicines and proprietary products. Herbal medicines also were, and continue to be, popular in consumer self-medication. Many products and remedies had no proven effectiveness (some even being considered quackery) according to medical science, yet consumers reported many beneficial effects. In the latter part of the twentieth century, self-medication has become a growing trend.[25]

Consumers generally view OTC medications, herbal remedies, and any substance they can obtain without a prescription as safer and easier to use as well as less expensive. There is also a fine line between self-care and quackery, depending on the approach or product that is being used. Many consumers hold fallacious or irrational health beliefs, and their health care practices reflect those misperceptions. For instance, one health survey found that 75% of consumers believe that extra vitamins are beneficial in slowing the aging process, curing the common cold, preventing allergies, and many other wonderful outcomes; 20% believed that arthritis and cancer are caused by vitamin or mineral deficiencies; and 12%

self-diagnosed their conditions of arthritis, asthma, and heart disease.[26]

Self-medication is the self-directed and self-administered use of a substance for a medical reason, either to treat or to prevent a health problem. The actual prevalence of self-medication in any population is not well known due to a lack of quality research on this subject. Most surveys suggest that about one-third of the general population engages in self-medication during any given period of time.[27,28] This rate may be much higher in the elderly.[29] The elderly may spend as much as one-third of their total expenditures for medications on nonprescription drugs.

In the most recent representative survey of health care practices in the U.S., the health problems reported most often by adults over a two-week period of time were: being overweight, upset stomach, muscle aches, minor eye problems, headache, and minor fatigue.[27] Of the top 23 most commonly cited problems, three of them had something to do with weight (with one of them explicitly stated as "overindulgence in food"), three were indicative of possible psychiatric conditions or mental symptoms related to a failure to adapt to tension and stress (minor fatigue, anxiety, and sleeping problems), and a number of symptoms (upset stomach, back problems, headache) could be related in part to the person's inability to cope with the everyday pressures of life.

Even more interesting were the reported actions taken in response to these symptom states.[27] Almost one-third of all problems were not treated (30%), or a nonmedicinal home remedy was used (16%). Nonprescription medications were used in 38% of the problem incidents. These figures are similar to those in other studies of health, illness, and self-treatment practices.[28,30]

Vignette 3. Self-Medicating with Nonprescription Medications and Herbal Medicine Products

> Mike has had trouble sleeping the past few months, possibly due to all of the changes occurring at his new job, so he asks his pharmacist to recommend a nonprescription sleep aid. Mike tries one of the products for a couple of nights, but it does not seem to work very well. After seeing an ad on television, he chooses a different product, which does help him get

to sleep, but makes him groggy and hung over the next morning. He stops using it and decides that the pharmacy does not have any useful sleep aids. While talking to a Russian friend, he learns that an herbal medicine, valerian, has been used for centuries in Europe as a sleep aid. Mike picks up some valerian tea at a health food store and tries it. He discovers that it does help him get to sleep and has no side effects, so he decides to use it every night.

Using Nonprescription Medications for Symptom Alleviation

There are different reasons for engaging in self-medicating behaviors. The dominant reason is that the individual seeks relief from, amelioration of, or even an end to a specific symptom or set of symptoms. Related to this is the desire to alleviate an illness or condition, which is more broadly defined in the patient's mind, but the symptoms may be vague or combined in a more implicit description of "feeling sick." People also self-medicate in the absence of specific symptom states or illness as a preventive measure. In this case, they seek to avoid getting sick, deter what is in their minds an impending disease, stay healthy, or even improve their physical and mental states of well-being beyond "normal." In as many as 30% of situations, self-medication may occur for reasons of preventive care.

In general, four key factors influence decisions to self-medicate: (1) one's past experiences and social knowledge, (2) perceived or anticipated symptom states, (3) advertising and promotion, and (4) the lay referral network. These factors can be closely interrelated in many situations, or they may interact to create a consumer mind-set about health, illness, and drug therapies as applied in specific circumstances.

Additional factors have been shown through research to be associated consistently with a greater tendency to self-medicate. The elderly, especially people over 75 years of age, and the very poor tend to self-medicate more than people of other age groups or socioeconomic classes. In some studies, women have been shown to self-medicate more than men, but often this is based more on shopping behavior; they are in the pharmacy buying drug products for any number of family members, but they are not ill themselves.

Some research also indicates that people living in urban settings tend to self-medicate more than people in rural settings, and whites tend to self-medicate more than members of other racial groups.

How do consumers make their decisions and why do they make the ones that they do? The process an individual might follow generally in making decisions about self-diagnosis and self-treatment is outlined in Figure 11.5. This abstract, highly rational and logical, yet theoretical, model should be viewed as a guide to the

FIGURE 11.5. Self-Medication Process

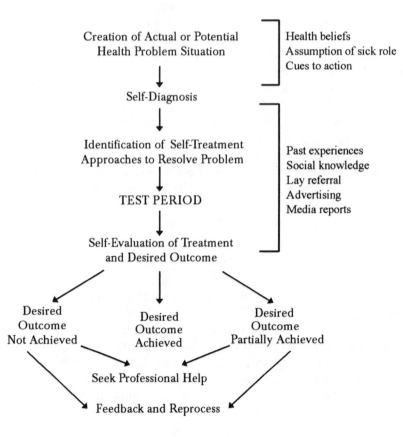

range of possible steps, and choices made at each step, that consumers may consider in their process of making a specific decision.

The process of self-medication, once the need for a drug is perceived, includes the steps of selecting a specific medicinal product, choosing the regimen, obtaining and preparing the product, and administering the product until the desired outcome is achieved or a decision is made to pursue an alternative treatment approach. The test period for any type of drug or nondrug self-treatment approach will vary in length of time depending upon the type of therapy chosen and the desired outcome. In those situations when self-medication is directed at real symptoms or illness, most patients expect to experience results from their drug therapy rather quickly. Some studies have shown that some type of effect should occur in four to six hours or the patient will change his or her mind, proceed to review the whole process, and make a new choice for treatment. As long as there is some effect, however, most patients will self-medicate for at least three days. Some patients (about 10% to 15%) will self-medicate for over two weeks.

If the desired outcome is achieved, self-treatment will end. Sometimes, however, the fear of symptom recurrence or other factors will encourage self-treatment beyond the necessary length of time. If the desired outcome is not achieved, or only partially achieved, the person may make additional decisions, such as seeking professional help or starting the whole decision-making process again. This reprocessing of the health problem situation may include any or all of the main steps, depending on feedback from the previous round of decision making.

Self-Medicating for Personal and Social Problems

When are changes in physical and mental states of being interpreted as symptoms, leading to a consideration of being ill? Are certain conditions such as pregnancy, child abuse, ordinary stress and tension ("pressures of life"), small breasts, thinning hair, mental fatigue, and many other personal and social life problems really diseases? Who determines which states of being are diseases to be addressed by medical care systems and public health efforts and which are problems and concerns better addressed by the family unit, social welfare agencies, religious groups, local government and law enforcement, or society in general?

There are many different definitions of what constitutes a disease or illness, and these may vary among patients, health professionals, health care corporations, health insurance companies, pharmaceutical manufacturers, and society in general. While "disease" is usually viewed as the specific negative change in physiological or psychological functioning, "illness" is viewed more in the cultural context in which it occurs. "Illness" is really a social phenomenon composed of meanings employed by those involved to make sense of the experience.[9,31]

Medicalization is the process of redefining or relabeling a personal or social "life" problem as a medical condition, thus necessitating involvement in and control by the health care system. More broadly, it can be seen as the imposition or substitution of medical care, and the biomedical model of health and illness, for care or activities that were previously nonmedical in nature. Many aspects of life are influenced or dominated by medicine, to the point that some have argued that we are seeing the medicalization of everything in life during the latter half of the twentieth century.[32]

The medicalization process can be positive or negative in terms of outcomes. Medicalization can prevent the balanced assessment of medicinal and nonmedicinal therapies to prevent or treat illness. Instead of attempting interpersonal, social, or structural changes to alleviate symptom states, diseases, and their causes, medications are prescribed and used for symptomatic relief.[33] Medicinal agents usually are the least expensive and most accessible form of health care technology that can be provided to a patient. Medications also are viewed as pure and potent, and their effects are quite detectable, which assures their efficacy in patients' minds.[34,35]

Using Herbal Remedies and Homeopathic Medicines

Vignette 4. Seeking Nontraditional Therapies: Homeopathic Medicines

> Kathy is a nurse who has arthritis. She believes in a holistic approach to her health and does not like the strong medicines that her physician has been prescribing for her condition. The

medications she is taking produce a number of side effects, and they do not seem to be very effective in alleviating her symptoms. She has read about homeopathic medicines and found journal articles reporting clinical trials of homeopathic products that demonstrated their effectiveness in treating arthritis. Kathy locates a pharmacy that carries a variety of homeopathic medicines, but the pharmacist has been stocking these products only because of consumer demand and does not really know why they work or how to use them. Kathy chooses a number of homeopathic products from the shelf, selects some reference books on homeopathy, and decides to treat her arthritis by herself.

As consumers seek less traditional approaches to solving their health problems, many of them will turn to herbal remedies and homeopathic medicines. While both of these approaches focus on the use of substances from plant materials, the basic premise of each is different.[36]

Herbal medicine is a traditional form of medical healing in many cultures, including most Western societies. Herbal remedies can be an integral part of a comprehensive therapeutic approach designed by a healer, or they can be simply a product used by a consumer. Conventional herbal therapy is based on a holistic, or whole person, approach to health care. After an assessment, the healer arrives at a diagnosis and prescribes treatment, which could include life-style changes, dieting and exercise, nonmedicinal treatment strategies, and the use of herbal medicines. Herbal treatments can consist of using individual plant species or, more commonly, combinations of different plants that have been found through the healer's experience to provide some type of therapeutic benefit.

Homeopathic medicine has a well-defined conceptual basis behind its use of plant substances. Homeopathy focuses on the whole person, viewing symptoms of illness as signs of the body's reactive forces at work in eliminating disease. Homeopathic treatment consists of stimulating these reactive forces. Unlike allopathic (traditional Western) medicine, which treats symptoms with a drug or procedure that produces opposite effects, homeopathic medicine treats symptoms with a drug, device, or procedure that will *produce or enhance* the symptom's effects. Homeopathic products consist of natural substances and plant material that are diluted to very low concentrations so that they are strong enough to stimulate the body's natural forces in healing, but not so strong (what pharmacists would consider the therapeutic dose for allopathic medicines or synthetic pharmaceuticals) as to prevent or limit the body's response to the illness.

Consumers using herbal and homeopathic medicines face unique barriers in the use of these products. These nontraditional therapies are not as well studied as prescription and nonprescription medications, and quality information on their safety and efficacy is often lacking. Although herbalists and homeopathic physicians are not considered part of the traditional health care delivery system, their remedies are available more widely than pharmaceutical products. Such products can be obtained easily and used, but without guidance from or supervision of the trained professionals. Consumers often consult written references rather than seek someone who is knowledgeable about these treatment approaches. Also, consumers often do not tell their allopathic physician or pharmacist that they are using herbal or homeopathic products because they fear their health professional will reprimand them or refuse to continue treating them.

Using Illicit Substances as Medications

Vignette 5. Self-Medicating with Illicit Substances

Bob has anxiety and often feels depressed. While smoking marijuana recreationally, he discovers that it helps alleviate some of his symptoms and makes him feel better. Even though marijuana is an illegal drug and is costly and sometimes difficult to obtain, he chooses to self-medicate with it. With continued use, he also discovers that it works well to stimulate his appetite, reduce nausea, and help him sleep at night. Bob is soon using marijuana every day. While Bob finds that it is still effective in alleviating symptoms, he notices that he is sleepy a lot during the daytime, he does not think clearly, and his work performance is declining. As he worries about what to do, his symptoms seem to get stronger, so he chooses more potent marijuana and smokes more of it.

Most pharmacists and other health professionals would not consider illicit substances as medications, but many consumers use these substances for medical reasons. Virtually every drug that is currently illegal in the U.S., including heroin, cannabis (marijuana), and psychedelic agents (e.g., LSD, psilocybin, mescaline), is used therapeutically in some other country or society as a legitimate, approved medication. One exception to the U.S. position was the approval of Marinol (dronabinol), an oral version of one of the psychoactive components of cannabis. Dronabinol is approved for the management of nausea after chemotherapy and anorexia in patients with AIDS. Clearly even the U.S. FDA has recognized the medicinal value of some illegal drugs. Considering these substances solely as drugs of abuse with no practical therapeutic benefits can obscure pharmacists' thinking with regard to how the drugs can be used by consumers. It can also hinder pharmacists' ability to counsel these drug consumers, to provide them with relatively objective information, and to assist them in preventing or resolving drug-use problems.

The idea that most people with chemical dependencies may have begun their drug use with a medical reason in mind is relatively new, but the basis for this idea is quite old. Recall that before the drug control legislation of the twentieth century, all known pharmacologically active substances were readily available for direct consumer use. Heroin, cannabis, cocaine, chloroform, and many other drugs were used in the nineteenth century in both patent medicines and ethical pharmaceuticals. However, in this century, most health professionals and others have focused exclusively on the misuse of these drugs rather than on their potential therapeutic benefits.

In the early 1980s, psychiatrists treating chemically dependent people hypothesized that many people who use alcohol, marijuana, cocaine and other stimulants, and certain opiates, such as heroin, are doing so because they view themselves as being sick in some way.[37] In essence, they are self-medicating, or using those substances not to "get high," but for medical reasons. For example, they feel depressed, so they drink alcoholic beverages to alter or mask those symptoms; they use cocaine to "lift their spirits" or to "give themselves energy." For instance, they may feel anxious, panicky, or unable to get to sleep at night, so they smoke marijuana, drink alcohol, or use a sedative to calm down. When one considers the pharmacological action of many substances, particularly psychoactive ones, it should be easy to identify many other situations in which people would use *any* drug for the particular effects that they think it will provide.

This self-medication hypothesis of addiction has fostered better treatment approaches for chemically dependent people. It also presents us with an important idea: consumers will use any active substance that they think or know will treat their perceived ills, regardless of the legal status, societal approval, "abuse" potential, sources of supply and information, or how anyone else (e.g., relatives, friends, health professionals, the media, or government regulators) feels about the drug. The process of using illicit substances as medications is a variation of the general process of consumer self-medication. The major differences include the legal status of the drug, which will influence its availability and price, and the lack of quality control measures in its production and packaging.

PROVIDING PHARMACEUTICAL CARE TO MEDICATION CONSUMERS

Many people view consumer medication use, especially self-medication, as beneficial and positive. Self-care is a way of supplementing medical care received from health professionals and can be an alternative to using health professionals and the health care delivery system. Self-medication, in general, may be one way of controlling the skyrocketing costs of health care, and, for many patients, it is an alternative form of hope, especially if hope for a cure has not been provided by health professionals. Finally, self-care is a way for patients to gain more control over or responsibility for their health.

There are, of course, disadvantages to engaging in self-medication, as well as many possible negative consequences. Self-care can lead to misdiagnosis and mistreatment of the symptoms, illness, and underlying condition. The use of specific drug products may be palliative in nature, leading the user to think that the cause of the symptom or illness is being corrected or cured when it is not. Improper patterns of self-medication may lead to side effects, interactions with other drugs, and many other medication use problems. It also represents a way of introducing various medications into the household that could become poisons—especially in the hands of children.

A lack of adequate knowledge about health, illness, and drug therapy is the most significant weakness of consumers who engage in self-medication activities. This often results in an irrational form of decision making that is based more on the highly variable information provided by relatives and friends, potentially biased information from promotional campaigns, incomplete or unclear media reports, and other information from less-than-accurate sources.

Role of the Pharmacist

Pharmacists and other health professionals should be prepared to counsel and inform patients who wish to treat themselves for minor, self-limiting conditions. Although it might be difficult to know who needs counseling, for what conditions, and which educational technique is the best in a given situation, the cornerstone to any counseling effort is the provision of high-quality drug information.

Other techniques have been developed for counseling patients about self-medication (Table 11.4).[38] Medication information may be provided in a structured manner through media, such as public service announcements (PSAs) on local radio or television, a regular column in a community or neighborhood newspaper, a drug information program on local cable-access television, or drug information sheets as a part of local mailings, posters, or billboards. Structured medication education programs should be developed for specific groups of patients or people in the community (e.g., elderly people, pregnant and breast-feeding women, people with specific health problems). Computer-assisted instruction can also be implemented in some settings. Knowledge of computer programming can lead to the development of drug IQ quizzes, games, and information modules for patients. These items can be used by patients waiting in

TABLE 11.4. Strategies for Improving Consumer Choice of Medications

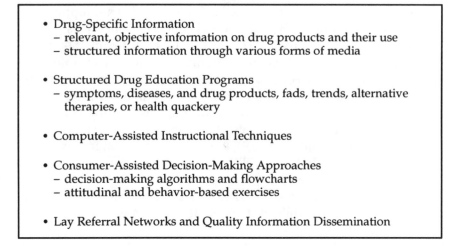

- Drug-Specific Information
  - relevant, objective information on drug products and their use
  - structured information through various forms of media

- Structured Drug Education Programs
  - symptoms, diseases, and drug products, fads, trends, alternative therapies, or health quackery

- Computer-Assisted Instructional Techniques

- Consumer-Assisted Decision-Making Approaches
  - decision-making algorithms and flowcharts
  - attitudinal and behavior-based exercises

- Lay Referral Networks and Quality Information Dissemination

the pharmacy for their prescription to be dispensed, or they can be rented or sold to patients for use at home.

The key to safe and effective self-medication practices is a rational and informed decision-making process, and some of the newer educational strategies have been developed to help patients make good choices. The use of algorithms or decision flowcharts has become popular. In this approach, a decision-tree is used to guide the patient through a series of Yes-No questions or a set number of alternate decisions to determine the best choice in terms of diagnosis and treatment. Also, attitudinal and behavioral games and exercises have been developed for use in structured medication education programs. Finally, the lay referral network, the circle of relatives and friends with whom the patient might consult for ideas, information, and actual products, can be employed in an active way to encourage and guide self-care in a positive direction.

Guidelines for Pharmacist Counseling of Consumers

As you know, pharmacists are obligated by law and ethics to ensure that patients understand why they are taking their prescription medication; to ensure that they know important facts about its use, such as administration, storage instructions, and common side effects; and to ensure that the medication selected is appropriate for the patient given his or her health and concurrent therapies. When patients self-medicate with OTC or other types of treatments, pharmacists still have an obligation to ensure that patients are informed and aware of the consequences of their decisions. When discussing self-selected therapies with patients, keep in mind the information that is important to *prescription* medication users: product name, administration, dosing, duration of use, precautions, side effects, contraindications, and storage.[39] Remind self-medicators that OTC medicines, especially Rx-to-OTC switch products, are just as powerful (and as dangerous) as prescription medicines. As their pharmacist, you can help them understand the benefits and potential dangers of self-care while still encouraging them to be responsible for their health.[40]

Guidance and accurate information from a health professional are essential to improving consumer medication behaviors. In practice, each person who wishes to self-medicate should be assessed to the same degree that prescription medication users are monitored. This would include taking a patient's medication use history, managing his or her medication use, and providing him or her with information and educational programs.

Each health professional should consider some additional guidelines when talking with current and prospective self-medicators.[22] Consumers should be taught to:

- Identify the problem as clearly as possible (e.g., symptom state, condition, illness prevention) and how to go about treating it, involving a health professional (pharmacist or physician).
- Get information, seek advice, and carefully read the labels of all medications they use.
- Know the potential dangers (i.e., side effects, adverse reactions, contraindications, effects from long-term use, and possibilities of misdiagnosis) of the medications they use in any treatment plan.

- Self-medicate for only a short period of time because most drug products intended for self-medication are meant only for symptomatic relief.
- Carefully consider promotional campaigns, reports in the media, and the general advice of other people who are not trained or do not have personal experience in this realm: to know the quality of their information source.

CONCLUSION

As the health care system changes, doctors lose prescribing autonomy, and patients demand more detailed information about both traditional and nontraditional disease therapies. Thus, the processes by which patients choose prescription and nonprescription medicines are changing. While pharmaceuticals are not purchased in exactly the same manner as high-involvement consumer goods, the *basic* purchase process (problem recognition, information search, evaluation of alternatives, and purchase) is identical. For patients, however, the nature of their problem involves more emotions when they are ill than when their television is broken, and drug information is less familiar and more confusing than a sales pitch for a new Magnavox. As such, pharmacists play an important role in facilitating appropriate medical treatment and self-care practices.

Pharmacists must be aware of the quality of health and drug information that patients and consumers receive from various sources, such as family members, the media, colleagues, other health care professionals, and proponents of alternative therapies. Understanding how and why consumers develop their health attitudes and why they behave in certain ways will help pharmacists counsel patients effectively. Finally, health care professionals must also respect the many types of nontraditional self-care alternatives—herbal remedies, holistic medications, and illicit substances—that patients will use. Many consider these alternative treatment choices effective, safe, and reasonable for managing illnesses. Providing objective, comprehensive, and accurate information is fundamental to ensuring that consumers are making optimal self-treatment choices.

REFERENCES

1. Howard JA, Sheth JN. The theory of buyer behavior. New York: Wiley, 1969.

2. Farley JU, Lehman DR. An overview of empirical applications of buyer behavior system models. Adv Consumer Res 1977;4:337.

3. Rossiter JR, Percy L. Advertising and promotion management. New York: McGraw-Hill Publishing, 1987.

4. Everett SE. Lay audience response to prescription drug advertising. J Advertis Res 1991;(Apr/May):43-9.

5. Widrick S, Fram E. Identifying negative products: do customers like to purchase your products? J Product Brand Manage 1992;1(1):43-50.

6. Svarstaad BL. Sociology of drugs in health care. In: Wertheimer AI, Smith MC, eds. Pharmacy practice: social and behavioral aspects. 3rd ed. Baltimore, MD: Williams & Wilkins, 1989.

7. Conrad P. The meaning of medications: another look at compliance. Soc Sci Med 1985;20:29-37.

8. Montagne M. Social knowledge and experiences with legal highs. J Drug Issues 1984;14:491-507.

9. Montagne M. The metaphorical nature of drugs and drug taking. Soc Sci Med 1988;26:417-24.

10. Montagne M, ed. Philosophies of drug giving and drug taking. J Drug Issues 1988;18:139-284.

11. Hartzema AG, Porta MS, Tilson HH. Pharmacoepidemiology: an introduction. 2nd ed. Cincinnati, OH: Harvey Whitney, 1991.

12. World Health Organization. The rational use of drugs. Geneva: WHO, 1987.

13. Kimmey JR. Consumer consciousness. Am Pharm 1971;11:328.

14. Cooper JD, ed. The efficacy of self-medication. Washington, DC: Interdisciplinary Communication Associates, 1973.

15. Koberstein W. Reach the people: the progress of direct-to-patient communications. Pharm Exec 1993;(Jul):36-58.

16. Masson A, Rubin PH. Warning: brief summaries may be hazardous to your health (and wealth) or the real issues in prescription drug advertising to consumers. J Pharm Market Manage 1986;1(2):29-43.

17. Waldholz M. Prescription-drug maker's ad stirs debate over marketing to public. Wall Street J 1987 Sept 22:35.

18. Pines WL. Trends in health care consumer communications. J Pharm Market Manage 1992;7(1):87-98.

19. Basara LR. Direct-to-consumer advertising: another perspective. Pharm Business 1993;4(1):15-8.

20. Madhavan S. Are we ready for direct-to-consumer advertising of prescription drugs? Pharm Business 1993;4(1):14+.

21. Ruby LA, Montagne M. Direct-to-consumer advertising: a case study of the Rogaine® campaign. J Pharm Market Manage 1991;6(2):21-32.

22. Basara LR, Montagne M. Searching for magic bullets: orphan drugs, consumer activism, and pharmaceutical development. Binghamton, NY: Pharmaceutical Products Press, 1994.

23. Young JH. The toadstool millionaires. Princeton, NJ: Princeton University Press, 1961.

24. Young JH. The medical messiahs. Princeton, NJ: Princeton University Press, 1967.

25. Coons SJ, McGhan WF. The role of drugs in self-care. J Drug Issues 1988;18:175.

26. U.S. House. Select Committee on Aging. Quackery: a $10 billion scandal (H.Doc. 98-262). Washington, DC: Government Printing Office, 1984.

27. Heller H. Health care practices and perceptions: a consumer survey of self-medication. Washington, DC: The Proprietary Association, 1984.

28. Knapp DA, Knapp DE. Decision-making and self-medication. Am J Hosp Pharm 1972;29:1004-7.

29. Montagne M. How to help the elderly self-medicate. US Pharm 1989;14(6):53+.

30. Dunnell K, Cartwright A. Medicine takers, prescribers and hoarders. London: Routledge and Kegan Paul, 1972.

31. Montagne M, ed. Drug advertising and promotion. J Drug Issues 1992;22:195-480.

32. Conrad P, Schneider JW. Deviance and medicalization: from badness to sickness. St. Louis, MO: Mosby, 1980.

33. Lennard HL. Mystification and drug misuse. San Francisco, CA: Jossey-Bass, 1971.

34. Arluke A. Judging drugs: patients' conceptions of therapeutic efficacy in the treatment of arthritis. Hum Organiz 1980;39:84-8.

35. Montagne M. The culture of long-term tranquilliser users. In: Gabe J, ed. Understanding tranquilliser use: the role of the social sciences. London: Tavistock/Routledge, 1991:48-68.

36. Sobel D, ed. Ways of health. New York: Harcourt Brace Jovanovich, 1979.

37. Khantzian EJ. The self-medication hypothesis of addictive disorders. Am J Psych 1985;142:1259-64.

38. Montagne M. Drug Education. In: Gennaro AR, ed. Remington's pharmaceutical sciences. 19th ed. Easton, PA: Mack, 1995.

39. Martin S. What you need to know about OBRA '90. Am Pharm 1993;NS33(1):26-8.

40. Srnka QM. Implementing a self-care consulting practice. Am Pharm 1993;NS33(1):61-70.

ADDITIONAL READINGS

For the Instructor

1. Arluke A. Judging drugs: patients' conceptions of therapeutic efficacy in the treatment of arthritis. Hum Organiz 1980;39:84-8.

2. Basara LR. Direct-to-consumer advertising: today's issues and tomorrow's outlook. J Drug Issues 1992;22:317-30.

3. Basara LR. Practical considerations when evaluating direct-to-consumer advertising as a marketing strategy for prescription medications. Drug Info J 1994;28:461-70.

4. Brody H. Placebos and the philosophy of medicine: clinical, conceptual, and ethical issues. Chicago, IL: University of Chicago Press, 1977.

5. Bush PJ, Davidson FR. Medicines and "drugs": what do children think? Health Educ Q 1982;9:113-28.

6. Conrad P, Schneider JW. Deviance and medicalization: from badness to sickness. St. Louis, MO: Mosby, 1980.

7. Cooper JD, ed. The efficacy of self-medication. Washington, DC: Interdisciplinary Communication Associates, 1973.

8. Cooperstock R, Lennard HL. Some social meanings of tranquilizer use. Sociol Health Illness 1979;1:331-47.

9. Coulter HL. Homeopathic medicine. In: Sobel DS, ed. Ways of health. New York: Harcourt, Brace, Jovanovich, 1979:289-310.

10. Dunnell K, Cartwright A. Medicine takers, prescribers and hoarders. London: Routledge and Kegan Paul, 1972.

11. Hall D. Prescribing as social exchange. In: Mapes R, ed. Prescribing practices and drug usage. London: Croom Helm, 1980.

12. Hartzema AG, Porta MS, Tilson HH. Pharmacoepidemiology: an introduction. 2nd ed. Cincinnati, OH: Harvey Whitney, 1991.

13. Heller H. Health care practices and perceptions: a consumer survey of self-medication. Washington, DC: The Proprietary Association, 1984.

14. Helman CG. "Tonic," "fuel," and "food": social and symbolic aspects of the long-term use of psychotropic drugs. Soc Sci Med 1981;15:521-33.

15. Khantzian EJ. The self-medication hypothesis of addictive disorders. Am J Psych 1985;142:1259-64.

16. Knapp DA, Knapp DE. Decision-making and self-medication. Am J Hosp Pharm 1972;29:1004-7.

17. Lennard HL. Mystification and drug misuse. San Francisco, CA: Jossey-Bass, 1971.

18. Montagne M. Social knowledge and experiences with legal highs. J Drug Issues 1984;14:491-507.

19. Montagne M. Social exchange, values, and the marketing of pharmaceuticals. J Pharm Market Manage 1989;3(3):45-55.

20. Montagne M. The culture of long-term tranquillizer users. In: Gabe J, ed. Understanding tranquillizer use: the role of the social sciences. London: Tavistock/Routledge, 1991:48-68.

21. Montagne M, Bleidt BA. Social forces in the premature removal of drug products from the marketplace. Clin Res Pract Drug Reg Aff 1987;5:83-127.

22. Morgan JP, Kagan DV, eds. Society and medications: conflicting signals for prescribers and patients. Lexington, MA: Lexington Books, 1983.

23. Pellegrino E. Prescribing and drug ingestion, symbols and substances. Drug Intell Clin Pharm 1976;10:624-30.

24. Rhodes LA. "This will clear your mind": the use of metaphors for medication in psychiatric settings. Cult Med Psychiatry 1984;8:49-70.

25. Ruby LA, Montagne M. Direct-to-consumer advertising: a case study of the Rogaine® campaign. J Pharm Market Manage 1991;6(2):21-32.

26. Segal R, Hepler CD. Drug choice as a problem-solving process. Med Care 1985;23:967-76.

27. Svarstaad BL. Sociology of drugs in health care. In: Wertheimer AI, Smith MC, eds. Pharmacy practice: social and behavioral aspects. 3rd ed. Baltimore, MD: Williams & Wilkins, 1989.

28. Temin P. Taking your medicines: drug regulation in the United States. Cambridge, MA: Harvard University Press, 1980.

29. U.S. House. Select Committee on Aging. Quackery: a $10 billion scandal (H.Doc. 98-262). Washington, DC: Government Printing Office, 1984.

30. Veatch RM. Value foundations for drug use. J Drug Issues 1977;7:253-62.

31. World Health Organization. The rational use of drugs. Geneva: WHO, 1987.

For the Student

1. Altman L. Who goes first? The story of self-experimentation in medicine. New York: Random House, 1987.

2. Basara LR. Direct-to-consumer advertising: another perspective. Pharm Business 1993;4(1):15-8.

3. Basara LR, Montagne M. Searching for magic bullets: orphan drugs, consumer activism, and pharmaceutical development. Binghamton, NY: Pharmaceutical Products Press, 1994.

4. Conrad P. The meaning of medications: another look at compliance. Soc Sci Med 1985;20:29-37.

5. Coons SJ, McGhan WF. The role of drugs in self-care. J Drug Issues 1988;18:175.

6. Everett SE. Lay audience response to prescription drug advertising. J Advertis Res 1991;(Apr/May):43-9.

7. Farley JU, Lehman DR. An overview of empirical applications of buyer behavior system models. Adv Consumer Res 1977;4:337.

8. Howard JA, Sheth JN. The theory of buyer behavior. New York: Wiley, 1969.

9. Kimmey JR. Consumer consciousness. Am Pharm 1971;11:328.

10. Madhavan S. Are we ready for direct-to-consumer advertising of prescription drugs? Pharm Business 1993;4(1):14+.

11. Martin S. What you need to know about OBRA '90. Am Pharm 1993;NS33(1):26-8.

12. Masson A, Rubin PH. Warning: brief summaries may be hazardous to your health (and wealth) or the real issues in prescription drug advertising to consumers. J Pharm Market Manage 1986;1(2):29-43.

13. Montagne M. The metaphorical nature of drugs and drug taking. Soc Sci Med 1988;26:417-24.

14. Montagne M, ed. Philosophies of drug giving and drug taking. J Drug Issues 1988;18:139-284.

15. Montagne M. How to help the elderly self-medicate. US Pharm 1989;14(6):53+.

16. Montagne M, ed. Drug advertising and promotion. J Drug Issues 1992;22:195-480.

17. Montagne M. Drug Education. In: Gennaro AR, ed. Remington's pharmaceutical sciences. 19th ed. Easton, PA: Mack, 1995.

18. Pines WL. Trends in health care consumer communications. J Pharm Market Manage 1992;7(1):87-98.

19. Rossiter JR, Percy L. Advertising and promotion management. New York: McGraw-Hill Publishing, 1987.

20. Silverman M, Lee PR. Pills, profits and politics. Berkeley, CA: University of California Press, 1974.

21. Silverman M, Lee P, Lydecker M. Prescriptions for death. Berkeley: University of California Press, 1982.

22. Srnka QM. Implementing a self-care consulting practice. Am Pharm 1993;NS33(1):61-70.

23. Vener AM, Krupka LR. Over-the-counter drug advertising in gender-oriented popular magazines. J Drug Educ 1986;16:367-81.

24. Vickery DM, Fries JF. Take care of yourself. Reading, MA: Addison-Wesley, 1976.

25. Waldholz M. Prescription-drug maker's ad stirs debate over marketing to public. Wall Street J 1987 Sept 22:35.

26. Wells HG. Tono-Bungay. New York: Dodd, Mead and Co., 1908.

27. Widrick S, Fram E. Identifying negative products: do customers like to purchase your products? J Product Brand Manage 1992;1(1):43-50.

28. Young JH. The toadstool millionaires. Princeton, NJ: Princeton University Press, 1961.

29. Young JH. The medical messiahs. Princeton, NJ: Princeton University Press, 1967.

SECTION E:
MEDICATION-TAKING BEHAVIOR

Chapter 12

Determinants of Medication Use

Mickey C. Smith

INTRODUCTION

Traditionally and historically, pharmacists have derived most of their livelihood directly or indirectly from the use of medications by the patients and customers they serve. It should be obvious, then, that we have had considerable stake in assuring that everyone who needed medicine actually received it. Today there is recognition (in the form of compensation/remuneration) of the role of the pharmacist in attempting to assure not only that patients with real medical needs have access to appropriate medication but also that medication is *not* used when it is not needed. Obviously, the foregoing is an oversimplification, but perhaps it will serve to introduce the topic of this chapter. In some ways, a chapter on the determinants of medication use is superfluous, as virtually every chapter in this book deals directly or indirectly with some aspect of the subject. Nevertheless, it should be worthwhile to spend a little time tying some of the threads.

How does it happen that people use medicines? There are many reasons to study and understand the determinants of medication use, for such behavior is not as simple and straightforward as logic might suggest. That same logic would tell one that an intelligent, rational individual suffering from disease, illness, sickness, or any combination of these who knows the existence of appropriate medication and has access to that medication would take it. The facts, however, suggest that there is many a slip between the medication cup and the lip.

Who cares?

When medication is used inappropriately–too much or too little–it is important to the patient's well-being, certainly. But there are others who have a stake in the process, too: the patient's family, his/her physician, any third-party payer that might be involved, the manufacturer of the medication, regulatory agencies such as the Food and Drug Administration, and most assuredly the practicing pharmacist involved. It is important to discover the reasons that people *do* and *do not* use medication properly for, as in the isolation of an infective bacterium, once the cause is discovered, action can be taken.

What do we know and what can be done about it? Let us respond with a collection of rather curious research data.

THE CASE OF LAXATIVES

As part of a large study of medication use among a sample of noninstitutionalized elderly, Juergens, Smith, and Sharpe focused especially on the use of laxatives.[1] Although older people appear to use laxatives in greater proportions and more frequently than the rest of the population, not all use them. That is consistent with the findings of the study under discussion.

Data from this study produced a profile of elderly laxative users that suggests that they tended to live alone in a rural setting and reported using a single pharmacy that did not deliver medication as a source of over-the-counter (OTC) drugs. They were not likely to ask a pharmacist questions about OTC medicines and perceived the availability of pharmacy services as low. Laxative use was equally likely among men and women. In terms of health perceptions, the use of OTC laxatives was associated with lower perceptions of general health and a higher degree of perceived morbidity. Users also were more likely to report signs of anxiety and use more prescription medication than did nonusers of laxatives. Interesting, but is a pharmacist supposed to care?

Gerbino and Gans describe the management of patients with chronic constipation as "one of the most frustrating problems in geriatric medicine."[2] Especially troublesome are patients who are laxative dependent or who abuse laxatives. Cummings has estimated that women constitute 90% of this group.[3]

Poe and Holloway suggest that laxatives are "the most overused and abused of all drugs."[4] They posited that advertising might be a major reason but also suggested custom and the lay referral system as factors. "Hundreds of thousands, if not millions, of people have adopted laxatives lore as a matter of faith. For many people the use of laxatives has been a lifelong custom, passed on from generation to generation."[4] They suggest that "isolation and depression cause people to become self-conscious" and therefore constipated. Certainly, other explanations suggest themselves. What kind of cooking and eating habits are practiced by people (whatever their ages) who live alone? Maybe they really need a laxative, even though a better diet is the treatment of choice.

Eve and Friedsam reported on a sample of more than 8,000 north Texans age 60 and older and their use of vitamins and laxatives.[5] Vitamins were selected as evidence of preventive health behavior and laxatives as evidence of curative behavior. Although the methods of analysis were somewhat different and each study collected data not duplicated in the other, there were some striking similarities in results between that study and the one reported here with regard to laxative use. Eve and Friedsam found laxative use to be positively related to age, feelings of loneliness, being single, being dissatisfied with social interaction, having less education, having lower income, having less access to medical care, and having a lower level of perceived health. Although not identical with the results of the current studies, none of these results is inconsistent with it.

The picture of the laxative user as a relatively lonely, socially isolated, "sick" person is reinforced by the findings of this study. Eve and Friedsam suggest as explanation:

- Poor health contributes to social isolation through lack of opportunity.
- Social isolation creates hypochondriasis, with self-medication for the somatic complaints to other solutions.

They do not suggest which is the more compelling.

Laxatives, indeed the entire process of evacuation, have been trivialized through jokes and television commercials. Yet laxatives are serious medicine. When used regularly, even daily, laxatives can have adverse physiological consequences, result in changes in drug effects, and, through fluid and electrolyte loss, affect or aggravate other problems.

It is apparent that long-term, chronic use of laxatives may be a problem, although data are sparse. In a Florida study among more than 3,000 elderly, nearly 15% reported using laxatives. Further, nearly two-thirds of the laxatives had been used by the respondents for longer than two years, and nearly half were used on a daily basis.[6]

More than 25 years ago, Ernest Dichter, then "guru" of motivation research, found deep meaning in constipation:

> In a psychological sense, irregular elimination is disturbing because the body is not functioning; control has been lost over it. It is not illness, but it is disquieting; it resembles impotence. On the other hand, bowel movements are accompanied by essentially pleasant feelings that ally it to sexual gratification. There is in elimination, as in sex, a relationship of tension and relief. However, as with sex, there are also feelings of inhibition, guilt, sin, and embarrassment.[7]

(We are not certain how to *act* on these observations, but we wanted to share them with our readers.)

A 1969 national survey jointly sponsored by seven different government agencies shed some light on laxative use patterns.[8] Among the findings:

- Forty-two percent of the respondents over the age of 65 (compared with 31% of the total sample) agreed with the statement, "People should do something regularly to help with bowel movements."

- Forty-seven percent of those over 65 (compared with 30% of the total sample) "[e]ver (did) anything to help with bowel movements."
- Fourteen percent of those over 65 (compared with 5% of the total) did something daily to help with bowel movements.
- People over 65 were nearly twice as likely as the total sample to overrely on bowel movement aids (i.e., do something daily or nearly daily to help with bowel movements, not on the advice of a physician).
- Although age breakdowns were not provided for this item, it appeared that the overreliers did not view daily laxative use as treatment but as something more akin to hygiene.

The results of the studies referred to above suggest that not all of the reasons for the use of laxatives are necessarily medical or physiological. The same could be said for virtually all medications. Discovery of the reasons for medication use and appropriate action following discovery make a worthy future professional agenda for pharmacists–individually and collectively. We will suggest a framework for doing that and provide examples in the remainder of this chapter. One other observation. If the reader believes that the laxative example is a trivial one, he or she should begin now to consider the importance of medications from the viewpoint of the *patient* and not that of a health professional.

ANALYZING THE MEDICATION USE PROCESS

The Framework

The framework for analysis flows from the work of Anderson and Newman, who propose a model of medical care utilization consisting of a three-stage sequence of predisposing, enabling, and need for care variables that interact in various ways to produce a continuum of utilization rates for the several components of medical care.[9] Our adaptation of the Anderson-Newman model is shown in general form in Figure 12.1, which requires some comment. First, the variables identified are exemplary rather than exhaustive.

FIGURE 12.1. Model for the Study of Determinants of Medication Use

	Predisposing factors →	Enabling factors →	Need for care factors →	Utilization decisions →
Patient Variables	Doctor Pharmacist Age Sex Race Prior Experience Marital status Media exposure Family Psychographics Residence	Doctor Pharmacist Relationship Family Education Symptoms Income Residence Third-party coverage	Diagnosis Symptoms Perceived morbidity	Use of drug (versus other therapy) Use of a specific drug Use of a specific drug product Acquire the drug Whether and how to administer the drug Discontinue use
Medication Variables	Cost Legal Status		Legal status Dosage form Dosage schedule Therapeutic class Effectiveness Safety	
Delivery System Variables	Pharmacy services Marketing practices Medical practices Third-party coverage Institutionalization	Pharmacy services Marketing practices Medical practices Third-party coverage Institutionalization	Diagnosis Marketing practices Medical practices Third-party coverage Institutionalization	

Source: Adapted from Anderson, R, Newman, J. Societal and individual determinants of medical care utilization. *Milbank Mem Fund Q*, 1973;51:91-124.

Indeed, variable identification is an ongoing process. Second, some variables are identified as more than one kind of factor. For example, third-party coverage can be both a predisposing and an enabling factor. In the first instance, the patient who knows that his medication costs will be covered in whole or in part by a third party may be predisposed to think of medication when a health problem arises. Then, clearly, the existence of such coverage will enable some patients to acquire medication who might not otherwise be able to afford it. Finally, although logic and professional intuition suggest some of the relationships shown, not all have been demonstrated empirically.

Utilization Decisions

Let us begin the discussion of the model at its *end*—utilization decisions. Data on medication use ultimately reflect the results of these decisions. It is notable that the first decision that must be made is that of the patient (or his surrogate) that something is wrong. Nothing happens in the field of medication use until that decision is made.

The first drug-related decision is comparatively simple: Will medication help or is some other action (diet, exercise, surgery) indicated? Nearly all serious research suggests that people confronted with some kind of symptom are more likely *not* to use medications than to use them. Studies by the Nonprescription Drug Manufacturers Association have found that people with everyday health problems used a prescription drug only 16% of the time and an OTC drug only 35% of the time.

What of those who decide to use a medication? They must still decide whether to self-medicate (with an OTC product or with an on-hand prescription drug) or to enter the medical care system, leading, in the majority of cases, to the use of a prescription drug. In either of these cases, someone—or some combination of persons— must still decide which drug to use, which product containing that drug to use, how to use the drug, and when to stop using the drug. Entangled with these decisions and complicating them are such factors as patient compliance, physician prescribing capabilities, coverage by managed care formularies, promotional activities of the manufacturers, and decision about need for care.

Need for Care

The most rational view of medications would suggest that sick people would use them and people who are sick would not. The fact is that not every sickness lends itself to easy diagnosis. Even though morbidity is comparatively highly correlated to medication use in most studies, the morbidity is frequently perceived and often not medically confirmed. Verbrugge recognizes this issue in her model of legal drug use (see Figure 12.2 and follow for detail in suggested readings).[10]

Enabling Factors

Many factors make it easier to use medications. (Their absence constitutes barriers to use, of course.) From the patient's perspective one must consider:

- The doctor-patient relationship and the pharmacist-patient relationship, as well as the patient's satisfaction with each.
- The family structure and medication-taking practices of the family members.
- The education and the socioeconomic class of the patient.
- Residence (i.e., rural or urban) as a measure of accessibility of care.
- Third-party coverage, especially as this factor interacts with income.

The medications themselves may vary in their effects on use. Variables to be considered include:

- Legal status: prescription, OTC, narcotics
- Brand-name versus generic issues
- Dosage forms that affect portability, compliance
- Comparative effectiveness
- Comparative safety
- Cost-benefit ratios and effects on such patient concerns as quality of life

The marketing activities of the pharmaceutical industry are just one group of variables in the medication delivery system that help

FIGURE 12.2. Model of Legal Drug Use

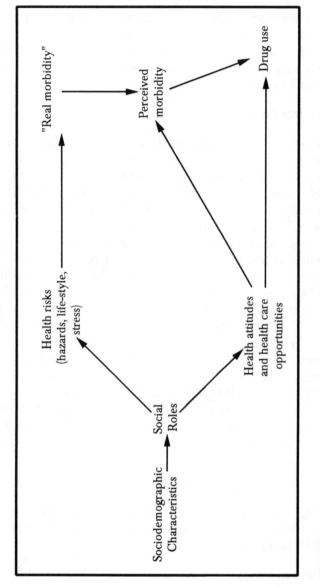

Source: Adapted from Verbrugge, LM. Sex differences in legal drug use. *J Soc Issues*, 1982;38:59-76.

determine whether medication will, in fact, be used. Some of these activities are directed at the prescribing physician (and at the pharmacist). The complexities of the simple act of prescribing are described in cogent fashion in the suggested reading by Hepler, Segal, and Freeman.

Predisposing Factors

In the list of predisposing factors are such mundane factors as the age and gender of the patient. Mundane perhaps, but definitely important. We have suggested readings to demonstrate that age (such as being a child or being an elder) and gender (especially being female) may make a difference in medication use that cannot be totally explained by illness experience.

There are many factors that may be predisposing but have not been explored and/or quantified through research. How great an effect do television commercials have? What are the effects of drug abuse education in the schools on the use of legitimate medications? Are some medications overvalued while others are thought of as something less than medicine? What has been the effect of sales of medications in grocery stores and convenience stores?

We cannot hope to explore all of the factors that go into determining use of medications here. Indeed, there is hardly room even to identify them. Rather, let us take another case study, the most widely prescribed class of medications in modern times, the minor tranquilizers. Even here, we can hope only to demonstrate, rather than explain, the complexities of the medication use process.

THE CASE OF MINOR TRANQUILIZERS

In 1982, the health research group of Ralph Nader's organization, Public Citizen, published a 108-page book entitled *Stopping Valium*.[11] The book charged, among other things, that:

- Thousands of people are addicted to Valium® and other benzodiazepines.
- Many of the 1.5 million people who have taken the drugs continuously for more than four months may be addicted and not know it.

• The drugs present special dangers to the elderly, to pregnant women, and to drivers.

In early 1983, it was announced that the recipient of the 1982 John Scott Medal Award for research was Dr. Leo H. Sternbach, the retired Roche chemist who invented Valium and its precursor, Librium®. Sternbach was in good company, as previous winners of the award included Madame Curie, Thomas Edison, and James Salk. How does it happen that a prestigious award is accorded the inventor of substances that a major consumer organization is committed to eliminating? Such an event is, in fact, a typical example of the ambivalence with which our society views the minor tranquilizers. The drugs are more than medicines. They are a social fact. Indeed, these drugs may be more social than medical. Both their use and the prescriptions for such use have been extensively studied and criticized.

PRESCRIBING STUDIES

A number of studies of prescribing patterns for psychotropic drugs have been conducted. Unfortunately, many of these studies failed to distinguish between major and minor tranquilizers, sedatives, antidepressants, and other drugs. Consequently, comparisons are often difficult. In this section, we review the reports of a few of the studies.

Shapiro and Baron were among the first to study the prescription of psychotropic drugs in a noninstitutionalized population.[12] Their study was conducted in 11 of the 32 medical groups that then participated in the Health Insurance Plan of Greater New York. Even then, the data showed women receiving psychotropic drugs at more than twice the rate of men. This class of drugs represented 12% of all prescriptions studied. Of all psychotropic drugs prescribed, 22% were minor tranquilizers. A notable finding, even in this early work, was that only one-third of the prescriptions for minor tranquilizers were written for diagnoses of mental, psychoneurotic, and personality disorders. The percentage so prescribed for all psychotropic drugs was only 18%.

Linn, in two related papers, examined attitudes of a sample of

235 Los Angeles physicians toward use of Librium in certain situations.[13,14] Physicians' general attitudes toward the use of medications were ascertained on the basis of their agreement with the following statements:

Statement	Percentage Agreeing
1. Certain medications are often very helpful in handling the social demands and stresses of everyday living.	79
2. A person should take pills only as a last resort.	17
3. A person is better off taking a sedative than missing a good night's sleep.	45
4. A person is better off taking a tranquilizer than going through the day tense and nervous.	57

Physicians in the sample were then asked to respond to a number of situations in which Librium was used. As a group, they were not in agreement over what constitutes legitimate use of the drug (Table 12.1).

In the second of these publications, Linn and Davis added data on physicians' preferred sources of drug information.[14] They reported that physicians preferring professional sources were significantly more likely to express conservative attitudes toward when drugs should be used than physicians preferring commercial sources. Physicians preferring professional sources were also significantly less likely to feel that medical advice from sources other than a physician was acceptable. Linn and Davis suggested that the medical profession contains "rather diverse philosophies of medication."

Williams has reviewed this use of "mental" drugs for "body" ailments and suggests three factors relating to why psychotropic (mainly minor tranquilizer) drugs are prescribed for physical illness:

1. Secondary properties of the drugs. Examples include reserpine in cardiovascular disease and diazepam as a muscle relaxant in arthritis/rheumatism.
2. Coexistence of physical and psychiatric disease. This is pur-

ported to be a function of (a) increased likelihood of physical morbidity among the psychiatrically ill, (b) increased likelihood of psychiatric morbidity among the physically ill, (c) existence of some individuals who are prone to both types of illness.

3. Psychic component of somatic disease. (The evidence cited in support of this type of use is reported to be weak.)[15]

It is, of course, more likely that a general practitioner will see the patient with some kind of mind/body problem.

Near the end of the first 25 years with the minor tranquilizers, T. Donald Rucker, noted health economist, essayed a quantitative opinion about the causes of "irrational" prescribing of minor tranquilizers.[16] His estimates, which appear in Table 12.2, reflect rather well the themes that have emerged, although the proportions hardly constitute a consensus.

UTILIZATION STUDIES

In addition to studies of prescribing and prescriptions, some researchers have used the patient as the unit of analysis, i.e., what proportion of a population used the drug under study? Parry was one of the first to present extensive data on the utilization rates for psychotropic drugs.[17] Parry noted that psychotropic drugs accounted for 14% of all prescriptions in the period 1963-1965 and pointed out that two-thirds of these prescriptions were refills, compared with about 50% for other drugs. "The preponderance of refills," Parry wrote, "[tended] to operate against any sharp decline of consumption."

In the first quarter century of the minor tranquilizers, only one truly national study of utilization was conducted. A number of researchers were involved in this project, which was conducted under the joint sponsorship of the National Institute of Mental Health (NIMH), the Social Research Group (George Washington University), and the Institute of Research in Social Behavior. The research resulted in a considerable number of publications, some of which contained similar data. Because this study was the only one of such magnitude in the 25-year history of the minor tranquilizers, it deserves special attention.

TABLE 12.1. Physician Evaluations of Legitimate Use of Librium from a Medical Point of View (n = 114)

Drug-Using Situation	Assessment of Legitimacy	(%)
A middle-aged housewife having marital trouble takes 15mg of Librium daily to settle her nerves.	Very legitimate Somewhat legitimate Not very legitimate Illegitimate	34 53 7 6 100
A college student takes 15mg of Librium occasionally when the stresses and demands of college life become too great.	Very legitimate Somewhat legitimate Not very legitimate Illegitimate	20 41 21 18 100
A physician takes 15mg of Librium occasionally when the stresses and demands of his practice become overbearing.	Very legitimate Somewhat legitimate Not very legitimate Illegitimate	17 38 25 20 100
A college student, highly anxious, takes 15mg of Librium daily to combat anxiety.	Very legitimate Somewhat legitimate Not very legitimate Illegitimate	22 31 25 22 100

Reprinted with permission from Linn, L. Physician characteristics and attitudes toward legitimate use of psychotherapeutic drugs. J Health Soc Behav, 1971;12:132-139.

TABLE 12.2. Factors Leading to Irrational Prescribing Classified by Quartile of Probable Level of Importance

Proportion of Problem		
50%	I.	• Promotional activities sponsored by pharmaceutical manufacturers that are designed, directly, or indirectly, to influence practitioner prescribing and dispensing • Limitations in the medical record system • Casual empiricism–practitioner interpretation of evidence regarding patient response to drug therapy
About 30%	II.	• Drug product proliferation with significant therapeutic contribution • Practitioner attitude toward treatment options in general and drug therapy in particular–the subjective component • Patient behavior–designed to influence prescribing patterns and subvert coordination of care
About 20%	III.	• Formal education received by practitioners • Practitioner postgraduate education (excluding promotional activities identified in I. but including all other inputs or lack thereof) • Practitioner proficiency in diagnosis and selection of optimum modality within treatment model
Nearly 10%	IV.	• Various characteristics of practice environment • Conflict of interest situations • Posture of professional associations toward drug use and drug industry

Source: Adapted from Rucker, T. Production and prescribing of minor tranquilizers: a macro view. Paper presented to the American Orthopsychiatric Association, 1980.

309

The comprehensive report on this collaborative study appeared in the *Archives of General Psychiatry* in June 1973.[18] The data were based on extensive personal interviews (60 to 90 minutes each) with 2,552 adults selected through "rigorous probability sampling methods" throughout the 48 contiguous states in late 1970 and early 1971.

Like many other studies, this one included a variety of psychotherapeutic agents. Because of our focus on the minor tranquilizers, and in the interest of space, we list some of the specific findings based on the survey relative to this class of drugs:

- Most current users felt they were helped by them.
- Women were more likely than men to say they would take a tranquilizer in advance of a possible unpleasant event.
- The United States would rank about in the middle of Western countries in level of use of minor tranquilizers, and the male-to-female use ratio is similar.
- About 5% of those using minor tranquilizers did so daily for at least two months (defined in the study as "high" use).
- Among users whose use level of minor tranquilizers was "high," the following characteristics appeared important:
  - Greater proportion in middle age
  - Greater proportion in the West
  - Greater proportion in lower socioeconomic class (similar to Parry's findings).
- There was little overlap between prescription and nonprescription tranquilizer use.

Using a symptom checklist and a social readjustment scale, it was possible to classify persons in terms of emotional distress and life crises. One of the principal findings was that 60% of the men and 70% of the women who used psychotherapeutic drugs in the year preceding the interview scored high on the psychic distress index or life crises index, or both indexes. Among the relatively small group of persons who used the drugs daily for two months or longer, the figures were somewhat higher: 70% of the men and 80% of the women.

A majority of the remaining drug users who were not classified as high on one index achieved a rating of at least moderate on the other. Only 2 to 3% of those who used drugs daily for two or more months were classified as low on both indexes, compared with 16% of persons who used no psychotherapeutic drugs in the preceding year.

Why these differences in prescribing and utilization of this fascinating group of drugs? The reasons discovered to date are many and affect and are affected by both the patient and his physician. Throughout their history, the minor tranquilizers, their use, and their prescription have been studied in efforts to determine reasons for their popularity. Various hypotheses, resulting mainly from research, are presented here.

In 1970 and 1972, Pflanz and colleagues conducted a survey in Hannover, West Germany, among 1,251 subjects.[19] They found that 14.7% of the men and 27.1% of the women took tranquilizers "regularly" and "presently." Consumption was higher in the upper and middle classes, particularly for men. Researchers also found a strong relationship between drug consumption and mental health (as assessed on a 22-item scale). Drug use was shown to be related to subjective ill health and unrelated to objective indicators of health status. The researchers saw the data as supporting a psychiatric medical model rather than a sociological one.

This last issue was at the core of the study. Pflanz and colleagues reviewed the literature of sociological studies of tranquilizer use and found that although a variety of cultural and social variations were reported, the one unanimous finding was a higher rate of use in women—and this finding was not dependent on the respective status of women in the various countries. In the Pflanz study, some of the findings were:

- Users were more "health conscious."
- Homemakers were not found to be heavier users than employed women.
- There was no relationship between drug use and role status inconsistency.
- Some 29 questions about "mental health" distinguished between users and nonusers, which was "true for mainly somatic symptoms, for psychosomatic symptoms, and for purely psychological symptoms" (perhaps most significant).

Pflanz and colleagues made a special effort to isolate the factors involved in the consistently higher use of tranquilizers by women. They found it appropriate to assume that women have more psychological symptoms and therefore take tranquilizers more often than men. They also found support for the hypothesis that women perceive and express more symptoms of ill mental health than men. They finally raised an interesting reverse question: Is the question why do women take more, or why do men take less? Again, the assumption appears to be that less is better.

According to Cooperstock and Lennard, the people who are most likely to receive prescriptions for minor tranquilizers fall into three groups:

1. Those who describe their problem to the physician in psychological or social terms.
2. Those who describe a physical complaint (insomnia, stomach pains, backache) but are discovered to experience "problems of living." (For women these tend to relate to structural strains in the nuclear family. For men they are more likely to be job-related.)
3. Those with chronic somatic illness for whom the drugs may be prescribed to diminish stress generally and in reaction to the illness particularly.[20]

They noted also that "since the legitimation of a personal problem as a 'disease' affords considerable advantages, it is readily 'traded' for a medical diagnosis."

Publication of such studies culminated in one of the strongest attacks on the increased prescribing of Valium, Librium, and other drugs. Waldron presented a plethora of arguments (some supported, others not supported) to the effect that these and many other drugs are often prescribed for nonmedical (i.e., social) reasons.[21] She referred to the "extremely rapid" increase in the use of Valium and Librium, pointing especially to the advance of Valium from its introduction in 1963 to a position as top seller in 1972. (In fact, that is not "rapid" by drug industry standards and suggests, instead, a steady rise based on positive experiences.)

Quoting drug advertisements for such conditions as a major discord in parent-child relations, she noted that they appear:

> To offer a resolution for a common and difficult dilemma for doctors, namely, how to respond to a patient who is distressed by psychological and social problems, given that both the doctor and the patient expect the doctor to do something to relieve the patient's distress in an appointment that averages less than 20 minutes.[21]

Such prescribing for social and psychological problems, Waldron noted, "appears to satisfy substantial needs of both doctor and patient." Indeed, even she agreed that the period of increasing use of Valium and Librium was also a period of increasing social problems.

Other researchers continued to search for determinants of tranquilizer use in a variety of places. Webb and Colette examined a hypothesis that urban crowding might be associated with a higher incidence of use of stress-alleviating drugs.[22] They found the reverse relationship and also that such drug use increases directly with the number of persons living alone.

Radelet set out "to identify health beliefs and social networks that distinguish users from nonusers."[23] His population consisted of only 181 university students (46 of whom used tranquilizers), but his findings were interesting. He found the users, in comparison with nonusers, to be:

- More reluctant to admit unpleasant feelings.
- More critical of OTC drug advertisements.
- More likely to have friends and relatives who use tranquilizers (especially the father).
- More anxious as measured by Spielberger's State-Trait Anxiety Inventory.

Radelet acknowledged that symptoms of anxiety play a role in tranquilizer use but argued that they are only part of the picture. He wrote:

> The extensive prevalence of tranquilizer use in American Society is therefore not indicative simply of high levels of anxiety, but also is associated with a general cultural frame-

work that tends to define anxiety as an individual biophysical reality that warrants and necessitates drug treatment.

Such cultural definitions of medical situations present dilemmas for the prescribing physician.

PRESCRIBING DILEMMAS

In 1980, Dr. Louis Lasagna published a collection of point-counterpoint papers under the broad title *Controversies in Therapeutics*. Dr. John Morgan drew the challenge of the position that perhaps our society is not overmedicated. The benzodiazepines were one class of drugs that he discussed in a cogent and pragmatic fashion:

> Everyone seems to know that we are an overmedicated society, and one imagines editors of publications small and large, common and arcane, swallowing benzodiazepines while rushing to meet a deadline with an article on the overmedicated society.

Lasagna's book exemplifies well the proposition that physicians face many dilemmas in their choice of drug therapy. Further, the use of the term controversies in the title illustrates the divergence of opinion that often exists. Among the controversial types of therapy were beta blockers, antacids, aspirin, opium derivatives, and, of course, benzodiazepines.

Morgan's Chapter 6 specifically addressed the issue of Valium use. He noted that Valium has an ample supply of both critics and defenders, and that, curiously, both sides tend to buttress their arguments with the same findings and data. Some of Morgan's comments offer a springboard for our brief discussion of the dilemmas that the minor tranquilizers have presented to the prescribing physician:

- One reason we stumble over questions of "appropriate" drug utility is our lack of understanding of the cultural context of drug use. We are asked to define a "proper" level of medication in society when we do not understand the social utility of powerful chemicals.
- It is presumptuous—even preposterous—to define appropriateness solely from clinical view.

- Until we make more progress in defining the social utility of drugs and the social utility of health and well-being, it behooves us to be a bit more humble about appropriateness.
- Diagnosis and conventional medical thought have essentially nothing to do with the usual use of benzodiazepines.

Morgan refers to a "somewhat tattered army of academics" caught between those who argue–forcefully–both sides of the overmedication issue. Equally tattered, certainly, must be the practicing physician who is trying to make prescribing decisions while awaiting some resolution of the conflict between the political and medical leadership. The foregoing surely demonstrates some of the physicians' dilemmas, a few of which will be reviewed briefly here.

"Do Something" Dilemma

In his book *Profession of Medicine*, sociologist Eliot Friedson argued for the activist orientation of the medical practitioner: "The aim of the practitioner is not knowledge but action. Successful action is preferred, but action with very little chance of success is to be preferred over no action at all."[24] He further noted that sociologist Talcott Parsons had observed that activism is, in fact, a basic part of the American value system.

Physicians, themselves action oriented, are thus faced with an "anxious" patient who also expects something to be done. Further, given the notoriety of the minor tranquilizers, the patient is likely to know that something *can* be done. On what basis can physicians withhold antianxiety therapy?

Definition/Diagnosis Dilemma

Although there are paper and pencil anxiety tests and clinical algorithms for diagnosis of clinical anxiety, these are relatively recent developments. Physicians have had no mental sphygmomanometer with which to measure mental pressure as with blood pressure.

A recurring medical problem is the difficulty in diagnosing and treating depression. Although this may seem removed from the present subject, it is not. Depression is apparently often misdiagnosed and consequently treated as anxiety. Kline has called depression the "most misunderstood of all major diseases," advising that "any case of anxiety that does not respond to appropriate treatment within one or two months should be reconsidered as possibly having an underlying depression. Many patients with such depression develop anxiety *secondarily* and do not respond to treatment until the primary depression is relieved."[25]

Information Dilemma

How much of the information supplied by the drug manufacturers should physicians accept? How much should they believe? What are their alternatives?

Clearly, commercial information is biased, although the Food and Drug Administration (FDA) has considerable authority to require a certain balance and to correct misleading messages. In any case, the information is palatable, convenient, and omnipresent. Further, the "scientific" information is often conflicting and sometimes unscientific.

Medicalization Dilemma

In 1971, Lewis reported that nearly two-thirds of his (nonscientific) sample of physician respondents felt that other physicians were prescribing too many tranquilizers.[26] He made clear, through the example of a Valium advertisement, his belief that a major reason for overprescribing was the "medicalization of human problems," noting that "once daily living is defined as disease, how logical it is for us to attempt to treat that disease."

Again, if one looks to Friedson's exposition on medicine, one finds an explanation—if not an excuse—for the medicalization of human problems. As he observes (and supports by argument), "The medical profession has first claim to jurisdiction over the label of illness and anything to which it may be attached, irrespective of its capacity to deal with it effectively." In fact, people come to physicians because there is anxiety in their lives. By the simple act of seeking medical assistance, the patient has medicalized the problem. Of course, by treating the patient, the physician confirms that this medicalization was appropriate. If the drug works, even if only briefly, the actions of both patient and physician are reinforced. The drugs have worked; criticism of the effectiveness of the benzodiazepines, at least, has been rare.

One may argue (and many do) that the drugs do not cure the anxiety and do not eliminate the cause. True enough, but the same can be said of antihypertensive medication, antiarthritics, and even aspirin. Should the patient be denied relief of mental discomfort when it is caused by problems of daily living?

Unless and until patients are provided with an acceptable alternative to the physician in dealing with their personal problems, as long as patients continue to seek relief from this source, it is unrealistic to expect the physician to turn them away because they have a problem with which he or she should not deal.

Society has medicalized human problems, it appears. Medicine has, perhaps, been an accessory, and the pharmaceutical industry, certainly, has provided both with the means. To expect either of the latter parties to do, or have done, otherwise bespeaks a considerable naiveté.

Friedson pointed out that "while medicine is hardly independent of the society in which it exists, by becoming a vehicle for society's values it came to play a major role in the forming and shaping of the social meanings imbued with such a role."[24] He argued that physicians became "moral entrepreneurs," seeing "mental illness" in cases in which the layperson has come to the physician for help. If the physician eschews treatment (with drugs), the patient is likely to be disappointed (the activism value), and the physician becomes equally a moral entrepreneur, saying, in effect, "You were wrong to come to me. Pull yourself together and get your life in order." However correct that judgment may be, the personal burden on the physician can be enormous.

LOOK AHEAD

Our struggle to deal with the role of the minor tranquilizers, very real in its own right, is also part of several other social concerns, including the future role of the physician. Many of the issues that these drugs have caused to surface have yet to be resolved. Each of us will be involved in that resolution. Mary Davis has described the kind of problem we face:

The role of the physician, according to Webster, is to "treat disease"; a physician is one "skilled in the art of healing." This seems a clear enough statement, one which should occasion little controversy, and yet many people today are confused as to exactly what duties a physician is to perform. It is widely accepted that one's life-style and environment may be etiological factors in diseases such as ulcers and hypertension (as well, of course, as many psychiatric diseases). Does it then become the physician's responsibility to modify the life-style or environment of the patient? Each physician must, finally, answer this question for himself, deciding where lies the dividing line between medical practice and unwarranted interference in another's life. Society, however, has not yet drawn this line clearly; until it does so, physicians and patients alike will continue to be confused regarding the physician's role.[27]

It is an intentional oversimplification to describe the minor tranquilizers as drugs used to treat the symptoms of minor mental irregularity—not so different from a laxative, antacid, or cold tablet, except that they affect the mind. Although certainly not innocuous, they are also not as dramatically dangerous as, for example, lysergic acid diethylamide (LSD). Yet we have dealt with these drugs in an emotional, sometimes irrational, fashion. Kline said, "There seems to be an icebox built somewhere into our skulls which is the seat of cold logic and rational thought, and next to it is a hothouse of irrational, affect-based beliefs," and the two do not interact.[28] Indeed, that seems often to have been the case with our views of the minor tranquilizers.

But what next?

Nearly 15 years ago, Kline suggested that the year 2000 would see drugs that:

- Provide safe, short-acting intoxication
- Prolong or shorten memory
- Provoke or relieve guilt
- Control affect and aggression
- Shorten or extend experienced time

These and other predictions were not idle futuristic exercises but serious predictions of a knowledgeable scientist.

Surely it is time to review our performance with the comparatively simple drugs we have had for a quarter century or more to discover how better to deal with those that are to come, especially if pharmacists are to exercise a greater and more responsible role in the rational use of medication.

SUMMARY

The title of this chapter, "Determinants of Medication Use," is truly just a synonym for the more contemporary term, pharmacoepidemiology. This field is growing and evolving and offers great promise in addressing socioeconomic issues in medication therapy.

Pharmacoepidemiological research to date has been much more frequently aimed at the *nature* of medication use than at the determinants. The majority of publications are descriptive in nature: who takes or prescribes what and how often or how much. This is not a negative criticism. As in any kind of epidemiological study, incidence and prevalence data, over time, are a necessary prerequisite to studies of cause.

REFERENCES

1. Juergens J, Smith MC, Sharpe TR. Determinants of nonprescription laxative use among an elderly population. J Soc Admin Pharm 1984;2:174-9.

2. Gerbino PP, Gans JA. Antacids and laxatives for symptomatic relief in the elderly. J Am Geriatr Soc 1982;30(Suppl):581-7.

3. Cummings JH. Laxative abuse. Gut 1979;15:578.

4. Poe WD, Holloway DA. Drugs and the aged. New York: McGraw-Hill, 1980.

5. Eve SB, Friedsam HJ. Factors influencing older persons' use of nonprescription medicines for prevention and cure. Paper presented to the XII International Congress of Gerontology, Hamburg, 1981.

6. Stewart RB. Laxative use among an elderly population: a report from the Dunedin Program. Contemp Pharm Pract 1982;5:166-9.

7. Dichter E. Handbook of consumer motivations. New York: McGraw-Hill, 1964.

8. Anon. A study of health practice and opinions. Final report, FDA contract. Philadelphia: National Analysts, 1972.

9. Anderson R, Newman J. Societal and individual determinants of medical care utilization. Milbank Mem Fund Q 1973;51:91-124.

10. Verbrugge LM. Sex differences in legal drug use. J Soc Issues 1986;38:59-76.

11. Bargmann E, Wolfe S, Levin J. Stopping Valium. Washington, DC: The Public Citizen Health Research Group, 1982.

12. Shapiro S, Baron S. Prescriptions for psychotropic drugs in a non-institutional population. Public Health Rep 1961;76:481-8.

13. Linn L. Physician characteristics and attitudes toward legitimate use of psychotherapeutic drugs. J Health Soc Behav 1971;12:132-9.

14. Linn L, Davis M. Physicians' orientation toward the legitimacy of drug use and their preferred source of new drug information. Soc Sci Med 1972;6:199-203.

15. Williams P. Physical ill-health and psychotropic drug prescription–a review. Psych Med 1978;8:683-93.

16. Rucker TD. Production and prescribing of minor tranquilizers: a macro view. Paper presented to the American Orthopsychiatric Association, 1980.

17. Parry H. Use of psychotropic drugs in U.S. adults. Public Health Rep 1968;83:799-810.

18. Parry H, Manheimer D, Mellinger G, Balter M. National patterns of psychotherapeutic drug use. Arch Gen Psychiatry 1973;238:769-83.

19. Pflanz M, Basler H, Schwoon D. The use of tranquilizing drugs by a middle aged population in a West German city. J Health Soc Behav 1972;18:194-205.

20. Cooperstock R, Lennard H. Some social meanings of tranquilizer use. Soc Health Illness 1979;1:331-47.

21. Waldron I. Increased prescribing of Valium, Librium and other drugs. Int J Health Ser 1977;1:37-62.

22. Webb S, Colette J. Urban ecological and household correlates of stress–alleviative drug use. Am Behav Scientist 1975;18:750-70.

23. Radelet M. Health beliefs, social networks, and tranquilizer use. J Health Soc Behav 1981;22:165-73.

24. Friedson E. Profession of medicine. New York: Dodd, Mead, 1970.

25. Kline N. Antidepressant medications. JAMA 1976;227:1158-60.

26. Lewis D. The physician's contribution to the drug abuse scene. Tufts Med Alum Bull 1971;30:36-9.

27. Davis M. Disease and its treatment. Comp Psych 1977;18:231-3.

28. Kline N. Psychotropic drugs in the year 2000. Springfield, IL: Charles C. Thomas, 1971.

SUGGESTED READINGS

Predisposing Factors

1. Jackson JD, et al. An investigation of prescribed and nonprescribed medicine use behavior within the household context. Soc Sci Med 1982;16:2009-15.

2. Kaufert PA, Gilber P. The context of menopause: psychotropic drug use and menopausal status. Soc Sci Med 1986;23:747-55.

Enabling Factors

1. Juergens JP, Smith MC, Sharpe TR. Mothers' anxiety and medication use in rural families on welfare. Am J Hosp Pharm 1983;40:103-6.
2. Fincham JE, Wertheimer AI. Using the Health Belief Model to predict initial drug therapy. Soc Sci Med 1985;20:101-5.
3. Hadsall RS, Norwood GJ. Differences in drug usage patterns between Medicaid and private-pay population. Med Market Media 1977;12:52-7.

Need for Care Factors

1. Smith MC, Sharpe TR, Banahan BF. Patient response to symptoms. J Clin Hosp Pharm 1981;6:267-76.
2. Anon. Health care practices and perceptions. Washington, DC: Harry Heller Research Corporation (for the Proprietary Association), 1984.

General

1. Hepler CD, Segal R, Freeman RA. How physicians choose the drugs they prescribe. Top Hosp Pharm Manage 1981;1:23-44.
2. Bush PJ, Davidson FR. Medicines and "drugs": what do children think? Health Educ Q 1982;9:113-28.
3. Smith MC, Sharpe TR. Medication use among a non-institutionalized rural elderly population. J Clin Exper Gerontol 1984;6:207-30.
4. Verbrugge LM. Sex differences in legal drug use. J Soc Issues 1982;38:59-76.

ADDITIONAL READINGS

For the Instructor

1. Higginbotham N, Streiner DL. The social science contribution to pharmacoepidemiology. J Clin Epidemiol 1991;44:73S-82S. This paper is so conceptually rich that it could form the basis for an entire course. Students can also be profitably referred to the *Journal of Pharmacoepidemiology* and to the pharmacoepidemiology section of *DICP (Drug Intelligence and Clinical Pharmacy)*.

For the Student

1. Jackson JD, et al. An investigation of prescribed and nonprescribed medicine use behavior within the household context. Soc Sci Med 1982;16:2009-15.
2. Kaufert PA, Gilber P. The context of menopause: psychotropic drug use and menopausal status. Soc Sci Med 1986;23:747-55.
3. Juergens JP, Smith MC, Sharpe TR. Mothers' anxiety and medication use in rural families on welfare. Am J Hosp Pharm 1983;40:103-6.
4. Fincham JE, Wertheimer AI. Using the health belief model to predict initial drug therapy. Soc Sci Med 1985;20:101-5.

5. Hadsall RS, Norwood GJ. Differences in drug usage patterns between Medicaid and private-pay population. Med Market Media 1977;12:52-7.

6. Smith MC, Sharpe TR, Banahan BF. Patient response to symptoms. J Clin Hosp Pharm 1981;6:267-76.

7. Hutchinson RC, Lewis RK, Hatoum HT. Inconsistencies in the drug use process. DICP Ann Pharmacother 1990;24:633-6.

8. Lowy FH. Realities of drug use in society. Drug Safety 1990;5:155-9.

Chapter 13

Predicting and Detecting Noncompliance

Mickey C. Smith

INTRODUCTION

The whole idea of noncompliance with a prescription medication regimen simply flies in the face of logic or intuition. Why would anyone spend the time and money necessary to see a physician, obtain a diagnosis and prescription, and then either not have the prescription filled at all (initial noncompliance) or not take it properly and retrieve the authorized refills? The fact is that such behavior is widespread, and the explanation for noncompliance and its subsequent correction is one of the most important professional challenges in pharmacy practice.

NONCOMPLIANCE AS A BEHAVIORAL DISEASE

Scores of reasons for noncompliance have been identified through research. In many cases, it can be traced to failure to communicate by the physician and/or pharmacist. Often, however, the problem resides with the patient. When this is true, it may be a combination of ignorance, inability to understand or remember, sensory problems, economic barriers, psychological factors, or sometimes simple "contrariness."

Dr. Richard Levy of the National Pharmaceutical Council has proposed the analogy of noncompliance as a "behavioral disease." Use of a disease model could be helpful in understanding and "treating" noncompliance. Certainly it is a disease worthy of prevention or treatment, as some estimate the overall costs of noncompliance to be more than $100 billion annually.

Noncompliance with medications has many features of a disease:

- Various risk factors have been demonstrated for the condition.
- Depending on numerous patient- and disease-related variables, noncompliance is associated with important variations in severity, morbidity, and mortality.
- Triage is necessary to identify those patients in greatest need for treatment for noncompliance.
- Iatrogenic noncompliance is an important aspect of this "disease," and validated screens are available to identify physicians who need to improve their communication skills.
- Some cases of noncompliance are "curable," but some are probably not.
- Noncompliance is a public health problem and, accordingly, prevention is a goal.

Patients having the noncompliance disease can be detected by general screening using validated demographic, sociographic, and psychographic predictors and questionnaires. The behavior of noncompliant patients and iatrogenic physicians, identified by these means, can be subjected to detailed examination. Differential diagnosis of noncompliant behavior in patients can be performed by computerized monitoring of refill patterns and more specifically by analysis of daily pill-taking behavior using microelectronic monitoring devices.

Interventions to treat the disease of noncompliance, like those to treat any other illness, must be tailored to the needs and circumstances of the individual patient and should be based on underlying causes. Specific interventions with validated effectiveness can be selected. Noncompliance with treatment for chronic disease is itself a chronic disease and needs sustained or periodic attention at appropriate intervals.

Importantly, interventions to improve compliance must be subjected to cost-benefit analysis to determine if and to what extent the cost savings resulting from improved compliance exceed the costs of the program. Outcomes studies are required to assess the effect of interventions not only on compliance behavior but also on the overall health of the patient and total treatment costs. A scientific approach to the noncompliance disease will require well-developed information on each of the dimensions discussed above: risk factors and predictors, differential diagnosis in the individual patient and doctor, effective and specific interventions to improve compliance, and cost-benefit analysis and outcomes studies.

Logic suggests that the most efficient approach to the compliance problem is to predict accurately those patients at highest risk and then to use the most specific and effective means to prevent this behavior. Choice of intervention should be matched to the type of illness, personality type, and social circumstances of the patient. This is a critical challenge to individual pharmacists and to pharmacy as a profession, but it is well worth the effort. It is far more efficient to focus our compliance-enhancing efforts on the minority of patients most at risk than to waste resources on one or more shotgun approaches.

So how do we identify those most likely to be noncompliant or predict noncompliant behavior?

PREDICTING NONCOMPLIANCE

The good news is that a great deal of study has been devoted to noncompliance and much has been learned. That is also the bad news! Many studies have been published, but no consensus has been achieved. As a consequence, interventions have resulted in mixed success. Nevertheless, some general risk factors for noncompliance have emerged.

More than 250 social, economic, medical, and behavioral factors have been found to affect compliance.[1] Studies to date on determinants of noncompliance have addressed the following types of variables:

- Demographic, such as gender, age, income.
- Sociographic, such as family stability, support, and size.
- Psychographic, such as attitudes, self-image, and locus of control.

Some of these patient characteristics may be disease-specific (e.g., those affecting cognitive ability), and some are subject to change by professional interventions. But all have value in alerting the concerned pharmacist to the potential for noncompliance. It is not far-fetched to suggest that a compliance profile should be a part of every patient's medical and pharmacy record. "Compliance risk" might very well appear right next to "allergic to penicillin," for this is truly an important factor in both the selection of the medication and the likelihood of a favorable outcome. Before that step is taken, however, much more progress in integrating our knowledge of compliance science and patient behavior will be necessary.

The need for additional synthesis for research results is demonstrated by the information in Table 13.1 which illustrates the diversity of factors shown in a selection of recent reports to be predictive of good or poor compliance. Both patient-related and physician-related predictors of compliance are reported.

For the pharmacist, the information derived from research and selected for inclusion in Table 13.1 may be interesting, but it hardly points to any action plan. For example, in the study of children with recent onset of diabetes, such factors as self-esteem, perceived competence, and social functioning were found to be reasonably good predictors. Obviously, it is not practical for pharmacists to make assessments such as these (although the information may be available in certain clinic and institutional settings). So what can *pharmacists* do to predict noncompliant behavior? Some have suggested that a measure of potential for noncompliance be included in initial contacts with patients. A short questionnaire taking only a few minutes to complete in the pharmacy would enable the motivated pharmacist to take the steps necessary to encourage compliance.

JUST ASK THEM!
(IN AN ORGANIZED WAY, OF COURSE)

General factors associated with poor compliance, discovered through research or incorporated into models, have only limited utility in identifying the noncompliant patient (e.g., a patient with a complex medication regimen probably warrants special attention). But if efficiency is to be built into programs to improve compliance, some means of predicting the likelihood of compliance in an individual patient is necessary. Fortunately, some progress has been

TABLE 13.1. Predictors of Compliance

Conditions/Medications	Predictors	Reference
Diabetes, hypertension, pulmonary disease	Patient satisfaction with care	Nagy and Wolfe, 1984 [2]
Diabetes	Belief in own ability to follow regimen	McCaul, Glascow, and Schafer, 1987 [3]
Diabetes (pediatric)	High self-esteem, perceived competence	Jacobson et al., 1987 [4]
Hypertension	Expectancy of internal control of health, medication knowledge, social support	Stanton, 1987 [5]
Antibiotics (short-term)	Medication knowledge, dose schedule	Cockburn et al., 1987 [6]
Influenza vaccination (elderly)	Worry about effects of injection results in noncompliance	Carter et al., 1986 [7]
Acne medication (OTC)	Initial health beliefs and social anxiety	Flanders and McNamara, 1984 [8]
Contraception (adolescent)	Self-pay, frequent intercourse	Litt and Cuskey, 1980 [9]
Hypertension	Positive: physicians who build patient confidence, or "medicalize" the condition. Negative: physicians who overdramatize the condition	Consoli and Safar, 1988 [10]
HIV seropositive	Professional and social support, distance from treatment, lack of isolation	Morse et al., 1991 [11]
Asthma	Ability of physician to speak patient's language (Spanish)	Manson, 1988 [12]

made in that direction. And, as sometimes happens, the solution may be surprisingly simple: ask the patient if she is a good complier.

Moriskey and colleagues have reported success in the use of a four-item scale as a predictor of future medication compliance as well as a measure of current compliance.[13] The scale was shown to be reliable and a valid predictor of compliance and blood pressure control. The four questions in the scale could be quickly administered to most patients:

1. Do you ever forget to take your medicine?
2. Are you careless at times about taking your medicine?
3. When you feel better do you sometimes stop taking your medicine?
4. Sometimes if you feel worse when you take the medicine, do you stop taking it?

Hogan and colleagues developed a 30-item scale that tested well on unidimensionality and reliability.[14] Use of the scale accurately assigned 89% of a sample of 150 schizophrenic patients to compliant and noncompliant groupings. The authors provided considerable food for thought in their speculation that compliers and noncompliers may differ in their awareness of internal body sensations, attitudes, feelings, and emotions. Thus, strategies designed merely to inform may be ineffective with patients who are unaware of these internal cues.

This particular scale focused on the experience with the medication, and a careful analysis showed that how the patient *feels* on the medication, rather than what the patient knows or believes about the medication, is a better predictor of compliance. For pharmacists, this suggests that a simple, "How are you doing on this medicine?" may be a very important question.

The ease of self-assessment is underscored by the work of Dr. Iris Litt, a physician with a research program in compliance. She used the two simple scales shown in Tables 13.2 and 13.3 to predict compliance with an oral contraceptive regimen for a period of six months. Use of the Post Medication Compliance Scale (Table 13.2)

TABLE 13.2. Four-Point Self-Assessment of Past Medication Compliance

Which of the following best describes you:
1. "I take my medicines if I feel that I need them. It varies from day to day."
2. "I sometimes go several days without taking my medicine because I forgot or am very busy."
3. "I rarely miss taking my medicines."
4. "I *never* forget to take my medicines."

Source: Litt, 1985[15]

TABLE 13.3. Vignettes on Compliance: Nancy and Sue

Nancy and Sue think it is important to take care not to get pregnant. But Nancy is *forgetful*, gets too involved in other things, and finds it is very difficult. By contrast, Sue is very *careful* and *systematic* about taking her pill or doing whatever else is required. She never forgets.

Whom are you like? CIRCLE ONE NUMBER.

1. Exactly or a lot like Nancy
2. Somewhat like Nancy
3. Halfway between Nancy and Sue
4. Somewhat like Sue
5. Exactly or a lot like Sue

Source: Litt, 1985[15]

accurately predicted compliance behavior 75% of the time. Use of the vignettes also proved helpful, as 64% of the compliant patients had rated themselves as "like Sue."

Just a few additional examples should illustrate that simple, but carefully planned, patient inquiry can be a significant help to the pharmacist in identifying patients most at risk of noncompliant behavior and the compromised outcomes that accompany it.

- Cromer and colleagues studied adolescents and their iron therapy (Stuartinic®) and found only 67% compliant.[16] Among the factors predicting compliance was patient's prediction of his/her own compliance at first visit.
- Edwards and Pathy found that a "good" score on the Stockton Geriatric Rating Scale was highly correlated with "good" compliance, additional evidence of the value of paper and pencil measures.[17]

THE HEALTH BELIEF MODEL CAN HELP

A frequently used, intuitively attractive, validated framework for understanding and predicting compliance behavior has been the Health Belief Model. This simple but powerful concept proposes that compliance is a function of:

- How serious and how likely are the consequences of noncompliance as perceived by the patient?
- How likely is it that something bad will happen, and how bad would it be?
- How beneficial will compliance be, and what real or perceived barriers to compliance exist?
- How much better will I be if I take my medicine, and is that worth the cost or risk?

In addition, the model acknowledges the importance of social, psychological, economic, and structural factors in determining compliance. It is, of course, the interaction of all of these variables that results in the behavior of an individual patient.

Using the Health Belief Model, Fincham and Wertheimer studied the phenomenon of initial drug therapy defaulting (never picking up the prescription) in an HMO population.[18] They analyzed factors predictive of compliance with medications and found more than a dozen to be significant. The two most important, however, were (1) lack of belief by patients in the benefits of care and (2) patients' acknowledgement that they often did not receive needed information on how to take newly prescribed medicines.

DETECTING NONCOMPLIANCE

Surely the reader will agree that it is far better to predict and then prevent noncompliant behavior by the patient. When this is not possible or when interventions fail, it is desirable to detect noncompliance. Indeed, there are specific situations where compliance monitoring goes beyond the best interests of the patient—tuberculosis treatment, alcoholism treatment, psychiatric outpatient programs.

Because of the keen interest in assessing levels of compliance, we will discuss some of the methods available. A list of the more popular methods is included as Table 13.4 and discussed in the following sections.

Indirect Methods

Self-reports and interviews with patients are the most common and simplest methods of attempting to determine compliance with

TABLE 13.4. Methods of Assessing Compliance

Methods	Comments
Indirect Assessments	
• Patient self-reports	Patients may lie or forget
• Therapeutic outcome	Only accurate if therapy itself is effective and correlated highly with compliance
• Physician estimate	Usually inaccurate, biased toward compliance
• Pharmacist estimate	Accurate for timeliness of refills, little basis for judgment otherwise
• Pill and bottle counts	Subject to patient falsification, do not reflect day-to-day variations
• Mechanical monitors (record of number and time sequence when bottles are opened)	Do not necessarily measure actual use
• Family interviews	Reasonably reliable if family member is involved in the therapy
Direct Assessments	
• Blood/serum assays	Accurate as regards blood levels, but assays not available for all drugs; invasive, time-consuming, expensive
• Urine assays	Time of collection important; must know the absorption and excretion pattern of the drug
• Tracer/marker methods	Usually used in urine assays; if added to marketed drugs, tracer requires FDA approval. Use of tracers in a portion of doses may allow more accurate measure.

therapy. However, studies have demonstrated that even the most skilled and highly refined interviewing techniques substantially overestimate medication compliance. Even Hippocrates remarked, "Patients often lie when they state they have taken certain medicines." Spot checks at the doctor's office are simply unreliable. Because the visit serves as a reminder (or threat) that enhances compliance for the previsit interval, the spot check at that visit would give fallaciously high results, inaccurate for the time period

between visits. Urquhart has referred to this temporary enhancement of compliance as the toothbrush effect (people are particularly
likely to brush their teeth before going to the dentist).[19] For those
who do not like this analogy, an alternative title might be white-coat
compliance, derived from the recent identification of white-coat
hypertension as blood pressure that rises in the doctor's presence.[20]

Diaries and various kinds of charts and records have been used to
track compliance. In these cases, the patient must remember not
only to take his or her medicine, but also to record the event,
introducing yet another potential error. On the other hand, use of
these techniques may have the effect of reminding some patients to
take their medicine. Thus, these methods may not be without value
in enhancing compliance, but their validity as a measure of compliance is highly questionable.

Pill counts are another common method used to measure compliance and are frequently used in clinical drug studies. A patient's
adherence to a medication regimen can be assessed by the difference between the number of dosage units initially dispensed and the
number remaining in the container on the patient's return visit or
during an unscheduled home visit. However, pill dumping (attempts
by patients to misrepresent their compliance by discarding medication) is common, and several studies have shown that return counts
grossly overestimate actual compliance rates. It is generally held
that pill counts no longer represent an effective measure of compliance.

Patient records are another source of compliance information. A
prescription refilled ten days after the patient should have run out of
medicine is certain evidence of some kind of noncompliance.
Tracking of initial filling of prescriptions is also possible using
copies generated when the prescription is issued. This technique
can be difficult and costly unless the delivery system is a closed one
(e.g., an HMO).

One device used to monitor compliance with solid dose therapy
used two clear plastic blister sheets, each containing 21 blisters
filled with the patient's medications. A self-adhering paper with
loops of conductive "wires" in the same pattern as the blisters was
then placed over the open face of each of the sheets of blisters to
form blister packs. The packs were connected to electronic compo-

nents and placed inside an easily opened plastic case. Every 15 minutes, the battery-operated electronic memory sent an impulse through each loop of conductive material. If a dose of medication had been removed (that is, if the paper covering a blister was torn), the electrical impulse failed to return to the electronic memory and the 15-minute interval during which this occurred was recorded. After the patient returned the compliance monitor, the data were collected with a microcomputer. Again, it is not possible to be sure that a patient actually ingests medication, but such a device provides insights into the timing of patients' actions. It can also detect white-coat compliance.

Direct Assessments

Assays of the patient's body fluids and similar techniques have the advantage of objectivity and reasonable accuracy, at least as regards the tests themselves. In some situations, such tests offer the best means for compliance assessment; they are not foolproof, however. Biological markers and tracer compounds indicate patient compliance over an extended period. Tracer compounds—small amounts of agents such as phenobarbital and digoxin—can be added to drugs and measured in biological fluids as pharmacological indicators of compliance. Both phenobarbital and digoxin have long half-lives and thus have the advantage of indicating compliance for the preceding few weeks rather than days. With either drug, good compliance cannot be simulated by ingesting a few doses immediately before assessment. In addition, phenobarbital has been shown to display relatively little interindividual and intraindividual variation in plasma concentration after allowances are made for age and weight. Other tracers used have included bromide, riboflavin, phenol red, and radioactive isotopes.

Compliance has also been measured through determination of drug concentrations in patients' biological fluids. However, the usefulness of data on drug concentrations in biological fluids is limited because:

1. Concentrations of drugs are affected by individual differences in absorption, distribution, metabolism, and excretion, and low

or erratic drug concentrations are not necessarily an indication of noncompliance.

2. Drug concentrations do not provide data on the timing of doses consumed.
3. Brief intake of rapidly cleared drugs before testing can produce results that show adequate drug concentrations, erroneously suggesting regular medication use.

Backes and Schentag have described some of the pitfalls in relying too heavily on pharmacokinetic methods of compliance assessment:[21]

Several different types of noncompliance exist. The drug concentrations used to assess compliance may be affected differently by different types of noncompliance. Practitioners should be aware of several possible patient scenarios that may occur in which patient noncompliance may affect pharmacokinetics. A patient may be "consistently" noncompliant (i.e., they may invariably eliminate a midday dose). This may result in a subtherapeutic plasma concentration, which would prompt a physician to increase the patient's dose. If the patient remains consistently noncompliant and continues to miss one daily dose, he or she may achieve a new serum concentration in the therapeutic range because the new dose now compensates precisely for his or her noncompliance. If, however, the patient suddenly becomes fully compliant with the prescribed dose, the plasma concentration may increase and approach toxicity. If the patient were receiving a medicine with nonlinear kinetics (e.g., phenytoin), a slight increase in prescribed dose in addition to a sudden increase in compliance may result in a dramatic increase in serum drug concentration and cause toxicity. As the reason for the sudden onset of toxicity seldom will be clear, the change in compliance is not usually considered as the cause of the problem.

Examples of drugs where biochemical methods of monitoring compliance have been used are shown in Table 13.5.

Computerized or electronic monitoring devices are the most recent—and in many ways most reliable—methods of assessing com-

TABLE 13.5. Assessing Compliance by Biochemical Means

Medication	Methods Used
Acetylsalicylic acid	Biochemical—urine
Antidepressants	Plasma concentrations
Aminosalicylic acid	Biochemical—urine
Atropine	Biochemical—urine
Barbiturates	Biochemical—serum
Digoxin	Biochemical—serum
Ethosuximide	Biochemical—serum, plasma
Flufenamic acid	Fluorescence—urinary metabolites
Imipramine	Biochemical—urine, plasma
Iron	Biochemical—stool
Isoniazid	Biochemical—urine
Mefenamic acid	Fluorescence—urinary metabolites
Mesantoin	Biochemical—serum
Paracetamol	Biochemical—urine
Para-aminosalicylic acid and/or isoniazid	Urine assay and/or isoniazid
Penicillin	Urine assay
Phenobarbital	Biochemical—serum
Phenothiazines	Biochemical—urine
Phenylbutazone	Biochemical—urinary metabolites
Phenytoin	Biochemical—serum, plasma
	Gas or liquid chromatography—serum
	Biochemical—urinary metabolite
Rifampicin	Bioassay—urine
Thiazides	Urine assay

Sources: Backes and Schentag, 1991[21]; Sackett, 1976[22]

pliance. A list of some of the better-known devices is given in Table 13.6. An example of the value of electronic monitoring is provided by the data in Table 13.7.

John Urquhart, developer of the device used in this study, has provided insightful comments on the value of the information provided in this example:

> The record is quite different from the once-daily regimen prescribed. There were episodes of acute pulmonary congestion on the second and third Sundays, both requiring emergency care, and both immediately preceded by two days without any record of medicine having been taken. The electronic

record showed that the medication event on the third Sunday occurred after the patient called for the ambulance.

The patient made these choices without asking or telling anyone. Her history at the time revealed nothing of this, so her doctors thought that her heart failure had gotten worse and treated her accordingly. Her doctors did not have this record at the time they were treating her. Today they could, but this was one of the very first uses of the electronic dosing monitors, and at that time it required a special computer to retrieve the data. Thus, her dosing record became evident only later.

What do we learn from it? The data tell us that the patient took the prescribed number of doses–actually one extra–but with very poor dose-timing. The patient was well-managed except when dosing lapsed for more than two days–a conclusion consistent with prior knowledge about diuretic actions in patients with heart failure. But how should we answer the question: Was the medicine effective?

We see two serious episodes of fluid retention in a patient prescribed a medicine meant to prevent fluid retention and its complications. If we didn't know the dosing record, *we would conclude that the medicine was ineffective* [emphasis added]. But we do know the dosing record. If we consider only the number of doses she took, we would also say that the medicine was *effective*. The data on dose-timing show the sequence of lapsed dosing and then clinical signs of fluid retention. The time sequence supports the conclusion that the medicine was effective, when taken. It also suggests the hypothesis that one day without dosing is forgiven, but two or more days without dosing leads to serious trouble.[19]

The device used in this example, the MEMS (APREX Corporation, Fremont, California) is an adaptation of a U.S. standard polypropylene vial. Within a childproof closure is a microelectronic circuit for recording day and time of each opening and closing. Data are retrieved by connecting the medication container to a microcomputer communication port. Openings of the medication container are recorded as presumptive doses. Compliance data are obtained as lists of the dates and times of individual container

TABLE 13.6. Electronic Monitoring Devices

Device/Developer	Use
Unit Dose Monitor DVA Medical Center St. Louis, Missouri	Records removal of tablets from blister pack
Eye Drop Monitor Washington University St. Louis, Missouri	Records inversion of bottle to dispense liquid medication
Nebulizer Chronolog Advanced Technology Products Lakewood, Colorado	Records spray of aerosol medication
Pill Box Monitor Clinical Research Center Harrow, United Kingdom	Records opening of box to remove tablets or capsules
Medication Event Monitoring System (MEMS) APREX Corp. Fremont, California	Records opening of bottle for removal of tablets or capsules
MEMS Reader Display-Printer (RDP) APREX Corp. Fremont, California	Dosing data retrieval, display and printout from MEMS units
Pill Ring Monitor APREX Corp. Fremont, California and Johnson & Johnson Co. New Brunswick, New Jersey	Records removal of each oral contraceptive tablet
Glucometer Ames/Miles Elkhart, Indiana	Records time of test and glucose level
Mini Doc Professional Medication Consultants-AB Uppsala, Sweden	Interactive electronic diary for notation of events by patient

openings and closings, the duration of openings, and the time since the previous opening. From the time pattern of openings, information is yielded about patients' compliance with the prescribed daily drug regimen. Compliance is defined as openings recorded during the period divided by prescribed number of doses during the period

TABLE 13.7. Variable Dosing with Once-Daily Diuretic by a Patient with Moderately Severe Congestive Heart Failure*

Monday	Tuesday	Wednesday	Thursday	Friday	Saturday	Sunday
3	1	1	0	2	1	1
2	1	2	1	0	0	0**
4	1	1	2	0	0	1**
2	0	1	1	1	0	1

* Each number represents the number of daily medication events, measured by an electronically monitored medicine container that records time and date each time the container is opened and closed. On the assumption that each medication event (a cycle of opening and closing of the container) represents a dose taken, the numbers indicate the number of doses of a prescribed once-daily diuretic (hydrochlorothiazide plus triamterene) taken each day by the patient. Each successive row of numbers is a successive week.

** The asterisks on the second and third Sundays of the four-week period indicate episodes of acute pulmonary congestion and dyspnea requiring emergency treatment.

Source: Urquhart, 1992.[19] Reprinted with permission.

multiplied by 100 and expressed as a percentage. Even here, of course, there is no assurance that the patient actually took the medication.

In an ideal situation, it would be possible to measure medication intake *directly* so that a patient would be considered compliant only if a dose were taken within the time interval specified by the drug regimen. This specificity of measurement has been performed in only one published account. Norell, in studying compliance of glaucoma patients with eye drops, developed an automatic eyedropper that provided continuous measurement of when and how many times the dropper was used.[23]

The Ideal Detection Method

The ideal detection method would measure compliance at the time and place of the medication-taking (or other treatment) event and, therefore, would possess perfect sensitivity (the proportion of patients with imperfect compliance identified by the measure) and specificity (the proportion of patients with perfect compliance identified by the measure). Although direct observation of the patient would come closest to satisfying this definition, this method, of course, is not practical.

The noncompliant patient can be a patient who does not take a single tablet, takes part of his or her medication each day, or stops treatment for a week or two. Usually, this information is left unspecified in the literature. The patient thinks that, when asked by the physician, he or she cannot answer, "I did not take the medication as directed," without damaging the physician's confidence in him or her. With this approach of simple questioning, therefore, we overestimate compliance, even when questioning is noninquisitorial.

The last word is not in on the value of patient self-reports as a measure of noncompliance. Patients *do* sometimes lie, and they frequently forget. But that does not mean that self-reports of compliance have no value. One study involving adolescents and their compliance with oral contraceptive regimens found that self-reports were consistent 90% of the time with serum tests in detection of compliance levels.[24]

In a study in England, the patient was asked, simply, "Many people have difficulty taking medications. Did you ever miss your digoxin?" Use of serum levels indicated a high degree of reliability from responses to this simple inquiry.[25] Yet another study found patients' own estimates to be superior even to biochemical testing in predicting compliance. (Of course, pill counts have their own reliability shortcomings.)

WHY ALL THE FUSS ABOUT COMPLIANCE?

An advertisement that appeared in a medical journal shows a physician observing, "When my patients don't return, I assume the therapy is working." On the facing page, one of those patients says, "I couldn't tell my doctor his migraine therapy didn't work." The text of the ad cites data indicating that nearly half of all migraine sufferers have given up on their physicians either because of failure to improve or because of side effects of the medication prescribed. This example is illustrative of the complexity, subtlety, and impor-

tance of the compliance problem. Among the potential consequences of this particular situation with migraine are:

- Physician misjudgment of the effectiveness of the prescribed
  therapy, in this case probably resulting in repeats of this scenario with the next medication prescribed for this patient.
- Loss of income to the pharmacist.
- Loss of confidence by the patient in the efficacy of medications and perhaps in the skill of the physician.
- Continued migraine attacks with continued erosion of the
  patient's quality of life.
- Loss of patient productivity. (One estimate, cited in the ad,
  was that annual lost productivity from migraine attacks fell in
  the $6-17 billion range.)
- Cost of other therapies, including OTC medicines, used by the
  patient to "try to cope," but to no avail.

There are literally hundreds of reports describing the rates of
noncompliance among patients with various medical conditions. A
sampling of chronic conditions and the consequences of noncompliance are shown in Table 13.8.

In all of these studies, the noncompliance behavior studied was
undercompliance, i.e., the process of taking a medication at a level
or for a duration less than that intended by the prescriber. Of course,

TABLE 13.8. Rates and Possible Consequences of Noncompliance with
Medication Regimens for Important Diseases

Condition	Rate of Compliance	Possible Consequences
Epilepsy	30-50%	Relapse
Arthritis	55-71%	Condition worsens
Hypertension	40%	Hospitalization
Diabetes	40-50%	Loss of control
Contraception (pill)	8%	Unwanted pregnancy
Asthma	20%	Attacks, hospitalization (?)
Alcoholism	48-56%	Relapse, hospitalization
Organ transplant	18%	Rejection, death
Anticoagulants	30%	Bleeding, hospitalization
Estrogen deficiency	57%	Symptoms, osteoporosis

for patients taking chronic medications, this practice eventually results in fewer refills.

Economic Consequences

Some of the most important economic figures associated with drug noncompliance are those reflecting the consequences in terms of increased hospital admissions. Only a few of these can be cited here.

- McEvoy and colleagues compared groups of noncompliant and compliant relapsed–and therefore hospitalized–schizophrenic patients. The noncompliant patients were found to have had a gradual onset of the determinant episode, to have been committed involuntarily, and to remain in the hospital longer.[26]
- In an Israeli study, Levy and colleagues identified 2.9% of nearly 1,200 hospital admissions as having been principally caused by noncompliance.[27]
- Kelly and Scott described a project to improve medication compliance among a group of outpatients with chronic mental disorders.[28] Compliance did improve, and at the end of six months, 33% of the better compliers were in the hospital compared with 45% of a control group whose compliance did not improve.
- Col and colleagues reviewed the records of and interviewed 315 consecutive elderly patients admitted to an acute-care hospital.[29] They determined that 11.4% of the admissions could be traced directly to some form of noncompliance. Total cost of these admissions was $77,000 ($2,150 each).
- Green used a retrospective chart review to compare community mental health center patients having three or more hospitalizations in an 18-month period with a matched group of patients without such hospitalizations.[30] Noncompliance with medications was associated with frequent hospitalization in 92% of the patients.

Some of the most dramatic data on the costs of noncompliance among the elderly come from Oregon. In this study by Strandberg, functional assessment profiles (ability to take care of oneself) of nursing home residents and of people successfully living at home were compared.[31] The single characteristic that best distinguished individuals who were in a nursing home from those who were not was found to be managing medications. It was more important than the actual health condition. Indeed, 60% of those placed at extreme risk of nursing home placement had no equally serious impairment other than their inability to manage their own medication. A more conservative view, removing other impairments that were lower on the risk scale, still left nearly 23% who had no high-risk problem other than medication management.

Much more research on this issue is urgently needed, but if even 10% of nursing home admissions are related to compliance problems, this represents an annual cost of at least $5 billion (based on a national figure of $50 billion annually for nursing home care).

CONCLUSION

Despite the enormity and complexity of the noncompliance problem, numerous studies have suggested remedies that may be operable in the real world. Additionally, the high cost of noncompliance is beginning to attract the attention of various stakeholders who wish to improve compliance rates.

Pharmacists are in the best possible position to predict, prevent, detect, and correct noncompliance problems. (Another chapter of this book deals with prevention and correction.) Given the medical and economic consequences of compliance deficits, the medical, professional, and financial incentives to take the initiative should be obvious.

REFERENCES

1. Fincham JE, Wertheimer AI. Correlates and predictors of self-medication attitudes of initial drug therapy defaulters. J Soc Admin Pharm 1985;3:10-7.

2. Nagy VT, Wolfe GR. Cognitive predictors of compliance in chronic disease patients. Med Care 1984;22:912-21.

3. McCaul KD, Glasgow RE, Schafer LC. Diabetes regimen behaviors: predicting adherence. Med Care 1987;25:868-81.

4. Jacobson AM, Hauser ST, Wolfsdorf JI, Houlihan J, Milley JE, Herskowitz RD. Psychologic predictors of compliance in children with recent onset of diabetes mellitus. J Pediatr 1987;110:805-11.

5. Stanton AL. Determinants of adherence to medical regimens by hypertensive patients. J Behav Med 1987;10:377-94.

6. Cockburn J, Gibberd RW, Reid AL, Sanson-Fisher RW. Determinants of noncompliance with short term antibiotic regimens. Br Med J 1987;295:814-8.

7. Carter WB, Beach LR, Inui TS, Kirscht JP, Prodzinski JC. Developing and testing a decision model for predicting influenza vaccination compliance. Health Serv Res 1986;20:897-932.

8. Flanders P, McNamara JR. Prediction of compliance with an over-the-counter acne medication. J Psychol 1984;118:31-6.

9. Litt IF, Cuskey WR. Compliance with medical regimens during adolescence. Pediatr Clin North Am 1980;27:3-15.

10. Consoli SM, Safar ME. Predictive value of the patient's psychological profile and the type of patient/practitioner relationship in compliance with antihypertensive treatment. Arch Mal Coeur 1988;81(Supp HTA):145-50.

11. Morse EV, Simon PM, Coburn M, Hyslop N, Greenspan D, Balson PM. Determinants of subject compliance within an experimental anti-HIV drug protocol. Soc Sci Med 1991;32:1161-7.

12. Manson A. Language concordance as a determinant of patient compliance and emergency room use in patients with asthma. Med Care 1988;26:1119-28.

13. Moriskey DE, Green LW, Levine DM. Concurrent and predictive validity of a self-reported measure of medication adherence. Med Care 1986;24:67-74.

14. Hogan TP, Awad AG, Eastwood R. A self-report scale predictive of drug compliance in schizophrenics: reliability and discriminative validity. Psychol Med 1983;13:177-83.

15. Litt IF. Know thyself–adolescents' self-assessment of compliance behavior. Pediatrics 1985;75:693-6.

16. Cromer BA, Steinberg K, Gardner L, Thornton D, Shannon B. Psychosocial determinants of compliance in adolescents with iron deficiency. Am J Dis Child 1989;143:55-8.

17. Edwards M, Pathy MSJ. Drug counseling in the elderly and predicting compliance. Practitioner 1984;228:291-300.

18. Fincham JE, Wertheimer AI. Using the Health Belief Model to predict initial drug therapy defaulting. Soc Sci Med 1985;20:101-5.

19. Urquhart J. Time to take our medicines, seriously. Maastricht, The Netherlands: University of Limburg, 1992.

20. Feinstein AR. On white-coat effects and the electronic monitoring of compliance. Arch Intern Med 1990;150:1377-8.

21. Backes JM, JJ Schentag. Partial compliance as a source of variance in pharmacokinetics and therapeutic drug monitoring. In: Cramer JA, Spilker B, eds. Patient compliance in medical practice and clinical trials. New York: Raven Press, 1991.

22. Sackett DL. The magnitude of compliance and noncompliance. In: Sackett DL, Haynes RB. Compliance with therapeutic regimens. Baltimore, MD: Johns Hopkins University Press, 1976:9-25.

23. Norell SE. Monitoring compliance with pilocarpine therapy. Am J Oph-thalmol 1981;92:727-31.

24. Jay MS, DuRant RH, Shoffitt T, Linder CW, Litt IF. Effect of peer counsel-ors on adolescent compliance in use of oral contraceptives. Pediatrics 1984;73: 126-31.

25. Gilbert JR, Evans CE, Haynes RB, Tugwell P. Predicting compliance with a regimen of digoxin therapy in family practice. Can Med Assoc J 1980;123: 119-22.

26. McEvoy JP, Howe AC, Hogarty GE. Differences in the nature of relapse and subsequent inpatient course between medication-compliant and noncompliant schizophrenic patients. J Nervous Mental Dis 1984;172:412-7.

27. Levy M, Mermelstein L, Hemo D. Medical admissions due to noncom-pliance with drug therapy. Int J Clin Pharmacol Ther Toxicol 1982;20:600-4.

28. Kelly GR, Scott JE. Medication compliance and health education among outpatients with chronic mental disorders. Med Care 1990;28:1181-97.

29. Col N, Fanale JE, Kronholm P. The role of medication noncompliance and adverse drug reactions in hospitalizations of the elderly. Arch Intern Med 1990;150:841-8.

30. Green JH. Frequent rehospitalization and noncompliance with treatment. Hosp Community Psychiatry 1988;39:963-6.

31. Strandberg LR. Drugs as a reason for nursing home admissions. Am Health Care Assoc J 1984;10(4):20.

ADDITIONAL READINGS

For the Instructor

1. American Association of Retired Persons. Prescription drugs: a survey of consumer use, attitudes and behavior. Washington, DC: AARP, 1984.

2. American College of Physicians. Therapeutic substitution and formulary systems. Ann Intern Med 1990;113:160-3.

3. American College of Physicians. Physicians and the pharmaceutical indus-try. Ann Intern Med 1990;112:624-6.

4. Bauman AE, Crain AR, Dunsmore J, Browne G, Allen DH, Vanderberg R. Removing barriers to effective self-management of asthma. Patient Educ Counsel 1989;14:217-26.

5. Barofski I, Sugarbaker PH. Determinants of patient nonparticipation in randomized clinical trials for the treatment of carcinoma. Cancer Clin Trials 1979;2:237-46.

6. Barth RT, Vertinsky P, Yang C-F. Some sociobehavioral and other determi-nants of compliance: a voluntary health service campaign. Hum Relations 1979;32:781-92.

7. Bloom BS. The medical, social, and economic implications of disease. In: van Eimeren W, Horisberger B, eds. Socioeconomic evaluation of drug therapy. New York: Springer-Verlag, 1988.

8. Bolton MB, Tilley BC, Kuder J, Reeves T, Schultz LR. The cost and effectiveness of an education program for adults who have asthma. J Gen Intern Med 1991;6:401-7.

9. Branche GC, Batts JM, Dowdy VM, Field LS, Francis CK. Improving compliance in an inner-city hypertensive patient population. Am J Med 1991; 91(1A):37S-41S.

10. Brittain E, Wittes J. Factorial designs in clinical trials: the effects on noncompliance and subadditivity. Stat Med 1989;8:161-71.

11. Clark LT. Improving compliance and increasing control of hypertension: needs of special hypertensive populations. Am Heart J 1991:664-9.

12. Cram DL Jr, Maesner AT, Witmore DM. Medication refill clinics: the Veterans Administration medical center experience. J Pharm Pract 1992;5(1):12-21.

13. Elixhauser A, Eisen SA, Romeis JC, Homan SM. The effects of monitoring and feedback on compliance. Med Care 1990;28:882-93.

14. Fincham JE, Wertheimer AI. Factors affecting self-medication attitudes of initially noncompliant patients in a health maintenance organization. Soc Pharmacol 1988;2:309-26.

15. Fincham JE, Wertheimer AI. Correlates and predictors of self-medication attitudes of initial drug therapy defaulters. J Soc Admin Pharm 1985;3:10-7.

16. Flanders P, McNamara JR. Prediction of compliance with an over-the-counter acne medication. J Psychol 1984;118:31-6.

17. Forstrom MJ, Ried LD, Stergachis AS, Corliss DA. Effect of a clinical pharmacist program on the cost of hypertension treatment in an HMO family practice clinic. Ann Pharmacother 1990;24:304-9.

18. Friedman IM, Litt IF, King DR, Henson R, Holtzman D, Halverson D. Compliance with anticonvulsant therapy by epileptic youth. J Adolesc Health Care 1987;7(1):12-7.

19. Garrity TF, Lawson EJ. Patient-physician communication as a determinant of medication misuse in older, minority women. J Drug Issues 1989;19:245-59.

20. Gilbert JR, Evans CE, Haynes RB, Tugwell P. Predicting compliance with a regimen of digoxin therapy in family practice. Can Med Assoc J 1980;123: 119-22.

21. Glik DC, Steadman MS, Michels PJ, Mallin R. Antihypertensive regimen and quality of life in a disadvantaged population. J Fam Pract 1990;30:143-52.

22. Goldberg SC, Schooler NR, Hogarty GE, Roper M. Prediction of relapse in schizophrenic outpatients treated by drug and sociotherapy. Arch Gen Psychiatry 1977;34:171-84.

23. Goldman AI, Holcomb R, Perry HM, Schnaper HW, Fitz AE, Forhlich ED. Can dropout and other noncompliance be minimized in a clinical trial? Controlled Clin Trials 1982;3:75-89.

24. Gordis L, Markowitz M, Lilienfeld AM. Studies in the epidemiology and preventability of rheumatic fever: IV. A quantitative determination of compliance in children on oral penicillin prophylaxis. Pediatrics 1969;43:173-82.

25. Grey MJ, Genel M, Tamboriane WV. Psychosocial adjustment of latency-aged diabetics: determinants and relationship to control. Pediatrics 1980;65: 69-73.

26. Gundlach CA, Ziqubu TT, Souney PF. Esmolol DUE: an examination of audit criteria and study outcomes. Hosp Form 1990;25:61-6.

27. Hansaen SO, Tuominen R. Economic evaluation and the societal impact of medicine. J Soc Admin Pharm 1987;4:81-5.

28. Harlan WR. Economic considerations that influence health policy and research. Hypertension 1989;13(5):I-158-I-163.

29. Haynes RB. A critical review of the "determinants" of patient compliance with therapeutic regimens. Sackett DL, Haynes RB, eds, Compliance with therapeutic regimens. Baltimore, MD: Johns Hopkins University Press, 1976:26-39.

30. Hemminki E, Brambilla DJ, McKinlay SM, Posner JG. Use of estrogens among middle-aged Massachusetts women. DICP Ann Pharmacother 1991;25: 418-23.

31. Hermoni D, Friedman M, Morel D, Mankuta D, Sivan A, Porter B. Effects of a health activist course on knowledge and awareness of antibiotic use. Fam Pract 1989;6(1):27-32.

32. Hillman AL, Bloom BS. Economic effects of prophylactic use of misoprostol to prevent gastric ulcer in patients taking nonsteroidal anti-inflammatory drugs. Arch Intern Med 1989;149:2061-5.

33. Hindi-Alexander MC. Asthma education programs: their role in asthma morbidity and mortality. J Allergy Clin Immunol 1987;80:492-4.

34. Hoge SK, Appelbaum PS, Lowlor T, Beck JC, Litman R, Greer A, Gutheil TG, Kaplan E. A prospective, multicenter study of patients' refusal of antipsychotic medication. Arch Gen Psychiatry 1990;47:949-56.

35. Isaksson H, Danielsson M, Rosenhamer G, Konarski-Svensson JC, Ostergren J. Characteristics of patients resistant to antihypertensive drug therapy. J Intern Med 1991;229:421-6.

36. Inui TS, Yourtee EL, Williamson JW. Improved outcomes in hypertension after physician tutorials. Ann Intern Med 1976;54:646-51.

37. Inui TS, Carter WB, Kukull WA, Haigh VH. Outcome-based doctor-patient interaction analysis: I. Comparison of techniques. Med Care 1982;20: 535-49.

38. Jacobson AM, Hauser ST, Wolfsdorf JI, Houlihan J, Milley JE, Herskowitz RD. Psychologic preditors of compliance in children with recent onset of diabetes mellitus. J Pediatr 1987;110:805-11.

39. Jay S, Litt IF, DuRant RH. Compliance with therapeutic regimens. J Adolesc Health Care 1984;5(2):124-36.

40. Jay S, DuRant RH, Litt IF, Linder CW, Shoffitt T. Riboflavin, self-report, and serum norethindrone: comparison of their use as indicators of adolescent compliance with oral contraceptives. Am J Dis Child 1984;138:70-3.

41. Joglekar M, Mohanaruban K, Bayer AJ, Pathy MSJ. Can old people on oral anticoagulants be safely managed as out-patients? Postgrad Med J 1988;64:775-7.

42. Jones EF, Forrest JD. Contraceptive failure rates based on the 1988 NSFG. Fam Plann Perspect 1992;24(1):12-9.

43. Kaplan SH, Greenfield S, Ware JE. Assessing the effects of physician-patient interactions on the outcomes of chronic disease. Med Care 1989;27:S110-27.

44. Kelly GR, Scott JE. Medication compliance and health education among outpatients with chronic mental disorders. Med Care 1990;28:1181-97.

45. Kessler DA, Pines WL. The federal regulation of prescription drug advertising and promotion. JAMA 1990;264:2409-15.

46. Lauper P. The socioeconomic study program on Nitroderm TTS. In: van Eimeren W, Horisberger B, eds. Socioeconomic evaluation of drug therapy. New York: Springer-Verlag, 1988.

47. Leppik IE. How to get patients with epilepsy to take their medication: the problem of noncompliance. Postgrad Med 1990;88:253-6.

48. Levy RA. Reductions in health care costs due to the introduction of new pharmaceuticals. Pharm Business 1991;(Feb):9-15.

49. Levy RA, Smith DL. Clinical differences among nonsteroidal antiinflammatory drugs: implications for therapeutic substitution in ambulatory patients. Ann Pharmacother 1989;23:76-85.

50. Levy RA. Failure to refill prescriptions: incidence, reasons, and remedies. In: Cramer JA, Spilker B, eds. Patient compliance in medical practice and clinical trials. New York: Raven Press, 1991.

51. Lindren-Furmaga EM, Schuna AA, Wolfe NL, Goodfriend TL. Cost of switching hypertensive patients from enalapril maleate to lisinopril. Am J Hosp Pharm 1991;48:276-9.

52. Lipton HL. Elderly patients and their pills: the role of compliance in safe and effective drug use. Pride Institute J 1989;8(1):26-31.

53. Litt IF, Cuskey WR. Compliance with salicylate therapy in adolescents with juvenile rheumatoid arthritis. Am J Dis Child 1981;135:434-6.

54. Litt IF, Glader L. Follow-up of adolescents previously studied for contraceptive compliance. Adolesc Health Care 1987;8:349-51.

55. Litt IF, Cuskey WR. Compliance with medical regimens during adolescence. Pediatr Clin North Am 1980;27:3-15.

56. Maling TJB, Kawachi I. Minimum effective dosage in the drug treatment of hypertension: a cost-effective strategy for prescribers. N Z Med J 1990;103:231-3.

57. McElnay JC, Thompson J. Dispensing of medicines in compliance packs. Int Pharm J 1992;6(1):10-4.

58. McKenney JM, Munroe WP, Wright JT. Impact of an electronic medication compliance aid on long-term blood pressure control. J Clin Pharmacol 1992;32:277-83.

59. Montano DE. Predicting and understanding influenza vaccination behavior: alternatives to the Health Belief Model. Med Care 1986;24:438-50.

60. Myers MJ, Gerbino PP. Aging well: meeting the drug needs of the elderly. Pharm Exec 1988;8(4):60-6.

61. Nagasawa M, Smith MC, Barnes JH. Meta-analysis of correlates of diabetes patients' compliance with prescribed medications. Diabetes Educ 1989; 16(3):192-200.

62. Neal WW. Reducing costs and improving compliance. Am J Cardiol 1989;63:17B-20B.

63. Neel EU, Jay S, Litt IF. The relationship of self-concept and autonomy to oral contraceptive compliance among adolescent females. J Adolesc Health Care 1985;6:445-7.

64. Neel EU, Litt IF, Jay MS. Side effects and compliance with low- and conventional-dose oral contraceptives among adolescents. J Adolesc Health Care 1987;8:327-9.

65. Nissinen A, Tuomilehto J, Pekkanen J, Enlund H, Gunther A. Drug treatment of high blood pressure in the community experience in eastern Finland. J Hum Hypertension 1989;3:165-71.

66. Pilkington MA, Dolinsky D. Selecting alternate drug therapies. Med Care 1991;29:152-65.

67. Powell BJ, Penick EC, Liskow BI, Rice AS, McKnelly W. Lithium compliance in alcoholic males: a six-month follow-up study. Addict Behav 1986;11: 135-40.

68. Rimer BK, Glanz K, Lerman C. Contributions of public health to patient compliance. J Com Health 1991;16:225-4.

69. Rovelli M, Palmeri D, Vossler E, Bartus S, Hull D, Schweizer R. Noncompliance in organ transplant recipients. Transplant Proc 1989;21:833-4.

70. Sackett DL, Haynes RB. Compliance with therapeutic regimens. Baltimore, MD: Johns Hopkins University Press, 1976:9-25.

71. Saunders E. Tailoring treatment to minority patients. Am J Med 1990; 88(3B):21S-23S.

72. Schmeider RE, Rockstroh JK, Messerli FH. Antihypertnesive therapy: to stop or not to stop? JAMA 1991;265:1566-71.

73. Schooler NR. Maintenance medication for schizophrenia: strategies for dose reduction. Schizophrenia Bull 1991;17:311-24.

74. Sclar DA, et al. Effect of health education on the utilization of antihypertensive therapy: a prospective trial among HMO enrollees. Primary Cardiol 1992;(Suppl 1):24-9.

75. Sclar DA, et al. Effect of health education on the utilization of HMO services: a prospective trial among patients with hypertension. Primary Cardiol 1992;(Suppl 1):30-5.

76. Sclar DA. Patient compliance: pharmacy's potential impact and economic incentive. Proceedings: Pharmacy Economics/Third-Party Programs, Montreal, Quebec, 1986.

77. Shepard DS, Foster SB, Stason WB, Solomon HS, McArdle PJ, Gallagher SS. Cost-effectiveness of interventions to improve compliance with anti-hypertensive therapy. Prev Med 1979;8:229.

78. Shulman NB, Martinez B, Brogan D, Carr AA, Miles CG. Financial cost as an obstacle to hypertension therapy. Am J Public Health 1986;76:1105-8.

79. Sintonen H, Alander V. Comparing the cost-effectiveness of drug regimens in the treatment of duodenal ulcers. J Health Econ 1990;9:85-101.

80. Steele DJ, Jackson TC, Gutmann MC. Have you been taking your pills? The adherence-monitoring sequence in the medical interview. J Fam Pract 1990;30:294-9.

81. Steiner JF, Fihn SD, Koepsell TD, Blair B, Kelleher K, D'Alessandro D. Clinical predictors of treatment reduction in hypertensive patients. J Gen Intern Med 1990;5:203-10.

82. Strogatz DS, Earp JAL. The determinants of dropping out of care among hypertensive patients receiving a behavioral intervention. Med Care 1983; 21(Suppl 10):970-80.

83. Sullivan SD, Kreling DH, Hazlet TK. Noncompliance with medication regimens and subsequent hospitalizations: a literature analysis and cost of hospitalization estimate. J Res Pharm Econ 1990;2(2):19-33.

84. Taubman AH, King JT, Weisbuch JB. Noncompliance in initial prescription filling. Apothecary 1975;9(10):14+.

85. Uhlmann RF, Inui TS, Pecoraro RE, Carter WB. Relationship of patient request fulfillment to compliance, glycemic control, and other health care outcomes in insulin-dependent diabetes. J Gen Intern Med 1988;3:458-63.

86. Ulmer RA. Editorial: patient noncompliance and health care costs. J Compliance Health Care 1987;2(1):3-5.

87. Uzoma CU, Feldman RHL. Psychosocial factors influencing inner city black diabetic patients' adherence with insulin. Health Educ Q 1989;20:29-32.

88. Wartman SA, Morlock LL, Malitz FE, Palm EA. Patient understanding and satisfaction as predictors of compliance. Med Care 1983;21:886-91.

89. Weintaub M, Taves DR, Hasday JD, Mushlin AI, Lockwood DH. Determinants of response to anorexiant. Clin Pharmacol Ther 1981;30:528-33.

90. Wilson W, Ary DV, Biglan A, Glasgow RE, Toobert DJ, Campbell DR. Psychosocial predictors of self-care behaviors (compliance) and glycemic control in non-insulin-dependent diabetes mellitus. Diabetes Care 1986;9:614-22.

91. Windsor RA, Bailey WC, Richards JM, Manzella B, Soong SJ, Brooks M. Evaluation of the efficacy and cost effectiveness of health education methods to increase medication adherence among adults with asthma. Am J Public Health 1990;80:1519-21.

92. Winkler R, Underwood P, Fatovich B, James R, Gray D. A clinical trial of a self-care approach to the management of chronic headache in general practice. Soc Sci Med 1989;29:213-9.

93. Worthen DM. Patient compliance and the usefulness product of timolol. Surv Ophthalmol 1979;23:403-8.

For the Student

1. Becker MH, Maiman LA. Sociobehavioral determinants of compliance with health and medical care recommendations. Med Care 1975;13:10-24.

2. Bond WS, Husser DA. Detection methods and strategies for improving medication compliance. Am J Hosp Pharm 1991;48:1978-88.

3. Boyd JR, Covington TR, Stanaszek WF, Coussons RT. Drug defaulting. Part I: Determinants of compliance. Am J Hosp Pharm 1974;31:362-7.

4. Enlund H, Poston JW. Impact of patient noncompliance on drug costs. J Soc Admin Pharm 1987;4:105-11.

5. Fedder DO. The pharmacist and cardiovascular risk reduction. US Pharm 1990;48(Suppl):4-8.

6. Fincham JE, Wertheimer AI. Correlates and predictors of self-medication attitudes of initial drug therapy defaulters. J Soc Admin Pharm 1985;3:10-7.

7. Fincham JE, Wertheimer AI. Elderly patient initial noncompliance: the drugs and the reasons. J Geriatr Drug Ther 1988;2(4):53-62.

8. Fincham JE, Wertheimer AI. Using the Health Belief Model to predict initial drug therapy defaulting. Soc Sci Med 1985;20:101-5.

9. Friedman IM, Litt IF. Adolescents' compliance with therapeutic regimens. J Adolesc Health Care 1987;8:52-67.

10. Hammel RW, Williams PO. Do patients receive prescribed medication? J Am Pharm Assoc 1964;NS4:331-7.

11. Hood JC, Murphy JE. Patient noncompliance can lead to hospital readmissions. Hospitals 1978;52:79-84.

12. Jay MS, DuRant RH, Litt IF. Female adolescents' compliance with contraceptive regimens. Pediatr Clin North Am 1989;36:731-46.

13. Litt IF, Cusky WR, Rudd S. Identifying adolescents in risk for noncompliance with contraceptive therapy. J Pediatr 1980;96:742-5.

14. Maronde RF, Chan LS, Larsen FJ, Strandberg LR, Laventurier MF, Sullivan SR. Underutilization of antihypertensive drugs and associated hospitalization. Med Care 1989;27:1159-66.

15. McCaffrey DJ, Smith MC, Banahan BF. Why prescriptions go unclaimed. US Pharm 1993;8:58-65.

16. Schering Laboratories. Improving patient compliance: is there a pharmacist in the house? Schering Report XIV. Kenilworth, NJ: Schering Laboratories, 1992.

17. Urquhart J, Chevalley C. Impact of unrecognized dosing errors on the cost and effectiveness of pharmaceuticals. Drug Info J 1988;22:363-78.

Chapter 14

Explaining and Changing Noncompliant Behavior

Dale B. Christensen

Listen to me, listen clearly please,
For I can tell you how to cure your sneeze.
Why oh why won't you comply?
I tell you to, that's for sure.
Why can't you be more mature?
I wish I could say to you,
"Don't you know what's the right thing to do?"

THE SIGNIFICANCE
OF THE NONCOMPLIANCE PROBLEM

Many observers have noted that it makes no sense for patients to go to the effort of seeking the care of a physician and getting a prescription for a diagnosed condition only to not bother to have it dispensed or to not follow the directions once prescribed. Similarly, it makes no sense for physicians to prescribe, for drug companies to promote drugs as cures for disease, for pharmacists to dispense, or for health insurers to pay for drug therapy if it is not consumed properly. If the drug does not fulfill expectations, why should anyone provide or pay for prescriptions? Noncompliance commands so much attention precisely because drug therapy is widely regarded as cost-effective and noncompliance is the only barrier to accomplishing this goal.

One way to frame our discussion of compliance is to relate it to a patient's expected health outcomes. Our working assumption is that good compliance leads to good outcomes and poor compliance leads to poor outcomes (Figure 14.1, Cells A and D). However, we should recognize that other alternatives are also possible, for example, good compliance but poor outcomes and poor compliance but good outcomes (Cells B and C). This is particularly true for disease conditions that are difficult to diagnose (e.g., depression) or require individualized treatments.

How common is noncompliance (Cells C and D combined), and how commonly does noncompliance lead to poor health outcomes (Cell D)? From all indications, the noncompliance "problem" is indeed significant. Available evidence suggests between 3% and 7% of patients do not have original prescriptions dispensed.[1-3] Of those who have them dispensed, it is commonly reported that only 30 to 60% of patients take their medicines as directed and that 4 to 35% misuse their medicines in a manner that can pose a serious threat to health.[4-6] It may be surprising to note that noncompliance is not particularly better

FIGURE 14.1. Compliance and Health Outcomes

HEALTH OUTCOMES

COMPLIANCE	Good	Poor
Good	**A** *Ideal*	**B** *Misdiagnose* *Misprescribe* *Misdose*
Poor	**C** *Misdiagnose* *Misprescribe* *Misdose* *Self-regulation*	**D** *Usual focus* *of concern*

even among patients with chronic conditions, for which drug therapy is particularly important, as illustrated in Table 14.1.

The adverse health consequences and costs associated with treating disease sequelae related to noncompliance are substantial.[7-10] For example, it is estimated that 5.5% of all hospital admissions (1.94 million in 1986) were due to noncompliance.[11] This was equal to an estimated $8.5 billion, or 1.7% of total health care dollars, in 1986. The elderly are one group in society particularly vulnerable to health problems from noncompliance.[12] While the elderly are not necessarily more noncompliant than other age groups, they typically have more drugs and dosage times to manage. The elderly are at greater risk for adverse health effects because of multiple chronic diseases, more complex drug therapies, and higher sensitivity to the effects of drug problems. For example, Col and colleagues found failure to comply with medication regimens was the reason for admission of 11% of elderly patients consecutively admitted to an acute care hospital.[13]

Achieving compliance in patients is no easy task. It first requires us as health professionals to understand the reasons why patients do not comply and then to determine effective strategies for intervention. In this chapter, we attempt to explain noncompliance in behavioral terms. We then turn attention to how noncompliant behavior can be changed in real-world practice environments, and we present strategies for changing behavior.

TABLE 14.1. Compliance Rates for Specific Chronic Conditions

Condition	Noncompliance Rate
asthma	20%
arthritis	55-71%
contraception	8%
diabetes	40-50%
epilepsy	30-50%
hypertension	40%

Source: Emerging issues in pharmaceutical cost containment. *Nat. Pharm. Council*, 1992.

EXPLAINING NONCOMPLIANT BEHAVIOR

To us as health professionals, noncompliance makes no rational sense. It is difficult to understand noncompliant behavior, let alone devise a strategy to change a patient's thinking. The reasons patients offer to explain why they did not take medicines as prescribed are varied. In the study of hospitalized elderly cited earlier, for example, the most commonly stated reasons for noncompliance (usually in the form of undercompliance) were:[13]

Forgetfulness	39.6%
Side effects	17.7%
Drug perceived as not necessary	12.5%
Confusion	11.5%
Cost	10.4%

Respondents often give multiple reasons for being noncompliant, and to make matters worse, the reasons given may not reflect the true reason. For over 20 years, sociologists and psychologists have worked to understand these reasons and have attempted to explain compliance behavior. Scientists first attempted to identify patient, disease, or situational factors related to compliance. Several of the factors investigated are listed in Table 14.2.

The next step for scientists was to develop theories or models to explain patient behavior that incorporated these factors. Several models emerged that are particularly useful in providing a framework to better understand patient behavior and patient responses to compliance-enhancement messages. They also shed light on our own drug-taking behavior. The most useful of the models and concepts developed to date are briefly described below.

Health Belief Model (HBM)

The HBM is perhaps the most well known and compelling of the behavioral models because it is intuitive and easily understood. The HBM, briefly described in another chapter, first and foremost reminds us that all health actions a patient undertakes are based on perceptions rather than actuality. A key to understanding patients, then, is to identify their perceptions and beliefs. The model, as

TABLE 14.2. Factors Related to Noncompliance

Factor	Comment
Nature of drug	perceived importance of the drug is more important than the drug itself.
Medication regimen, number of different drugs, doses	indicators of the degree of complexity and disruption in the patient's normal routine.
Treatment period	longer = higher likelihood of compliance.
Patient characteristics: age, sex, education, family family status	no consistent predictor. The elderly are sometimes found to be less compliant, but they also have more complex regimens.
Type of illness	perceived severity is more important than the actual illness.
Patient-practitioner relationship	the amount of information transmitted and the degree of trust are important factors.

originally developed, states that a person is motivated to some action by his or her *perceived threat to health*.[14] This threat is determined, in turn, by each patient's:

- Perceived value of good health.
- Perceived susceptibility or resusceptibility to a particular disease.
- Perceived seriousness of a disease condition.

Faced with a perceived threat to health, the person evaluates alternative health actions that offer the potential for reducing this threat. The workings of the model are best explained with an example.

Example 1: Mr. Smith

Mr. Smith, a 45-year-old male, is newly diagnosed with high blood pressure. According to the HBM, Mr. Smith's motivation to bring his blood pressure under control will be directly proportional to how great a threat to his health high blood pressure represents.

If, for example, he prides himself on being in optimal health and good physical condition, he will be more motivated. Similarly, if he believes his high blood pressure is likely to lead to heart or circulatory problems, and if he believes these problems to be serious, he will be more motivated. Among the alternative courses of action Mr. Smith may consider as a means to reduce this perceived threat are: do nothing and hope the threat disappears, change his diet, exercise more, try to relax more, and take prescribed medications.

The HBM further states that Mr. Smith will evaluate each course of action, consciously or subconsciously, in terms of benefits versus costs. The benefits of each course of action (e.g., lowered blood pressure, peace of mind, feeling better) are weighed against its costs.

Costs are considered in economic as well as in other terms. For example, there are the time and effort costs to keep an appointment with the physician, out-of-pocket costs and time costs to have the prescription filled, the "cost" of inconvenience in remembering to take the medication as prescribed, and the "cost" of enduring any unpleasant side effects. When one considers that Mr. Smith may not feel any better (or he may feel even worse) with blood pressure controlled, we begin to understand why he may not optimally comply.

The process of weighing the advantages and disadvantages of alternative courses of action can be vexing. For example, should Mr. Smith initiate drug therapy now or commit to a program of exercise and strict diet? If he initiates drug therapy, is it really necessary to comply 100%, or will 80% do? How can one justify taking medicines if they only make one feel worse? Sometimes the patient finds himself on the horns of a dilemma, so perplexed by the pros and cons of each choice that a decision does not come easily. In these cases, the model indicates that a cue or stimulus is needed to precipitate a decision. A cue may be a newspaper article showing a link between compliance and hypertension and stroke, for example, or it may be finding out that an acquaintance experienced a heart attack believed to be linked to hypertension.

How well does this model explain behavior? In studies of the HBM, researchers have found that while the overall model is not particularly predictive of health behavior, individual components are. For example, Janz and Becker reviewed several studies of preventive health behaviors and sick role behaviors through 1984.[15] They found perceived barriers (e.g., costs) to be the most powerful of all the HBM dimensions. More recent studies have confirmed this general conclusion.[16] In a study of compliance behavior during a chemotherapy prevention trial, Sauer and colleagues found the importance of HBM components were related to the type of non-compliant behavior.[17] Overcompliance was significantly related to perceived benefits of drug therapy and severity of disease, while undercompliance was related to perceived barriers to therapy. Fincham and Wertheimer have identified lack of strong belief in the benefits of medical care for symptoms or illness and inadequate information from the prescriber on how to take a newly prescribed drug as prominent reasons why patients do not have new prescriptions dispensed.[18]

The HBM has been modified by several researchers in an attempt to improve its explanatory powers. This author, for example, has suggested that a more dynamic model of drug-taking compliance applies.[19,20] This modification recognized that a person's perceptions change over time in response to information and advice from health professionals and close family friends as well as from the experiences of drug taking. The initial encounter with the prescriber is seen as a learning process for the patient wherein the information received is compared with the patient's own experiences and beliefs. If the information is consistent, the patient is much more likely to follow the advice of the prescriber. Similar reasoning applies to information received from the pharmacist.

Returning to our patient, perhaps Mr. Smith doubts the diagnosis (because he believes his last blood pressure reading was artificially high) or the need for drug therapy (because he believes his blood pressure is not sufficiently high to warrant treatment). In this case, his commitment to comply is much more tenuous. Reinforcement of messages about the drug from the physician, nurse, or pharmacist may help to persuade him to comply for a trial period, but ultimately, the decision to continue using the drug will be based on his perception of the drug-taking experience and his reassessment of his blood pressure. Cues derived from drug taking are interpreted and serve as reinforcers for continued compliance. For example, if

Mr. Smith's blood pressure is indeed lowered after taking the drug for a short time, compliance is reinforced. On the other hand, side effects (e.g., cough, lack of energy) serve as negative reinforcers. In this manner, the patient is constantly reassessing the benefits versus the costs of continued compliance behavior. The pharmacist can help by providing information or other assistance that adds to the benefits side of the ledger or reduces the perceived cost side of the ledger.

A closely related model, the Theory of Reasoned Action (TRA), builds upon the HBM by recognizing the importance of social influences on the individual.[21] For example, it recognizes that a person's behavioral intention is also affected by the health beliefs of other persons important in the patient's life (e.g., "significant others"), as well as by the patient's motivation to comply with these beliefs. The model also distinguishes between behavioral intentions and actual behavior, recognizing that certain barriers may prevent one from carrying out one's intentions.

Returning to the case of Mr. Smith, we would add that if his family, coworkers, and employer have a strong belief that hypertension is serious business and if Mr. Smith is generally inclined to follow the collective advice of these individuals, his commitment and intent to control his blood pressure will be comparatively strong.

With these added variables, it is not surprising that the TRA has been shown to be a better predictor of intent to comply and compliance behavior. In a study of females with urinary tract infections, Ried and Christensen found that two HBM variables (perceived benefits and perceived barriers) alone explained 10% of the variance in compliance, while three added TRA variables (belief strength, outcome evaluation, and behavioral intention) explained an additional 19% of the variance.[16] It is apparent that even this model leaves a considerable amount of variance unexplained. One problem in testing models of compliance behavior is that patients' beliefs and perceptions are usually measured at one point in time, while compliance is often assessed over a period of time.

Let us restate the factors related to noncompliance from a different perspective with these models in mind (Table 14.3). Note that these factors are not the same as patients' often-stated reasons for

TABLE 14.3. Factors Influencing Compliance Based on the Health Beliefs Model and Theory of Reasoned Action

---

General health perceptions

- How strongly does the patient believe that optimal health is important?
- To what extent does the patient follow preventive health practices?
- What is the patient's prior illness experience?

Perceived susceptibility

- What is the likelihood that the illness symptoms will get worse if nothing is done?
- How susceptible does the patient feel to a worsened illness condition?

Perceived seriousness

- How seriously does the patient regard this illness episode?
- How serious is a worsened condition to which the patient feels vulnerable?
- How disruptive of normal social roles is this patient's condition?
- What degree of physical discomfort does the patient attribute to this condition?

Treatment benefits

- What is the expected action (benefit) of each drug? *e.g., prevention, symptom relief, cure.*
- How strong is the patient's belief that the drug will work as intended?
- How well is the patient informed about how the medicine is to be used and stored?

Treatment costs

- What are perceived costs of not following the prescribed regimen? *e.g., prolonged recovery, discomfort, likelihood of condition worsening.*
- What are the economic costs? *e.g., $$ out of pocket costs, travel costs*
- Time costs: *e.g.,*
  - *Waiting time at physician's office, at pharmacy*
  - *Drug-taking inconvenience costs: how disruptive is the drug regimen to the patient's daily life?*
  - *Complexity of regimen (number of drugs, doses, dose times)*
- Drug side effects: e.g.,
  - *Does the patient link the physiological effects to taking the drug?*
  - *How unpleasant are these effects?*

Normative expectations and beliefs

- How do the patient's family and close friends regard this illness situation and how important do they think it is to follow the prescribed therapy.
- How do the patient's physician and pharmacist regard this illness and the need for careful compliance?

Motivation to comply

- How likely is the patient to follow the advice of family and friends?
- How likely is the patient to follow the advice of a prescriber or pharmacist?

Cues:

- What cues are likely to be most meaningful to the patient?

---

noncompliance. However, they can be used to interpret those reasons. For example, "simply forgetting" most likely reflects perceptions that the disease condition is not serious, that perceived expected benefits of drug therapy are low, or that the costs of drug taking are high vis-à-vis benefits.

Self-Regulation

Examining drug-taking compliance from the point of view of the patient suggests that noncompliance may not even be the appropriate term. Indeed, there is evidence to suggest that deliberate, rather than passive or forgetful noncompliance, is involved. In these cases, patients may, in fact, be engaging in a form of self-regulation of their illness situation. Self-regulation is generally defined as the way people control and adapt their thoughts, emotions, and behavior to changing situations and personal outcomes.[22]

Despite what we may wish to believe, acting against the doctor's orders does not necessarily lead to poorer health outcomes. Weintraub used the phrase "intelligent noncompliance" to describe patient self-directed behavior.[23] While most of the compliance literature suggests that noncompliant patients are usually worse off, this is not always the case (Figure 14.1, Cell C). One study involving hypertensive patients made this point. During the course of a project aimed at initiating a step-down program among patients with well-controlled blood pressure, Steiner and colleagues found that among control group patients, 24% appropriately initiated self-regulatory behavior (i.e., obtained fewer medications than they were prescribed) while continuing to maintain control of their blood pressure.[24]

To better understand this phenomenon, it may be helpful to view illness as a source of stress and a disruption in our life-style. We seek to cope with this stress by cognitively and emotionally adapting to our illness and treatment regimen. Taylor has suggested that we tend to adapt by (1) seeking to derive meaning from our illness and drug-taking experience, (2) seeking to restore some sense of self-esteem, and (3) seeking to regain some mastery and control over our environment.[25] Seen in this regard, drug-taking itself can be perceived as either stress producing or stress reducing. Drug-consumption behavior that is willfully consistent *or* willfully inconsistent with a prescriber's intentions may be undertaken as a way to reduce stress. To the individual, the action is appropriate and rational. To better illustrate these factors and forces, consider a second case.

Example 2: Mr. Jones

Mr. Jones was diagnosed with epilepsy four years ago. He calls in a refill at your pharmacy. As you prepare the prescription, you notice he has only been about 80% compliant since his last refill some four months ago. You know he will be in for his prescription soon. Why do you suppose he, of all persons, would not be fully compliant?

Epilepsy is a disease condition most would regard as stressful. An illuminating study of patients with epilepsy was conducted by Conrad, who sought to determine when and why these patients, in particular, would be noncompliant with their drug therapy.[26] His findings were most revealing. First, a surprising 42% of patients admitted self-regulating their seizure medication. Second, noncompliant patients were not a discrete or easily identifiable group. In terms of their general attitudes about drug-taking, they viewed their medication as a reminder that they had epilepsy. Seizures were a cue that the disease was getting better or worse. So, too, was a prescribed increase or decrease in dosage. A stated *instrumental* reason for taking medications was to control seizures, and stated *psychological* reasons were to reduce worry and to lead a normal life. Reasons for altering their schedule (and thereby exercising self-control) were to avoid side effects, to test for the progress of their disorder and the continued need to take their medication, and to assert control over their perceived dependence on the drug or their health practitioner. Conrad speculated that self-regulation is probably common for many term chronic illnesses but that the degree and amount of self-regulation may differ.

The "Take-Away" Message

First, to understand the noncompliant patient is to understand ourselves. Our drug-taking behaviors, whether compliant or noncompliant, are better viewed as self-regulation behaviors. We tend to act according to our own best self-interests (as we define them). Our actions are understandable and (to a degree) predictable. We act to bring or restore normalcy and homeostasis to our lives. We act to reduce stress posed by fears and threats and disruptions to our daily routine. We formulate our actions based on information provided to us about our disease and the drugs prescribed and based on cues derived from our own illness and drug-taking experience. We make decisions with information we have at hand. If our personal database were loaded differently or if our personal health perceptions were different, we would probably exhibit different actions. Putting our pharmacist hats back on, we as pharmacists ought to be able to build upon these notions as we develop strategies to influence compliance behavior.

CHANGING NONCOMPLIANT BEHAVIOR

As previously discussed, sociologists and psychologists have long recognized that understanding the patient's point of view is most important in understanding and influencing behavior. As Stimson notes, any strategy directed at changing patient behavior must recognize that the patient makes the decision to act or not act in a medically appropriate way.[27] Physicians and pharmacists must forego paternalistic notions that compliance can be achieved by merely telling patients what to do. A much better approach is to recognize that you as a pharmacist are but one source of information for the patient, who will make an independent decision whether, and to what extent, to comply with drug therapy. A second message is that you must provide information relevant to attainment of the health goals the patient defines as important. Figure 14.2 illustrates the sources of influence on patient compliance decision-making processes.

Ideally, the patient would have a long and continuing relationship with the prescriber and pharmacist and would be actively involved in treatment decisions.[22] If such a relationship exists, it sets the stage and a framework for effective interventions. Strategies can be of two general types: educational or behavioral. Each has motivational components, and it is the motivational nature of the message—often a hidden message—that affects the success of intervention strategies. Messages about drug therapy go through several different phases before being acted upon, as illustrated by Figure 14.3.

The pharmacist as communicator must assure information transmitted is understandable, rational, helpful, and easy to remember. Usually, multifaceted approaches work best. A variety of strategies for changing behavior have been tried. These include stressing the importance of drug therapy, monitoring drug use closely, and using reminder aids. It turns out that no matter what strategy is used, it is better than no strategy at all. This was shown in a study by Myers and Calvert, who demonstrated that verbal and written information about prescribed antidepressant medicines improved compliance with the medication by an "attention-placebo" effect, rather than a cognitive effect.[28] Further, information about beneficial effects led to fewer reported side effects than either information about side effects or no information at all.

TYPES OF NONCOMPLIANT BEHAVIOR

There are a number of ways to categorize noncompliant behaviors. One such categorization, useful from a practitioner's perspective, is the following:

- *Nonstarters:* patients who, for a variety of reasons, do not get initial prescriptions filled.

- *Inappropriate users:* patients who take prescribed medications routinely but in the wrong manner, for example, at the wrong time.
- *Abrupt quitters:* patients who prematurely terminate prescribed treatment.
- *Taperers:* patients who take prescriptions less frequently than prescribed.
- *Intermittent users:* patients who take prescribed medicines only intermittently, perhaps in response to recurrence of symptoms.

In each case, it is possible to speculate about specific reasons for this behavior. For example, there may be misperceptions of:

1. The purpose of taking the medicine.
2. How to take the medicine properly.
3. The value of taking, or continuing to take, the medication.

Here are some "how to's" based on what we know about compliance behavior.

1. *Optimize drug therapy by simplifying and consolidating it whenever possible.* It is well established that lower compliance rates are associated with more drug-taking inconveniences, such as dosage intervals and different drugs to be taken each day.[29-31] Compliance rates fall off dramatically beyond two doses per day. Any appropriate reduction in the number of different drugs in the patient's regimen is also helpful in that the possibility of drug interactions, therapeutic duplication, or side effects are likely to be commensurately reduced.

2. *Provide complete and useful drug use information to the patient.* Talking to the patient works. Further, it should be the pharmacist, rather than the clerk or technician, who interacts with the patient. Largely as a result of the Omnibus Budget Reconciliation Act of 1990 (OBRA 90), this is now a practice requirement in most states for new prescriptions. However, to monitor patient response optimally, pharmacists should directly interact with patients for refills as well. The simple act of the pharmacist, rather than the clerk, handing the medication to the patient has been reported to improve compliance by 25%.[32] By merely directing attention to the

FIGURE 14.2. Patient Information Processing and Compliance Decision Making

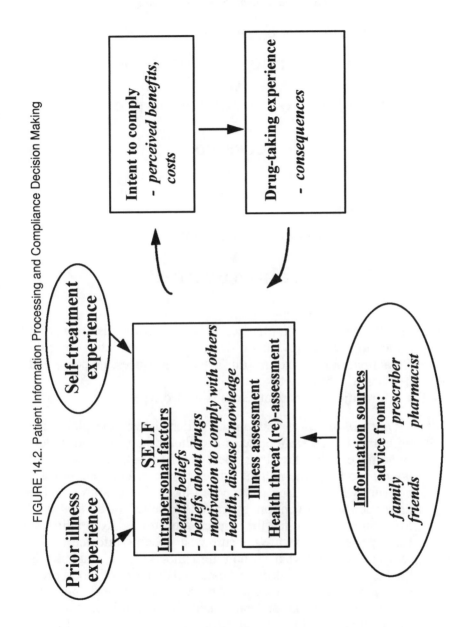

FIGURE 14.3. Patient Drug-Related Information Processing

PRESCRIPTION MESSAGE SENT

⬇

RECIEVED AND COMPREHENDED

⬇

ACCEPTED

⬇

RETAINED

⬇

INTENDED TO COMPLY

⬇

COMPLIED WITH

drug and the need for compliance, the pharmacist adds a sense of importance to the therapy, again demonstrating an "attention-placebo" reaction. In HBM terms, this activity enhances the perceived expected benefits of drug taking.

The information the patient needs is of two types:

- *"What" information:* When the patient leaves the pharmacy with a new prescription, he or she should have a good understanding of what the drug does, why it was prescribed, and how it works in the patient's disease state. Many patients are not sure what the drug is supposed to do. For example, is the drug used for prevention, symptomatic relief, or cure? When Mr. Smith received his first prescription for an ACE inhibitor for hypertension, we are not sure what specific information he was provided or what he now remembers. If he understands the benefits of drug taking to be not only control of blood pressure and relief of headache or occasional dizziness, but also prevention of stroke or heart attack, his commitment to take the medication will likely be stronger. Other "what" information should more directly address the perceived cost of drug taking. What are the possible side effects? Are they serious?

Will they go away? How can they be managed? What should the patient do if they occur?

- *"How to" information:* This includes how the drug should be taken, and when. What times during the day should the medication be taken? What about meal times? What should be done if a dose is missed? What other drugs should be avoided and why?

3. *Provide compliance-enhancement packaging or devices.* Given the busy status of most everyone's lives, anything offered in the form of cues or memory aids can be most helpful. Cues can greatly help the patient to fulfill his or her own drug-taking intentions. Encouraging the patient to engage "significant others" is one way. Another is to offer the patient one of a multitude of compliance-enhancement devices. Some common devices include:

- *"Medi-set" type containers:* These are rectangular plastic boxes designed to hold dosage units for one or more medications. Each individual box can be used to hold all doses for a single dosage time or a single day. Other examples are electronic pillboxes with alarms.
- *Counter caps:* These are caps that fit over the top of a prescription vial. Each time the cap is removed, the dial ratchets forward to a new number (representing dosage intervals per day or days of the week).
- *Calendars:* Easily produced by some software programs, these provide the patient with a paper form to check off when doses are taken.
- *"Bingo" cards:* Disposable blister or bubble packs mounted on an 8 1/2 × 11-piece of cardboard contain prescribed medications for each dosage time. Usually, each card contains a single medication.

In addition to compliance-enhancement devices, other types of reminders have been developed to help patients remember to obtain refills. Providing patients on chronic medications with postcard or telephone refill reminders has proven quite successful. Sclar and colleagues, for example, have demonstrated that patients on antihypertensive therapy who were provided with telephone and mail reminders and a newsletter acquired significantly higher days' supply of medication than patients not provided this service.[33] Windsor and colleagues have similarly demonstrated the effectiveness of a multiple intervention strategy for patients with asthma, consisting of a self-help guide, one-to-one instruction, and reinforcement telephone calls.[34]

In recent years, there has been a dramatic increase in the number and variety of commercially available patient compliance-enhancement programs and devices. For example, there are now software programs that, when connected to a pharmacy's existing computer system, use the actual voice of the pharmacist to deliver a message that a target prescription needs refilling.[35] Recognizing the importance of compliance–to company sales as well as to patients–a number of companies have developed and now sponsor compliance-enhancement programs. They are specifically directed to enhancing compliance of a specific product, and most involve multi-intervention techniques, such as postcard or telephone reminders. For the practicing pharmacist, a list of many of these offerings appears in an excellent review article by Berg and colleagues.[35] Often they are offered to practicing pharmacists, who are asked either to provide these services to identified patients or to enroll them for later contact.

4. *Monitor patient drug-taking experiences.* After the initial prescription, the pharmacist's task should be to follow up and monitor patient response to therapy. The following are suggested activities for the pharmacist. Each is briefly discussed:

- Assess compliance (query + pill counts)
- Monitor drug effects and disease state:
  - Patient experiences
  - Physiological signs
- Link compliance to disease status
- Advise patients accordingly

Assessing and monitoring is best accomplished at scheduled or actual refill times. Among the various methods of assessing compliance (discussed in another chapter), the most practical for the practicing pharmacist is to use a combination of prescription refill rate monitoring and brief patient query. Most pharmacy computer systems are capable of displaying information indicating how long it has been since the prescription was last refilled. Knowing the last refill date, the quantity supplied, and the days' supply or daily dose, the pharmacist can assess how many days late—or early—the patient is for a refill. This is sometimes known as the "medication possession ratio." Helpful computer algorithms have also been developed for this purpose and now exist on many pharmacy computer systems. One caution, however, is that the "days' supply" field in the prescription database may not be filled in accurately by pharmacists or may not be filled in at all. Pharmacists in practice are urged to fill in this field accurately and consistently so that compliance calculations can be more accurate.

If this calculation or computer message indicates the patient is not compliant, it is a clear signal for some type of planned intervention with the patient. The initial focus of inquiry should be to determine if the patient was, in fact, noncompliant. There may be several reasons for a false positive alert. Perhaps the patient obtained an extra supply of the medication as samples from the prescriber. Second, there may be a discrepancy between the pharmacist records of the prescribed dose (affecting days' supply) and what the physician intended or what the patient recalls he was told. For example, suppose Mr. Smith's blood pressure was under control and he was instructed to reduce the dose, but this message was not transmitted to the pharmacist.

A number of specific questions have been developed for practitioners' use in assessing compliance, knowledge, and experiences taking prescribed medications. The following single question has been used successfully: "Most people have trouble remembering to take their medicine. Do you have trouble remembering to take yours?"[36, 37] Note how this question attempts to elicit a positive response by communicating a normative standard—that other patients have difficulty and that it is OK to admit having such difficulty yourself. Responses to this question, when asked using a structured interview format, have been found to correlate highly with pill count compliance.[37]

The four questions developed by Morisky and colleagues and presented in another chapter, are also useful in assessing compliance:[38]

1. Do you ever forget to take your medicine?
2. Are you careless at times about taking your medicine?
3. When you feel better do you sometimes stop taking your medicine?
4. Sometimes if you feel worse when you take the medicine, do you stop taking it?

Another approach, developed by the Indian Health Service, is designed to assess patient knowledge and drug-taking experience more comprehensively.[39] It uses structured open-ended questions to identify specific potential drug-taking problems in a manner that allows the pharmacist to address them specifically. The questions are shown in Table 14.4.

COMPLIANCE ENHANCEMENT IN TOMORROW'S WORLD OF PHARMACY PRACTICE

Here are some observations concerning the importance of compliance:

- Improved compliance is a recognized win-win situation for health professionals, patients, third parties, and product manufacturers. All parties have a vested interest in the appropriate use of drugs, and all should work collaboratively to achieve it.
- With few exceptions, compliance is not currently being detected or improved systematically by pharmacists or by anyone else.
- Health reform is changing health professional roles and treatment expectations.
- Pharmacists are logical choices to be actively involved in compliance-enhancement activities.
- However, pharmacy is a profession undergoing considerable change.
- Effective use of pharmacists will involve a mixture of training and incentives.

The pharmacist is the logical choice to be engaged in compliance-enhancement services for patients. Yet there are many

TABLE 14.4. Prime Questions for Interactive Patient Counseling

---

New prescriptions:
  "What did your doctor tell you the medication is for?"
  "How did your doctor tell you to take the medication?"
  "What did your doctor tell you to expect?"

Verification:
  "Just to be sure I haven't left anything out, please tell me how you are going to use this medication?"

Refills: "Show and tell."
  "What do you take the medication for?"
  "How do you take it?"
  "What kind of problems are you having?"

EXPANDING THE PRIME QUESTIONS:

  "What did your doctor tell you the medicine is for?"
    "What problem or symptom is it supposed to help?"
    "What is it supposed to do?"

  "How did your doctor tell you to take the medication?"
    "How often did your doctor say to take it?"
    "How long are you to continue taking it?"
    "What did your doctor say to do when you miss a dose?"
    "How should you store this medication?"
    "What does three times a day mean to you?"

  "What did your doctor tell you to expect?"
    "What good effects are you supposed to expect?"
    "How will you know if the medication is working?"
    "What bad effects did your doctor tell you to watch for?"
    "What should you do if a bad reaction occurs?"
    "What precautions are you to take while on this medication?"
    "How will you know if it's not working?"
    "What are you to do if the medication doesn't work?"

---

barriers to optimal performance of this role. Some typical comments we hear from pharmacists are those expressed below:

"This wasn't the job I was trained to do."
"Why get involved? As a practicing pharmacist, I am only paid to dispense. I simply do not have the time, and no one expects me to do this."

"What can I as a pharmacist do about noncompliance? It is the patient's decision and the patient's business, not mine."

Let us briefly address each concern.

"This wasn't the job I was trained to do." The simple truth is that pharmacy practice standards are changing dramatically. It is now increasingly recognized that merely providing prescription products to patients is an inadequate basis to justify pharmacists' continued professional role. Largely as a result of health reform and legislation, professional activities and responsibilities are expanding in areas consistent with the concepts of pharmaceutical care.[40] For example, OBRA 90 requirements have resulted in new state regulations requiring pharmacists to offer to counsel patients each time a new prescription is dispensed. Educators across the country are vigorously integrating the concepts of pharmaceutical care into all practice-related course work, and schools of pharmacy are moving to adopt the PharmD as the sole entry-level degree.

"Why get involved? As a practicing pharmacist, I am only paid to dispense. I simply do not have the time, and no one expects me to do this." There is no question that the fee-for-service reimbursement system provides disincentives to practice pharmaceutical care. The percentage of prescriptions paid for under third-party programs has dramatically increased. Third-party prescription services are often offered to the lowest bidder, which means low dispensing fees and minimal services to patients. Further exacerbating the problem is a pharmacist shortage in many geographic areas. At a time when pharmacist salaries are at an all-time high, lower third-party dispensing fees mean pharmacists must work harder to generate the same level of revenue. There are strong incentives to make the dispensing process more efficient through use of technicians and technology, as exhibited by the use of automatic dispensing machines and mail-order services.

On the positive side, there is a growing recognition among health benefit managers that the economic goal is not to obtain the lowest cost per-prescription price, but to achieve appropriate drug utilization. There is increased realization that appropriate use of prescription drugs can positively affect total health care utilization and that higher drug utilization may be cost-effective. Disease management, in which attention is directed to appropriate utilization of all services necessary to manage a specific disease state optimally, is emerging as the new focus of attention of health benefits managers. Because most chronic diseases involve use of prescription drugs, the pharmacist has an opportunity to refocus his or her professional efforts toward a patient care goal. As of this writing, we are witnessing the emergence of different reimbursement plans to provide more appropriate incentives toward practicing pharmaceutical care, including compliance enhancement. One prominent national example is the incentive package offered to pharmacists by PAID-Medco through its Coordinated Care Network.[41] The program offers additional compensation opportunities to pharmacists in the areas of formulary adherence, therapeutic and generic interchange, disease management, and drug-taking compliance. Beginning in 1994, the disease management focus initially targeted patients with asthma and chronic obstructive pulmonary disease, with plans to expand to other diseases such as osteoporosis, diabetes, gastrointestinal problems, and smoking cessation. The compliance program provides incentives for pharmacists to improve patient compliance for target drugs used in chronic diseases.

"What can I as a pharmacist do about noncompliance? It's the patient's business, not mine." Two additional barriers must be overcome in providing compliance enhancing services. One is the pharmacist's view of his or her professional role, and the other is the patient's view. If the pharmacist adopts a disease management role as opposed to a dispensing role, providing compliance-enhancement services comes naturally. If the patient expects assistance from the pharmacist in managing the drug therapy portion of his or her disease management plan, the services become easier to offer. This transition in expectations is evolving but will require continued education and communication of the pharmacist and patient. Perhaps more to the point, compliance-enhancement services are not imposed; rather, they are offered as a service to assist the patient in better controlling his or her disease state.

"How do I do it effectively and efficiently?" We identified several "how to's" above. To be sure, to engage in all of them for each patient is not practical. The following are essential points about compliance-enhancement strategies.

- Provide privacy and uninterrupted time. A person's health is very personal and should be treated confidentially. Time spent may be brief but needs to be focused and uninterrupted.
- Keep it simple. Cognitive processes are different for all of us. The elderly in particular need simple, straightforward messages. You will likely see the patient again. Follow-up visits are the best time to supplement initial messages.
- Tailored messages work best. Relate the message to what the patient already knows. Adapt strategies to the patient's lifestyle.
- Multiple intervention strategies work better than single ones. Offer verbal and written information as well as reminder aids.
- Develop a consistent strategy. Develop a standardized approach to addressing compliance in your pharmacy.

CONCLUDING THOUGHTS

Managed care entities are increasingly dominating the health care marketplace, and their influence on pharmacists' professional practice will be increasingly felt in future years. In general, directors of managed care programs are oriented toward a population-based perspective and toward optimizing the use of professional resources. Drug-taking compliance is increasingly being recognized as an important issue. But the manner in which managed care programs address it differs considerably from that of the practicing pharmacist. A manager might well ask:

- Are resources directed toward optimizing compliance prudently and efficiently deployed?
- Given that not all diseases need equally fervent attention to compliance, what diseases should be targeted?
- What types of patients are likely to become noncompliant, and how can patients at highest risk be identified?

Intervention strategies:

- What is the most cost-effective way to identify noncompliance problems?

- Which interventions work best to improve compliance?
- Where should compliance-enhancement strategies take place (e.g., at the time of prescribing [prescriber or nurse, pharmacist], at the time of dispensing [pharmacist], after dispensing [telephoned follow-up by pharmacist, nurse, physicians' office, or by an outside party])?

The message is that compliance is increasingly recognized as important, but established mechanisms to achieve optimal compliance levels are not yet in place. Therein lies the opportunity–and the challenge–for pharmacists.

REFERENCES

1. The forgetful patient. Schering report IX. Kenilworth, NJ: Schering Laboratories, 1987.

2. Taubman AH, King JT, Weisbuch JB. Noncompliance in initial prescription filling. Apothecary 1975;9:14.

3. Hammel RW, Williams PO. Do patients receive prescribed medication? J Am Pharm Assoc 1964;NS4:331-7.

4. Eraker SA, Kirscht JP, Becker MH. Understanding and improving patient compliance. Ann Intern Med 1984;100:258-68.

5. Stewart RB, Caranasos GJ. Medication compliance in the elderly. Med Clin North Am 1989;73:1551-63.

6. Stewart RB, Cluff LE. Commentary: a review of medication errors and compliance in ambulant patients. Clin Pharmacol Ther 1972;13:463-8.

7. McKenney TJ, Harrison WL. Drug related hospital admissions. Am J Hosp Pharm 1976;33:792-5.

8. Cross FS, Long MW, Banner AS, Snider DE. Rifampin-isoniazid therapy of alcoholic and nonalcoholic tuberculosis patients in a U.S. Public Health Service Cooperative trial. Am Rev Resp Dis 1980;122:349-53.

9. Fox W. Compliance of patients and physicians: experience and lessons learned from tuberculosis. Brit Med J 1983;287:33-5.

10. Ettenger RR, Rosenthal JT, Marik JL, et al. Improved cadaveric renal transplant outcome in children. Pediatr Nephrol 1991;5:137-42.

11. Sullivan SD, Kreling DH, Hazlet TK. Noncompliance with medication regimens and subsequent hospitalizations: a literature analysis and cost of hospitalization estimate. J Res Pharm Econ 1990;2(2):19-33.

12. Weintraub M. Compliance in the elderly. Clin Geriatr Med 1990;6:445-52.

13. Col K, Fanale JE, Kronhom P. The role of medication noncompliance and adverse drug reactions in hospitalizations of the elderly. Arch Intern Med 1990;150:841-5.

14. Rosenstock IM. Historical origins of the Health Belief Model. Health Educ Monogr 1974;2:328-35.

15. Janz NK, Becker MH. The Health Belief Model: a decade later. Health Educ Q 1984;2(1):1-47.

16. Ried LD, Christensen DB. A psychosocial perspective in the explanation of patient drug-taking behavior. Soc Sci Med 1988;3:70-7.

17. Sauer KA, Coons SJ, Bootman JL, Moon TE. The relationship between health beliefs and compliance in a clinical trial. J Soc Admin Pharm 1987;4:41-7.

18. Fincham JE, Wertheimer AI. Using the Health Belief Model to predict initial drug therapy defaulting. Soc Sci Med 1985;20:101-5.

19. Christensen DB. Understanding patient drug-taking compliance. J Soc Admin Pharm 1985;3:70-7.

20. Christensen DB. Drug-taking compliance. A review and synthesis. Health Serv Res 1978;13:171-87.

21. Fishbein M, Ajzen I. Belief, attitude, intention and behavior: an introduction to theory and research. Reading, MA: Addison-Wesley, 1975.

22. Mellins RB, Evans D, Zimmerman B, Clark NM. Patient compliance: are we wasting our time and don't know it? Am Rev Resp Dis 1992;46:1376-7.

23. Weintraub M. A different view of patient compliance in the elderly. In: Vestal RE, ed. Drug treatment in the elderly. Balgowlah, Australia: ADIS Health Sciences Press, 1984:43-50.

24. Steiner JF, Fihn SD, Blair B, Inui TS. Appropriate reductions in compliance among well controlled hypertensive patients. J Clin Epidemiol 1991;44: 1361-71.

25. Taylor S. Adjustment to threatening events: a theory of cognitive adaptation. Am Psychol 1983;38:1161-73.

26. Conrad P. The meaning of medications: another look at compliance. Soc Sci Med 1985;20:29-37.

27. Stimson GV. Obeying the doctor's orders: a view from the other side. Soc Sci Med 1974;8:97.

28. Myers ED, Calvert EJ. Information, compliance, and side effects: a study of patients on antidepressant medication. Br J Clin Pharmacol 1984;17:21-5.

29. Eisen SA, Miller DK, Woodward RS, Spitznagel E, Przybeck TR. The effect of prescribed daily dose frequency on patient medication compliance. Arch Intern Med 1990;150:1881-4.

30. Cramer JA, Mattson RH, Prevey ML, Scheyer RD, Ouellette VL. How often is medication taken as prescribed? A novel assessment technique. JAMA 1989;261:3273-7.

31. Kruse W, Eggert-Kruse W, Rampmaier J, Runnebaum B, Weber E. Dosage frequency and drug-compliance behaviour–a comparative study on compliance with a medication to be taken twice or four times daily. Eur J Clin Pharmacol 1991;41:589-92.

32. Schering report XIV. Improving patient compliance: is there a pharmacist in the house? Kenilworth, NJ: Schering Laboratories, 1992.

33. Sclar DA, Chin A, Skaer TL, Okamoto MP, Nakahiro RK, Gill MA. Effect of health education on the utilization of antihypertensive therapy: a prospective trial among HMO enrollees. Primary Cardiol 1992;(Suppl 1):24-35.

34. Windsor RA, Bailey WC, Richards JM, Manzella B, Soong SJ, Brooks M. Evaluation of the efficacy and cost effectiveness of health education methods of increased medication adherence among adult patients with asthma. Am J Public Health 1990;80:1519-21.

35. Berg JS, Dischler J, Wagner DJ, Raia JJ, Palmer-Shevlin N. Medication compliance: a healthcare problem. Ann Pharmacother 1993;27(Suppl):3-19.

36. Sackett DL, Snow JS. The magnitude and measurement of compliance. In: Haynes RB, Taylor DW, Sackett DL. Compliance in health care. Baltimore: Johns Hopkins Press, 1979.

37. Haynes RB, Taylor W, Sackett DL, Gibson ES, Bernholz MA, Mukherjee J. Can simple clinical measurements detect compliance? Hypertension 1980;2: 757-64.

38. Morisky DE, Green LW, Levine DM. Concurrent and predictive validity of a self-reported measure of medication adherence. Med Care 1986;24:67-74.

39. Gardner M, Boyce RW, Herrier RN. Pharmacist-patient consultation program: an interactive approach to verify patient understanding. An educational unit developed by the U.S. Public Health Service, Indian Health Service in cooperation with the University of Arizona School of Pharmacy.

40. Hepler CD, Strand LM. Opportunities and responsibilities in pharmaceutical care. Am J Hosp Pharm 1990;47:533-41.

41. Conlan MF. APhA delegates embrace PAID, reject Clinton at convention. Drug Top 1994;138(4):15-6.

ADDITIONAL READINGS

1. Cramer JA. Identifying and improving compliance patterns: a composite plan for health care providers. In: Cramer JA, Spilker B, eds. Patient compliance in medical practice and clinical trials. New York: Raven Press, Ltd., 1991.

2. Task Force for Compliance. Noncompliance with medications–an economic tragedy with important implications for health care reform. Baltimore, MD: Task Force for Compliance, 1994.

3. Berg JS, Dischler J, Wagner DJ, Raia JJ, Palmer-Shevlin N. Medication compliance: a heathcare problem. Ann Pharmacother 1993;27(Suppl):3-19.

4. Leventhal H, Cameron L. Behavioral theories and the problem of compliance. Patient Educ Counsel 1987;10:117-38.

5. Feldman JA, DeTullio PL. Medication noncompliance: an issue to consider in the drug selection process. Hosp Form 1994;29:204-11.

SECTION F:
OUTCOMES
OF PHARMACEUTICAL CARE

Chapter 15

Clinical Outcomes

Charles E. Daniels

INTRODUCTION

Inherent in the professional responsibility of the pharmacist is the monitoring of the outcomes of medication use in patients.* Pharmacists have been involved in identifying and monitoring clinical outcomes for many years. It is only recently that this function has been so titled and has taken on such a high profile. The present interest in clinical outcomes monitoring comes in conjunction with increasing interest in providing defined progressive service programs. It also comes at the same time that many pharmacists search for reimbursement for cognitive services.

The purpose of this chapter is to discuss the concept of clinical

---

*Editors' Note: The focus of this text is the social and behavior aspects of pharmaceutical care. Nevertheless, we have asked Dr. Daniels to provide a short statement on the more traditional clinical outcomes before the presentations on economic and humanistic outcomes.

outcomes vis-à-vis broader outcome measures, briefly review applicable pharmacy clinical outcomes literature, and consider its relevance in the practice of pharmacy.

DEFINITION OF CLINICAL OUTCOMES

Traditional clinical monitoring parameters used by pharmacists have been simple objective measures of a biological and clinical nature. These include serum drug levels, blood pressure monitoring, blood glucose monitoring, cholesterol levels, and a long list of other measures. Pharmacists have adopted these simple measures in part because the measures were in general use by the medical community and also because they offered a measure which has often been considered to be an outcome. Many of them are actually indicators of structure or process.[1] They in fact have only marginal application to the global outcome and thus have limited "real" value.

In the broadest sense, clinical outcome measures are not easily defined. Outcome is a relative term based upon the natural clinical course of a disease, the intended therapeutic objective of any treatments, and the patient's expectation of the impact of disease and therapy.

There is not a single definition of clinical outcomes. Donabedian defines outcomes as "changes in the current or future health status of patients that occur as a result of antecedent medical care."[2] Lohr defines outcomes as the end results of medical care: palliation, control of illness, cure, or rehabilitation.[3]

MONITORING OUTCOMES

Kozma describes the ECHO (*economic, clinical, humanistic, outcomes*) model for monitoring outcomes. This model relies on clinical outcomes such as death, morbidity, emergency room use, etc., to play a greater role in the decision-making process.[4] The model asks that the user consider more than simply the most pure pharmaceutical outcome. For instance, drug serum levels may be optimized with a four-times-daily regimen. However, compliance studies suggest that a drug that can be taken twice daily will be used more consistently by more patients. As the objective is to reach the best outcome, the selection of the twice daily dose may be the better choice in the long run. The ECHO model relies on the user to consider the more global clinical outcomes in the decision-making process. This may often result in a different choice for the best course of action.

Pharmacy has a small but helpful body of literature regarding the topic of outcomes monitoring. Pain is a debilitating symptom of many diseases and is often poorly treated because it is not well monitored. There are now a number of pain assessment question-naires which are being regularly used to help quantify pain for purposes of pharmacological and nonpharmacological treatment. These include the McGill Pain Questionnaire, the Mood Adjective Checklist, and the Profile of Mood States.[5-7] Each of these instru-ments can be used by the pharmacist in monitoring progress in patients whose disease includes continuous or intermittent pain.

Outcome measures for oncology patients are certainly different from those for patients undergoing treatment for psychiatric disease or any other type of disease. Therefore, the intended clinical out-come must be determined by the pharmacist and the patient, in conjunction with other health professionals who are involved in the care of the patient. Simple clinical outcome measures that have been used include tumor mass and survival time. There a number of more complex measures of overall well-being or satisfaction that are coming into more widespread use in conjunction with oncology trials and treatment. These include the Functional Living Index-Cancer (FLIC) and the Cancer Inventory of Problem Situations (CIPS).[8] The pharmacist regularly working with oncology patients must identify which of these instruments is most applicable to his or her practice and incorporate this broader view of outcomes.

Rector reviewed the development of a pharmacist outcome assessment for heart failure patients.[9] He notes that there are sub-stantial differences between the traditional medical monitoring tools for heart disease and the patient's subjective determination of his or her status. He goes on to say that there are multiple compo-nents to the development of an assessment tool for these patients. An important component is the determination of how the disease has affected the patient's life. This includes consideration of the physical, mental, social, and economic impact of the disease.

A few pharmacists have begun to use standardized health monitoring questionnaires as part of their routine monitoring of patients. Wagner and colleagues describe a program whereby patients in a neurology clinic routinely complete a broad health survey which is then used in patient assessment.[10] The same assessment tool is regularly completed by patients in a Boston dialysis unit.[11] In both cases, the patient is asked to complete the instrument as part of the routine workup. This information is then used by the pharmacist and other health professionals who provide the ongoing care to assure that the most global outcomes are being addressed along with the routine therapy issues.

DISEASE STATE MANAGEMENT

Disease management is a concept that has recently come into the spotlight as a component of health care systems.[12] Disease state management is a specialized application of clinical outcome monitoring activities. Chronic diseases are characterized by alternating periods of better or poorer health. This type of disease has historically been frustrating for patient and provider. It has also been expensive for the payer. Disease state management relies upon a predesignated program to optimize health services through a combination of diagnostic and therapeutic resources. Pharmacists have become key players in the disease state management programs of many managed care organizations. The first of the many planned disease state management programs appears to be for asthma.[13] Other likely candidates for disease management programs are ulcers, depression, and arthritis.

Ilersich and colleagues reported the work of pharmacists who monitored the outcome of patients on antibiotic therapy.[14] Of 37 patients, only 26 could be monitored through the course of therapy. This typifies another perspective on the problems associated with clinical outcome monitoring, namely, consistent application of the planned monitoring program. Whatever the outcome monitoring to be done, it must be adopted and used consistently by the pharmacists involved in the program. This will require that a system be put in place to easily track which patients are being monitored by what method. The selection of the outcome measures to be used must be relevant to the disease state.

CONCLUSION

There is promise that the future will bring more scientific evidence to the application of outcomes to the clinical setting. This type of information has been used to a limited degree in formulary decision making and development of treatment guidelines.[15] However, at present, outcomes research is still in a chaotic stage. While over 7,000 articles on outcomes appeared in the medical literature from 1990 to 1994, there is still no generally accepted paradigm of what to measure and how to do so.

REFERENCES

1. Gouveia WA, Chapman MM. The outcomes of patient care. Am J Health-Syst Pharm 1995;52(Suppl 3):S11-S15.

2. Donabedian A. Evaluating the quality of medical care. Milbank Mem Fund Q 1966;44:166-206.

3. Lohr KN. Outcomes measurement: concepts and questions. Inquiry 1988; 25:37-50.

4. Kozma CM, Reeder CE, Schulz RM. Economic, clinical, and humanistic outcomes: a planning model for pharmacoeconomic research. Clin Ther 1993;15: 1121-32.

5. Melzak R. The McGill Pain Questionnaire: major properties and scoring methods. Pain 1975;1:277-99.

6. MacLochlon JFC. A short adjective checklist for depression. J Clin Psych 32:195-7.

7. McNair DM. Profile of mood states. San Diego, CA: Educational and Industrial Testing Service, 1971.

8. Lindley C. Outcome assessment: functional status measures and quality of life as therapeutic endpoints in oncology. Top Hosp Pharm Manage 1990;10(2): 54-63.

9. Rector TS. Outcome assessment: functional status measures as therapeutic endpoints for heart failure. Top Hosp Pharm Manage 1990;10(2):37-43.

10. Wagner AK, Bungay KM, Bromfield EB, Ehrenberg BL. The outcomes of patient care. Am J Health-Syst Pharm 1995;52(Suppl 3):S29-S31.

11. Chapman MM, Meyer KB. Assessing health status in a dialysis clinic: the outcomes of patient care. Am J Health-Syst Pharm 1995;52(Suppl 3):S31-S32.

12. Petersen C. Disease state management. Managed Healthcare 1995;(May): 45-8.

13. Troy TN. Asthma monographs give MCOs the tools necessary to battle a costly disease. Managed Healthcare 1995;(May):48-9.

14. Ilersich AL, Rovers JP, Einarson TR. A pilot study of process and outcome assessment in antibiotic therapy. Can J Hosp Pharm 1991;44:251-8.

15. Schrogie JJ, Nash DB. Relationship between practice guidelines, formulary management, and pharmacoeconomic studies. Top Hosp Pharm Manage 1994;13(4):38-46.

ADDITIONAL READINGS

For the Instructor

1. The importance of outcomes research for pharmacy. Am Pharm 1993;NS33(12)(Suppl).

2. Kozma CM. Outcomes research and pharmacy practice. Am Pharm 1995;NS35(7):35-41.

3. Yuen E, Johnson N. Severity of illness and outcomes research. P & T 1995;(Sept):594-9.

4. Ilersich AL, Rovers JP, Einarson TR. A pilot study of process and outcome assessment in antibiotic therapy. Can J Hosp Pharm 1991;44:251-8.

5. Perfetto EM, Epstein RS, Baldwin R. What are the necessary research methodology components for high quality, scientifically rigorous pharmaceutical outcomes research? In: Patient outcomes of pharmaceutical interventions: a scientific foundation for the future. Washington, DC: American Pharmaceutical Association, 1994.

For the Student

1. Bjornson DC, Hiner WO Jr, Potyk RP, Nelson BA, et al. Effect of pharmacists on health care outcomes in hospitalized patients. Am J Hosp Pharm 1993;50: 1875-84.

2. Bungay KM, Wagner AK. Introduction to pharmacist participation in measuring and monitoring patients' health-related quality of life. Am J Health-Syst Pharm 1995;52(Suppl 3):S19-S23.

3. Rector TS. Outcome assessment: functional status measures as therapeutic endpoints for heart failure. Top Hosp Pharm Manage 1990;10(2):37-43.

4. Meade V. Helping pharmacists provide disease-based pharmaceutical care. Am Pharm 1995;NS35(3):45-8.

5. Gurwitz JH, Noonan JP, Soumerai SB. Reducing the use of H_2-receptor antagonists in the long-term setting. J Am Geriatr Soc 1992;40:359-64.

6. Petersen C. Disease state management. Managed Healthcare 1995;(May): 45-52.

Chapter 16

Economic Outcomes

Brian G. Ortmeier

INTRODUCTION

✱The paradigm for pharmacy practice has shifted from that of dispensing medication to evaluating medication use, ensuring appropriateness of care, and monitoring the outcomes of pharmaceutical interventions.✱Pharmacists were prompted, in part, to take a more active role in patient-focused care by the Omnibus Budget Reconciliation Act of 1990 (OBRA 90). This act mandates, among certain circumstances, the conduction of prospective drug use reviews and patient counseling. In addition, in 1993, seven national pharmacy organizations and the National Consumer League joined efforts to form the Coalition for Consumer Access to Pharmaceutical Care. A primary objective of the coalition is to "create a cost-effective medication and pharmacy services delivery and financing system that enhances patient quality of life."[1]

In 1970, medical care expenditures accounted for 7.5% of the United States gross domestic product (GDP). By 1992, the ratio had increased to 13.9%, and it is projected to reach 19.8% by the year 2015.[2] Although pharmaceuticals account for only 7% of total medical care expenditures, pharmaceutical budgets have been under increased scrutiny as a way to contain costs in many health care sectors. To restrain this growth in health care costs, an increased emphasis has been placed on employing all available health care resources as efficiently and effectively as possible.[3] Pharmacists must be able to evaluate the clinical and economic impact pharmaceuticals have on patient care, the patient's health status, and on the entire health care delivery system.[4,5]

To assist in the explanation of the impact health care professionals have on a patient, a theoretical model was developed which relates disease and certain outcome indicators to assist health care practitioners in decision making about medical care interventions. The economic, clinical, and humanistic outcomes model, called the ECHO model, provides a mechanism by which outcomes of treatment and services can assist the health care practitioner in making utilization decisions.[6] This chapter concentrates on the economic component of outcomes research.

Economic evaluation is a process by which the costs and consequences of an intervention, be it a pharmaceutical, medical device, or service, are assessed. The outcome of an economic evaluation is to determine the influence such an intervention has on health care resource utilization, the patient, and/or society. Maynard states that the objectives of health economic studies are: "(1) to prioritize expenditures and determine decisions about who will live and in what degree of pain and discomfort; and (2) to inform medical audit so that efficient decision makers can be identified and inefficient decision makers be reformed or dismissed."[7]

PERSPECTIVE

The first step in conducting an economic evaluation is to identify the perspective from which the result will be viewed. In an economic analysis, the perspective from which the study was conducted and the audience for which it is intended become very important. While the benefit or costs associated with a service or product may be important from one point of view, they may be of little interest to an individual with another viewpoint. There are many possible perspectives to consider when conducting an economic analysis. Examples would be patients, health professionals, health care institutions, third-party payers, and society. Table 16.1 summarizes how costs and outcomes differ in terms of importance based upon the perspective of the individual.

As an example, suppose an economic analysis shows that two antibiotic therapies provide the same therapeutic benefit. Let's fur-

TABLE 16.1. Costs and Consequences Relative to Various Perspectives

Perspective	Relevant Costs	Relevant Consequences
Patient	Out-of-pocket costs (copayments, deductibles) Lost income Transportation	Therapeutic effectiveness Adverse events Quality of life
Health Practitioner	Hospitalization costs (inpatient and outpatient) Pharmaceutical costs Personnel Supplies	Therapeutic effectiveness Adverse events
Hospitals	Costs incurred during hospital stay Treatment of adverse events and complications	Therapeutic effectiveness Adverse events
Third-Party Payers	Hospitalization costs (inpatient and outpatient) Pharmaceutical costs Nursing home care	None
Society	All possible costs, including lost productivity	All possible consequences, including quality of life and extended life years

ther suppose that one therapy, Therapy A, is significantly less expensive than the other, Therapy B. However, the less costly Therapy A has a higher incidence of gastrointestinal (GI) upset than the more costly alternative Therapy B. The perspective of a third-party payer may be to include the less costly Therapy A on its formulary over the more expensive Therapy B. However, from the perspective of the patient–whose out-of-pocket costs are minimal–the more costly alternative, Therapy B, would be more attractive due to the decreased incidence of GI upset.

Obviously, it is important to determine from whose viewpoint the economic analysis will be conducted. One must be careful not to be too restrictive in the viewpoint, such as selecting only an institution's point of view, which would limit the evaluation's generalizability to other populations. Although a societal perspective may be too broad and encompassing, Drummond and colleagues recommend that a broader viewpoint should always be a consideration, especially in conjunction with a more limited one.[8]

DEFINING COSTS AND CONSEQUENCES

As noted above, the initial step in conducting or reviewing an economic analysis is to determine from whose perspective the study is or was undertaken. This step is important in that it helps us identify the scope of relevant costs and consequences to include in the economic analysis. The next step is to delineate the costs and consequences that are relevant to the intended audience.

Costs

Costs for economic analyses are usually divided into four groups: direct medical costs, direct nonmedical costs, indirect costs, and intangible costs. Direct medical costs involve expenses incurred directly pertaining to medical care for an illness by an institution, third-party payer, or patient. Examples of direct medical costs include medication costs, physician fees and office visits, laboratory tests and diagnostic procedures, and treatment for adverse events.

Direct nonmedical costs are primarily related to out-of-pocket expenses incurred by the patient due to the illness. An example would be transportation to the clinic or hospital for treatments or procedures. These expenses are related to the disease, but are not *medical* expenses.

Indirect costs are generally considered when a societal perspective is taken for the study. These costs are associated with morbidities and mortalities of the disease. Examples of indirect costs are lost productivity and decreased earning power due to morbidities and lost earnings due to premature death. Another indirect cost would be future costs caused by a longer life expectancy. Due to the nature of these expenses, indirect costs are very difficult to capture and to assign a value.[9-13] However, third-party payers are starting to pay closer attention to them. As part of their offering to employers, third-party payers are expressing an interest in showing that a decrease in indirect costs, such as absenteeism, is due to the medical care provided in their health care plan. This decrease in lost productivity may lead to increased productivity, potentially giving the third-party payer a marketing edge over its competitors.

Intangible costs are those that result from psychological factors such as stress, pain due to the illness, and side effects of treatments. Like indirect costs, intangible costs are usually considered when a societal perspective is taken. Intangible costs are also similar to indirect costs in that they are very difficult to assess and assign a value. Table 16.2 summarizes the more frequent items found in each cost category that might be considered for inclusion in an economic evaluation.

Benefits/Consequences

Similar to costs, benefits or consequences derived from an intervention can be divided into three categories: direct benefits, indirect benefits, and intangible benefits. Direct benefits would include a reduction or avoidance of future illness and its associated treatment costs, a reduction of insurance payments, and a reduction of ser-

TABLE 16.2. Costs

Cost Category	Costs
Direct Costs	Outpatient costs • Medications • Physician office visits • Home health care • Social services Inpatient costs • Hospitalization • Medications • Laboratory and diagnostic tests • Consultations • Adverse event treatment • Failure/relapse treatment
Direct Nonmedical Costs	Cost borne by the patient such as: • Transportation • Accommodations for family • Meals
Indirect Costs	Loss of work days Income foregone due to death
Intangible Costs	Quality of life Psychological factors (stress, fear, pain, and suffering)

vices and products used (e.g., decreased number of hospitalizations or decreased number of laboratory tests ordered). Indirect benefits are those that would be experienced through avoidance of physical disability or death, resulting in increased employee productivity. Intangible benefits include avoidance of pain and improvement in quality of life.

COST/BENEFIT VALUATION

Assigning Costs

Once all costs and benefits have been identified for inclusion in the economic evaluation, it is time to assign a dollar value. Determining direct costs is fairly straightforward. However, it is important to note that a difference exists between costs and charges in an analysis. What the actual cost of the intervention or treatment cost is and what is actually charged to the payer are two separate amounts in most instances. One commonly used cost is average wholesale price (AWP), which is a standard list price for a pharmaceutical product, available to all purchasers. Rarely, though, is AWP paid because most contracts for pharmaceuticals are negotiated at a price less than AWP. However, if a researcher uses actual charge data for a particular institution in the economic analysis, the results would not be generalizable because they are used on values unique to that particular institution.

Assigning a dollar value to benefits and indirect costs is even more difficult. There are two main approaches to placing value on benefits and indirect costs. They are: (1) the human capital approach and (2) the willingness-to-pay method.

In the human capital approach, the value of human work and life is calculated by the economic productivity of the individual. This method allows us to equate actual market income of an individual or group of individuals with lost resources and profit related to an illness. A criticism of the human capital method is that often the market price does not represent the true value of human productivity. In addition, it is biased toward middle-aged, educated individuals with an above-average income and discriminates against children, older adults, and full-time homemakers.

A second approach to valuing benefits and indirect costs is the willingness-to-pay method. The value of benefits and indirect costs in this case are determined by what the individual is willing to pay for preventing negative or receiving positive effects of an intervention. One difficulty with this method is that often it is difficult to differentiate between the price people are willing to pay and the price they are actually able to pay.

Discounting

Costs and benefits of an intervention generally arise at different times and are often spread over an extended period of time. Discounting is a method that takes into account the effect time has on costs and outcomes. Discounting should be considered when costs are to be incurred in the future or benefits are to be experienced in the future to reflect the current, present value of the costs or benefits. Future costs and benefits should be valued lower than present costs and benefits because a dollar invested today is worth more than a dollar invested in the future. Future values should first be discounted and then incorporated into the economic analysis.

There is continuing debate over whether health outcomes or benefits and costs should be discounted and over what discount rate should be used. Selection of a discount rate varies significantly in the published literature. Most recently, economists have recommended uniform discount rates ranging from 3 to 6%. The use of a standardized discount rate becomes very important when making comparisons between interventions that occur over varying periods of time. A simple discount equation multiplies future costs by a discount factor:

$$PV = FC \times (1 + DR)^{-n}$$

where *PV* is the present value, *FC* is the future costs, *DR* is the discount rate used, and *n* is the year cost will be incurred in the future.[14]

For example, let's assume that we have a program that will be implemented over a period of four years at a cost of $1,000 per year. Let's assume further that these costs are realized at the end of each year. Using the above equation, the unadjusted cost over the four-year period would be $4,000. With a discount rate of 6%, the cost incurred would be $3,465 ($943, $889, $840, and $792, for each year, respectively).

Sensitivity Analysis

As one can deduce from the previous sections, there is a measure of uncertainty about whether the correct value was placed on a cost or benefit, an outcome was appropriately estimated, or the correct discount rate was used. To address this uncertainty, researchers conduct a sensitivity analysis. In a sensitivity analysis, an economic evaluation is reanalyzed using different estimates of the areas of which the researcher is most uncertain. For instance, if the original analysis used a discount rate of 3%, a sensitivity analysis would reanalyze the costs, possibly over a range of 2 to 5%. If the results were not significantly different from the original analysis, the researcher would be fairly confident that the values obtained in the original analysis were robust. If, however, the results were significantly different, the researcher would have to reevaluate the methods used to select the particular rate, ensuring that the rate used was the one most appropriate for the analysis. A similar reanalysis can be conducted on any assigned value of concern.

TYPES OF ECONOMIC ANALYSES

Five methods of analysis are associated with an economic evaluation. These methods are: (1) cost-of-illness, (2) cost-minimization, (3) cost-benefit, (4) cost-effectiveness, and (5) cost-utility analysis.

Cost-of-Illness Analysis

A cost-of-illness analysis involves the evaluation of the overall economic impact a disease has on a particular patient population. This type of evaluation should include all costs and consequences of treating that disease and entails a simple summation of costs and consequences associated with the disease. In a cost-of-illness analysis, no separate comparison of alternative treatment approaches is involved.

There are several advantages to performing a cost-of-illness study prior to initiating further economic evaluation. The cost-of-illness evaluation allows the researcher to collect and assess disease-specific data, providing a definition of the illness, its epidemiology and potential outcomes, and the consequences associated with the disease.

An example of a cost-of-illness study is a large multicenter survey conducted to collect health care resource utilization data for people infected with human immunodeficiency virus (HIV).[15] The analysis was conducted from the perspective of the health care provider. Costs included in this cost-of-illness study were inpatient hospitalization, outpatient visits, home health care, medication costs, and long-term care. It was estimated that the lifetime cost of treating a person with HIV from the time of infection until death would be approximately $119,000.

Another example of a cost-of-illness analysis of persons with HIV was conducted from the perspective of a business firm.[16] The costs included in this study were medical, disability, and life insurance; pension; and employee replacement costs. The average five-year cost to a business firm for an HIV-infected employee was estimated at $17,000. As this example demonstrates, the perspective used defined which costs were important for inclusion in the evaluation. This change in perspective led to a significantly different conclusion from the previous cost analysis.

Cost-Minimization Analysis

A cost-minimization analysis is used to examine the costs associated with two or more alternatives that are clinically equivalent in terms of outcomes or consequences. Equivalency must be established prior to conducting a cost-minimization analysis. Equivalency should be demonstrated not only for clinical efficacy, but also in terms of incidence and type of adverse events. When equivalency has been demonstrated, the analysis focuses on finding the least expensive alternative, usually based on acquisition and administration costs. The formula for a cost-minimization analysis is:

Costs (in dollars) of Intervention A <, =, or > Costs (in dollars) of Intervention B[17]

An example of a cost-minimization analysis would be the evaluation of two pharmacological treatments with the same pharmaceutical agent but different routes of administration. Both have the same effectiveness in terms of cure rates. The difference, however, may lie in the administration route, the costs associated with the method of administration, and the amount of medication administered. One type of delivery may be more labor intensive and costly. The alternative administration method may be less labor intensive at a similar cost. Thus, if the outcome is the same regardless of method of administration, the recommended therapy would be the method that is less labor intensive.

Another example of a cost-minimization analysis would be when there are two alternative treatment approaches. In one instance, the patient receives therapy in the hospital. In the second instance, the patient receives the same therapy at home. In either instance, the ultimate outcome in regard to effectiveness and incidence of adverse events is the same. The home therapy alternative would be the preferred therapy because it provides the least costly alternative, avoiding more expensive hospitalizations.

Cost-Benefit Analysis

An economic evaluation where all costs and outcomes or consequences are expressed in monetary terms is a cost-benefit analysis. A cost-benefit analysis answers the question: Will the benefits of a program exceed the cost of implementing it? Typically, the result of a cost-benefit analysis is stated in the form of a ratio of costs to benefit. It can, however, also be reported as a net benefit dollar amount, subtracting total costs from total dollar benefit. The two equations commonly used for cost-benefit analysis are:

$$\text{Cost-Benefit Ratio} = \text{Costs (\$)/Benefits (\$)}$$
$$\text{Net Benefit} = \text{Benefits (\$)} - \text{Costs (\$)}$$

Cost-benefit analyses are typically conducted when two competing therapies or programs have different outcomes. The goal is to identify the program or intervention with the highest net benefit. An advantage to performing a cost-benefit analysis is that different treatments or programs with different outcomes can be compared because they are all termed in the same units, dollars. It is an evaluation that can be used in situations where resources are limited and a decision must be made between two programs competing for the same resource.

For example, a small regional clinic has the resources to invest in only one of two new diagnostic devices, each used to diagnose a different disease. The decision to purchase one device over another could be based on the cost-benefit ratios associated with each device. For each alternative, the researcher would have to determine the actual cost of each device and the personnel time to run the device. The benefits might include morbidities and mortalities prevented due to early diagnosis of the disease and early intervention.

A disadvantage to cost-benefit analyses is that many consequences are difficult to value in monetary terms, for instance, the monetary value of pain and suffering and quality of life.[18,19] Another disadvantage is that many consequences *cannot* be valued in monetary terms; thus, there is a danger that these consequences will be excluded from the analysis.

An example of a cost-benefit analysis is one conducted on contraceptive therapy.[20] Costs and benefits associated with the use of four hormonal contraceptive therapies were identified and analyzed based on the cost per patient day of effective pregnancy prevention. Costs included in the study were medication cost, physician office visits, costs associated with the administration and removal of the contraceptive method, medication for treatment of adverse events, and contraceptive failure. Failure costs were further broken down into those associated with carrying the fetus to full-term versus those that chose to abort. Benefits included in the study were pregnancies prevented and protective effect for endometrial cancer. Dollar amounts were applied to all costs and benefits associated with each therapy. Table 16.3 summarizes the outcome of the analysis. As is seen, Therapy A had the highest net benefit. The differences in total benefits were due in part to varying efficacy rates of the therapy, discontinuation rates, and length of therapy.

Cost-Effectiveness Analysis

When costs and consequences are simultaneously measured— costs in monetary terms, consequences in terms of an obtained unit

TABLE 16.3. Summary of a Cost-Benefit Analysis of Four Hormonal Contraceptive Methods

Cost/Benefit (Per Patient Per Day)	Therapy ($) A	Therapy ($) B	Therapy ($) C	Therapy ($) D
Total Cost	0.88	0.96	1.08	1.78
Total Benefit	3.75	3.75	3.85	3.42
Net Benefit	2.87	2.79	2.77	1.64

of effectiveness—the result is a cost-effectiveness analysis.[14] This type of analysis differs from cost-benefit evaluations in that the health outcome or consequence in a cost-effectiveness analysis is measured in nonmonetary terms. The ratio typically used in cost-effectiveness is:

Cost-Effectiveness Ratio = Costs ($)/Therapeutic Effect (natural units)

Effectiveness is measured in some type of natural outcome unit. Examples include number of cases cured, reduced length of illness, gained working days, years of life saved, number of human lives saved, and cases of disease prevented.

A simple example would be in the treatment of noninsulin-dependent diabetes mellitus. Two drugs, A and B, lower plasma glucose levels, but Drug A may lower the plasma glucose on average 30mg/dl, while Drug B may lower plasma glucose by 40mg/dl, on average. If the drugs are similar in price, Drug B would be the more cost-effective agent.

Another example of a cost-effectiveness analysis is an evaluation of three antibiotics used to treat an infection. The cost of Therapy A is $50, the cost of Therapy B is $70, and the cost of Therapy C is $95. Effectiveness, as measured by cure rate, is 80% for Therapy A, 90% for Therapy B, and 92% for Therapy C. Based on 100 patients, the cost-effectiveness ratios for Therapies A, B, and C are $63, $78, and $103 per cure, respectively. Therapy A requires the lowest investment per cure. Based purely on the cost-effectiveness ratio, Therapy A would be the therapy of choice.

Consideration should also be given to the incremental cost difference between Therapy A and the other therapies. The incremental difference between Therapy A and Therapy B is $2,000 [($50 × 100) $ ($70 × 100)]. The incremental difference between Therapy A and Therapy C is $4,500 [($50 × 100) $ ($95 × 100)]. However, when compared to Therapy A, Therapy B would cure ten more patients for an additional cost of $200 per cure, and Therapy C would cure 12 more patients at an additional cost of $375 per cure. The difference in costs associated with an increase of one more unit, in this case a cure, is called the marginal cost of therapy. The point of this additional cost per cure must be discussed by the people responsible for making formulary or treatment decisions to determine whether the additional benefit is "worth" the additional cost.

There are a couple of disadvantages to performing a cost-effectiveness analysis. One disadvantage is that only interventions with identical clinical endpoints can be compared. For instance, a cost-effectiveness analysis could not be performed on two pharmacological therapies where one clinical endpoint might be reduction in mmHg for hypertension and the other a reduction in mg/dl total cholesterol. Another disadvantage is that there is an implicit assumption that all outcomes are equivalent, that is, the years of lives saved, as a unit of effectiveness, are valued the same. One way to address the latter disadvantage is to perform a cost-utility analysis.

Cost-Utility Analysis

A cost-utility analysis is defined as a method where costs are measured in terms of dollar amounts and consequences are measured in terms of quality of life, willingness to pay, or preference for one treatment or intervention over another.[14] A cost-utility analysis measures the quality of a health outcome derived from an intervention–the change in a patient's health status/outcome–and assigns a dollar value per unit of utility. The formula for calculating a cost-utility ratio is:

Cost-Utility Ratio = Costs ($)/Utilities (e.g., QALY)

In many ways, cost-utility and cost-effectiveness analyses are similar. Both measure costs in terms of dollar amounts, and both are concerned about treatment consequences. They differ, however, in how those treatment outcomes are measured. Cost-effectiveness analyses measure health effects in natural units (e.g., number of cures, lives saved) and are expressed in cost per unit of effect. Cost-utility analyses measure health effects in quality-adjusted life years (QALYs) gained.[14] The result of a cost-utility analysis is expressed as cost per quality-adjusted life years gained.

Quality of life can be determined through a wide variety of health status scales. Examples include the Medical Outcomes Study Short Form 36 (SF-36), the Sickness Impact Profile (SIP), and the Quality of Well-Being Scale (QWB). Another chapter provides an in-depth discussion on quality of life outcome measures.

Quality-adjusted life years are the number of years at full health compared to the number of years as actually experienced with the patient's illness. A year of perfect health would equal 1.0 QALY, and death would equal 0.0 QALYs. This value is then adjusted based on the actual quality of life of the life years gained due to an intervention. For example, four years of perfect health would be valued at 4.0 QALYs, whereas four years of living with a disease such as cancer, with each year having a utility of 0.3, would be valued at 1.2 QALYs.

Utilities can be determined by three methods: (1) through estimation, (2) through the literature, and (3) through actual measurement. Clinicians, investigators, or an expert panel can provide estimates of the health utility of a particular disease. Increasingly, research on utility values for specific disease states are being published. Finally, the optimal method of determining health utilities is through actual measurement. Three instruments are commonly used to measure utility values: rating scale, standard gamble, and time trade-off.[8]

A cost-utility analysis has the advantage of being the only economic analysis that can determine the impact an intervention has on the quality of a patient's life. The major disadvantage associated with this type of evaluation is the lack of utility measure standardization.

CRITICAL EVALUATION
OF THE HEALTH ECONOMIC LITERATURE

Economics applied to the health sciences is not new. Pharmacy and pharmaceuticals have used several cost analytical approaches for many years. However, the application of various types of economic analyses in inappropriate ways and a wide variation in the definition of various economic terms has led to confusion in this rapidly developing science.[21,22] In response to the wide variation in application and interpretation of these analytical tools, several countries have established economic standards for evaluating pharmaceuticals. Australia requires that economic analyses be included in submissions for national drug formulary acceptance following set guidelines.[23,24] In Canada and Italy, similar guidelines are being considered for adoption.[25]

A variety of evaluative aids have been published to assist the individual charged with evaluating economic literature for health care policy decision making.[8,26,27] The contents of the lists are similar and are summarized as follows: (1) Was a research question proposed in an answerable form, with the alternatives compared and perspective clearly specified? (2) Were all appropriate, competing alternatives identified and included in the analysis? (3) Were all relevant costs and consequences, based on the perspective, included in the study and measured in an appropriate manner? (4) Was the appropriate economic evaluation used? (5) Was discounting of costs and consequences used when appropriate? (6) Was a sensitivity analysis performed? (7) Were the conclusions appropriate based on the analysis?

APPLICATIONS FOR HEALTH ECONOMIC RESEARCH

Once the economic research has been conducted, the conclusions generated can be used in a variety of ways by decision makers within the health care system. When applied correctly, this information can assist in employing limited and valuable health care resources in the most efficient and effective way possible. The information garnered can be instrumental in formulary decisions for inclusion, exclusion, or use of a pharmacological agent on a restricted basis. In addition, information provided by economic analyses can be used to make difficult decisions in regard to selecting services to be offered by a health care institution. As an example, a question a pharmacy director may ask is: What impact will a pharmacokinetic service have on health care expenditures? Potential outcomes to be measured may be pharmaceutical costs, length of stay, and costs of adverse events due to over- or underutilized pharmacological agents.

The disease modeling involved in economic research can also assist the health care practitioner in developing a rational care plan based on expected clinical, economic, and humanistic outcomes. But above all, it is imperative that quality patient care be maintained as further financial constraints are placed on the health care budget.

REFERENCES

1. Coalition for Consumer Access to Pharmaceutical Care. An outpatient pharmacy services benefit in a reformed health care system. Am J Hosp Pharm 1993;50:1465.

2. Warshawsky MJ. Projections of health care expenditures as a share of the GDP: actuarial and macroeconomic approaches. Health Sci Res 1994;29:293-313.

3. Freund DA, Dittus RS. Principles of pharmacoeconomic analysis of drug therapy. Pharmacoeconomics 1992;1:20-32.

4. Reeder CE. Economic outcomes and contemporary pharmacy practice. Am Pharm 1993;NS33(Suppl 12):S3-S6.

5. Cooper JW. Clinical outcomes research in pharmacy practice. Am Pharm 1993;NS33(Suppl 12):S7-S13.

6. Kozma CM, Reeder CE, Schulz RM. Economic, clinical and humanistic outcomes: a planning model for pharmacoeconomic research. Clin Ther 1993;15: 1121-32.

7. Maynard A. The design of future cost-benefit studies. Am Heart J 1990;119:761-5.

8. Drummond MF, Stoddart GL, Torrance GW. Methods for the economic evaluation of health care programmes. Oxford: Oxford Medical Publications, 1987.

9. McGhan WF. Pharmacoeconomics and the evaluation of drugs and services. Hosp Form 1993;28:365-78.

10. Henry DA. The Australian guidelines for subsidisation of pharmaceuticals. Pharmacoeconomics 1992;2:422-4.

11. Gerard K, Donaldson C, Maynard AK. The cost of diabetes. Diabetic Med 1989;6:164-70.

12. Williams A. Economics of coronary artery bypass grafting. Brit Med J 1985;291:326-9.

13. Drummond MF. Cost-of-illness studies: a major headache? Pharmacoeconomics 1992;2:1-4.

14. Bootman JL, Townsend RJ, McGhan WF. Principles of pharmacoeconomics. Cincinnati, OH: Harvey Whitney Books Company, 1991.

15. Hellinger FJ. The lifetime cost of treating a person with HIV. JAMA 1993;270:474-8.

16. Farnham PG, Gorsky RD. Costs to business for an HIV-infected worker. Inquiry 1994;31(1):76-88.

17. McGhan WF. Pharmacoeconomics and the evaluation of drugs and services. Hosp Form 1993;28:365-78.

18. Weinstein MC, Fineberg HV, Elstein AS, et al. Clinical decision analysis. Philadelphia, PA: W. B. Saunders, 1980.

19. Phelps CE, Mushlin AI. On the (near) equivalence of cost-effectiveness and cost-benefit analyses. Int J Technol Assess Health Care 1991;7:12-21.

20. Ortmeier BG, Sauer KA, Langley PC, Bealmear BK. A cost-benefit analysis of four hormonal contraceptive methods. Clin Ther 1994;16:707-13.

21. Gaspari KC. The use and misuse of cost-effectiveness analysis. Soc Sci Med 1983;17:1043-6.

22. Lee JT, Sanchez LA. Interpretation of "cost-effective" and soundness of economic evaluations in the pharmacy literature. Am J Hosp Pharm 1991;48: 2622-7.

23. Commonwealth of Australia. Guidelines for the pharmaceutical industry on preparation of submissions to the Pharmaceutical Benefits Advisory Committee: including submissions involving economic analysis. Canberra, Australia: Department of Health, Housing and Community Services, 1990.

24. Henry D. Economic analysis as an aid to subsidization decisions: the development of Australian guidelines for pharmaceuticals. Pharmacoeconomics 1992;1:54-67.

25. Canadian Coordinating Office for Health Technology Assessment. Guidelines for economic evaluation of pharmaceuticals. 1st ed. Ottawa, Canada: CCOHTA, 1994.

26. Thiel KA. Assumptions, subjective measures and variables: ten principles for analyzing a pharmacoeconomic study. Outcomes Measure Manage 1994;5(3):2.

27. Sacristan JA, Soto J, Galende I. Evaluation of pharmacoeconomic studies: utilization of a checklist. Ann Pharmacother 1993;27:1126-33.

Chapter 17

Humanistic Outcomes

Stephen Joel Coons
Jeffrey A. Johnson

INTRODUCTION

There is a pressing need in the United States (and most other countries) to maximize the net health benefit derived from the utilization of limited health care resources. However, a major obstacle is the lack of critical information as to what value is received for the tremendous amount of resources expended on medical care. As Maynard has stated, it is commonplace in health care "for policy to be designed and executed in a data free environment!"[1] This lack of critical information as to the outcomes produced by medical care is an obstacle to optimal health care decision making at all levels.

In 1988, Arnold Relman stated that the United States was entering the third era of modern medical care: the Era of Assessment and Accountability.[2] This era follows the Era of Expansion (i.e., the late 1940s through the 1960s) and the Era of Cost Containment (i.e., the 1970s and 1980s). During the Era of Expansion, in an effort to improve access to medical care, a great number of hospitals and health care facilities were built and/or better equipped through federal legislation such as the Hill-Burton Act of 1946 and the Hill-Harris Act of 1964. In addition, financial access was increased by the rapid growth of private health insurance and through the enactment of Medicare and Medicaid legislation in 1965. During the Era of Cost Containment, containing costs was an explicit goal of most of the stakeholders in health care. Nevertheless, no significant cost containment appeared to have occurred. As a matter of fact, the

United States experienced some of the largest annual increases in the rate of growth of health care expenditures during the Era of Cost Containment. The excesses or failures of these earlier eras have led to the Era of Assessment and Accountability. An essential element of this third era is the growing consensus that health outcomes data is essential for the determination of the value of medical care interventions to individuals, systems, and societies.

The Outcomes Movement

Although the implicit objective of medical care is to improve health outcomes, there is minimal evidence of the true effectiveness of many current health care practices.[3] In addition, measures of the overall quality of the U.S. health care system, such as access to primary health care, health indicators (e.g., infant mortality and life expectancy), and public satisfaction in relation to costs, provide evidence that the U.S. trails behind countries that spend significantly less on medical care.[4]

Outcome is one of the three components of the conceptual framework articulated by Donabedian for assessing and assuring the quality of health care: structure, process, and outcome.[5] Traditionally, the approach to evaluating health care has emphasized the structure and processes involved in medical care delivery rather than the outcomes. However, the principal stakeholders in health care–payers, providers, regulators, accrediting bodies, manufacturers, and patients–are placing increasing emphasis on the outcomes that medical care products and services produce.[6] For example, the Joint Commission on the Accreditation of Healthcare Organizations has shifted toward outcome (i.e., "clinical indicator") measurement.[7]

Types of Outcomes

There are a number of ways of describing or characterizing the types of outcomes that result from medical care interventions. One classic list of health outcomes is called the five D's. It includes death, disease, disability, discomfort, and dissatisfaction. The five D's capture a wide range of outcomes that could be used in assessing the quality of medical care.[7] However, the negativity of the terms does not reflect the truly positive impact that medical interventions can have.

A more comprehensive (and more positively worded) conceptual framework has been proposed, the ECHO Model, which classifies outcomes into three categories: *e*conomic, *c*linical, and *h*umanistic *o*utcomes.[8] The model is more comprehensive in that it embodies the five *D*'s within the clinical and humanistic outcomes domains and provides an added economic outcomes dimension. As described by Kozma and colleagues, clinical outcomes are the "medical events that occur as a result of the condition or its treatment." Economic outcomes are the "direct, indirect, and intangible costs compared with the consequences of medical treatment alternatives." The primary patient-reported or humanistic outcomes are health-related quality of life and patient satisfaction. Clinical and economic outcomes associated with pharmaceutical care have been addressed in earlier chapters of this book.

As stated by Ellwood, outcomes research is "designed to help patients, payers, and providers make rational medical care-related choices based on better insight into the effect of these choices on the patient's life."[9] Essential elements of outcomes research are the assessment of patient function and well-being, or health-related quality of life, and patient satisfaction. This chapter focuses on quality of life and patient satisfaction as outcomes or endpoints of the use of pharmaceutical products and services, or pharmaceutical care. In this chapter, we depart somewhat from an aspect of the often-repeated definition of pharmaceutical care, i.e., "the responsible provision of drug therapy for the purpose of achieving definite outcomes that improve a patient's quality of life," in that we will be referring to quality of life as an outcome itself, not the result of an outcome.[10]

QUALITY OF LIFE

Definition

As discussed above, one of the essential elements of outcomes research is the assessment of patient health-related quality of life. However, there is no consensus on the definition of quality of life or its overall conceptual framework.[11] In the literature, the term quality of life has been used in a variety of ways. Kaplan and Bush proposed that studies of health outcomes use the term health-related quality of life to distinguish health effects from the effects of job satisfaction, environment, and other factors that affect perceived life quality.[12] Only health outcomes are discussed in this chapter, so the terms quality of life and health-related quality of life are used interchangeably. In addition, another term that will be used interchangeably with health-related quality of life is health status.

Quality of life, like other aspects of the human experience, is hard to define. In much of the medical research literature, explicit definitions of quality of life are rare. However, some authors have provided definitions. For example, Schron and Shumaker define quality of life as "a multidimensional concept referring to a person's total well-being including his or her psychological, social, and physical health status."[13] Patrick and Erickson propose that quality of life is "the value assigned to duration of life as modified by the impairments, functional states, perceptions, and social opportunities that are influenced by disease, injury, treatment, or policy."[14] Although the two definitions differ in certain respects, a conceptual characteristic they share is the multidimensionality of quality of life. While the terminology may vary with the author, commonly measured dimensions of health-related quality of life include:

- Physical health
- Emotional health
- Social and role functioning
- Perceptions of general well-being
- Disease- and/or treatment-related symptomatology.

Significance of Quality of Life as an Outcome

For medical care providers, quality of life is increasingly being viewed as a therapeutic endpoint. An overriding factor leading to this view has been the gradual shift in the focus of primary medical care from limiting mortality to limiting morbidity and the patient-reported impact of that morbidity. This is because the pattern of illness in our society has shifted from mostly acute disease to one in which chronic conditions predominate. In the early part of this century, many individuals died from infectious diseases for which cures (e.g., antibiotics) and/or effective preventive measures (e.g., increased sanitation, vaccines) were unavailable or underutilized. Although there currently remain many diseases that may shorten life expectancy, it is more likely that a disease will have adverse health consequences leading to dysfunction and decreased well-being. For those conditions that shorten life expectancy and for which there are no cures, managing symptoms and maintaining function are the primary objectives of medical care. Medications are used extensively as a means of maintaining and/or enhancing patients' quality of life.

Because therapeutic interventions such as medications can enhance as well as decrease quality of life, pharmacists and other medical care providers must strive to achieve enhanced quality of life as an outcome of therapy. Although it must be assumed that quality of life has always played an implicit role in the provision of health care, it has not always been viewed as equal in importance to the more clinical or biological outcome parameters (e.g., blood pressure, serum cholesterol, blood glucose level). The subjective nature of quality of life assessment has made many people uneasy with it as a measure of the patient outcomes produced by medical treatment.[15] However, there is growing awareness that, in certain diseases, quality of life may be the most important health outcome to consider in assessing treatment.[16] The logic behind the increasing attention to quality of life includes:

- The goal of therapy is to make people feel better.
- Physiological measures may change without people feeling better.
- People may feel better without measurable change in physiological values.
- There may be trade-offs between positive treatment effects and adverse events.
- There is an increasing emphasis on patient-focused care.[17]

Types of Quality of Life Measures

Hundreds of health-related quality of life instruments have been developed.[18-22] The following is a simple taxonomy of quality of life measures:

- Disease-specific instruments
- Generic/general instruments
  - Health profiles
  - Utility-based instruments

Disease-Specific Instruments

Disease-specific instruments are intended to provide greater detail concerning particular outcomes, in terms of functioning and well-being, that may be uniquely associated with a condition or its treatment. Guyatt and colleagues provide a further distinction in the types of specific instruments available, i.e., disease specific (e.g., arthritis, diabetes), population specific (e.g., frail elderly), function specific (e.g., sexual functioning), and condition or problem specific (e.g., pain).[23] A few selected examples of disease-specific instruments are shown in Table 17.1.

Generic/General Instruments

Some investigators believe that all conditions have a general effect on quality of life and that the purpose of quality of life assessment should be not only to identify clinical information relevant to a specific disease but also to determine the impact of the condition on general function and well-being. A concern is that by focusing too specifically on clinical correlates of a disease or condition, the gen-

TABLE 17.1. Selected Disease-Specific Quality of Life Instruments

■ *Arthritis Impact Measurement Scale (AIMS)*
■ *Asthma Quality of Life Questionnaire (AQLQ)*
■ *Diabetes Quality of Life (DQOL)*
■ *Functional Living Index Cancer (FLIC)*
■ *Quality of Life in Epilepsy (QOLIE)*
■ *HIV Overview of Problems--Evaluation System (HOPES)*

Source: Berzon et al., 1995[22]

eral or overall impact is overlooked. In studies involving pharmacotherapy, the use of a generic instrument and a specific instrument
may be the best approach. The generic instrument will provide a
more holistic outcome score and allow comparability across other
disease states in which it has been used. An appropriately selected
specific instrument will provide more detailed clinical information
regarding expected changes in the particular patient population. The
development of the Kidney Disease Quality of Life (KDQOL)
instrument will be discussed later in this chapter. The KDQOL is an
example of an instrument that contains a generic core (i.e., the
RAND 36-item Health Survey 1.0) plus disease-specific items.

Health Profiles. Health profiles provide an array of scores representing individual dimensions or domains of quality of life or health
status. An advantage of a health profile is that it provides multiple
outcome scores that may be useful to clinicians and/or researchers
who are attempting to measure differential effects of a condition or
its treatment on various quality of life domains. A commonly used
profile instrument is the Medical Outcomes Study Short Form-36
(SF-36). The SF-36 is also known and distributed as the RAND
36-Item Health Survey 1.0, a copy of which is provided in Appendix I. The SF-36/RAND 36-Item Health Survey 1.0 grew out of the
Medical Outcomes Study (MOS) conducted by researchers from
the RAND Corporation.[24, 25] The instrument includes the nine
health concepts or dimensions that are shown in Table 17.2. The

TABLE 17.2. Dimensions of RAND 36-Item Health Survey 1.0/SF-36 Scales

- General health perceptions
- Physical functioning
- Social functioning
- Role limitations attributed to physical problems
- Role limitations attributed to emotional problems
- Bodily pain
- General mental health
- Energy/fatigue (vitality)
- Health transition

Source: Ware and Sherbourne, 1992; Hays, Sherbourne, and Mazel, 1993[24,25]

SF-36/RAND 36-Item Health Survey 1.0 can be administered by a trained interviewer (face to face or via telephone) or be self-administered. It has many advantages. For example, it is brief (the instrument takes approximately five to ten minutes to complete), and there is substantial evidence for its reliability and validity.

Utility-Based Instruments. Utility-based quality of life instruments incorporate the measurement of specific patient health states along with an adjustment for the preferences (e.g., utilities) for the health state. The preferences are empirically measured or assigned through a variety of procedures, including visual analog scales, the time trade-off technique, and standard gamble.[26] The outcome scores on utility-based instruments range from 0.0 to 1.0 and represent the quality of life associated with death and perfect health, respectively. This makes utility-based measures useful in pharmacoeconomic research, particularly cost-utility analysis.[27] Cost-utility analysis involves comparing the costs of an intervention (e.g., a medication) to its outcomes, with outcomes expressed in units such as quality-adjusted life years (QALYs) gained.[28]

QALYs gained is an outcome measure that incorporates both quantity and quality of life. This can be a key outcome measure, especially in diseases such as cancer where the treatment itself can have a significant impact on patient functioning and well-being. A study by Smith and colleagues illustrates the importance of adjusting length of life or survival for quality of life.[29] The authors compared the incremental costs and outcomes associated with surgery plus adjuvant chemotherapy versus surgery alone in colon cancer patients. Their results indicated that 2.4 unadjusted years of life were gained from the addition of the chemotherapeutic regimen. However, after adjusting for quality of life, only 0.4 QALYs were gained. The calculated cost per life year gained was $2,916, while the cost per QALY gained was $17,500. Maximizing the potential of pharmaceutical care will require that pharmacists understand and be able to address the economic *and* quality of life implications of therapeutic decisions, such as those made in regard to adjuvant chemotherapy.

Examples of utility-based measures include the Quality of Well-Being Scale (QWB), the Health Utilities Index (HUI), and the Euro-Qol.[14,30,31]

PATIENT SATISFACTION

Definition

The other humanistic outcome that should be considered when evaluating the outcomes of pharmaceutical care is patient satisfaction. As with quality of life, patient satisfaction is a multidimensional concept that often is not explicitly defined in the empirical literature. It is implicitly defined by the way in which its measurement is operationalized. As reported by Cleary and McNeil, the most commonly measured dimensions of patient satisfaction are:

- Personal aspects of care
- Technical quality of care
- Accessibility and availability of care
- Continuity of care
- Convenience
- Physical setting
- Financial considerations
- Efficacy[32]

Rossiter and colleagues define satisfaction as the extent to which individual needs and wants are met.[33] They assert that, in health care, patient satisfaction is linked to attitudes toward the medical care system, as well as expectations and perceptions regarding the quantity and quality of the care received. Pascoe defined patient satisfaction "as a health care recipient's reaction to salient aspects of the context, process, and result of their service experience."[34] He contends that satisfaction includes two interrelated psychological components: (1) a cognitively based evaluation, or grading, of the structure, process, and outcomes of services and (2) an emotional, or affective, reaction to the structure, process, and outcomes of services.

Significance of Patient Satisfaction as an Outcome

As observed by Davies and Ware, patient satisfaction has emerged as a critical outcome of medical care because of the increasing emphasis on patients as consumers in the medical marketplace.[35] Patient satisfaction is viewed as an integral component of the measurement of medical care quality.[32,36,37] Donabedian has stated that "achieving and producing health and satisfaction, as defined for its individual members by a particular society or subculture, is the ultimate validator of the quality of care."[38]

The significance of patient satisfaction is reflected in its correlates.[39] Patient satisfaction (or dissatisfaction) has been associated with willingness to initiate malpractice litigation, switching medical care providers, compliance with medical recommendations, and disenrollment from prepaid health plans.[5,36,40-42] The extent to which a health care delivery system satisfies its consumers is considered a major determinant of the viability of that system.[37]

Measuring Patient Satisfaction

The importance of patient satisfaction as an outcome became recognized in the mid-1970s, at which time Ware and colleagues developed the Patient Satisfaction Questionnaire (PSQ).[43] This instrument has become one of the most widely used measures of satisfaction with medical care.[37] The initial instrument consisted of 80 items and was intended to be used for planning, administration, and evaluation of medical care delivery systems.[44] Subsequent development work led to a 68-item version called the PSQ-II, a 43-item short form, and the PSQ-III, consisting of 50 items.[37] Versions of the PSQ have been administered in the RAND Health Insurance Experiment, the RAND Medical Outcomes Study, and various national surveys.

More recently, a short-form version of the PSQ that contains only 18 items has been developed.[39] A copy of the PSQ-18 is provided in Appendix II. The brevity of this instrument is a meaningful advantage in minimizing respondent burden. In addition, the PSQ-18 retains each of the seven dimensions of satisfaction with medical care measured by the PSQ-III without significantly decreasing measured reliability. The dimensions of patient satisfaction measured by the PSQ-18 are listed in Table 17.3.

QUALITY OF LIFE AND PATIENT SATISFACTION: MEASUREMENT ISSUES

Humanistic outcomes are closely linked to individual attitudes and beliefs relating to health and health care. As such, the measure-

TABLE 17.3. PSQ-18 Dimensions/Scales

- General satisfaction
- Technical quality
- Interpersonal manner
- Communication
- Financial Aspects
- Time Spent with Doctor
- Accessibility and convenience

Source: Marshall and Hays, 1994[39]

ment of these outcomes is inherently subjective. A number of methodological issues must be considered when evaluating existing health outcomes measurement and/or deciding on the appropriate instrument to use when designing a study or assessment. A thorough review of these issues is not within the scope or objectives of this chapter. More in-depth reviews of methodological considerations are available in the literature.[23,37,44-49] The discussion in this section will focus on psychometric properties of instruments that are essential for the successful measurement of humanistic outcomes (and other psychological constructs) and practical issues regarding administration of the instruments.

Psychometric Properties

Of particular concern are the psychometric properties of a chosen instrument. Psychometrics refers to the measurement of psychological constructs, such as quality of life or patient satisfaction. It is concerned with the development and testing of instruments in such a way that we can have confidence in the measurement made. Psychometric properties include the reliability and validity of measurements. In this chapter, we discuss different instruments that have been used to measure quality of life or patient satisfaction and address their use in terms of these psychometric properties. In this section, we review reliability and validity as they pertain to the measurement of humanistic outcomes.

Reliability of the Measure

Reliability refers to the consistency, stability, or reproducibility of scores obtained on different administrations of a measure when all pertinent conditions remain relatively unchanged. The two reliability assessment methods most often discussed in the health-status literature are test-retest and internal consistency. However, as opposed to a measure of a trait that is assumed to be constant over the course of time, there can be a problem in attempting to use test-retest methods to assess the reliability of measures of humanistic outcomes. For example, quality of life is not assumed to be constant over the course of time. In fact, most clinical studies attempt to assess how health status and/or quality of life changes. Test-retest reliability estimates may have little value in evaluating measures that are designed to assess a dynamic process.

Internal consistency, which indicates the extent to which an instrument is free of random error, is an aspect of reliability assessment that can be very important to outcome measures. The degree of internal consistency is typically indicated by coefficient alpha.[50] It has been proposed that values above 0.90 are needed for making comparisons between individuals and above 0.50 for comparisons between groups.[51,52] Inter-rater agreement, although not as critical with self-administered instruments, is a consideration when obtaining data through interviews or direct observation. The assessment of reliability is a complex topic that is reviewed in detail in the psychometrics literature.[53,54]

Validity of the Measure

Validity defines the range of inferences that are justifiable on the basis of a score or measure, i.e., is the instrument really measuring what it is supposed to be measuring. Validity is not absolute, but is relative to the domain under study. Three basic types of validity commonly considered are criterion, content, and construct. Criterion validity is achieved when a new measure corresponds to an established measure or observation that accurately reflects the phenomenon of interest. By definition, the criterion must be a superior measure of the phenomenon if it is to serve as a comparative norm.[55] However, in quality of life or patient satisfaction assessment, gold standards or criterion measures against which a new measure can be compared rarely exist. Content validity, which is infrequently tested statistically, refers to how adequately the sampling of questions reflects the aims of the measure. Construct validity refers to the relationship between measures purporting to measure the same underlying theoretical construct (i.e., convergent evidence) or purporting to measure a different construct (i.e., discriminant evidence). For example, convergent evidence for the validity of role performance items is established by showing associations between the responses to the items and observed verifiable functioning. Evidence for the construct validity of other aspects of the measure might be established through comparisons with physiological measures, organ pathology, or clinical signs. Construct validity is not absolute. We do not say that a measure is "valid," but rather support its validity through research findings.

Another aspect of a measure that supports its validity is responsiveness or sensitivity to change. Responsiveness is the ability or power of the measure to detect clinically important change when it occurs.[56] Although some authors have suggested that responsiveness is a psychometric property of a measure that is distinct from validity, others argue that responsiveness is an aspect of validity rather than a separate property.[47,53,57,58] We include it as an indication of a measure's validity.

Administration

A practical aspect of the measurement of quality of life and patient satisfaction that was referred to in the discussion of the PSQ-18 is length of the instrument or the administration time involved. Instruments should be as brief as possible without significantly compromising the validity and reliability of the measurement. The longer an instrument is, the greater the respondent burden, which can lead to unwillingness/refusal to complete the instrument or to incomplete responses.

Another issue is the means of administration.[59] Many quality of life and patient satisfaction measures can be administered in different ways. The primary modes of administration are: (1) interviewer administered, either in person or over the telephone, and (2) self-administered. Also used, but not recommended, are surrogate responders, i.e., using a health care provider, family member, or friend to respond for the subject when the subject is unable to complete the instrument. Because humanistic outcomes are, by definition, patient reported, it is essential that patients have the opportunity to provide their perspective on the impact of medical care on their lives and on their level of satisfaction with the delivery of that care. Their perspective is likely to be quite different from that of an outside observer. For example, Jachuck and colleagues, in an assessment of the quality of life of hypertensive patients, found differences among the assessments provided by the patients, their spouses (or close companions), and their physicians.[60]

HUMANISTIC OUTCOMES AND PHARMACEUTICAL CARE

Quality of Life and Pharmaceutical Care

Information on the impact of pharmaceutical products and services on quality of life can provide additional data for making health care policy and clinical decisions. The inclusion of quality of life data in economic analyses of alternative interventions is becoming increasingly prevalent and involves the comparison of the costs and the outcomes of the interventions. It is no longer acceptable to make a selection between two medications of equal clinical efficacy based on the acquisition costs alone. One of the medications may be more appropriate because it produces better outcomes relative to its cost than the alternative.

Quality of life assessment can assist by providing documentation of differential patient outcomes. Pharmacists on pharmacy and therapeutics committees should be prepared to assist other committee members in incorporating existing quality of life data into the formulary or practice guideline decision-making process. A study by Croog and associates was one of the first in a growing body of literature reporting the quality of life impact of antihypertensive agents.[61-64] This is just one area of pharmacotherapy that has received a great deal of attention in regard to quality of life implications.

Quality of life as an input to clinical decision making at the patient level is also very important. For example, alternative treatments may have equal efficacy based on traditional clinical parameters (e.g., blood pressure reduction) but produce very different effects on the patient's quality of life. Thus, a provider's selection among competing alternatives may hinge upon documented differential impact on quality of life. A perceived decrease in quality of life attributed by the patient to an adverse effect of the drug may lead to a decrease in adherence to the medication regimen.[65] However, empirical evidence regarding the relationship between quality of life/health status and patient compliance is lacking.

In an attempt to explore that relationship, Hays and colleagues examined the association between self-reported adherence to medical recommendations and health outcomes over time for patients in the Medical Outcomes Study with one or more chronic conditions (i.e., diabetes, hypertension, congestive heart failure, recent myocardial infarction, and/or depressive symptoms).[66] The humanistic outcomes measured were physical, role, and social functioning; energy/fatigue; pain; emotional well-being; and general health perceptions. For this analysis, two composite indices of functioning and well-being (i.e., physical health and mental health) were produced from these scales/dimensions. The findings were mixed; specifically, adherence to medication recommendations was associated with negative effects on physical health for both insulin-using diabetic patients and depressed patients. The authors suggest that the perceptions of impaired physical health among those who complied with the medications may have been due to side effects. Compliance with medication regimens is a complex issue. For example, several studies have found higher levels of compliance to be related to better outcomes regardless of whether the drug was active or a placebo.[67,68] Compliance is discussed in much greater detail in another chapter of this book.

As described by Smith, when a patient takes a medication, there are four possible quality of life outcomes: (1) quality of life is improved, (2) quality of life is maintained as a direct result of the medication, (3) quality of life is decreased, or (4) quality of life remains unchanged.[69] When medication-related quality of life data are available in the literature from clinical trials or other studies, the pharmacist should be aware of and communicate the data to the prescriber and the patient. When the information is not available, the pharmacist should take an active role in monitoring the quality of life impact at the individual patient level.

Published research on the quality of life impact of pharmaceutical care is hard to find. This should change over the next few years. One study being conducted currently is assessing the clinical, economic, and humanistic outcomes associated with the provision of pharmaceutical care in renal dialysis units. The primary goal of the project is to improve dialysis patient care outcomes through pharmacist involvement in optimizing the dose of erythropoietin and other medications. A quality of life instrument, the Kidney Disease Quality of Life (KDQOL) Instrument, was developed specifically for this project.[70] In the instrument development process, focus groups with patients and staff were conducted to identify areas of particular concern for dialysis patients. A preliminary version of the instrument was pretested on a small sample of patients and then revised. The revised instrument includes the RAND 36-item Health Survey 1.0 as the generic core, supplemented with multi-item scales targeted at particular concerns of individuals with kidney disease.

Pharmacists have a tremendous opportunity to play an important role in optimizing the patient-reported or humanistic outcomes of medication therapy. Smith and colleagues have provided a list of action items to empower pharmacists to become actively engaged in the consideration of quality of life as an outcome of pharmaceutical care (Table 17.4).[71]

Patient Satisfaction and Pharmaceutical Care

Satisfaction with pharmacy services in general, and with pharmaceutical care specifically, has received limited attention. Like quality of life and general patient satisfaction, patient satisfaction with pharmacy services is a multidimensional concept. Some researchers have measured satisfaction with single items or questions.[72-74] However, this approach has been criticized for not reflecting, or capturing, the multidimensionality of patient satisfaction.[44,46]

TABLE 17.4. Consideration of Quality of Life: What Pharmacists Can Do

- Educate themselves and the entire pharmacy staff about the concept of medication-related quality of life and its importance.
- Translate the concept into terms that are meaningful to patients and physicians and educate them through newsletters, talks at local civic group meetings, senior citizen centers, and the like.
- Be alert for medication-induced patient complaints that may go unrecognized by the patient.
- Communicate medication-related patient complaints to the physician and suggest alternatives.
- Encourage patients to discuss any perceived negative effects of therapy with the pharmacist or the physician.
- Explain to patients what they can and cannot expect from their therapy and help them weigh the benefits against any negative effects.
- When a patient receives a new medication, explore whether the therapy is likely to interfere with important aspects of his or her life-style.
- Offer suggestions on how to minimize the impact of the negative effects of the therapy on the patient's quality of life.
- Include important life-style characteristics such as hobbies and occupation in patients' medication profiles.
- Alert the physician to potential and perceived conflicts between patient life-style and medication effects.

Source: Smith, Juergens, and Jack, 1991[71]

Fincham and Wertheimer examined the factors that affect patient satisfaction with pharmacy services in a health maintenance organization (HMO).[72] The significance of this topic stems from the increasing influence of HMOs on the delivery of medical care and the resultant impact on the provision of pharmacy services. Members of a large midwestern HMO, which operated in-house pharmacies as well as utilizing a closed panel of out-of-plan pharmacies, were surveyed. Variables shown to be predictive of higher levels of satisfaction with pharmacy services were satisfaction with the HMO in general, convenience of prescription filling, self-assessed positive health status, communication between the patient and provider, and a view of prescription drugs as being inexpensive. Variables that were not significantly related to satisfaction with pharmacy services included age, sex, length of membership with the HMO, and income.

Fincham and Wertheimer suggested that the results indicated areas to which HMOs, pharmacy directors, and pharmacies should direct attention to maintain patient satisfaction with pharmacy services. It is important to note, however, that some of the identified areas were not specifically related to pharmacy services but were more general (e.g., overall satisfaction with HMO) or were related by the patient (e.g., self-assessed health). To measure satisfaction with pharmaceutical care, patients must also be questioned specifically about the services provided directly by the pharmacist.

MacKeigan and Larson developed and provided evidence for the validity of a multidimensional instrument to measure satisfaction with pharmacy services and recently provided further validation of a subsequent version of the instrument.[46,75] This instrument was adapted from the PSQ but modified to include personal referent wording (e.g., *I am confident that the pharmacist dispenses all prescriptions correctly* versus *Pharmacists usually dispense all prescriptions correctly*) and a reduced number of items.[46]

A second version of the Satisfaction with Pharmacy Services Questionnaire, resulting from the initial administration and analysis of 45 items in nine scales, consists of 33 items comprising seven scales that tap the dimensions of satisfaction shown in Table 17.5.[75] The seven scales demonstrated adequate reliability in a telephone survey of a sample of patients who had recently purchased one or more prescriptions. Construct validity of the measurement was supported by association between satisfaction scores and variables such as type of pharmacy, prescription payment method, and level of medication use.

More recently, a 29-item version of the Satisfaction with Pharmacy Services Questionnaire was developed for use in a study called the Applied Pharmaceutical Care Project. This version, currently undergoing further validation, consists of a subset of five dimensions of satisfaction: four from the original instrument (i.e., explanation, consideration, technical competence, and general) and a new dimension, physical attributes. A copy of this instrument is provided in Appendix III.

TABLE 17.5. Satisfaction with Pharmacy Services Questionnaire Dimensions/ Scales

- Explanation
- Consideration
- Technical competence
- Financial aspects
- Accessibility
- Product availability
- General satisfaction

Source: Larson and MacKeigan, 1994[75]

Similar to the measurement of quality of life, pharmacists should also be concerned with the level of satisfaction achieved in the patients whom they serve. Researchers have attempted to develop instruments that provide valid and reliable measures of satisfaction with pharmacy services in general. Practitioners must begin to put these instruments to use in an effort to evaluate the impact of pharmaceutical care on measures of patient satisfaction.

SUMMARY AND CONCLUSION

The concept of quality of life has gained increasing attention in the evaluation of pharmaceutical products and services. Health-related quality of life is a health outcome that is gaining increasing attention. In fact, in certain diseases, quality of life may be the most important outcome to consider in assessing the effectiveness of health care interventions. Pharmacists, other health care practitioners, and policymakers must remember that efforts to increase quantity of life must not outstrip the ability to maintain or to improve quality of life.

Patient satisfaction has also been recognized as an important consideration in the assessment of the impact of health services. In the United States, where competition among health systems is increasing, the measurement of patient satisfaction will become even more important. Satisfaction with pharmacy services in particular has received limited attention. With the implementation of pharmaceutical care practices, it will be essential to assess the outcomes of these services in regard to patient satisfaction.

As might be expected in a very new scientific field, there are a considerable number of unresolved theoretical and methodological issues. One area that needs further exploration is the association between patient satisfaction and quality of life. However, there are some general concepts in the measurement of humanistic outcomes that should be carefully considered when designing a study, evaluating existing research, or evaluating new programs or services. This chapter has provided only a brief overview of the concepts in an effort to sensitize current or potential pharmaceutical care practitioners to the complexity of the area and to provide insight as to how these concepts can and should be incorporated into their practice.

Appendix I

RAND 36-ITEM HEALTH SURVEY 1.0

Scoring Instructions: To enable scoring, two tables appear after the *RAND 36-Item Health Survey 1.0* form in this appendix. The first step is to recode the items as directed in Table 1. After recoding, the second step is to combine the items into the scales as per Table 2. Scale scores are then calculated by averaging the recoded scores for all of the scale items. For example, recoded scores for items 21 and 22 would be added together and then divided by 2 to obtain the *Pain* scale score.

For additional information on the *RAND 36-Item Health Survey 1.0* contact:

RAND
1700 Main Street
P.O. Box 2138
Santa Monica, CA 90407-2138
310/451-7002 FAX: 310/451-6915

RAND 36-ITEM HEALTH SURVEY 1.0

QUESTIONNAIRE ITEMS

RAND HEALTH SCIENCE PROGRAM

1. In general, would you say your health is:

(Circle One Number)

Excellent	1
Very Good	2
Good	3
Fair	4
Poor	5

2. **Compared to one year ago,** how would you rate your health in general **now?**

(Circle One Number)

Much better now than one year ago	1
Somewhat better now than one year ago	2
About the same	3
Somewhat worse now than one year ago	4
Much worse now than one year ago	5

The following items are about activities you might do during a typical day. Does **your health now limit you** in these activities? If so, how much?

(Circle One Number on Each Line)

	Yes, Limited a Lot	Yes, Limited a Little	No, Not Limited at All
3. **Vigorous activities,** such as running, lifting heavy objects, participating in strenuous sports	1	2	3
4. **Moderate activities,** such as moving a table, pushing a vacuum cleaner, bowling, or playing golf	1	2	3
5. Lifting or carrying groceries	1	2	3
6. Climbing **several** flights of stairs ..	1	2	3
7. Climbing **one** flight of stairs	1	2	3
8. Bending, kneeling, or stooping	1	2	3
9. Walking **more than a mile**	1	2	3
10. Walking **several blocks**	1	2	3
11. Walking **one block**	1	2	3
12. Bathing or dressing yourself	1	2	3

RAND HEALTH SCIENCE PROGRAM

During the **past 4 weeks,** have you had any of the following problems with your work or other regular daily activities **as a result of your physical health?**

(Circle One Number on Each Line)

	Yes	No
13. Cut down the **amount of time** you spent on work or other activities	1	2
14. **Accomplished less** than you would like	1	2
15. Were limited in the **kind** of work or other activities	1	2
16. Had **difficulty** performing the work or other activities (for example, it took extra effort)	1	2

During the **past 4 weeks,** have you had any of the following problems with your work or other regular daily activities **as a result of any emotional problems** (such as feeling depressed or anxious)?

(Circle One Number on Each Line)

	Yes	No
17. Cut down the **amount of time** you spent on work or other activities	1	2
18. **Accomplished less** than you would like	1	2
19. Didn't do work or other activities as **carefully** as usual	1	2

20. During the **past 4 weeks,** to what extent has your physical health or emotional problems interfered with your normal social activities with family, friends, neighbors, or groups?

(Circle One Number)

Not at all	1
Slightly	2
Moderately	3
Quite a bit	4
Extremely	5

21. How much **bodily** pain have you had during the **past 4 weeks?**

(Circle One Number)

None	1
Very mild	2
Mild	3
Moderate	4
Severe	5
Very severe	6

RAND HEALTH SCIENCE PROGRAM

22. During the **past 4 weeks,** how much did **pain** interfere with your normal work (including both work outside the home and housework)?

(Circle One Number)

Not at all	1
A little bit	2
Moderately	3
Quite a bit	4
Extremely	5

These questions are about how you feel and how things have been with you **during the past 4 weeks.** For each question, please give the one answer that comes closest to the way you have been feeling.

How much of the time during the **past 4 weeks...**

(Circle One Number on Each Line)

	All of the Time	Most of the Time	A Good Bit of the Time	Some of the Time	A Little of the Time	None of the Time
23. Did you feel full of pep?	1	2	3	4	5	6
24. Have you been a very nervous person?	1	2	3	4	5	6
25. Have you felt so down in the dumps that nothing could cheer you up?	1	2	3	4	5	6
26. Have you felt calm and peaceful?	1	2	3	4	5	6
27. Did you have a lot of energy?	1	2	3	4	5	6
28. Have you felt downhearted and blue?	1	2	3	4	5	6
29. Did you feel worn out?	1	2	3	4	5	6
30. Have you been a happy person?	1	2	3	4	5	6
31. Did you feel tired?	1	2	3	4	5	6

32. During the **past 4 weeks,** how much of the time has your **physical health or emotional problems** interfered with your social activities (like visiting with friends, relatives, etc.)?

(Circle One Number)

All of the time	1
Most of the time	2
Some of the time	3
A little of the time	4
None of the time	5

RAND HEALTH SCIENCE PROGRAM

How TRUE or FALSE is <u>each</u> of the following statements for you?

(Circle One Number on Each Line)

	Definitely <u>True</u>	Mostly <u>True</u>	Don't <u>Know</u>	Mostly <u>False</u>	Definitely <u>False</u>
33. I seem to get sick a little easier than other people	1	2	3	4	5
34. I am as healthy as anybody I know	1	2	3	4	5
35. I expect my health to get worse	1	2	3	4	5
36. My health is excellent	1	2	3	4	5

RAND HEALTH SERVICES PROGRAM

TABLE 1

STEP 1: RECODING ITEMS

ITEM NUMBERS	Change original response category [a]	To recoded value of:
1,2,20,22,34,36	1 ---------->	100
	2 ---------->	75
	3 ---------->	50
	4 ---------->	25
	5 ---------->	0
3,4,5,6,7,8,9,10,11,12	1 ---------->	0
	2 ---------->	50
	3 ---------->	100
13,14,15,16,17,18,19	1 ---------->	0
	2 ---------->	100
21,23,26,27,30	1 ---------->	100
	2 ---------->	80
	3 ---------->	60
	4 ---------->	40
	5 ---------->	20
	6 ---------->	0
24,25,28,29,31	1 ---------->	0
	2 ---------->	20
	3 ---------->	40
	4 ---------->	60
	5 ---------->	80
	6 ---------->	100
32,33,35	1 ---------->	0
	2 ---------->	25
	3 ---------->	50
	4 ---------->	75
	5 ---------->	100

RAND HEALTH SERVICES PROGRAM

(a) Precoded response choices as printed in the questionnaire

TABLE 2

STEP 2: AVERAGING ITEMS TO FORM SCALES

Scale	Number of Items	After Recoding per Table 1, Average the Following Items:
Physical Functioning	10	3,4,5,6,7,8,9,10,11,12
Role limitations due to physical health	4	13,4,15,16
Role limitations due to emotional problems	3	17,18,19
Energy/fatigue	4	23,27,29,31
Emotional well-being	5	24,25,26,28,30
Social functioning	2	20,32
Pain	2	21,22
General Health	5	1,33,34,15,36

Appendix II

SHORT-FORM PATIENT SATISFACTION QUESTIONNAIRE (PSQ-18)

Scoring Instructions: The PSQ-18 yields separate scores for each of seven different subscales. General Satisfaction (Items 3 and 17); Technical Quality (Items 2, 4, 6, and 14); Interpersonal Manner (Items 10 and 11); Communication (Items 1 and 13); Financial Aspects (Items 5 and 7); Time Spent with Doctor (Items 12 and 15); Accessibility and Convenience (Items 8, 9, 16, and 18). Some PSQ-18 items are worded so that agreement reflects satisfaction with medical care, whereas other items are worded so that agreement reflects dissatisfaction with medical care. All items should be scored so that high scores reflect satisfaction with medical care (see Table 1). After item scoring, items within the same subscale should be averaged together to create the 7 subscale scores. It is recommended that items left blank by respondents (missing data) be ignored when calculating scale scores. In other words, scale scores represent the average for all items in the scale that were answered.

For additional information on the *PSQ-18* contact:

RAND
1700 Main Street
P.O. Box 2138
Santa Monica, CA 90407-2138
310/451-7002 FAX: 310/451-6915

Reprinted by permission of RAND.

Patient Satisfaction Questionnaire

SHORT-FORM PATIENT SATISFACTION QUESTIONNAIRE (PSQ-18)

These next questions are about how you feel about the
medical care you receive.

On the following pages are some things people say about medical care. Please read each one carefully, keeping in mind the medical care you are receiving now. (If you have not received care recently, think about what you would expect if you needed care today.) We are interested in your feelings, good and bad, about the medical care you have received.

How strongly do you AGREE or DISAGREE with each of the following statements?

(Circle One Number on Each Line)

	Strongly Agree	Agree	Uncertain	Disagree	Strongly Disagree
1. Doctors are good about explaining the reason for medical tests	1	2	3	4	5
2. I think my doctor's office has everything needed to provide complete medical care	1	2	3	4	5
3. The medical care I have been receiving is just about perfect	1	2	3	4	5
4. Sometimes doctors make me wonder if their diagnosis is correct	1	2	3	4	5
5. I feel confident that I can get the medical care I need without being set back financially	1	2	3	4	5
6. When I go for medical care, they are careful to check everything when treating and examining me	1	2	3	4	5
7. I have to pay for more of my medical care than I can afford	1	2	3	4	5
8. I have easy access to the medical specialists I need . . .	1	2	3	4	5

Patient Satisfaction Questionnaire

How strongly do you AGREE or DISAGREE with <u>each</u> of the following statements?

(Circle One Number on Each Line)

	Strongly Agree	Agree	Uncertain	Disagree	Strongly Disagree
9. Where I get medical care, people have to wait too long for emergency treatment	1	2	3	4	5
10. Doctors act too businesslike and impersonal toward me . . .	1	2	3	4	5
11. My doctors treat me in a very friendly and courteous manner.	1	2	3	4	5
12. Those who provide my medical care sometimes hurry too much when they treat me	1	2	3	4	5
13. Doctors sometimes ignore what I tell them	1	2	3	4	5
14. I have some doubts about the ability of the doctors who treat me.	1	2	3	4	5
15. Doctors usually spend plenty of time with me	1	2	3	4	5
16. I find it hard to get an appointment for medical care right away.	1	2	3	4	5
17. I am dissatisfied with some things about the medical care I receive	1	2	3	4	5
18. I am able to get medical care whenever I need it	1	2	3	4	5

TABLE 1

STEP 1: SCORING ITEMS

Item Numbers	Original Response Value		Scored Value
1,2,3,5,6,8,11,15,18	1	---------->	5
	2	---------->	4
	3	---------->	3
	4	---------->	2
	5	---------->	1
4,7,9,10,12,13,14,16,17	1	---------->	1
	2	---------->	2
	3	---------->	3
	4	---------->	4
	5	---------->	5

TABLE 2

STEP 2: CREATING SCALE SCORES

Scale	Average These Items
General Satisfaction	3, 17
Technical Quality	2, 4, 6, 14
Interpersonal Manner	10, 11
Communication	1, 13
Financial Aspects	5, 7
Time Spent with Doctor	12, 15
Accessibility and Convenience	8, 9, 16, 18

Note. Items within each scale are averaged after scoring as shown in Table 1.

Appendix III

SATISFACTION WITH PHARMACY SERVICES QUESTIONNAIRE
(Applied Pharmaceutical Care Project Version)

Scoring Instructions: Unfavorably worded items should be reverse coded such that higher item scores indicate higher levels of satisfaction. This recoding can be accomplished by subtracting the raw item score from 6 (e.g., for item 14, a raw score of 2 would be recoded as 6-2 = 4). Total scores are then calculated by summing item scores within each dimension. Mean dimension scores are calculated across individuals. To allow comparability across dimensions, prorated mean scores, expressed as a percentage, are calculated by dividing the mean dimension score by the maximum possible score and multiplied by 100.

This instrument was developed by Linda D. MacKeigan, Ph.D. and Lon N. Larson, Ph.D. The development and validation of earlier versions of the instrument have been reported (MacKeigan and Larson, 1989; Larson and MacKeigan, 1994). Additional information on scoring the instrument can be found in those two publications. Reprinted with permission.

Satisfaction with Pharmacy Services Questionnaire

On the following pages are some statements about your pharmacy services. Please read each one carefully, keeping in mind the pharmacy that you presently go to. On the line next to each statement circle the number for the opinion which is closest to your own view.

For example, if you strongly agreed with the following statement, you would circle the number 1 as indicated below:

	Strongly Agree	Agree	No Opinion	Disagree	Strongly Disagree
Parking is a problem when I go to the pharmacy.	1	2	3	4	5

Some statements look similar to others, but each statement is different. You should answer each statement by itself. This is not a test of what you know. There are no right or wrong answers. We are only interested in your opinions or best impression. Please circle only one number for each statement.

	Strongly Agree	Agree	No Opinion	Disagree	Strongly Disagree
1. The pharmacist spends as much time as is necessary with me.	1	2	3	4	5
2. The pharmacist usually explains the possible side effects that a new medication may cause.	1	2	3	4	5
3. If I have a question about my prescription, the pharmacist is always available to help me.	1	2	3	4	5
4. There is no place where I can have a private conversation with the pharmacist.	1	2	3	4	5
5. The pharmacist knows how to explain things in a way that I understand.	1	2	3	4	5
6. The pharmacy area is as clean as any medical office.	1	2	3	4	5
7. The pharmacist is not as thorough as he/she should be.	1	2	3	4	5
8. The pharmacy services that I receive are just about perfect.	1	2	3	4	5
9. My prescriptions are always filled promptly.	1	2	3	4	5

Satisfaction with Pharmacy Services Questionnaire

	Strongly Agree	Agree	No Opinion	Disagree	Strongly Disagree
10. When I get a prescription filled, the pharmacist makes sure that I understand how to take the medication.	1	2	3	4	5
11. The pharmacy has a comfortable waiting area.	1	2	3	4	5
12. Sometimes the pharmacist does not spend enough time with me.	1	2	3	4	5
13. The pharmacy staff should be more friendly.	1	2	3	4	5
14. The pharmacist hardly ever explains what my medication does.	1	2	3	4	5
15. I am confident that the pharmacist dispenses all prescriptions correctly.	1	2	3	4	5
16. There are things about the pharmacy services that could be better.	1	2	3	4	5
17. The pharmacy staff always looks professional.	1	2	3	4	5
18. The pharmacist should do more to keep people from having problems with their medications.	1	2	3	4	5
19. I usually have to wait a long time when I get a prescription filled.	1	2	3	4	5
20. I sometimes have trouble understanding what the pharmacist means when he or she talks to me about my medication.	1	2	3	4	5
21. I sometimes wonder about the accuracy of the prescriptions that the pharmacist dispenses.	1	2	3	4	5
22. The pharmacy staff seems to have a genuine interest in me as a person.	1	2	3	4	5

Satisfaction with Pharmacy Services Questionnaire

	Strongly Agree	Agree	No Opinion	Disagree	Strongly Disagree
23. There are so many distractions in the pharmacy area that you often can't get good service.	1	2	3	4	5
24. I have some complaints about the pharmacy services.	1	2	3	4	5
25. The pharmacy staff are always courteous and respectful.	1	2	3	4	5
26. The pharmacist is always thorough.	1	2	3	4	5
27. I am very satisfied with the pharmacy services.	1	2	3	4	5
28. The pharmacist often does not tell how to take my prescription medication.	1	2	3	4	5
29. The prescription area is too cluttered with signs and merchandise.	1	2	3	4	5

Satisfaction with Pharmacy Services Questionnaire

TABLE 1

Dimensions

Dimension	Favorably Worded Items	Unfavorably Worded Items	Matched Pairs
Explanation	2, 3, 5, 10	14, 20, 28	5, 20; 10, 28
Consideration	1, 9, 22, 25	12, 13, 18, 19	1, 12; 9, 19
Technical competence	15, 26	7, 21	7, 26, 15, 21
Physical attributes	6, 11, 17	4, 23, 29	—
General	8, 27	16, 24	—

REFERENCES

1. Maynard A. Developing the health care market. Econ J 1991;(Sept): 1277-86.

2. Relman AS. Assessment and accountability. N Engl J Med 1988;319: 1220-1221.

3. Roper WL, Winkenwerder W, Hackbarth GM, Krakauer H. Effectiveness in health care: an initiative to evaluate and improve medical practice. N Engl J Med 1988;319:1197-1202.

4. Starfield B. Primary care and health: a cross-national comparison. JAMA 1991;266:2268-71.

5. Donabedian A. Explorations in quality assessment and monitoring. Vol. I: The definition of quality and approaches to its assessment. Ann Arbor, MI: Health Administration Press, 1980.

6. Zitter M. Outcomes assessment: true customer focus comes to health care. Med Interface 1992;(May):32-7.

7. Lohr KN. Outcome measurement: concepts and questions. Inquiry 1988;25:37-50.

8. Kozma CM, Reeder CE, Schulz RM. Economic, clinical, and humanistic outcomes: a planning model for pharmacoeconomic research. Clin Ther 1993;15: 1121-32.

9. Ellwood PM. Outcomes management: a technology of patient experience. N Engl J Med 1988;318:1549-56.

10. Hepler CD, Strand LM. Opportunities and responsibilities in pharmaceutical care. Am J Hosp Pharm 1990;47:533-43.

11. Stewart AL. Conceptual and methodologic issues in defining quality of life: state of the art. Prog Cardiovasc Nurs 1992;7(1):3-11.

12. Kaplan RM, Bush JW. Health-related quality of life measurement for evaluation research and policy analysis. Health Psychol 1982;1:61-80.

13. Schron EB, Shumaker SA. The integration of health quality of life in clinical research: experience from cardiovascular clinical trials. Prog Cardiovasc Nurs 1992;7(2):21-28.

14. Patrick DL, Erickson P. Health status and health policy: allocating resources to health care. New York: Oxford University Press, 1993.

15. Schipper H, Clinch J, Powell V. Definitions and conceptual issues. In: Spilker B, ed. Quality of life assessments in clinical trials. New York: Raven Press, 1990:11-24.

16. Staquet M, Aaronson NK, Ahmedzai, S et al. Editorial: health-related quality of life research. Qual Life Res 1992;1:3.

17. Torrance GW, Feeny DH. Cost-utility workshop. Presented at Eli Lilly and Company, Indianapolis, IN, December 6, 1993.

18. Spilker B, Molinek FR, Johnston KA, Simpson RL, Tilson HH. Quality of life bibliography and indexes. Med Care 1990;28(Suppl 12):DS1-DS77.

19. Spilker B, White WSA, Simpson RL, Tilson HH. Quality of life bibliography and indexes: 1990 update. J Clin Res Pharmacoepidemiol 1992a;6:87-156.

20. Spilker B, Simpson RL, Tilson HH. Quality of life bibliography and indexes: 1991 update. J Clin Res Pharmacoepidemiol 1992b;6:205-66.

21. Berzon RA, Simeon GP, Simpson RL, Tilson HH. Quality of life bibliography and indexes: 1992 update. J Clin Res Drug Develop 1993;7:203-42.

22. Berzon RA, Simeon GP, Simpson RL, Donnelly MA, Tilson HH. Quality of life bibliography and indexes: 1993 update. Qual Life Res 1995;4:53-74.

23. Guyatt GH, Feeny DH, Patrick DL. Measuring health-related quality of life. Ann Intern Med 1993;118:622-9.

24. Ware JE Jr, Sherbourne CD. The MOS 36-Item Short-Form Health Survey (SF-36): I. Conceptual framework and item selection. Med Care 1992;30:473-83.

25. Hays RD, Sherbourne CD, Mazel RM. The RAND 36-Item Health Survey 1.0. Health Econ 1993b;2:217-27.

26. Revicki DA. Relationships between health utility and psychometric health status measures. Med Care 1992;30:MS274-MS282.

27. Kongpatanakul S, Strom BL. Quality of life, health status and clinical drug research. PharmacoEconomics 1992;2:8-14.

28. Coons SJ, Kaplan RM. Cost-utility analysis. In: Bootman JL, Townsend RJ, McGhan WF, eds. Principles of pharmacoeconomics. 2nd ed. Cincinnati, OH: Harvey Whitney Books Company, in press.

29. Smith RD, Hall J, Gurney H, Harnett PR. A cost-utility approach to the use of 5-fluorouracil and levamisole as adjuvant chemotherapy for Dukes' C colonic carcinoma. Med J Aust 1993;158:319-22.

30. Kaplan RM, Anderson JP. The general health policy model: an integrated approach. In: Spilker B, ed. Quality of life assessments in clinical trials. New York: Raven Press, 1990:131-49.

31. Essink-Bot ML, Stouthard MEA, Bonsel GJ. Generalizability of valuations on health states collected with the EuroQol-questionnaire. Health Econ 1993;2:237-46.

32. Cleary PD, McNeil BJ. Patient Satisfaction as an indicator of quality care. Inquiry 1988;25:25-36.

33. Rossiter LF, Langwell K, Wan TTH, Rivnyak M. Patient satisfaction among elderly enrollees and disenrollees in Medicare health maintenance organizations. JAMA 1989;262:57-63.

34. Pascoe GC. Patient satisfaction in primary health care: a literature review and analysis. Eval Prog Plan 1983;6:185-210.

35. Davies AR, Ware JE Jr. Involving consumers in quality of care assessment. Health Aff 1988;7(1):33-48.

36. Ware JE Jr, Davies AR. Behavioral consequences of consumer dissatisfaction with medical care. Eval Prog Plan 1983;6:291-7.

37. Marshall GN, Hays RD, Sherbourne CD, Wells KB. The structure of patient satisfaction with outpatient medical care. Psychol Assess 1993;3:477-83.

38. Donabedian A. Evaluating the quality of medical care. Milbank Mem Fund Q 1966;44:166-203.

39. Marshall GN, Hays RD. The Patient Satisfaction Questionnaire Short-Form (PSQ-18). Santa Monica, CA: RAND, 1994.

40. Vaccarino JM. Malpractice: the problem in perspective. JAMA 1977;238:861-3.

41. Marquis MS, Davies AR, Ware JE. Patient satisfaction and change in medical care provider: a longitudinal study. Med Care 1983;21:821-9.

42. Sherbourne CD, Hays RD, Ordway L, DiMatteo MR, Kravitz R. Antecedents of adherence to medical recommendations: results from the Medical Outcomes Study. J Behav Med 1992;15:447-68.

43. Ware JE, Snyder MK, Wright WR. Development and validation of scales to measure patient satisfaction with medical care services. Vol. I, Part A: Review of literature, overview of methods, and results regarding construction of scales. (NTIS Pub. No. PB 288-329). Springfield, VA: National Technical Information Service, 1976.

44. Ware JE Jr, Snyder MK, Wright WR, Davies AR. Defining and measuring patient satisfaction with medical care. Eval Prog Plan 1983;6:247-63.

45. Ware JE Jr. Standards for validating health measures: definition and content. J Chronic Dis 1987;40:473-80.

46. MacKeigan LD, Larson LN. Development and validation of an instrument to measure patient satisfaction with pharmacy services. Med Care 1989;27: 522-36.

47. Coons SJ, Kaplan RM. Assessing health-related quality of life: application to drug therapy. Clin Ther 1992;14:850-8.

48. Pathak DS, MacKeigan LD. Assessment of quality of life and health status: selected observations. J Res Pharm Econ 1992;4(4):31-52.

49. Smith ND. Quality of life studies from the perspective of an FDA reviewing statistician. Drug Info J 1993;27:617-23.

50. Cronbach LJ. Coefficient alpha and the internal structure of tests. Psychometrika 1951;16:297.

51. Nunally J. Psychometric theory. 2nd ed. New York: McGraw-Hill, 1978.

52. Helmstadter GC. Principles of psychological measurement. New York: Appleton-Century-Crofts, 1964.

53. Hays RD, Anderson R, Revicki D. Psychometric considerations in evaluating health-related quality of life measures. Qual Life Res 1993a;2:441-9.

54. Kaplan RM, Saccuzzo DP. Psychological testing: principles, applications, and issues. 3rd ed. Pacific Grove, CA: Brooks/Cole, 1993.

55. Kaplan RM, Bush JW, Berry CC. Health status: types of validity and the Index of Well-Being. Health Serv Res 1976;11:478-507.

56. Jaeschke R, Guyatt GH. How to develop and evaluate a new quality of life instrument. In: Spilker B, ed. Quality of life assessments in clinical trials. New York: Raven Press, 1990:47-57.

57. Guyatt G, Walter S, Norman G. Measuring change over time: assessing the usefulness of evaluative instruments. J Chronic Dis 1987;40:171-8.

58. Hays RD, Hadorn D. Responsiveness to change: an aspect of validity, not a separate dimension. Q Life Res 1992;1:73-5.

59. Cook DJ, Guyatt GH, Juniper E, Griffith L, McIlroy W, Willan A, Jaeschke R, Epstein R. Interviewer versus self-administered questionnaires in developing a

disease-specific, health-related quality of life instrument for asthma. J Clin Epidemiol 1993;46:529-34.

60. Jachuck SJ, Brierly H, Jachuck S, Wilcox PM. The effect of hypotensive drugs on the quality of life. J R Coll Gen Pract 1982;32:103-5.

61. Croog SH, Levine S, Testa MA, Brown B, Bulpitt CJ, Jenkins CD. The effects of antihypertensive therapy on the quality of life. N Engl J Med 1986;314: 1657-64.

62. Hollenberg NK, Testa M, Williams GH. Quality-of-life as a therapeutic end-point: and analysis of therapeutic trials in hypertension. PharmacoEconomics 1991;6:83-93.

63. Beto JA, Bansal VK. Quality of life in the treatment of hypertension: a metaanalysis of clinical trials. Am J Hyperten 1992;5:125-33.

64. Bulpitt CJ, Fletcher AE. Quality-of-life instruments in hypertension. PharmacoEconomics 1994; 6:523-35.

65. Curb JD, Borhani NO, Blaszkanski RTP, et al. Long-term surveillance for adverse effects of antihypertensive drugs. JAMA 1985;253:3263-8.

66. Hays RD, Kravitz RL, Mazel RM, Sherbourne CD, DiMatteo MR, Rogers WH, Greenfield S. The impact of patient adherence on health outcomes for patients with chronic disease in the Medical Outcomes Study. J Behav Med 1994;17:347-60.

67. Horwitz RI, Viscoli CM, Berkman L, Donaldson RM, Horwitz SM, Murray CJ, Ransohoff DF, Sindelar J. Treatment adherence and risk of death after a myocardial infarction. Lancet 1990;336:542-5.

68. Gallagher EJ, Viscoli CM, Horwitz RI. The relationship of treatment adherence to risk of death after myocardial infarction in women. JAMA 1993; 270:742-4.

69. Smith M. Medication, quality of life and compliance: the role of the pharmacist. PharmacoEconomics 1992:1;225-30.

70. Hays RD, Kallich JD, Mapes DL, Coons SJ, Carter WB. Development of the Kidney Disease Quality of Life (KDQOL™) instrument. Qual Life Res 1994;3:329-38.

71. Smith M, Juergens J, Jack W. Medication and the quality of life. Am Pharm 1991;NS31:275-81.

72. Fincham JE, Wertheimer AI. Predictors of patient satisfaction with pharmacy services in a health maintenance organization. J Pharm Market Manage 1987;2(2):73-88.

73. Arneson DL, Jacobs EW, Scott, DM, Murray WJ. Patronage motives of community pharmacy patrons. J Pharm Market Manage 1989;4(2):3-22.

74. Smith HA, Coons SJ. Patronage factors and consumer satisfaction with sources of prescription purchases. J Pharm Market Manage 1990;4(3):61-81.

75. Larson LN, MacKeigan LD. Further validation of an instrument to measure patient satisfaction with pharmacy services. J Pharm Market Manage 1994; 8(1):125-39.

ADDITIONAL READINGS

For the Instructor

1. Cleary PD, McNeil BJ. Patient satisfaction as an indicator of quality care. Inquiry 1988;25:25-36.

2. Ellwood PM. Outcomes management: a technology of patient experience. N Engl J Med 1988;318:1549-56.

3. Fincham JE, Wertheimer AI. Predictors of patient satisfaction with pharmacy services in a health maintenance organization. J Pharm Market Manage 1987;2(2):73-88.

4. Froberg DG, Kane RL. Methodology for measuring health state preferences I: measurement strategies. J Clin Epidemiol 1989;42:345-52.

5. Froberg DG, Kane RL. Methodology for measuring health state preferences II: scaling methods. J Clin Epidemiol 1989;42:459-71.

6. Froberg DG, Kane RL. Methodology for measuring health state preferences III: population and context effects. J Clin Epidemiol 1989;42:585-92.

7. Froberg DG, Kane RL. Methodology for measuring health state preferences IV: progress and a research agenda. J Clin Epidemiol 1989;42:675-85.

8. Guyatt GH, Feeny DH, Patrick DL. Measuring health-related quality of life. Ann Intern Med 1993;118:622-9.

9. Larson LN, MacKeigan LD. Further validation of an instrument to measure patient satisfaction with pharmacy services. J Pharm Market Manage 1994;8(1):125-39.

10. MacKeigan LD, Larson LN. Development and validation of an instrument to measure patient satisfaction with pharmacy services. Med Care 1989;27:522-36.

11. MacKeigan LD, Pathak D. Overview of health-related quality-of-life measures. Am J Hosp Pharm 1992;49:2236-45.

12. Marshall GN, Hays RD, Sherbourne CD, Wells KB. The structure of patient satisfaction with outpatient medical care. Psychol Assess 1993;3:477-83.

13. Relman AS. Assessment and accountability. N Engl J Med 1988;319:1220-1.

14. Ware JE Jr, Sherbourne CD. The MOS 36-Item Short-Form Health Survey (SF-36): I. conceptual framework and item selection. Med Care 1992;30:473-83.

15. Ware JE Jr. Conceptualizing and measuring generic health outcomes. Cancer 1991; 67:774-9.

16. Ware JE Jr, Snyder MK, Wright WR, Davies AR. Defining and measuring patient satisfaction with medical care. Eval Prog Plan 1983;6:247-63.

For the Student

1. Cleary PD, McNeil BJ. Patient satisfaction as an indicator of quality care. Inquiry 1988;25:25-36.

2. Ellwood PM. Outcomes management: a technology of patient experience. N Engl J Med 1988;318:1549-56.

3. Guyatt GH, Feeny DH, Patrick DL. Measuring health-related quality of life. Ann Intern Med 1993;118:622-9.

4. MacKeigan LD, Larson LN. Development and validation of an instrument to measure patient satisfaction with pharmacy services. Med Care 1989;27: 522-36.

5. MacKeigan LD, Pathak D. Overview of health-related quality-of-life measures. Am J Hosp Pharm 1992;49:2236-45.

6. Marshall GN, Hays RD, Sherbourne CD, Wells KB. The structure of patient satisfaction with outpatient medical care. Psychol Assess 1993;3:477-83.

7. Ware JE Jr, Sherbourne CD. The MOS 36-Item Short-Form Health Survey (SF-36): I. Conceptual framework and item selection. Med Care 1992;30:473-83.

8. Ware JE Jr, Snyder MK, Wright WR, Davies AR. Defining and measuring patient satisfaction with medical care. Eval Prog Plan 1983;6:247-63.

PART II:
SPECIAL CLASSES OF PATIENTS

Chapter 18

Children and Medicines

Patricia J. Bush

INTRODUCTION

Relative to health and illness behavior, children traditionally have been viewed as passive, with adults making decisions for them and providing information about them to health professionals. However, a moment's reflection leads to the conclusion that children not only take medicines themselves but also are aware that medicines are stored at home, and they observe medicine use by family members and medicine promotion by the media on a regular basis. In addition, every trip to the drugstore and food market exposes children to the sight of medicines and sometimes to the sight of a pharmacist. It would be extremely unlikely, in view of this exposure, that children would not be forming beliefs and attitudes about medicines. Moreover, these beliefs and attitudes should change as children gain more experience and become more skilled at interpreting what they observe.

Interviews with mothers in Washington, DC, revealed that the average mother, irrespective of the age of her own child, believes that 12 years is the age that the average child should be able to take a medicine for common health problems, such as a headache or sore throat, without asking an adult. This response raises several questions. What should that average child know about medicines before assuming this responsibility? Do children know enough to take medicines on their own? How do children learn about medicines? Who should teach children about medicines? Do some children, perhaps children with chronic health problems, have more responsibility for medicine use than other children of the same age?

The answers to these questions remain largely unknown. There is certainly no consensus among health professionals or health educators about the need for children to be educated about medicines, who should assume the responsibility for education, or what forms such education should take. However, it is a goal of this chapter to review what is known about children and medicines to produce some pragmatic suggestions for practitioners.

We begin with a caution. Adults, perceiving a need and with the best of intentions, have a tendency to rush in and teach children what they think the children should know. Most of these adults have no idea of what the target children's existing knowledge and behaviors are in the particular area in which the need is perceived, and most of the well-meaning adults have no knowledge of how children process information at the different cognitive developmental stages through which they progress. Certainly, health professionals are well-imbued with the notion that children are something other than little adults. But where medicines are concerned, this notion has not led them much farther than to formulate for calculating pediatric doses and to knowledge about how drug metabolism is different in children, especially very young children.

The fact that children are not simply little adults is just as important in communications. It is not just that children have less experience than adults, it is that children's thinking processes develop through stages and are *qualitatively* different from those of adults. For reasons of differences relating to knowledge, experience, autonomy, and developmental levels, children may view and understand health, illness, and treatment differently from adults. Thus, children should be treated differently from adults with regard to pharmaceutical care, and programs intended for teaching children about medicines should take into account what children already know and do.

The remainder of this chapter presents information about how children learn; children's autonomy, knowledge, and attitudes relative to medicine use; factors associated with children's medicine-related beliefs and behaviors; and finally, suggestions about how health professionals can help children learn to use medicines wisely before they bear the responsibility for doing so.

HOW CHILDREN LEARN

Cognitive Developmental Stages

Adults not only know more than children, but they are also capable of processing what they know in a more complex way. Children progress through four stages as they develop these more complex cognitive skills. The theory of cognitive developmental stage is credited to Jean Piaget.[1] Although the sequence of progression through the stages is the same, there is individual variation in the rate of progression, and the progression has been shown to vary with the specific topic area. For the area of health in particular, even some adults are not operating in the final stage.

A value in being aware of the cognitive developmental stages of children is in appreciating that adults cannot provide children with information about health and illness and expect them to infer appropriate behavior. Moreover, in any given situation of health education addressed at a particular behavior, an adult cannot predict what a child will perceive. Knowledge of Cognitive Development Theory may not help you to predict what a child thinks, but at least you should not be surprised when a seven-year-old child who can correctly name a number of "bad" drugs (i.e., abusable substances) will then tell you that you get them at the drugstore. That is perfectly logical. Another lovely example occurred when a first grader was asked how likely she would be to take something special if she had trouble falling asleep. Her response, indicated by pointing to the largest bar on a graph: "Very likely." She was then asked, "What would you take?" Her response: "A teddy bear."

Stage 1: Sensory Motor

The first stage, lasting from birth to about two years, is known as the sensory motor stage. In this stage, the child learns through interacting with the environment, and cannot recall or imagine an object or person that is not present. Thus, learning about medicines does not (cannot) occur during this stage.

Stage 2: Preoperational

The second stage, roughly from two to seven years, is known as the preoperational stage. During this stage, children begin to be able to recall past events, to understand symbols, and to use mental imaging. However, links from cause to effect are not understood, and magical thinking is often used to explain events. Children in this stage are often "yea sayers." Here is an example from an interview with a four-year-old child, illustrating both "yea saying" and magical thinking. "Did you ride an elephant to school today?" "Yes." "Do you have wings?" "Yes." "How do you know you have wings?" "The elephant told me."

Clearly, it is wise to avoid a question format that calls for a yes/no response when trying to get information from children in the preoperational stage. In addition, children at this stage are sensitive to what they perceive the interviewer wants to hear and will respond to please the interviewer. Thus, the interviewer must take great care not to provide either verbal or body indicators of approval or disapproval to the young child.

Stage 3: Concrete Operational

From age seven to twelve years, children's thinking becomes more logical and systematic. Children become problem solvers and are able to focus on several aspects of the same situation. They understand the difference between permanency and change. However, they use whatever is at hand within their sphere of knowledge and experience to explain an event, however illogical it may seem to an adult. For example, a seven-year-old urban boy, when asked, "Is that good, being on drugs?" responded, "No, you can mess up your mind and then you die." But, when next asked, "So what happens . . . ?" the boy said, " . . . and then the police catch you and then they take you to the hospital and they gonna have to break your head open to get all the drugs out."[2] This seven-year-old shows a child entering the concrete operational stage and just beginning to apply logic and reasoning to events in his life. A sensitive adult, understanding this developmental process, can understand the child's logic. In this case, if drugs "mess up the mind," the child reasons that the mind is in the head and knows that hospitals fix people. Thus, it follows, with compelling logic, that the head must be broken open and the drugs removed to unburden the mind from its "messed up" state. A sensitive adult can use this information to help the child acquire an understanding that will lead to a more appropriate conclusion.

Stage 4: Formal Operational

This stage is considered to be from 12 years on. Children become more capable of hypothetical and abstract thought as they enter adolescence, but for formal operational thinking to occur in a particular area, attention and motivation are required. Thus, although individuals develop the capacity for this type of thinking—a capacity fully developed by late adolescence—the associated skills are not always applied. Understanding causal processes in health and illness is one of those content areas that seems to lag in some individuals through lack of interest or motivation, but certainly a better understanding of related processes and the relationships between internal and external factors is acquired. Someone trying to get information about health and illness from an adolescent (or an adult, for that matter) or to educate the adolescent cannot simply assume that the adolescent (or adult) is operating at this most advanced cognitive developmental stage.

Theories and Models

Three conceptual systems predominate in explaining how children learn. One of these, Cognitive Development Theory (CDT), as explained above, emphasizes the role of developmental processes that influence children's understanding. CDT has influenced studies of children's health beliefs and understanding of illness-related processes.[3-6] Behavioral Intention Theory (BIT) is more often used for adults but is attractive for children because it emphasizes the influence of reference group norms and focusing on specific behaviors rather than inferences and abstractions for which children are often not cognitively prepared. Moreover, BIT posits that a behavioral intention is the best predictor of an actual behavior. Social Cognitive Theory (SCT), a revision of Social Learning Theory (SLT), predominates.[7] According to SCT, behaviors are acquired and shaped through attention, retention, production, and motivation operating in three domains: personal, behavioral, and environmental. Personal factors include the child's own value system, and expectations derived from observation and experience; behavioral factors include performance skills; environmental factors include modeling and expressed opinions of peers, family, and media.

The Children's Health Belief Model (CHBM) was adapted from the Health Belief Model (HBM) to explain children's expectations of taking medicines by including aspects of children's learning theories.[8] The original Health Belief Model (HBM), first hypothesized by Rosenstock,[9] focused on preventive health behaviors. The HBM was later explored for its application in sick role behavior, especially compliance with physician directives.[10-13] The reformulated version placed greater emphasis on motivations, prior experience, and interpersonal relationships.

The CHBM (Table 18.1) is consistent with Gochman's[14] recommendation to place children's health behavior within their personal and social context, a context that recognizes the relationship of children's health behavior to their personal attributes (i.e., their beliefs, expectations, motives, and other cognitive elements) and recognizes that these personal attributes are influenced by peers, families, and other social groups. This view of children's health behaviors supported the inclusion of the influence of the child's primary caregiver in the CHBM and also cognitive and psychological attributes that change with age and experience, such as knowledge, risk taking, and perception of control over health status (health locus of control).

The CHBM was very successful in predicting urban elementary schoolchildren's expectation to take medicines for five common problems: cold, fever, upset stomach, trouble sleeping, and nervousness.[8, 15] The child's perception of illness severity and benefit of taking medicines were the two best predictors. Two other variables—concern with the illnesses and perceived vulnerability to the illnesses—had weaker relationships. From a developmental perspective, two findings were particularly noteworthy. Mothers' attitudes influenced their children's expectations of taking medicines, but the children's own cognitions and attitudes appeared to develop independently of their mothers. The primary message was that mothers'

TABLE 18.1. A Children's Health Belief Model

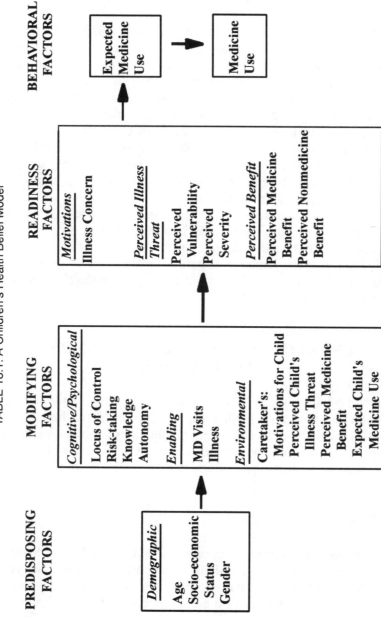

health beliefs are correlated with those of their children but that this effect is relatively small when compared to other developmental influences associated with the child. In fact, expected medicine use was not correlated between mothers and their children. Developmental cognitive and affective variables, particularly health locus of control, were significantly–but not strongly–related to the mothers' beliefs and yet played a significant role in children's health belief systems. Children's own motivational variables, influenced by their cognitive and affective processes, were more important in determining their expectations to take medicines than their mother's values, beliefs, or expectations.

An additional finding was that the variables in the CHBM were surprisingly stable among the children measured.[15] Not only were most of the variables correlated over three years, but also expected medicine use among children in grades three to seven was predicted from variables measured three years earlier, when the children were in grades kindergarten to four.

CHILDREN'S MEDICINE-RELATED BELIEFS AND BEHAVIORS

Factors likely to influence children's beliefs, attitudes, and expectations regarding medicines, in addition to the beliefs, attitudes, and behaviors of caretakers, include frequency of the child's use of medicines and environmental exposure. Some or all of these may interact. Cultural beliefs and practices may have an effect as well.

Medicine Use Frequency

Medicine use is common among children. In the United States, a recent survey found that one-half of the children under three years had been given an OTC (over-the-counter) medicine in the previous 30 days.[16] Although rates of medicine use–both OTC and prescription–tend to decrease after infancy and to begin to rise for females as they reach reproductive age, all surveys lead to the conclusion that medicine use is a common activity for most school-age children. Surveys consistently find that from about one-third to about two-fifths of elementary school-age children have used one or more medicines (including vitamins and minerals) in a 48-hour period.[17-18] Mothers' reports of their children's use tend to be higher than the children's reports, with the primary discrepancy attributable to vitamin and mineral use.[17] The higher incidence reported by the mothers may reflect their reporting what the child was supposed to have done and not what the child actually did.

Medicines in the Child's Household

The number of medicines stored in the household may also influence children's medicine-related beliefs and expectations as well as behaviors. The number of medicines stored may vary widely among subpopulations. For example, a household inventory in Chapel Hill, North Carolina, found 24 medicines stored in the average home where at least one Caucasian child lived, whereas in Washington, DC, a comparable survey found seven medicines in the average home where at least two African-American children lived.[19] In an earlier study in Ohio, the average number of stored medicines per household was close to that found in the Chapel Hill sample.[20] In a replication of the household inventory used in seven European countries, the average number varied from eight in the location in Finland to 24 in the location in Yugoslavia.[21] The number of medicines per person in the households varied from two in Finland to six in Madrid and Yugoslavia; seven was the average number of medicines per person in Chapel Hill, compared with only 1.4 per person in Washington, DC.

Children's Medicine Knowledge

A medicine knowledge index containing ten items believed to include those fundamental to appropriate use of medicines was administered to urban elementary schoolchildren.[17] The results from children in the third, fifth, and seventh grades are shown in Table 18.2. Only in seventh grade did the average child answer more than half of the questions correctly. There was considerable variation among the items in terms of the percentage of children

TABLE 18.2. Children's Medicine Knowledge Index (percentage of children responding correctly)

ITEMS	GRADE		
	3 (%)	5 (%)	7 (%)
If you were sick, which would help you more...			
a big pill or a little pill?			
no difference/it depends	23	14	35
a good- or a bad-tasting medicine?			
no difference/it depends	28	22	46
medicine from the drug store or the same thing from the grocery store?			
no difference/it depends	16	7	40
could the same medicine be different colors?			
yes	50	54	78
would medicine a doctor told you to take always make you feel better?			
no	66	63	76
What is a doctor's prescription?			
correct	60	85	92
What is the difference between medicines you pick out yourself and those you get with a doctor's prescription?			
correct	38	58	82
Can the same medicine be good for people and also bad for people?			
correct	64	63	75
Are drugs the same as medicines?			
yes/sometimes	36	45	56
Mean score on sum of correct items	3.8	4.2	5.9

Source: Adapted from Ref. 22.

who answered correctly, but the total scores were positively correlated with the children's grade. The seventh graders scored highest on every item, but for half of the items, the third graders performed as well or better than the fifth graders. In general, the children scored best on questions relating to prescriptions and poorest on questions relating to the relationship between efficacy (i.e., how well a medicine works) and medicine characteristics such as size, color, taste, and place acquired.

Even six-year-old children are aware of brand names. In the District of Columbia, 20 kindergarten children were asked to name the medicine they would take for a cold. Among the responses were 18 different brand names plus vitamins and others described by form, taste, or color, e.g., cherry-tasting medicine.[17]

Children's Medicine Autonomy

Areas of inquiry in a medicine autonomy scale include taking medicine without asking an adult, accessibility of medicines in the household, medicine purchase, giving medicine to other children, and taking medicine to school.[17,22] Except for general questions about household accessibility, when elementary schoolchildren were administered this scale, no child's response to any question was counted as affirmative unless the child named a specific medicine or the age at which he or she first engaged in the behavior. For example, relative to taking medicine without asking an adult for a headache, the response was not counted as affirmative unless the child named a medicine in response to the open-ended question asking, "What would you do if you were home alone and had a headache?" There was a second opportunity to score affirmatively; if the child responded, "I would call an adult," the child was then asked, "What would you do if you couldn't reach an adult?"

The responses to the 14-item Medicine Autonomy Scale are shown in Table 18.3. As expected, for every item, seventh-grade students indicated more autonomy than younger children. The item receiving the most positive responses was household accessibility. It was noted above that the average mother said that 12 years was the age when the average child could take a medicine for a headache independently; however, about one-fifth of third and fifth graders indicated they would do so, although most of them were younger than 12 years.

From 14 to 29% of the children in grades three, five, and seven indicated they had purchased a medicine independently, and 38 to 44% of the children said they had picked up a prescription. That children do purchase OTC medicines and pick up prescriptions without being accompanied by an adult was confirmed by visiting

TABLE 18.3. Children's Autonomy in Medicine Use

AUTONOMY ITEMS	GRADE		
	3 (%)	5 (%)	7 (%)
Had access to household medicines	77	83	100
All household medicines are accessible	45	75	78
Asked for the last medicine taken	67	72	78
Got household medicines for self	73	83	89
Got household medicines for self independently	22	33	45
Got household medicines for others	62	62	71
Got household medicines for others independently	9	24	36
Took medicine independently	23	28	51
Took medicine independently for headache	20	21	45
Took medicine independently for sore throat	17	28	33
Gave medicine to another child independently	9	9	25
Purchased medicine independently	14	29	29
Picked up prescription independently	38	34	44
Had medicine at school on day of interview	6	5	9

Source: Adapted from Ref. 22.

all of the commercial establishments within a half-mile radius of the schools attended by the children and by informally interviewing the managers or pharmacists on duty. Only in liquor stores was it unlikely that a child might buy an OTC medicine.

The majority of the children got medicines for themselves or others and had asked for medicine. Most or all of the household medicines are physically accessible to most children by age six or seven years, a finding replicated in Chapel Hill and in Europe (locations in Greece, Italy, Spain, Germany, Finland, Yugoslavia, and the Netherlands). Moreover, medicines are kept in more than one place in the average household.[21] Children also take medicines to school. In Washington, DC, 8.3% of the children had 1 or more medicines at school on the day they were interviewed.[17] Interestingly, when the mothers of these same children were interviewed later by phone, half of them said their child *never* takes a medicine to school.

Children believe themselves to be involved in their treatment. In a different study, one involving northern Virginia students, fewer than 16% of preadolescents thought they did not have a role in treatment of their last episode of injury or illness.[23] A majority of the students had taken a medicine, usually an OTC, for this last episode.

Media Exposure

In the mid-1970s, it was documented that the average child saw 1,139 drug commercials on TV annually.[24] This number is likely to have increased, as the number of hours children devote to TV watching increased through the last decade. The impact of such exposure is largely unknown. Much of the debate has concerned the impact on use of abusable substances rather than on beliefs, attitudes, and behaviors relative to medicines.[25] However, numerous studies have found that children of all ages are susceptible to media messages.

To assess the impact of TV commercials on children, fifth- and sixth-grade students were asked to watch commercials in two schools, one with and one without a program wherein children participate in decisions involving their own health care.[26] The students were asked to record information about six messages involving health. The students in the school with the program more frequently reported on commercials dealing with sedatives and tranquilizers and gastrointestinal medication, whereas the students in the school without the program reported more frequently on advertisements involving cold symptom relief. The proportion of children reporting they believed the messages was higher in the school without the program. Moreover, children in the school without the program took more vitamins and tonics and used fewer hygiene products, such as toothpaste.

In another study conducted in the mid-1970s among 673 children eight to twelve years old, exposure to TV was moderately related to children's proprietary medicine-related beliefs, affect, intentions, and request behavior.[27] In a later study involving children the same age, the children were found to have negative attitudes toward OTC drug advertising in general, but this general attitude was found to be unrelated to their attitudes toward specific brand-name drugs, about which most children seemed uncertain.[28]

Children's recognition of OTCs was investigated by showing seven-year-old and ten-year-old children 23 medicines that had been advertised frequently on TV during the previous month.[29] Children were asked if they had ever seen the medicine before, and if yes, where. To evaluate the validity of the source named, 18 generic drugs and drugs in the same category but not advertised were shown to the children. The average number of medicines recognized was 5.5 for the younger children and 7.9 for the older children. The source mentioned most often was TV or advertising (mean = 2.9 drugs for younger children and 4.1 drugs for older children), followed by "the store" for the younger children (1.7) and the home for the older children (2.1). The validity test revealed very few instances of a child claiming to recognize one of the unadvertised drugs.

Unfortunately, there are no longitudinal studies tracking children's exposure to advertising and later beliefs, attitudes, and behaviors. The question is, should advertising of medicines be banned? Results from earlier studies on the issue conflict, but these studies have not tested the influence of beliefs on the benefits of legal medicines.[25] It is very possible that drug advertising induces "irrational" perceptions of medicine benefits without increasing knowledge, use, and acquisition skills, or understanding of the limitations of medicines.

Interrelationships

Children who know the most about medicines and who feel the most in control of their health are less likely than other children to perceive that medicines will help them for common health problems and are less likely to expect to take medicines for common health problems.[8,30] Perceived benefit of medicines was measured by asking children to indicate by pointing to a bar on a graph how likely a medicine would be to help them for a series of common health problems. Expected medicine use was measured similarly by asking children how likely they would be to take "something special" for the same common health problems, and next, to say what they would take. Health locus of control was measured using the nine-item Children's Health Locus of Control Scale.[31] The concept of the health locus of control is that people vary according to how much control they perceive they have over their health. Beliefs range from one extreme that individuals feel they can affect the direction of their own health (internal) to the other extreme of helplessness, that good or bad health is due to factors outside of the person's control, good or bad luck, or others more powerful or knowledgeable such as parents or physicians.

Negative correlations between the Perceived Benefit of Medicines Scale and both the Medicine Knowledge Index and the Children's Health Locus of Control Scale were found among four populations of children, two in Spain and two in the United States.[17,30] In addition, in one of the U.S. studies, perceived benefit of medicines was strongly positively associated with expected medicine use.

Because there is little or no information given to children in schools about legal medicines, it was proposed in a hypothesized model tested among the Chapel Hill children that their knowledge and understanding of medicines are products of their development and the amount of influence from the environment in the form of parents and advertising. The more internal a child's health locus of control was, the more knowledgeable he or she was about medicines. The ten-year-old children were more knowledgeable than the seven-year-olds. The ten-year-olds tended to answer correctly on roughly one question more than the seven-year-olds. Moreover, the parent's level of education contributed to the child's knowledge. Next it was found that a one-unit (10%) increase in the predicted value on the ten-item Medicine Knowledge Index led to more than a four-unit (10.3%) decrease in the 16-item Perceived Benefit of Medicines Scale (range 16-55). Children's recognition of advertised medicines did not show a statistically significant relationship with perceived benefit of medicines, nor did the number of children in the household or whether a parent was a health professional.

Because the negative relationship found between children's knowledge of medicines and children's perceived benefit of medicines has been shown in four locations, it would be interesting to learn if this relationship exists among children in other countries and whether such "cynicism" leads to improved medicine use in any location.

HELPING CHILDREN LEARN ABOUT MEDICINES

Considering the pervasiveness of medicine use in society, it is surprising that school health education has been restricted, almost completely, to prevention of poisoning among young children and to prevention of abusable substance use among older children. It is most certainly possible to educate children about medicines as early as the second grade. Children think about medicines and have already discovered the role of medicines in the repair of the body's malfunctions. They are ready to hear messages about what medicines are able to do (i.e., cure some diseases, prevent others, relieve symptoms) and that medicines may be harmful.

Children are not learning about medicines in school, and they are not learning about them from their health care providers either. This leaves families and the media as the main medicine "educators." Pediatricians usually seek some information from the child but are more than four times more likely to provide information on diagnosis and therapy to the accompanying adults. However, children are interested in and often anxious about their medical condition and retain some information better than adults. Why is it important to talk to the children? If for no other reason than the quality of communications to children affects compliance and treatment outcomes, probably through the mechanism of anxiety.[32,33]

It is not enough simply to have good intentions and to decide to educate children about medicines. Most adults do not know that children's concepts of health and illness are qualitatively different from adults and that they change dramatically with age as the children progress through cognitive development stages. It is not simply that adults have more knowledge. Most health providers are poor judges of what children at various levels can comprehend, but most health providers can be taught how to recognize the child's developmental level and to communicate at that level effectively. Making an effort to talk at the child's level is not condescension or "talking down," rather, it is communicating at a mutually comfortable level.

For communications to children, the first step is to make a commitment to do it, to do it consistently, and to not give up if not every child responds. The second step is to decide where. The third step is to prepare materials to augment communications and/or to give to the children.

One-on-One Communication

1. Immediately focus on the child.

Engaging the child immediately will reduce the probability that a parent will respond for the child and will help the child recognize that you are interested in him or her. Bring yourself to the child's eye level and make eye contact with the child.

2. Attempt to communicate at the child's developmental level.[23]

For example, here are developmentally different responses to the same question, "What makes a cold go away?" from three children, ages six, eight, and ten years. The six-year-old and eight-year-old both replied, "A medicine." The ten-year-old said, "By taking the right medication from your doctor." When asked, "What kind of medicine?" the six-year-old replied, "Medicine that helps you." The eight-year-old replied, "Tylenol, Dimetapp." The next question was, "How does that help?" The six-year-old responded, "It takes the coughing away so you won't have a cold anymore." The eight-year-old said, "It helps fight the germs away." The ten-year-old replied, "If you take it and it goes in your system and it clears away the fluids and stuff that is in your body, like the mucus in your nose or something when it is stuffed up."

Although there is a general correlation with age, children develop at different rates that vary with the subject. The skills may be present but not applied within a content area, or a child may appear more advanced during a transitional phase. For example, another eight-year-old exhibited formal operational thinking when she was asked about how a medicine would help cancer: "Well the medicine kind of helps your immune system and white blood cells and from what I've heard kind of sterilizes the germs."

For a child in the preoperational stage with otitis media, you might say, "This medicine will go into your body to make your ear feel better. It will only work if you take it three times every day. Your mom will help you to know when to take the medicine. Be sure to use up all the medicine, even if your ear feels better." For a child in the concrete operational stage, you might say, "This medicine will go into your body to help fight off the germs that are causing the infection in your ear. The medicine will only work if you take it three times a day until (date). If you don't take it that way, your ear infection is likely to come back. So keep taking it even if you feel better. Work with your mom so you both know you have taken the medicine at the right time." For a child in the formal operational stage, you might say, "This medicine goes into your system to help your immune system fight off the bacteria that are causing your ear infection. The medical name for your ear infection is otitis media. The medicine used to fight the bacteria is an antibiotic. Its name is (name). What time do you usually get up? You must take this antibiotic every eight hours, that is, three times a day at (times) for the next ten days. If you don't, the bacteria may not all be killed and your ear infection can return. Keep taking it until it is all gone, even if you think your ear infection is cured. Also, you will need to shake the medicine very well before taking it, and you must keep it in the refrigerator. Here's a special spoon so you will always get the right amount. Let's look at the label together. What does it say?" "What should you do if you miss a dose?"

3. Ask open-ended questions rather than those requiring only a "yes" or "no" response.

Ask follow-up questions to make sure the child understands and have the child repeat what you said. A child may parrot back a correct response or read the label correctly without real understanding.

4. Use simple declarative sentences for all children.

This is good advice when talking to adults as well. In fact, an advantage of using developmentally appropriate child-directed communications is that the accompanying adult is more likely to understand what you say. The goal is not to empower the child with total responsibility but to build a partnership between parent and child, to work with the parent to begin to grant autonomy coupled with proper patient education and information. Modeling an appropriate communication style will help the parent continue communicating with the child about medicines and other health issues and grant appropriate autonomy.

5. Ask the child if he or she has any questions to ask you.

Be a medicines educator. Try to empower the child to feel comfortable asking you questions about his or her health problem, the particular medicine, or medicines in general.

6. Augment oral communications with written material.

Average adults only recall about 30% of what they are told by a doctor. You should not expect more from children. Written materials can be reviewed at home after the hospital, clinic, or pharmacy visit. For antibiotics, for example, a calendar can be given to the child with times and dates. The child can be asked to cross off each dose and to bring the calendar to the next visit. The date for a follow-up visit can also be on the calendar.

Children with Special Needs

Chronically ill children need special attention. Many chronically ill children (e.g., children with asthma) are responsible for taking their own medicines.[22] Older children in a large family and urban children also may have more autonomy at an earlier age than other children.

Illustrated materials are particularly important for those chronically ill children who use devices such as inhalers, patches, or syringes to take their medicines.

Classroom Children

After selecting a target grade, your first task is to learn what these children already know and do relative to medicines. The second task is to set out learning objectives. This involves clarifying what you think these children should know. The third task is to prepare a teaching plan. For example, for third-grade children, you might decide that you want to address attributes of medicines that are unrelated to medicine efficacy such as color, taste, size, and proprietary name. Examples can be brought to the class. To illustrate that the same medicine can be different colors, two tablets of different colors that are bioequivalent can be shown along with two M&Ms that are different colors. The children understand that the M&Ms have the same "active" ingredient inside, and the analogy can be made with the tablets. Labels can be read to teach that the same thing can have different brand names and different prices. For older children, you may want to develop a lesson about antibiotics and the importance of compliance, with the lesson including the differences between viruses and bacteria.

Some particular points that you may want to make include how to understand and counteract media promotion of medicines and understanding that legal drugs may be as dangerous as illegal drugs and should not be taken without restraint.

Adolescents

Although the topic of adolescents and medicines has not been the subject of this chapter, some recommendations regarding communicating with adolescents can be made. Adolescents need privacy, and they need to know that their communications are confidential. Adolescents need to talk with adults who are absolutely comfortable talking with them about *any* subject, including abortion, acne, birth control, sexuality, steroids, and STDs (including HIV+ and AIDS). This must be done in a completely nonjudgmental way and without letting the adolescent perceive that you find his or her beliefs foolish, irrational, or amusing. Adolescents have a great deal of misinformation about how their bodies work and how they should be maintained. Many adolescents are distrustful of information provided by older persons. If you cannot be completely comfortable when talking with adolescents, you can at least provide appropriate written materials and referrals.

REFERENCES

1. Piaget J. The moral judgement of the child. Gabain M, trans. New York: Harcourt, Brace, World, 1932.

2. Bush PJ, Davidson FR. Medicines and "drugs": what do children think? Health Educ Q 1982;9:113-28.

3. Bibace R, Walsh ME. Development of children's concepts of illness. Pediatrics 1980;66:913-7.

4. Bibace R, Walsh ME. Children's conceptions of illness. In: Bibace R, Walsh ME, eds. Children's conceptions of health, illness, and bodily functions. San Francisco: Jossey-Bass, 1981:31-48.

5. Burbach DJ, Peterson L. Children's concepts of physical illness: a review and critique of the cognitive developmental literature. Health Psychol 1986;5:307-25.

6. Perrin E, Gerrity S. There's a demon in your belly. Pediatrics 1981;57:841-9.

7. Bandura A. Social foundations of thought and action: a social cognitive theory. Englewood Cliffs, NJ: Prentice-Hall, 1986.

8. Bush PJ, Iannotti RJ. A children's health belief model. Med Care 1990;28:69-86.

9. Rosenstock I. Why people use health services. Milbank Mem Fund Q 1966;44:94-124.

10. Rosenstock IM, Strecher VJ, Becker MH. Social learning theory and the Health Belief Model. Health Educ Q 1988;15:175-83.

11. Maiman LA, Becker MH. The Health Belief Model: origins and correlates in psychological theory. Health Educ Monogr 1974;2:336-53.

12. Becker MH, Maiman LA, Kirscht JP, Hefner DP, Drachman RH. Predictions of dietary compliance: a field experiment. J Health Soc Behav 1977;18:348-65.

13. Christensen DB. Understanding patient drug taking compliance. J Soc Admin Pharm 1985;32:70-7.

14. Gochman DS. Labels, systems and motives: some perspectives for future research and programs. Health Educ Q 1982;9:167-74.

15. Bush PJ, Iannotti RJ. The origins and stability of children's health beliefs relative to medicine use. Soc Sci Med 1988;27:345-55.

16. Kogan MD, Pappas G, Yu SM, Kotelchuk M. Over-the-counter medication use among U.S. preschool-age children. JAMA 1994;272:1025-30.

17. Bush PJ, Iannotti RJ, Davidson FR. A longitudinal study of children and medicines. In: Breimer DD, Speiser P, eds. Topics in pharmaceutical sciences. Amsterdam, NY: Elsevier Science Publishers, 1985:391-403.

18. Bush PJ, Rabin DL. Medicines: who's using them? J Am Pharm Assoc 1977;NS17(4):117-230.

19. Bush PJ, Moore TS, Almarsdottir AB. Medicines in the home: differences between two U.S. communities. Annual Meeting American Public Health Association, Washington, DC, November 1994.

20. Knapp DA, Knapp DE. Decision-making and self-medication: preliminary findings. Am J Hosp Pharm 1972;24:1004-12.

21. Sanz EJ, Bush PJ, Garcia M. Medicines at home: the contents of medicine cabinets in eight countries. In: Bush PJ, Trakas DJ, Sanz EJ, Wirsing R, Vaskilampi T, Prout A, eds. Children, medicines, and culture. Binghamton, NY: Pharmaceutical Products Press, 1996:77-104.

22. Iannotti RJ, Bush PJ. The development of autonomy in children's health behaviors. In: Susman EJ, Feagans LV, Ray W, eds. Emotion, cognition, health, and development in children and adolescents: a two-way street. Hillsdale, NJ: Lawrence Erlbaum, 1992:53-74.

23. O'Brien RW, Bush PJ. Helping children learn how to use medicines. Office Nurse 1993;6(3):14-9.

24. Choate R, Debevoise N. Caution! Keep this commercial out of reach of children! J Drug Issues 1976;6:91-8.

25. Almarsdottir AB, Bush PJ. The influence of drug advertising on children's drug use attitudes, and behaviors. J Drug Issues 1992;22:361-76.

26. Lewis CE, Lewis MA. Children's health-related decision making. Health Educ Q 1982;9:225-37.

27. Robertson TS, Rossiter JR, Gleason TC. Children's conception of medicine: the role of advertising. Adv Consumer Res 1978;515-7.

28. Riecken G, Yavas U. Children's general, product and brand-specific attitudes toward television commercials–implications for public policy and advertising strategy. Int J Advertis 1990;9:136-48.

29. Almarsdottir AB. Children's perceptions of the role of medicines in health, illness, and treatment–a triangulation of qualitative and quantitative methods [Dissertation]. Chapel Hill, NC: University of North Carolina, 1994.

30. Almarsdottir AB, Aramburuzabala P, Garcia M, Sanz EJ. Children's perceived benefit of medicines in Chapel Hill, Madrid, and Tenerife. In: Bush PJ, Trakas DJ, Sanz EJ, Wirsing R, Vaskilampi T, Prout A, eds. Children, medicines, and culture. Binghamton, NY: Pharmaceutical Products Press, 1996:127-171.

31. Bush PJ, Parcel GS, Davidson FR. Reliability of a shortened children's health locus of control scale. ERIC #ED 223354, 1982.

32. Iannotti RJ, Bush PJ. Toward a developmental theory of compliance. In: Krasnegor N, Epstein S, Johnson S, Yaffe S, eds. Developmental aspects of health compliance behavior. Hillsdale, NJ: Lawrence Erlbaum, 1992:59-76.

33. Pantell RH, Lewis CC. Physician communication with pediatric patients: a theoretical and empirical analysis. Adv Develop Behav Pediatrics 1986;7:65-119.

ADDITIONAL READINGS

For the Instructor

1. Bush PJ, Hardon AP. Towards rational medicine use: is there a role for children? Soc Sci Med 1990;9:1043-50.

2. Bush PJ, Iannotti RJ. A children's health belief model. Med Care 1990;28:69-86.

3. Bush PJ, Iannotti RJ, Davidson FR. A longitudinal study of children and medicines. In: Breimer DD, Speiser P, eds. Topics in pharmaceutical sciences. Amsterdam, NY: Elsevier Science Publishers, 1985:391-403.

4. Iannotti RJ, Bush PJ. Toward a developmental theory of compliance. In: Krasnegor N, Epstein S, Johnson S, Yaffe S, eds. Developmental aspects of health compliance behavior. Hillsdale, NJ: Lawrence Erlbaum, 1992:59-76.

5. Pantell RH, Lewis CC. Physician communication with pediatric patients: a theoretical and empirical analysis. Adv Develop Behav Pediatrics 1986;7:65-119.

For the Student

1. Almarsdottir AB, Bush PJ. The influence of drug advertising on children's drug use attitudes, and behaviors. J Drug Issues 1992;22:361-76.

2. Bush PJ, Davidson FR. Medicines and "drugs": what do children think? Health Educ Q 1982;9:113-28. Reprinted in: Gochman DS, Parcel GS, eds. Children's health beliefs and health behaviors. New York: Human Sciences Press, 1982. Abstracted in: Smith MC, Wertheimer AI, eds. Pharmacy practice: social and behavioral aspects. 3rd ed. Baltimore: University Park, 1989:250-2.

3. National Council on Patient Information and Education. Children and America's other drug problem: guidelines for improving prescription medicine use among children and teenagers. Washington, DC: NCPIE, 1989.

4. O'Brien RW, Bush PJ. Helping children learn how to use medicines. Office Nurse 1993;6(3):14-9.

Chapter 19

Adolescents and College Students

Tracy S. Portner

INTRODUCTION

Adolescents and college students are distinct groups of persons whose behavior may mimic that of adults but whose unique physical and social characteristics present a different picture. Adolescents–those in the ten-year-old to late teens age group–are, by definition, a group in transition. Preparing to enter the adult world, adolescents experience changing physical bodies and emotional responses. For this reason, the provision of pharmaceutical care to adolescents should reflect knowledge and recognition of the special patient needs that they may have.

Similarly, college students provide opportunities for special emphasis in pharmaceutical care. As young adults often experiencing a tremendous amount of freedom and number of choices away from parental role models, college students warrant special attention.

While logic insists that adolescents and college students would demand less attention from pharmacists and other health professionals due to the generally good health that most have, we find increasing evidence suggesting that adolescents and college students are active participants in the health care marketplace. Increased levels of independence, the beginning of new relationships, and opportunities for experimentation with alcohol, drugs, and sexual relationships place students in higher education at special risk of problems with mental and physical health. It is believed that interventions aimed at the time of the developmental transition from youth to adulthood can influence long-term behavior patterns.

473

According to government statistics, those persons 22 years of age and younger comprise more than 30% of the population of the United States. Of these, around 5% of the U.S. population, or 12 million people, are currently enrolled in college.[1] Two million of these are "conventional" students, that is, those 18 to 22 years of age enrolled full-time in school, representing about 10% of the total number of persons in the 18 to 22-year-old age group.[2] Projections for 1995 are for the total number of enrolled college students to continue to increase.[3]

Another 35 million persons fall in the age categories of 10 to 19 years of age.[4] This age group, the adolescents, comprise approximately 14% of the total U.S. population. Males in this age category slightly outnumber females, although the mix is approximately equal.

SOCIAL AND HEALTH ISSUES

A number of critical health issues face adolescents and college students. The United States Public Health Service has formulated National Health Objectives for the year 2000 including guidelines and recommendations on college-student health. Health status goals for healthy adolescents and young adults hope to reduce deaths by 20% relative to 1977, with special focus on motor vehicle injuries, alcohol, and drugs.[5]

The top five issues affecting college students through the year 2000 were identified in 1987 by medical professionals, health educators, and residence hall staff offering testimony at a meeting of the American College Health Association.[6] The issues identified were sexual health concerns; substance abuse; mental health concerns; issues surrounding food, such as eating disorders, chronic disease prevention, and general nutrition; and the efficient and wise use of health resources.

Sexual Behavior and Contraception

Over 90% of the experts identifying issues affecting the health of college students identified sexual health concerns as a primary threat to the physical and emotional health of young adults.[6] Sexually transmitted diseases, with heightened emphasis on AIDS, affect increasing numbers of college students. It is estimated that 86% of all sexually transmitted diseases occur among persons aged 15 to 29 years old.[7]

It is said that this age group possesses little understanding of the seriousness of many sexually transmitted diseases, and the lack of appreciation of the realities of contracting AIDS is frightening. A great deal of misinformation and ignorance about human sexuality, sexual behavior, and sexual problems exists.

Sexual Practices

Social acceptability of sexual intimacy is growing, and the number of sexually active adolescents and college students is staggering. Seidman and Reider, in a review of various surveys of American sexuality and the 1988-90 General Social Surveys, found that most American males have intercourse by 16 to 17 years of age, and females do so by 17 to 18 years of age.[8] In contrast to adults in the 25 to 59 years age group, where the norm is monogamous relationships, the majority of young adults aged 18 to 24 have multiple serial sex partners. In the American college student population, some 75% are sexually active, with those sexually active averaging more than two partners per year. One-third to one-half of this sexually active group will have six or more sexual partners before marriage.[9-11] Because having multiple sexual partners is associated with increased risk of contracting sexually transmitted diseases, this behavior indicates that a large proportion of young heterosexual persons is at considerable risk of contracting sexually transmitted diseases.

AIDS

It is estimated that two in every 1,000 American college students are infected with human immunodeficiency virus (HIV), and 0.4% of all cases of AIDS occurs in adolescents.[9,12] Despite research indicating that college students have a high level of knowledge about AIDS and that they claim to have changed their sexual behavior because of the threat of the disease, there appears to be no relationship between knowledge level and sexual behavior changes.[9,13-17]

There are gender differences in the incidence of AIDS and risk behavior associated with contracting the disease. The majority of those affected with the virus are men, despite recent increases in AIDS among women.[18] Unmarried women and men possess different motives, values, and attitudes surrounding premarital sex. College women are significantly less permissive, view sex in emotional rather than physical terms, and tend to score higher on measures of sexual responsibility and conventionality.[19-21] Given the preceding, women may be more likely to respond to increased knowledge of the AIDS threat through behavior changes.

Contraception

Attitudes toward contraception vary considerably in this population. Relatively few teenagers and young adults use contraception on a regular basis. In 1990, there were an estimated 1 million pregnancies in the U.S. among women ages 15 to 19 years. Most of these pregnancies (approximately 95%) are unintended, and 400,000 were terminated by abortions. The pregnancy rates have not changed significantly since 1980. However, the abortion rate appears to have declined.[22,23]

The condom and oral contraceptives are the most popular contraceptive methods used by sexually active adolescents. Hale and colleagues report that one-half of the college students studied used contraception, with oral contraceptives as the first choice and condoms second.[24] Thirty percent of the college students at one northeastern university reported that they used condoms all or almost every time they had sex.[25] In another study of 2,000 college students, 37% to 54% of students indicated that they always used some method of birth control. The method of choice was condoms alone for 56% of the men and 42% of the women.[26] However, fairly ineffective methods, such as periodic abstinence, coitus interruptus, and withdrawal before ejaculation are in use.[27]

The selection of birth control options may be even more of a puzzle in adolescents. One option that has potential for success in this patient group is levonorgestrel implants (Norplant®). One recent study looked at satisfaction and side effects during the first six months of use with this method and compared the results among adolescents and adults.[28] The most frequent reason adolescents gave for selecting Norplant was convenience. None of the adolescents in the study reported being upset by the insertion process or by its appearance in the arm. Some side effects were reported, including weight gain (38%), emotional disturbances (33%), and headaches (38%). Overall, most of the adolescents reported being satisfied with this method of birth control. This may be a suitable method of contraception in the age group due to its long duration of action and lack of compliance problems.

Substance Abuse

Addictive behaviors and the abuse of alcohol, drugs, tobacco, and food have been identified as the second greatest health risk for young adults.[6] Our nation's youth have experimented and gained considerable experience with alcohol and other social drugs in the last 25 years. Widespread abuse of alcohol and social drugs has become a major societal problem, and most students report the beginning of experimentation with alcohol and other drugs in their early to late teens.

Alcohol is the primary recreational drug for students as well as adults. In a national survey of high school seniors, 73% of high school students reported use of alcohol in the year preceding the survey.[29] The use of alcohol by most of the students compares with 31% reporting use of marijuana in the preceding year, 9% amphetamines, 8% inhalants, 7% psychedelics, and 4% cocaine. Data reporting use of substance in the month preceding the survey suggest that students are regular users. Over one-half of the students surveyed reported use of alcohol in the month preceding the survey, and use of marijuana (19%), inhalants (3%), and psychedelics (3%) to a lesser extent.

Use of alcohol increases in college students. Initial findings of the core alcohol and drug survey indicate that the average number of drinks consumed per week by college students across the nation is 5.0, with men reporting 7.5 drinks and women 3.2 drinks.[30] The good news is that binge drinking, the consumption of five or more drinks on a single occasion at least once in the two weeks prior to the survey, continues to decline slowly. In a survey of Indiana teenagers, the rates of binge drinking for 1995 were the lowest reported since 1991.[31]

Following a 14-year decline in use of marijuana among students from 1978 to 1992, marijuana use in 1993 climbed and continues to climb nationally.[29] From 1992 to 1995, the number of teenagers who had smoked marijuana at least once during the previous 12 months had doubled. One of two high school seniors have used illegal drugs at least once in their lifetime.

Use of inhalant drugs, such as nitrous oxide, volatile nitrites, and petroleum-based solvents, appears to be increasing. Eighth and ninth grade students are the most frequent users with use declining in higher grades.

The rate of cigarette smoking in children and adolescents has increased since 1993.[29] This increase was seen in most grades, and is particularly worrisome since daily cigarette smoking is the best statistical predictor of future use of alcohol and other drugs.[32] In children and adolescents, daily smokers are three times more likely to be heavy drinkers and 10 to 60 times more likely to be users of controlled substances as are nonsmokers.

Some gender differences do exist in substance abuse profiles. Based on a National Center for Health Statistics report, adolescent boys are more likely than girls of the same age to smoke, and white teenagers have higher smoking rates than minorities.[33] Teens from troubled families smoke the most. Although drinking patterns indicate that males drink more than females, the rate of increase is greatest in female students.

The medical and social implications of the recreational use of drugs by college students are staggering. The short-term health-related impact is reflected by an increase in student injuries, accidents, date rapes, car crashes, and suicide attempts. It has been estimated that 2 to 3% of college students will eventually lose their lives due to alcohol use.[34]

Mental Health Concerns

Stress, anxiety, pressure to achieve, fear of failure, low self-esteem, lack of social support, and superficial relationships are mental health concerns that affect a great number of college students. Many college students have underdeveloped skills in coping and communicating, come from family backgrounds that are problematic, experience depression, and in general have difficulty dealing with their personal lives. These problems may be manifested through physical illness, loneliness, and alienation. In fact, many of the other social and health concerns that have been discussed, such as alcohol and substance abuse, are probably related to poor mental health.[6]

Teenage suicide is perhaps the ultimate expression of these problems, and the rate at which it occurs is growing. In the 1950s, suicide rates for men of college age were under 10 per 100,000 persons and about 2.5 for women. These rates tripled for both sexes over the next 25 years.[35] Between 1980 and 1989, the suicide rate among teens in the United States increased by 18%. In some states, the rate more than doubled in that nine-year time period. In 1989, the state with the highest suicide rate was Alaska, with 9.6 deaths per 100,000 teens, while Hawaii reported the lowest rate of 2.9 teen deaths per 100,000.[36] Suicide is among the three leading causes of death for 15- to 19-year-olds and college-age persons in the general population.[35,37]

Problems with Food and Nutrition

In general, our nation's youth weigh more than their parents and exercise less often. Although obtaining adequate amounts of food is not a common problem for most youths, the quality of the content of that food, from a nutritional standpoint, continues to decline.

Binge-Purge Syndrome

One of the most common food problems encountered in young adults is bulimia, or binge-purge syndrome. This is a common eating disorder associated with significant medical complications. The disorder is characterized by continual patterns of binge eating followed by fasting and/or various methods of purging, including self-induced vomiting and abuse of laxatives and diuretics.[38] Preoccupation with weight and dieting and an obsessive fear of weight gain and body-image distortions are common among bulimics.

Most typically, efforts to control weight are the starting point of the binge-purge cycle. A teenager may begin a self-imposed highly restrictive diet that causes hunger. At some point, the hunger is relieved by bingeing behavior. Following the bingeing, the individ- ual experiences guilt and fear of gaining weight, which leads to self-induced vomiting. Hunger follows, and a cycle of behavior is established.

Most patients with bulimia report that the syndrome began in adolescence and was associated with attempts to regulate weight. Killen and colleagues report that about 13% of adolescents in their study of tenth-grade (15-year-old) high school students reported some form of purging behavior, including use of diet pills (Table 19.1).[39] These findings and those from several other studies suggest that in adolescent girls, vomiting occurs at a rate of 10 to 11%, laxative use at 4 to 7%, and diuretic use at 3 to 4%.[40,41]

As with adults, bingeing behavior is seen more frequently with females than males.[42] In the 15-year-olds studied, female purgers

TABLE 19.1. Frequency of Purging Behaviors and Diet Pill Use Among 15-Year-Old Males (M) and Females (F)

Frequency	Diet Pills M	F	Laxatives M	F	Diuretics M	F	Vomiting M	F
Never	96.2	91.7	94.3	93.2	97.6	96.3	95.0	89.4
Monthly or less	2.5	6.6	4.1	6.1	1.0	3.1	3.2	8.6
Weekly	0.5	0.5	0.6	0.5	0.6	0.1	0.8	1.1
Several times a week or more	0.8	1.2	1.1	0.2	0.8	0.4	1.0	0.9
Total yes responses (M vs F subjects)	3.8	8.3*	5.8	6.8	2.4	3.6**	5.0	10.6*

*   significant difference at p<.0005

** significant difference at p<.02

Source: Killen, J. et al. Self-induced vomiting and laxative and diuretic use among teenagers. *JAMA*, 1986;255(11):1448.

outnumbered male purgers by two to one. Heatherton and colleagues examined changes in eating disorder symptoms among college students from 1982 to 1992.[43] The estimated prevalence of bulimia nervosa decreased from 7.2% to 5.1% for women and from 1.1% to 0.4% for men. Binge eating, vomiting, and diuretic use declined during this period. However, the incidence and prevalence rates of bulimia in adolescents has steadily increased since 1982.[42] In general, both male and female purgers felt guiltier after eating large amounts of food, counted calories more often, dieted more frequently, and exercised less than nonpurgers.[35]

It is evident that a large number of adolescents use unhealthy strategies for controlling their weight. A growing national preoccupation with having the "perfect" (slender and fit) body and the typical psychological struggles of adolescence are a potentially serious combination for predisposal to eating disorders.

New and expanded roles for pharmacists may present increased opportunities for preventing the adoption of harmful weight control strategies by the younger age groups. Educational programs targeted for primary prevention of eating disorders and discussion of healthful weight control procedures are possible areas for attention. The programs should focus on the social influence mechanisms promoting unrealistic and unhealthy attitudes about body weight, and impart accurate knowledge and effective influence-resistance techniques.

Calcium Intake in Young Women

Of lesser short-term consequence–but an increasing problem in the young female–is lack of adequate calcium intake. Findings of low calcium intake by adolescent females and young adult women appear consistently.[44] Such patterns contribute to the low bone mass associated with late-life osteoporotic fractures in women. Because of these findings, efforts are being made to target young females, especially white females who suffer more hip fractures than other racial/ethnic groups.

Retail pharmacists may have the opportunity to observe behavior patterns that suggest eating disorders in their patients who are adolescents and college-age. Patient histories should include questions about over-the-counter medication use, and specific attention should be paid to youths who purchase laxatives, diet pills, or diuretics on a regular basis. Educational programs for younger patients should emphasize normalized eating habits and increased physical activity.

Efficient Use of Medical Resources

Expenditures for Care

According to government statistics, persons 25 years of age and younger spend about $86 per year on drugs and medical supplies–about one-fourth of the average annual total health care budget of $360.[45] Annual per capita expenditures increase for those with chronic conditions. It is estimated that approximately 31% of adolescents in the United States have one or more chronic conditions.[46] Looking beyond the patient, family members have the responsibility of caring, in part or in sum, for these patients.[46] As can be expected, the costs of care are tremendous.

Over one-third of the total hospital days for all children under the age of 15 years in the United States were associated with care of those with chronic conditions. Average per capita expenditures (in 1980) on care for disabled children were estimated to be $601 higher than expenditures for nondisabled children, with total direct costs of such care approaching $3 billion (in 1980).[47] Not included in these estimates are nonmedical costs such as transportation, clothing, special diets, and missed work for parents.

Expenses associated with illness in adolescents and children go beyond direct medical costs of medications and other related health care costs. Indirect costs of illness, defined as those costs due to loss of productivity, are significant. Males lose an average of 4.7 work days per year due to illness, while females lose an average of 5.9 work days.[48] While these data include days of personal illness for adults, one would expect, particularly in the case of the mother, that these losses include days of work missed to care for a sick child.

In those of nonworking age, this indirect (productivity) loss is often expressed as days of school lost. Data from 1990 indicate that males of school age (6 to 16 years of age) lost 4.3 days per year due to illness, and females lost an average of 5.0 days per year from school.[49]

SELF-CARE ACTIVITIES

An important component of medical consumerism is the increasing emphasis on self-care. Self-care has been defined by Levin as activities that individuals undertake to restore and promote health, prevent disease, and limit illness.[49] Self-care activities range from self-medication to taking care with diet and exercise, and are generally divided into three distinct general areas: (1) symptom-related self-care for acute problems, (2) symptom-related self-care for chronic problems, and (3) asymptomatic self-care involving alteration of life-style and risk factors. Although the latter two areas of self-care have received the most attention in the literature, the self-care of acute problems is gaining more attention.[50]

The practice of self-care is quite extensive, and the literature suggests that anywhere from 50% to 80% of all care is self-provided.[51] Self-care practices of college students appear to occur similarly. In a study to assess the incidence of self-care measures by college students prior to visiting a student health service, almost two-thirds of the subjects had engaged in self-care for the symptoms they had.[52] This is probably an underestimate of the extent of self-care activity because some self-care practices probably occurred that did not lead to a subsequent visit to the student health service.

Not unlike consumers in general, it is likely that students practice self-care with medications without adequate knowledge. The most recently published report appeared in 1987 and resulted from a study of 561 students at Michigan State University.[53] The researchers administered a standardized drug knowledge test and concluded that students deserve a grade no better than "C" based on scores.

Comments from a recent focus-group discussion with college students indicate that students may not realize the significance of using nonprescription medications appropriately.[54] The students expressed the belief that nonprescription medications were "relatively harmless or they wouldn't let you buy them without a prescription." The students indicated that the piece of information desired about nonprescription medications is what is the *most* (quantity) amount that they could safely take.

As stressed by Levin:

> Self-care is the essential ingredient in health care that needs to be complemented by necessary professional and technical resources. The task is to appreciate and strengthen beneficial self-care practices, encourage the replacement of inappropriate ones, and back up self-care with accessible (and acceptable) professional care.[55]

The point has been well made that college students may be an especially good group to target for education and assistance regarding self-care activities. As old attitudes and beliefs are dropped or reshaped and new attitudes and behaviors emerge, the opportunity exists to effect change in this group of young adults. This, in fact, was found by Coons and colleagues when they conducted a study to determine whether an intervention using self-care information would change college students' attitudes and beliefs about personal responsibility and involvement in health care.[52] Results indicated that those in the treatment group, who were exposed to general information about the benefits of personal responsibility for health as well as sections from a consumer-oriented health care book, had a positive change in attitudes toward active participation in health care.

In an experiment to determine the effect of a "dial-a-nurse" program on a college campus, nursing personnel were available by telephone to answer students' questions about medical problems and emergencies.[56] The program, which was successful, was aimed at enhancing self-care (if appropriate) and decreasing unnecessary utilization of health resources. In the program, most calls reflected diseases of the respiratory system, injury and poisonings, and diseases of the digestive system.

ILLNESS AND MEDICATION USE IN YOUNG ADULTS

A certain amount of "illness" exists even in typically well populations. Zeltzer and colleagues found that almost one-third of a group of 345 "healthy" adolescents reported currently being ill when questioned.[57] Illnesses reported included allergies or colds (27% of total) and asthma and sinus problems (2% each). Self-reported treatment for this group included allergy medications (10%), aspirin (1%), antibiotics (.75%), cold medications (.5%), and thyroid medication (.3%).

OTC Medication Use

It is not surprising to discover that the use of OTC medications by adolescents and young adults occurs at a relatively high rate. Common medication categories used by adolescents and young adults include analgesics, antacids, vitamins, laxatives, cough medicines, and skin ointments, with various types of analgesics and cough and cold preparations heading most lists. Some differences in frequencies of categories used by age groups is evident, however.

Krupka and colleagues found in 1982 that aspirin was the most prevalent substance used by children aged 9 to 11 years.[53] In a study of children in England, aspirin accounted for almost 15% of all drugs (both prescription and over-the-counter) used in any one week.[58] However, medications affecting the respiratory tract and coughs were the major product category used. In these children, prescription medication accounted for the majority (55%) of all medications used.

In contrast, over-the-counter medications were used with greater frequency than prescribed medications by college students in a two-week study reported by Busching and Bromley in 1975.[59] Aspirin was used by a greater percentage of respondents than other OTC medications, followed by medicinal creams; soaps; and shampoos, vitamins, and dandruff preparations.

Vener, Krupka, and Climo identified three classes of over-the-counter medications that accounted for over 90% of those consumed by users of nonprescribed medication in a study of undergraduate college students reported in 1981.[60] Forty-one percent had used analgesics, 36% had used vitamins, and 15% had used cold remedies. In a later study reported in 1991, students were relatively infrequent users of OTC products when the number of products consumed in a period of one week is examined. Approximately 40% of college students surveyed indicated that they take no OTC products in an average week, and 50% reported taking one to two different OTC products in an average week.[61]

Brand-Name Preferences

In one recent study, college students from across the nation were asked to name a brand name or type of over-the-counter medication most frequently used or purchased. Respondents indicated the most frequently used/taken product category was analgesic products, followed by cough and cold products.[61] Details of product categories and specific brand names cited appear in Table 19.2.

Appropriateness of OTC Product Selection

In 1980, Cafferata, Lach, and Reifler identified analgesics, cough and cold products, and topical antimicrobial products as the nonprescription medications most frequently used by college students.[62] The symptoms for which these products were used were headaches, skin cuts, colds, "runny" nose or sneezing, cough, and acne. Conditions for which the fewest number of nonprescription medications were used were drowsiness, insomnia, nervous tension, hemorrhoids, corns, and obesity. Student use of OTC products was evaluated by university health professional staff for appropriateness of use. The data indicated that self-medication for the more frequently treated symptoms (e.g., headaches, skin cuts, colds) was generally appropriate.

Misuse of OTC Medications

The public's belief that OTC medications are harmless leads many to misuse these products in a variety of ways. Silverman and Lee suggest that consumers overmedicate through their belief that if three tablets work well, then six tablets will work two times better.[63] This assumption has been voiced by college students in a focus group discussion. Students expressed a desire to know the "most amount" of medications they could take, suggesting that the label directions are not as important to OTC medication use as taking the maximum amount they can without harmful effects.[54]

Sources of OTC Products

College students purchase over-the-counter medications in a variety of locations. When asked to name the primary source of OTC products, college students listed grocery stores and pharmacies.[61] More than 40% indicated that they purchased their OTC products at grocery stores, with another one-third reporting pharmacies as their primary source.

SOURCES OF SELF-CARE
AND OTC PRODUCT INFORMATION

General Perceptions

In general, three distinct groups of sources of information about self-care and health-related activities have been found:

1. Formal sources such as family doctors, pharmacists, chiropractors, and natural therapists;
2. Informal sources such as friends and relatives;
3. Commercial and media sources such as health food shop personnel, fitness instructors, and newspaper articles.

Studies have found that young people may have different preferences for and opinions about health information sources than older adults.[64,65] In some studies, young people rated the reliability of commercial, media, and unorthodox formal sources higher than did other respondents. However, when asked to rate information sources as to accuracy of information provided, the responses were similar to those of other age groups. Formal information sources, such as physicians and pharmacists, were rated as more accurate than informal sources. In general, younger consumers appear to view the family doctor and pharmacist as the most reliable sources of health information, while TV advertisements, newspapers, and magazine articles were considered the least reliable.

Availability and access to information may greatly influence product selection. In product adoption decisions, college students

TABLE 19.2. Most Frequently Used or Purchased OTC Medications

Product Type or Brand Name	n[1]	Percentage (%)
Analgesics		
Analgesics/pain relievers	11	1.38
Non-aspirin pain relievers	9	1.13
Acetaminophen	2	0.25
Anacin 3	4	0.50
Panadol	2	0.25
Tylenol	288	36.09
Aspirin	109	13.66
Bayer Aspirin	6	0.75
Ibuprofen	12	1.50
Advil	153	19.17
Mediprin	1	0.13
Motrin IB	17	2.13
Nuprin	17	2.13
"Other" Branded Products		
Ascriptin	1	0.13
Bufferin	10	1.25
Excedrin	11	1.38
Midol PMS	7	0.88
Mobigesic	2	0.25
Pamprin	3	0.38
Total Analgesics	665	83.34
Cough/Cold		
Allergy Products	10	1.25
"Cold" Medicines	17	2.13
"Cough" Medicines	6	0.75
Antihistamines	5	0.63
Decongestants	5	0.63
Actifed	10	1.25
Benadryl	10	1.25
Chloroseptic	1	0.13
Chlor Trimeton	2	0.25
Co-Advil	2	0.25
Comtrex	3	0.38
Contac	1	0.13
Dimetapp	9	1.13
Drixoral	5	0.63
Niquil	2	0.25
Sine-Aid	4	0.50
Sinutab	3	0.38
Sucrets	3	0.38
Sudafed	9	1.13
Triaminic	4	0.50
Total Cough/Cold	111	13.93

Antacids	2	0.25
Maalox	4	0.50
Mylanta	4	0.50
Tums	3	0.38
Total Antacids	13	1.63
Other		
Anusol	1	0.13
Bronkaid Mist	2	0.25
Fergon	2	0.25
No Doz	2	0.25
Primatene Mist	1	0.13
Vitamin C	1	0.13
Total Other	9	1.14
Total	**798**	100.04[2]

[1]Responses to the question, "What brand name or type of over-the-counter (OTC) medication do you most frequently use or purchase?"
[2]Percentage does not total 100 due to rounding.

rated the advice of friends, packaging, and media greater than that of health professionals such as physicians, nurses, or pharmacists.[62] Approximately 60% of those sampled reported multiple influences on the selection of a drug product. However, when asked to rate the single most important influence, physicians and nurses were more likely than friends, relatives, pharmacist, packaging, and advertising to be rated the single most important influence.

Differentiating Information Sources

Different information sources present varying images to consumers. Results of a recent study suggest that formal (physicians and pharmacists) and informal (family, friends, TV ads) information sources were perceived differently by students.[65] Formal sources were identified as accurate but inconvenient. In contrast, informal sources were convenient sources of information but were perceived as less accurate.

Attributes of these information sources are perceived differently depending upon the situation for which the information is needed. In general, the family serves as a very important source of information for college students. The family is likely to be the information source used unless one wants to know the cause of the problem or unless it is important that relief be obtained quickly. In addition, if the situation is perceived as more threatening (and hence more risky) by the student, with longer lengths of symptom experience and more discomfort, formal sources of information are more likely to be contacted.

It appears that students will use those information sources that are most convenient as long as they believe that the source is capable of providing adequate information. As the need for more complex information becomes apparent, convenience and time are traded for access to "better" (and different) information.

Ratings of Information Importance

Table 19.3 lists responses of college students when asked to rate the importance of having various types of information about OTC products and minor health problems available to them. In general, the importance of various types of information that may be provided about OTC medications produced generally high ratings.[61]

TABLE 19.3. Importance of Information About OTC Medications

Information Type	n	Mean[1]	Standard Deviation
How much medication to take	785	6.28	1.21
The most amount of medication you can safely take	785	6.17	1.27
Side effects	785	6.10	1.28
How to take the medication (e.g.,with food, on an empty stomach)	785	5.84	1.43
When to take the medication	785	5.75	1.53
How much it costs	785	5.10	1.82
[1]Respondents rated the importance of having the information on a scale where 1 = not important and 7 = very important.			

"How much medication to take" and "the most amount of medication you can safely take" received the overall highest importance ratings. This is not surprising given student comments in a focus group discussion which provided insight into perceptions of OTC medications.[54] The students in the focus group indicated that OTC medications would not be allowed for sale if they were dangerous. Because they are not perceived as dangerous, the package dosage directions are sometimes disregarded. Dosages may be doubled or tripled to get a "better" effect.

HEALTH SERVICES UTILIZATION

As health care consumers, adolescents and college students should be viewed as full participants having unique needs. Tootelian and Gaedeke suggest that a highly specialized market exists for the provision of health care services to college students.[66]

Health Professionals

Students use the services of health care professionals on a relatively frequent basis, with a recent survey finding that more than two-thirds of those surveyed had been seen by a physician or nurse practitioner at least two times in the last two years.[66] About 30% indicated contact with a health professional five or more times in the last two years. While student health programs exist at most college campuses, students indicate that their health needs are not met completely by college health programs.

Pharmacist contact, however, may be much more infrequent than contact with other health professionals. A study of 2,791 high school students in rural Mississippi found that only 23% of the students surveyed said that they had spoken to a pharmacist about a health problem within the past year.[67]

Emergency Room

College students' use of other sources of medical care is less than that of the general population. In one study, college students used the emergency room at a rate of 41.7 visits per 1,000 students per semester, or 141 visits per 1,000 students per calendar year.[56] This is considerably lower than the 252 visits per 1,000 people per year reported for those aged 18 to 24 and the general public's emergency room use rate of 300 visits per 1,000 people per year.[68,69]

In addition, when the appropriateness of use of emergency room services is examined, college students perform well. McKillip found that 88% of college students' visits to emergency rooms were rated as emergencies and deemed an appropriate use of resources.[56] In contrast, research from 1980 on the general population found that only 67% of visits were emergencies that required medical treatment within two to 12 hours.[70] The diagnostic categories and weekly pattern of use of emergency rooms by college students were similar to those of the general public. Injuries were the primary reason for visits, and visits occurred more frequently on weekends than during the week.

ROLE OF THE FAMILY IN ILLNESS CARE

The Basic Unit in Health Care

The family is the primary unit in health and health care. Litman remarks that "the family constitutes the most important social context within which illness occurs and is resolved" and discusses the role family members play in the etiology, treatment, and recovery of patients.[71] The family, most frequently the mother, is involved in defining whether a family member is sick, validating that the person is indeed sick, and providing the impetus for professional care to be sought. Parental modeling behavior was found to influence the use of a medication product in children.[72] Predictably, 71% of the fifth and sixth graders studied reported using a particular product that their parents had used, and 90% of the children had not used the product when their parents had not.

Mothers play a crucial role in determining family decision-making styles, the utilization of health care services, and health beliefs held by the family. In addition, the mother plays a major role in the health care and health practices of the family. "Doctor Mom" is alive and well. In more than two-thirds of the cases, the decision to seek out some form of professional assistance for an ill family member generally rested with the wife-mother.[71] A more recent study found an even greater reliance on the mother for advice about minor, self-limiting illness or over-the-counter medications.[61] When asked to name the family member whom they would most likely ask for advice, respondents most often named "Mom" (78.5%), followed by "Dad" (11.1%).

The Role of Heredity

Genetics play a tremendous role in an individual's potential for disease. A familial link exists in those with diabetes, heart disease, and cholesterol levels that may be more important than any environmental factors in predicting disease.

The pharmacist can and should assist individuals in identifying familial risk factors. As a source of more frequent contact with patients than most other health professionals, pharmacists should include questions addressing family disease history in medication histories and issue periodic reminders of needed screenings. Diseases with a strong hereditary component include alcoholism, diabetes, hypercholesterolemia, heart disease, ovarian cancer, and breast cancer.

ADOLESCENTS AND CHRONIC ILLNESS

To most of us, children with a chronic illness present a more tragic picture than adults with similar afflictions. Any family member who requires the care and attention demanded by illness causes major changes in daily living for both the patient and the family. Those with chronic diseases have an extended, if not lifetime, involvement with the health care system. This involvement, if begun while in adolescence, creates significant physical, emotional, social, and economic issues for these individuals. Adolescents exhibit different illness behaviors than their younger or older counterparts, and their perspective is important in providing appropriate treatment and educational programs.

Incidence of Chronic Disease

It is estimated that 10 to 20% of children (ages 0 to 20 years) in the United States have chronic conditions, comprising 50% of pediatric practice.[73] The most prevalent diseases are asthma (estimated to affect 38 per 1,000 children), visual impairment (30 per 1,000), mental retardation (25 per 1,000), and hearing impairment (16 per 1,000). Less prevalent diseases include seizure disorders (3.5 per 1,000), arthritis (2.2 per 1,000), and diabetes (1.8 per 1,000). Approximately one of three U.S. adolescents have been reported to have one or more chronic conditions such as allergies, asthma, or severe headaches.[46] Certainly, each of these diseases differs in the effect of the illness on the patient, the patient's family, and the involvement with medical and pharmaceutical treatment required.

Unique Aspects of Chronic Illness

Chronic illness differs from acute physical conditions in several respects. First, chronic disease must be "managed" for long periods of time—it is treatable but not curable. Second, responsibility shifts from the physician to the patient, who is then the disease "manager." The patient (or patient's family) is the day-to-day manager who commits to perhaps a lifelong and expensive process of care.

A chronic disease in an adolescent may serve to enhance the formidable insecurities and stress experienced by healthy adolescents. Some conditions or treatments of a condition cause obvious disfigurement, a tremendous problem in a group where awareness of physical characteristics is heightened. Other conditions and/or treatments, such as anorexia or corticosteroids, can delay the onset of puberty. Doubts about the adequacy of one's body, perhaps exacerbated by lack of normal pubertal development, may occur as a result of chronic disease.

Psychological Effects of Illness of Adolescents

It seems reasonable to assume that adolescents who face life-threatening or lifelong disease would demonstrate higher anxiety levels and/or lower levels of self-esteem than healthy adolescents. However, this is not necessarily the case. Kellerman and colleagues found no increase in anxiety or decrease in self-esteem in a group of teenagers with chronic illness as compared with a group of healthy teenagers.[74] A later study by Wolman and colleagues investigated the emotional well-being of adolescents with chronic diseases. Adolescents with chronic conditions had lower scores regarding emotional well-being and worried more about dying soon and about school work. In addition, the chronically ill group exhibited poorer body image.[75]

Zeltzer studied the impact of illness on various quality of life components such as relationships within the family (siblings and parents); school and peer activities; independence and autonomy; perceptions of personal, social, and sexual functioning; future orientation; and effects of treatment.[57] Interestingly, a comparison of "healthy" adolescents and those with chronic diseases identified no significant differences in total illness impact scores. Adolescents in certain disease groups, such as oncology and rheumatology patients, expressed greater total impact than cardiac and diabetic patients.

Both "healthy" patients and those afflicted with chronic illnesses felt that freedom and popularity were the life areas most frequently disrupted. More than one-half of all adolescents agreed that illness disrupted relations with peers, siblings, and parents, as well as their own lives. While ill adolescents expressed more general school and treatment-related disruption, healthy adolescents indicated that popularity and peer activities suffered when they fell ill. Both groups, however, were similar in measures of positive future outlook.

It is important to note that physical change seems to affect female adolescents more than male adolescents. While female adolescents may believe that their looks are strongly related to social acceptance, males rely on other attributes (such as physical strength, athletic prowess, scholastic achievement) for such acceptance. Similar findings have found a stronger relationship in females than males between suicide intent and illness-related cosmetic change.[76]

In addition, the degree to which the disease is visible to others is a factor in adolescents' adjustment to illness. McAnarney and colleagues found that children with arthritis showed significantly poorer adjustment than others.[77] Specifically, those with arthritis but without visibly obvious disability, showed the poorest adjustment. In this situation, those with the least obvious problems (but with perhaps equal discomfort) may have experienced less acknowledgement of their disease from parents, teachers, friends, and others. Those who received the greatest support and services may have been those with the most obvious difficulties.

Figure 19.1 presents the particular problems of each illness group and lists those problem areas where the illness groups differ significantly from the healthy adolescents. Note the similarities expressed between adolescents with oncological and rheumatoid problems. In these groups, treatments were reported as problematic due to the effect upon looks and the problems that treatments cause. The diabetic group reported the least impact of illness upon life. Adolescents with diabetes actually expressed *less* illness-related peer disruption than healthy adolescents.

Similarities in responses between "healthy" and ill patients are probably the results of the relativity of illness. Because healthy adolescents are usually well, even minor illness may considerably disrupt normal routines. However, ill adolescents who frequently undergo treatment and are inconvenienced due to the effects of major illness may become accustomed to all but the most serious of

FIGURE 19.1. Impact of Illness in Healthy Adolescents and Those with Specific Diseases

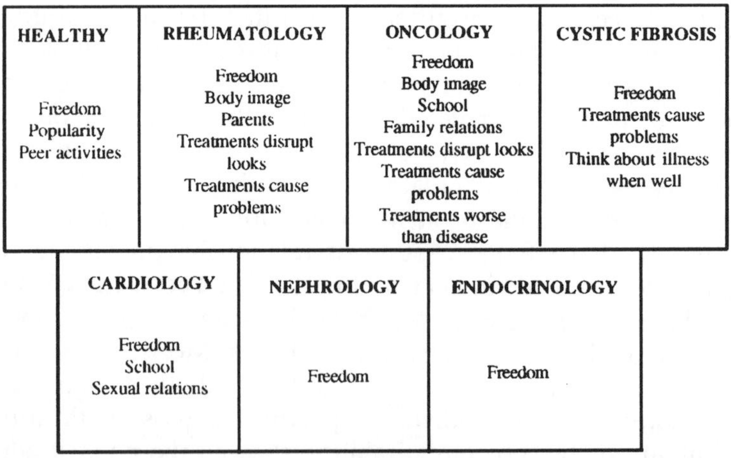

Source: Zeltzer L, Kellerman J, Ellenberg L, Dash J, Rigler D. Psychologic effects of illness in adolescence. II. Impact of illness in adolescents—crucial issues and copying styles. J Pediatrics 1980;97(1):132-138.

events. This premise is supported by Bedell and colleagues, who found more of a connection between life and stress and anxiety in healthy children than in chronically ill children.[78]

Adolescence is typically a time of experimentation and risky behavior. However, chronically ill patients may place themselves at greater risk through adoption of normal teenage behavior.[79] For example, preoccupation with the ideal body image causes many adolescents to try various diets. Such experimentation in a diabetic patient can have disastrous results. The risks of such experimentation should be part of patient education provided by health care providers and family members.

Pharmacists should be aware of the concerns these treatments raise and focus special educational and informational attention upon adolescents in these illness groups who have pharmaceutical needs. In this age group, careful explanation of medication therapy, the planned effects, and what to watch for in terms of improvement or worsening is important. Information and reassurance are good remedies for anxiety that may be present.

Access to Care

We would expect that chronically ill adolescents would have increased contact with their specialty providers. However, an apparent consequence of this specialty contact is isolation from primary physicians. In a study of 14 to 18-year-olds with chronic diseases, Carroll and colleagues found that 40% reported having no primary care providers. Most of these teenagers viewed their specialist as their personal physician, but few discussed "common" health problems with specialists.[80] This is interesting because 90% of the adolescents studied reported one or more health concerns of headache, acne, enuresis, insomnia, anxiety, and school problems.

The potential for chronically ill adolescents to have health concerns that are unaddressed is great. These patients should be encouraged to maintain an ongoing relationship with a primary care provider who can address general health needs. Of issue here is the need for health care providers to have continuous involvement with patients and their families.

Community pharmacists have historically been points of constant contact with patients with chronic illness. Knowledge of the patient's disease, disease history, and family characteristics; familiarity with primary care and specialty physicians providing care; and awareness of current and past medication use is vital to being a part of the patient's health care team.

COMPLIANCE IN ADOLESCENTS

As discussed in other chapters in this book, patient noncompliance with taking medication, keeping appointments, and other medical care behaviors is a significant problem that has great costs—both medical and economic—to society. As with adult noncompliance, a number of factors have been identified that affect regimen compliance in children and adolescents.

The demands of the regimen, the form of the disease, and the tasks involved in treatment for the disease are variables associated with compliance behavior. Compliance also varies with the age of the child, sex of the child, and family ethnicity. Characteristics of the mother have been predictive of compliance in children; however, characteristics of family functioning and peer pressures are thought to be more predictive of compliance in adolescents. In general, aspects of family and social *structure*, such as number of children in the family, education of parents, and single-parent versus two-parent households, were not likely to be related to compliance in adolescents. However, family *process* variables, such as family cohesiveness and the degree of interference of the regimen with family activities and social roles, were more likely to be related to compliance.[81]

In addition, it is believed that opportunity exists for family process and structure variables to inhibit and detract from compliance, but few opportunities exist to maximize compliance. Those negative aspects of family structure and process, perhaps where care and the care regimen cause restrictions on family activities or where the parents' relationship suffers, have the greatest effect on compliance. Therefore, efforts should focus on creating regimens that, where possible, have the least impact on the processes of family functioning.

MINORS' LEGAL RIGHTS TO CARE

A number of unique social issues related to ethics and legality arise when the patients in question are minors. Contraception, abortion, and the right to treatment of minor illnesses are examples of potential situations for pharmacist involvement, and conflicting issues may arise. As the pharmacist's role continues to expand beyond dispensing to therapeutic monitoring and protocol and disease management, a number of issues relating to minors' rights will become important.

It is obvious that today's adolescents are markedly different in attitude and behavior from their counterparts of 50 years ago. They are allowed much more freedom and are considered to have adult privileges such as access to charge accounts and credit cards without parental consent. Even colleges no longer consider their role to be "parental." These trends toward allowing more "adult" behaviors and decision making by adolescents have been addressed by the courts.

While adults are allowed to make decisions about their own health care, only recently have older minors been given choices that extend beyond parental control. In England in the 1600s, anyone under 21 years of age was considered the property of his or her father.[82] A physician who treated a son or daughter without the father's permission was likely to be sued because of interference with the father's right to control the child. The courts now allow adolescents much more decision-making autonomy than adolescents have had historically.

For over 200 years, the concept of the emancipated minor has been recognized in the courts in the Anglo-American system. An emancipated minor is one who is living on his or her own, is self-supporting, and is not subject to parental control. An emancipated minor may be dealt with medically as an adult, and no parental permission is required for treatment.[83] College students, minors in the military, and unmarried minor mothers are considered to be emancipated.

Treatment Statutes

Treatment statutes regarding minors came into effect in the 1960s and 1970s to counter concerns of medical providers about the legality of treating minors. These statutes are in place in most states and provide a specific age (usually 16 years but sometimes as young as 14 years) at which a minor may be considered completely independent for all health care purposes and may be provided treatment as if he or she were an adult. In states that have not enacted those statutes, a physician may treat any minor for venereal disease without parental knowledge. In most states, minors are able to receive certain services based on their own consent and without the consent of a parent or guardian. These situations include the following:

- Diagnosis, treatment, and counseling for sexually transmitted diseases (including HIV infection).
- Diagnosis, treatment, and counseling for substance abuse.
- Diagnosis, treatment, and counseling for emotional problems or mental disorders.
- Diagnosis, treatment, and counseling for contraception or pregnancy.[83]

In these instances where parents are not informed, it follows that parents would not receive a bill for the care received by the child. Thus, decisions for care may encompass legal, ethical, and economic components.

Mature Minors

A legal principle called the mature minor rule exists that calls for the age and maturity of the patient to be considered in each case, as well as the nature of the illness and the risks of therapy. The rule states that if a young person (age 14 or older) understands the nature of proposed treatment and its risks, the physician believes that the patient can give the same degree of informed consent as an adult patient, and the treatment does not involve very serious risks, the young person may validly consent to receiving it.[84] While this applies to clinic or office visits which usually do not entail major risks, elective admission to a hospital may be viewed differently by the hospital in that the courts have generally held that parents are not responsible for payment of the bill if nonemergency medical care is given without contacting parents and obtaining their consent. This holds true especially if the procedure is completely elective.

Specific Health Issues

Contraceptives

The U.S. Supreme Court first struck down state statutes making the prescription or use of contraceptives a criminal offense in 1965.[85] In 1977, the Court held that minors do have a right to privacy that extends to contraception after considering a New York statute that made it a criminal offense to sell nonprescription contraceptives to minors.[86]

Title X of the Public Health Service Act of 1970 has a provision stipulating that federally funded family planning agencies are required to provide services without regard to age or marital status.[87] However, the U.S. Department of Health and Human Services published regulations in January of 1983 requiring federally funded family planning facilities to notify a minor's parents within ten days of her visit to the facility that she had been given contraceptives. This "squeal rule" was later found unconstitutional by the courts.[88-90]

In general, it is the constitutional right of an adolescent to obtain contraception if requested by the adolescent. Although a private physician has the right to refuse to prescribe or provide contraception or a pharmacist may refuse to dispense or sell the product, a federally funded facility must respect the constitutional right of the individual.

Abortion

The courts have gone through several steps in deciding the abortion issue, from the Supreme Court's *Roe v. Wade* decision in 1973 which struck down state laws making abortion a criminal offense to state enactment of statutes that required parental consent before a minor could obtain an abortion. Currently, the Supreme Court has held that statutes requiring parental consent without any alternative procedures to determine the girl's capacity to decide for herself—thus giving the parents an absolute veto over this decision—were unconstitutional violations of the girl's right to privacy.[91,92] It appears that the Supreme holds that state statutes will be constitutionally acceptable if access to a judge will be provided if a girl does not wish to involve her family. The role of the judge is to determine if she is mature enough to make the decision about abortion for herself.

State statutes may, however, invoke another ruling, as in Tennessee, where a minor's parents or guardian must be informed two days prior to the operation that an abortion is to be performed upon such minor, "provided however, that the provisions of this section shall in no way be construed to mean provide for, or authorize parental objection, in any way, prevent or alter the decision of the minor to proceed with the abortion."[93]

While the dispensing of contraceptives to minors or the offering of advice regarding venereal disease has direct applicability to pharmacists, it has been more difficult to understand the relationship of the pharmacist to issues surrounding abortion. Abortion, which has historically involved an invasive procedure performed by a physician, required no pharmacist involvement except perhaps to dispense postprocedure antibiotics. On the horizon for pharmacy, however, is contemplation of the moral, social, and psychological issues associated with the dispensing of abortifacients.

Approved for use in some countries such as France and China, mifepristone, known as RU486, has not been approved for use in the United States. However, it is not unrealistic to suggest that some type of chemical abortifacient will gain approval in the United States. If this occurs, will the drug be distributed through typical pharmacy channels? If so, will some pharmacists refuse to dispense it?

Certainly, the pharmacist's involvement would increase if he controls the distribution of a product that has such social and moral overtones. As Smith suggests, "When a patient may request, a physician prescribe, and a pharmacist dispense an abortifacient agent, the pharmacist is no longer an interested bystander."[94]

Confidentiality of Information

In addition to the issue of whether a minor has the right to consent to treatment, confidentiality of patient information is an important part of the medical care system. Although most state laws allow the minor to consent to treatment in certain areas, this does not necessarily mean that the information can be withheld when requested by a parent or guardian.[95]

Maryland law states that anyone under 18 years of age may consent to treatment for or receive advice about venereal disease, just as an adult may.[96] Interestingly, another section of the law states that the attending physician may, but is not required to, provide information to the parent or guardian of the minor about the services provided, except as related to abortion.[97] This may occur without the consent of, or over the express objection of, the minor patient.

Some state codes simply do not address the issue of confidentiality of information. In Virginia, the law allows a minor to consent to "medical or health services needed to determine the presence of or to treat venereal disease or any other [reportable] infectious or contagious disease. . . ."[98] The law does not, however, contain any protection for the confidentiality of this information if requested by the parent or guardian.

The type of information requested may determine the confidentiality of the information. An increasing number of states have laws that absolutely prohibit disclosure of HIV-related information without the express written consent of the patient, even if the patient is a minor.

PHARMACISTS:
MAKE A SPECIAL EFFORT

Talk to the Patient!

Health care professionals may tend to ignore children and adolescents in the information provision part of the care process. Several studies have found that programs based upon providing information directly to the child (rather to the parent only) are beneficial and may serve to supply reassurance. Such programs are more effective with older patients (over seven years of age).[99] Such health professional-child interaction serves to foster a trusting relationship between the child and the information agent, as well as to reduce the child's anxiety by reducing uncertainty about what is going to happen.[100,101] Physicians are not the only health professionals who benefit patients by providing information to younger patients.[102] As pharmacists' role as information providers increases and perhaps becomes the predominant activity for the profession, the discussion of medication-related therapy with adolescents should not be ignored. Pharmacists should be prepared to communicate information in a format appropriate for these younger patients.

Continuity of Care

Continuity of care is a term being used to describe a desired relationship between health professionals and patients. Although most commonly thought of as one patient seeing the same physician for medical care, the concept is applicable to any health professional who has the opportunity to interact with the same patient or caregiver regularly. Benefits to the patient are positive effects on compliance, appointment keeping, and patient satisfaction.[103]

Pharmaceutical care calls for pharmacists to assume responsibility for patient outcomes. This responsibility cannot be fulfilled through episodic contact that is hit or miss, initiated by the patient only, or confined to the working environment of the pharmacist. It is conceivable and possible that pharmacists, in whatever work situation, would maintain contact with their patients, whether those patients are ambulatory or institutionalized. Pharmaceutical care should be continuous, as evidenced by follow-up phone calls, letters, and comprehensive record keeping.

The "doc-in-the-box" minor emergency clinics provide an example, on a limited basis, of good follow-up care. Many of these facilities have a system in place whereby any patient seen at the clinic is telephoned within 24 hours and asked about progress or any problems that they may be having. The person conducting the follow-up call may be the receptionist rather than the physician or nurse who provided care. However, the reason for the success of this practice and the increase in patient satisfaction is that patients perceive that they are receiving a level of care beyond what is expected. This type of service differentiates health care providers to patients.

Adolescents and college students (who are often away from their family environment) may be excellent patient groups to target for follow-up activities. Whereas the mother often serves as the intermediary with health professionals for children under seven years of age, adolescents often accept responsibility as the patient. If possible, continuity of care for adolescents, whether by personal or written communication, should be directed toward the patient and supplemented through the parent.

CONCLUSION

There is little question that adolescents and college students present considerable opportunities as a target group for health professionals. These patient groups differ significantly from their older and younger counterparts in social and physical characteristics, and the problems they incur are often unique to their age group.

Perhaps it is recognition of the social and health problems that should be considered first and foremost by pharmacists in focusing on this patient group. It is interesting that the majority of the most visible social and health threats to young adults' health are related to such complex issues as low self-esteem, peer pressure, underdeveloped communication skills, superficial relationships, and loneliness. Although these problems are not limited to the under 22-year-old age group, certainly this population can be more strongly influenced and their behavior more easily shaped than older groups.

For college students, the competitive nature of academic environments often creates and increases feelings of inferiority, insecurity, and emotional distress. The experience of being away from the family for the first time and the lack of previous experience with health-influencing factors make it extremely difficult for college students to make mature decisions–particularly for students who were raised in rural environments and find themselves in an urban university environment. In addition, economically disadvantaged students with little or no health education preparation from schools or parents may be especially challenged.

The special needs that pharmacists may fill for younger patients are not dramatically different than for older patient groups; however, the emphasis may differ. The prevalence of alcohol and substance abuse problems, eating disorders, and food abuse should come to mind when prescriptions are filled, medication therapy is monitored, and education is provided. Pharmacists should be the primary self-care advisor for this age group, which practices self-care but may not always practice it effectively. The opportunity to be a source of accessible, ongoing care is great.

Other niches for pharmacists? Nutrition and contraceptive counseling, the provision of information on stress management, and development of exercise centers are not far-fetched ideas for pharmacists who primarily want to be care providers. The wise pharmacist will remember that this group of patients (who may be a relatively small user group now) will grow into the parents of small children and will become the elderly population of the future. This is the time to influence future life-styles and professional relationships.

REFERENCES

1. Davis CD, Conant SB, Fordon K, Gold RL. Achieving the health objectives for the nation in higher education. J Am Coll Health 1986;35:1.

2. Hodgkinson HL. Demographics of education, kindergarten through graduate school, all one system. Boston, MA: Institute for Educational Leadership, 1985.

3. U.S. Center for Educational Statistics. Digest of educational statistics—annual. 1993.

4. U.S. Bureau of Census. Current population reports, 1990.

5. U.S. Department of Health and Human Services. Promoting health/preventing disease: objectives for the nation. 1980.

6. Guyton R, Corbin S, Zimmer C, et al. College students and national health objectives for the year 2000: a summary report. J Am Coll Health 1989;38:9-14.

7. Sexual behavior among high school students—United States, 1990. MMWR Morb Mortal Wkly Rep 1992;40:885-8.

8. Seidman SN, Snider RO. A review of sexual behavior in the United States. Am J Psychiatry 1994;151:330-41.

9. Tucker VL, Cho CT. AIDS in adolescents. Postgrad Med 1991;89(3): 49-53.

10. Baldwin JD, Baldwin JI. Factors affecting AIDS-related sexual risk taking behavior among college students. J Sex Res 1988;25(May):181-96.

11. Carroll L. Concern with AIDS and the sexual behavior of college students. J Marr Fam 1988;50:405-11.

12. Painter K. AIDS on college campuses. USA Today 1989, May 23:D1.

13. Hays HE, Hays JR. Students' knowledge of AIDS and sexual behavior. Psychol Rep 1992;71(2):649-50.

14. Manning DR, Barenberg N, Gallese L, Rick JC. College students' knowledge and health beliefs about AIDS: implications for education and prevention. J Am Coll Health 1989;37:254-9.

15. Winslow RW. Student knowledge of AIDS transmission. Sociol Soc Res 1988;72(Jan):110-3.

16. Keller ML. Why don't young adults protect themselves against sexual transmission of HIV? Possible answers to a complex question. AIDS Educ Prev 1993;5(3):220-33.

17. Thurman QC, Franklin KM. AIDS and college health; knowledge, threat, and prevention at a northeastern university. J Am Coll Health 1990;38:179-84.

18. Researchers see AIDS among leading killers of childbearing women. Providence (RI) J 1990;11(July):A-1+.

19. Jadack RA, Hyde JS, Keller ML. Gender and knowledge about HIV, risky sexual behavior, and safer sex practices. Res Nurs Health 1995;18(4):313-24.

20. Carroll J, Volk KD, Huyde JS. Differences between males and females in motives for engaging in sexual intercourse. Arch Sex Behav 1985;14(Apr):131-9.

21. Hendrick S, Hendrick C, Slapion-Foote MJ, Foote FH. Gender differences in sexual attitudes. J Pers Soc Psychol 1985;48:1630-42.

22. Spitz AM, Ventura SJ, Koonin LM, Strauss LT, Frye A, Heuser RL, Smith JC, Morris L, Smith S, Wingo P. Surveillance for pregnancy and birth rates among teenagers, by state–United States 1980 and 1990. MMWR–CDC Surveill Summ 1993;42(SS-6):1-27.

23. Guyer B, Strobino DM, Ventura SJ, Singh GK. Annual summary of vital statistics–1994. Pediatrics 1995;96(6):1029-1039.

24. Hale RW, Char DF, Nagy K, Stockert N. Seventeen-year review of sexual and contraceptive behavior on a college campus. Am J Obstet Gynecol 1993; 168(6 Pt 1):1833-8.

25. Carroll L. Gender, knowledge about AIDS, reported behavioral change, and the sexual behavior of college students. J Am Coll Health 1991;40:5-12.

26. Feigenbaum R, Weinstein E, Rosen E. College students' sexual attitudes and behaviors: implications for sexuality education. J Am Coll Health 1995; 44(3):112-18.

27. Creatsas GK. Sexuality: sexual activity and contraception during adolescence. Curr Opin Obstet Gynecol 1993;5:744-83.

28. Berenson AB, Wiemann CM. Patient satisfaction and side effects with levonorgestrel implant (Norplant) use in adolescents 18 years of age or younger. Pediatrics 1993;92:257-60.

29. Johnson LD, O'Malley PM, Bachman JG. National high school senior survey of alcohol and other drug use–preliminary data–1994. Washington, DC: National Clearinghouse on Alcohol and Drug Information, 1994.

30. Presley CA, Meilman PWE, Lyerla R. Development of the core alcohol and drug survey: initial findings and future directions. J Am Coll Health 1994;42:248-55.

31. Bailey WJ, Martin MH, Jimenez JM, Ding K, Holtsclaw MA, English C, Carroll A, Pearson W. Drug, tobacco, and alcohol use by Indiana children and adolescents. Bloomington, IN: Indiana Prevention Resource Center Institute for Drug Abuse, 1995.

32. Torabi MR. Cigarette smoking as a predictor of alcohol and other drug use by children and adolescents: evidence of the "gateway drug effect." J School Health 1993;63:302-306.

33. Krarrt S. The Marlboro Man is really a troubled teen. Am Demographics 1993;15(12):12-3.

34. Eigen LD. Alcohol practices, policies, and potentials of American colleges and universities: a white paper. Rockville, MD: Office for Substance Abuse Prevention, September 1991.

35. Schwartz AJ. Suicide and stress among college students. In Wallace HM, Patrick K, Parcel GS, Igoe JB, eds. Principles and practice of student health vol. 3. Oakland, California: Third Party Publishing Company, 1992.

36. Mulcahy C. Costs of treating teen suicide attempts soaring. National Underwriter Life/Health/Financial Services 1993;9(14):5.

37. CDC. Suicide among children, adolescents, and young adults–United States. 1980-1992. MMWR 1995;44(15):289-91.

38. American Psychiatric Association. Task Force on Nomenclature and Statistics. Diagnostic and statistical manual of mental disorders. 3rd ed. Washington, DC: American Psychiatric Association, 1980.

39. Killen JD, Taylor CB, Telch MJ, Saylor KE, Maron DJ, Robinson TN. Self-induced vomiting and laxative and diuretic use among teenagers. JAMA 1986;255:1447-9.

40. Halmi KA, Falk JR, Schwartz E. Binge-eating and vomiting: a survey of a college population. Psychol Med 1981;11:697-706.

41. Crowther JH, Post G, Zaynor L. The prevalence of bulimia and binge eating in adolescent girls. Int J Eat Disord 1985;4:29-42.

42. Felker KR, Stiver SC. The relationship of gender and family environment to eating disorder risk. Adolescence 1994;29(116):821-34.

43. Heatherton TF, Nichols P, Mahamedi F, Keel P. Body weight, dieting and eating disorder symptoms among college students, 1982-1992. Am J Psychiatry 1995;152(11):1623-9.

44. Anderson J. Calcium–a critical dietary nutrient. Brit Food J 1993; 95(4):22-4.

45. Bureau of Labor Statistics. Consumer Expenditure Survey–annual. Average annual expenditures per consumer unit for health care, 1984-1991.

46. Newacheck PW, McManua MA, Fox HB. Prevalence and incidence of chronic illness among adolescents. Am J Dis Child 1991;145(12):1367-73.

47. Butler JA, Budett P, McManus MA, et al. Health care expenditures for children with chronic illness. In: Hobbs N, Perrin J, eds. Issues in the care of children with chronic illness. San Francisco, CA: Jossey-Bass Inc. Publishers, 1985.

48. U.S. Center for Health Statistics. Vital and health statistics. Series 10, No. 181, 1992.

49. Levin L. Self care in health. Rev Public Health 1983;4:181-201.

50. Green KE, Moore SH. Attitudes toward self-care. Med Care 1980;18:872-7.

51. Elliott-Binns CP. An analysis of lay medicine. J Res Collection Pract 23:255.

52. Coons SJ, McGhan WF, Bootman JL, et al. The effect of self care information on health-related attitudes and beliefs of college students. J Am Coll Health 1989;38:121-4.

53. Krupka LR, Vener AM. Drug knowledge (prescription, O-T-C and social): young adult consumers at risk? J Drug Educ 1982;17(2):129-42.

54. Focus group discussion. Use of OTC medication and information sources. University of Mississippi. November 19, 1990.

55. Levin LS. Self care: toward fundamental changes in national strategies. Int J Health Educ 1981;24:219-28.

56. McKillip J, Courtney CL, Locasso R, Eckert P, Holly F. College students' use of emergency medical services. J Am Coll Health 1990;38:289-92.

57. Zeltzer L, Kellerman J, Ellenberg L, Dash J, Rigler D. Psychologic effects of illness in adolescents. II. Impact of illness in adolescents–crucial issues and coping styles. J Pediatrics 1980;97:132-8.

58. Rylance GW, Woods CG, Cullen RE, Rylance ME. Use of drugs by children. Brit Med J 1988;297:445-7.

59. Busching BC, Bromley DG. Sources of non-medicinal drug use: a test of the drug-oriented society explanation. J Health Soc Behav 1975;16:50-62.

60. Vener AM, Krupka LR, Climo JJ. Drugs (prescription, over-the-counter, social) and the young adult: use and attitudes. Int J Addict 1982;17:399-415.

61. Portner TS. Factors influencing college students' selection of a source of information about OTC medications [Dissertation]. University, MS: University of Mississippi, 1991.

62. Cafferata GL, Lach PA, Reifler CB. Use of over-the-counter and home remedies by college students. J Am Coll Health 1980;29:61-5.

63. Silverman M, Lee PR. Pills, politics, and the public purse. Berkeley, CA: University of California Press, 1974.

64. Worsley A. Perceived reliability of sources of health information. Health Educ Res 1989;4:367-76.

65. Portner TS, Smith MC. College students' perceptions of OTC information source characteristics. J Pharm Market Manage 1994;8(1):161-85.

66. Tootelian DH, Gaedeke RM. Marketing News 1990, July 23:26.

67. Gannon K. Few rural students report contact with pharmacists. Drug Top 1992;(Jul 6):39+.

68. Kennedy DW, Kennedy SL. Using importance-performance analysis for evaluating university health services. J Am Coll Health 1987;36:27-31.

69. National Center for Health Statistics. Unpublished data from the National Health Interview Survey, 1986.

70. Gillford MJ, Franaszek JB, Gibson G. Emergency physicians and patients' assessments: urgency of need for medical care. Ann Emerg Med 1980;9:502-7.

71. Litman TJ. The family as a basic unit in health and medical care: a social-behavioral overview. Soc Sci Med 1974;8:495-519.

72. Lewis CE, Lewis MA. The impact of television commercials on health-related beliefs and behaviors of children. Pediatrics 1974;53:431-5.

73. Gortmaker SL, Sappenfield W. Chronic childhood disorders: prevalence and impact. Pediatr Clin North Am 1984;31:3-18.

74. Kellerman J, Zeltzer L, Ellenberg L, et al. Psychologic effects of illness in adolescents. I. Anxiety, self-esteem, and perception of control. J Pediatrics 1980;97:126-31.

75. Wolman C, Resnick MD, Harris LJ, Blum RW. Emotional well-being among adolescents with and without chronic conditions. J Adolesc Health 1994; 15(3):199-204.

76. Weinberg S. Suicidal intent in adolescence: a hypothesis about the role of physical illness. J Pediatrics 1970;77:579.

77. McAnarney ER. Social maturation: a challenge for handicapped and chronically ill adolescents. J Adolesc Health Care 1985;6:90-101.

78. Bedell JR, Giordani B, Amour JL, et al. Life stress and the psychological and medical adjustment of chronically ill children. J Psychosom Res 1977;21:237.

79. Siegel DM. Adolescents and chronic illness. JAMA 1987;257:3396-9.

80. Carroll G, Massarelli E, Opzoomer A, et al. Adolescents with chronic disease: are they receiving comprehensive health care? J Adolesc Health Care 1983;4:261-6.

81. Baranowski T, Nader PR. Family health behavior. In: Turk DC, Kerns RD, eds. Health, illness, and families: a life-span perspective. New York: John Wiley and Sons, 1985:66-72.

82. Pollack F, Maitlant FW. The history of English law before the time of Edward I. 2nd ed. Cambridge: University Press; Boston, MA: Little, Brown & Company, 1903.

83. Holder AR. Minors' rights to consent to medical care. JAMA 1987;257: 3400-2.

84. Munson CF. Toward a standard of informed consent by the adolescent in medical treatment decisions. Dickenson Law Rev 1981;431.

85. *Griswold v. Connecticut*, 381 US 479 (1965).

86. *Carey v. Population Services International*, 431 US 678 (1977).

87. Public Health Service Act, 42 USC §§ 300-300a(8) (1970).

88. *Memphis Association for Planned Parenthood v. Schweiker*, DC Tenn Civit No. 83-2060, temporary injunction granted Feb. 24, 1983.

89. *State of New York v. Schweiker*, 557 FSupp 354 (SD NY 1983).

90. *Planned Parenthood of America v. Schweiker*, 559 FSupp 658 (DDC), *aff'd.* 712 F2d 650 (DC Cir 1983).

91. *City of Akron v. Akron Center for Reproductive Health*, 462 US 416 (1983).

92. *Planned Parenthood of Kansas City, Missouri v. Ashcroft*, 462 US 476 (1983).

93. Tn Code Ann § 39-15-202(f) (1991).

94. Smith M. The abortion pill. US Pharm 1977;2(9):79-80.

95. Confidentiality of information for care provided to a minor. Med Legal Lessons 1994;2(3):2-3.

96. Ann Code of Md § 2-102(c).

97. Ann Code of Md § 20-102(e).

98. Va Code Ann § 54.12969(D)(1).

99. Melamed BG, Siegel LJ. Behavioral medicine: practical applications in health care. New York: Springer, 1980.

100. Kendall PC, Watson D. Psychological preparation for stressful medical procedures. In: Prokop CA, Bradley LA, eds. Medical psychology. New York: Academic Press, 1981.

101. Clough F. The validation of meaning in illness-treatment situations. In: Hall D, Stacey M, eds. Beyond separation: further studies of children in hospital. London: Routledge and Kegan-Paul, 1979.

102. Skipper JK Jr, Leonard RC, Rhymers J. Child hospitalization and social interaction: an experimental study of mothers' feelings of stress, adaption, and satisfaction. Med Care 1968;6:496-506.

103. Haggerty RJ. The university and primary medical care. N Engl J Med 1969;281:416.

ADDITIONAL READINGS

For the Instructor

1. Baldwin HJ, Alberts KT. Joan Johnson and adolescent confidentiality. In: Smith M, Strauss S, Baldwin HJ, Alberts KT, eds. Pharmacy ethics. Binghamton, NY: Pharmaceutical Products Press, 1991:484-92.

2. Baranowski T, Nader PR. Family health behavior. In: Turk DC, Kerns RD, eds. Health, illness, and families: a life-span perspective. New York: John Wiley and Sons, 1985:66-72.

3. Coons SJ, McGhan WF, Bootman JL, Larson LN. The effect of self-care information on health-related attitudes and beliefs of college students. J Am Coll Health 1989;38:121-4.

4. Coons SJ, McGhan WF, Bootman JL. Self care practices of college students. J Am Coll Health 1989;37:170-3.

5. Gortmaker SL, Sappenfield W. Chronic childhood disorders: prevalence and impact. Pediatr Clin North Am 1984;31:3-18.

6. Guyton R, Corbin S, Zimmer C, et al. College students and national health objectives for the year 2000: a summary report. J Am Coll Health 1989;38:9-14.

7. Hale RW, Char DF, Nagy K, Stockert N. Seventeen-year review of sexual and contraceptive behavior on a college campus. Am J Obstet Gynecol 1993; 168(6 Pt 1):1833-8.

8. Holder AR. Minors' rights to consent to medical care. JAMA 1987;257: 3400-2.

9. Kellerman J, Zeltzer L, Ellenberg L, et al. Psychologic effects of illness in adolescents. I. Anxiety, self-esteem, and perception of control. J Pediatrics 1980;97:126-31.

10. Lewis CE, Lewis MA. The impact of television commercials on health-related beliefs and behaviors of children. Pediatrics 1974;53:431-5.

11. Litman TJ. The family as a basic unit in health and medical care: a social-behavioral overview. Soc Sci Med 1974;8:495-519.

12. Manning DR, Barenberg N, Gallese L, Rick JC. College students' knowledge and health beliefs about AIDS: implication for education and prevention. J Am Coll Health 1989;37:254-9.

13. Melamed BG, Siegel LJ. Behavioral medicine: practical applications in health care. New York: Springer, 1980.

14. Portner TS, Smith MC. College students' perceptions of OTC information source characteristics. J Pharm Market Manage 1994;8(1):161-85.

15. Presley CA, Meilman PWE, Lyerla R. Development of the core alcohol and drug survey: initial findings and future directions. J Am Coll Health 1994;42:248-55.

16. Promoting health/preventing disease: objectives for the nation. U.S. Department of Health and Human Services, 1980.

17. Seidman SN, Snider RO. A review of sexual behavior in the United States. Am J Psychiatry 1994;151:330-41.

18. Vener AM, Krupka LR, Climo JJ. Drugs (prescription, over-the-counter, social) and the young adult: use and attitudes. Int J Addict 1982;17:399-415.

19. Zeltzer L, Kellerman J, Ellenberg L, et al. Psychologic effects of illness in adolescents. II. Impact of illness in adolescents—crucial issues and coping styles. J Pediatrics 1980;97:132-38.

For the Student

1. Sexual behavior among high school students—United States, 1990. MMWR Morb Mortal Wkly Rep 1992;40:885-8.

2. Baldwin JD, Baldwin JI. Factors affecting AIDS-related sexual risk taking behavior among college students. J Sex Res 1988;25(May):181-96.

3. Baranowski T, Nader PR. Family health behavior. In: Turk DC, Kerns RD, eds. Health, illness, and families: a life-span perspective. New York: John Wiley and Sons, 1985:66-72.

4. Butler JA, Budett P, McManus MA, et al. Health care expenditures for children with chronic illness. In: Hobbs N, Perrin J, eds. Issues in the care of children with chronic illness. San Francisco, CA: Jossey-Bass Inc. Publishers, 1985.

5. Cafferata GL, Lach PA, Reifler CB. Use of over-the-counter and home remedies by college students. J Am Coll Health 1980;29:61-5.

6. Carroll G, Massarelli E, Opzoomer A, et al. Adolescents with chronic disease: are they receiving comprehensive health care? J Adolesc Health Care 1983;4:261-6.

7. Carroll J, Volk KD, Huyde JS. Differences between males and females in motives for engaging in sexual intercourse. Arch Sex Behav 1985;14(Apr):131-9.

8. Carroll L. Concern with AIDS and the sexual behavior of college students. J Marr Fam 1988;50:405-11.

9. Confidentiality of information for care provided to a minor. Med Legal Lessons 1994;2(3):2-3.

10. Coons SJ, McGhan WF, Bootman JL, et al. The effect of self care information on health-related attitudes and beliefs of college students. J Am Coll Health 1989;38:121-4.

11. Creatsas GK. Sexuality: sexual activity and contraception during adolescence. Curr Opin Obstet Gynecol 1993;5:744-83.

12. Davis CD, Conant SB, Fordon K, Gold RL. Achieving the health objectives for the nation in higher education. J Am Coll Health 1986;35:1.

13. Freimuth VS, Edgar T, Hammond SL. College students: awareness and interpretation of the AIDS risk. Sci Technol Hum Values 1987;12(Summer/Fall):37-40.

14. Guyton R, Corbin S, Zimmer C, O'Donnell M, Chervin DD, Sloane BC, Chamberlain MD. College students and national health objectives for the year 2000: a summary report. J Am Coll Health 1989;38:9-14.

15. Hale RW, Char DF, Nagy K, Stockert N. Seventeen-year review of sexual and contraceptive behavior on a college campus. Am J Obstet Gynecol 1993; 168(6 Pt 1):1833-8.

16. Hendrick S, Hendrick C, Slapion-Foote MJ, Foote FH. Gender differences in sexual attitudes. J Pers Soc Psychol 1985;48:1630-42.

17. Holder AR. Minors' rights to consent to medical care. JAMA 1987;257: 3400-2.

18. Johnson SB. The family and the child with chronic illness. In: Turk DC, Kerns RD, eds. Health, illness, and families: a life-span perspective. New York: John Wiley and Sons, 1985:220-6.

19. Kellerman J, Zeltzer L, Ellenberg L, et al. Psychologic effects of illness in adolescents. I. Anxiety, self-esteem, and perception of control. J Pediatrics 1980;97:126-31.

20. Killen JD, Taylor CB, Telch MJ, Saylor KE, Maron DJ, Robinson TN. Self-induced vomiting and laxative and diuretic use among teenagers. JAMA 1986;255:1447-9.

21. Krupka LR, Vener AM. Drug knowledge (prescription, O-T-C and social): young adult consumers at risk? J Drug Educ 1982;17(2):129-42.

22. Litman TJ. The family as a basic unit in health and medical care: a social-behavioral overview. Soc Sci Med 1974;8:495-519.

23. Melamed BG, Siegel LJ. Behavioral medicine: practical applications in health care. New York: Springer, 1980.

24. Portner TS, Smith MC. College students' perceptions of OTC information source characteristics. J Pharm Market Manage 1994;8(1):161-85.

25. Presley CA, Meilman PWE, Lyerla R. Development of the core alcohol and drug survey: initial findings and future directions. J Am Coll Health 1994;42:248-55.

26. Promoting health/preventing disease: objectives for the nation. U.S. Department of Health and Human Services, 1980.

27. Seidman SN, Snider RO. A review of sexual behavior in the United States. Am J Psychiatry 1994;151:330-41.

28. Self DR, Self RM. The adolescents: target market for prevention. Health Market Q 1990;3(4):125-39.

29. Siegel DM. Adolescents and chronic illness. JAMA 1987;257:3396-9.

30. Vener AM, Krupka LR, Climo JJ. Drugs (prescription, over-the-counter, social) and the young adult: use and attitudes. Int J Addict 1982;17:399-415.

31. Zeltzer L, Kellerman J, Ellenberg, et al. Psychologic effects of illness in adolescents. II. Impact of illness in adolescents—crucial issues and coping styles. J Pediatrics 1980;97:132-8.

Chapter 20

Ambulatory Elderly

Jack E. Fincham

INTRODUCTION

When the elderly are considered or depicted, more often than not a stereotype of an old, institutionalized individual immediately comes to mind. There are elderly who reside in nursing homes or other institutions; however, the number is very small in comparison with the number of elderly living on their own. Furthermore, the ambulatory elderly for the most part do very well on their own not only socially and financially but also from a health standpoint. Those who do poorly are definitely worth considering and focusing upon, but it is important not to lose sight of the fact that many elderly do well when left to their own devices.

The future success of the health care system in meeting the needs of the elderly will involve the restructuring of health care delivery and utilization processes. Part of this restructuring must include enhanced methods of delivery of health services to the elderly via means that prevent premature institutionalization, adverse drug-related episodes, unnecessary disease morbidity, and inappropriate use of various health care services.

Pharmacists can and must play a more active role in diminishing the occurrence of drug-induced morbidity and lack of therapeutic benefits due to regimens not aiding to reach of an appropriate therapeutic threshold. This enhanced role must include active yet appropriate interventions in drug-taking activities of select groups of subacutely ill elderly patients. Enhancements must include:

- Compliance enhancing strategies.
- More astute use by the elderly of self-medication agents and therapies.
- Proactive avoidance of adverse drug effects.
- Assistance in other health-related needs and necessities.

DEMOGRAPHICS

The United States has long thought of itself as a nation of youth; however, the country is now aging to a significant degree.[1] There can be no doubt of the impact that the aging of America has upon society, the economy of health care delivery, health professionals, and the elderly themselves. The U.S. population over the age or 65 is now ten times as large as it was in 1900.[2] In addition, the population over the age of 65 is rapidly increasing, with those over the age of 80 years increasing by 160% between 1980 and 1990, compared with a 10% increase in the total population.[2] By the year 2040, varying estimates place the number of individuals over the age of 85 at 12 to 40 million.[2] In the 1990s, Europe has the greatest share of the world's oldest people; however, 13% of the U.S. population is over the age of 65.[2] The U.S. has 12.8% of the octogenarians in the world. Only China has more individuals over the age of 80.[2]

Because of the rapid increase in the elderly population, the ways of defining the elderly are rapidly changing. Rather than simply referring to everyone over the age of 65 as "old," it is now acceptable to categorize the elderly as follows:

> *young old*–age 65-75 years
> *middle old*–age 75-85 years
> *old old*–age 85 years and older

The fastest growth segment of these three categories is the old old group. In 1995, it is estimated that there are 60,000 individuals age 100 or older. It is estimated that by the year 2035 there will be 1,000,000 individuals age 100 or older.

The elderly are often depicted as a deprived, disadvantaged, disabled group of patients with little or no hope for well-being. This distorted and ageism-related view of the elderly is counterproductive for both the elderly and the health provider alike.[3] There are startling differences in how individuals age. The health and well-being of the aged depend upon the services and empowerment provided to them to maintain an acceptable quality of life.[4]

Population Shifts

As the ambulatory elderly segment of our society expands, it is important to consider the impact this will have not only on the elderly but also on health care delivery. Because the elderly occupy 30% of beds for general medical services and consume a significant portion of ambulatory care resources utilized in the health care system, where the elderly reside and how population shifts are occurring are important to note.[5]

Geographic

The elderly are less likely to change residence than other age groups.[6] The majority of those who do move, move to another residence in the same state. Nine states account for 52% of the elderly population: California, Florida, New York, Pennsylvania, Texas, Illinois, Ohio, Michigan, and New Jersey. Although this concentration of the elderly in a small minority of states is important to consider, the urban/rural breakdown of the elderly population must also be examined. Approximately 25% of the elderly live in nonmetropolitan areas of the U.S.[7] The elderly comprise 15% of the nonmetropolitan and 12% of the metropolitan population. The rural elderly experience higher levels of poverty and poorer health and functional status than their urban counterparts.[8]

Aging Segments

One-third of Americans born between 1946 and 1964 are now entering middle age.[9] The baby boom generation will now become the senior boom generation. Thus, the elderly segment of our population will continue to explode in the years to come. Furthermore, the elderly will become a much more diverse and heterogeneous group. The influence of Latin, Hispanic, and Asian segments on the

makeup of the elderly population will expand.[9] The elderly population of the next several decades will have fewer white members and
be less English-speaking.[9] A growing assortment of languages,
customs, and racial characteristics will affect laws, benefits, and
services. The current circumstances of many of the elderly of today
reflect discrimination from earlier times. Minority elders are dispro-
portionately at risk of poverty, malnutrition, and poor health and are
more likely to live in substandard housing.[10]

Current Concerns

Even though each day there is a net gain of 1,000 new people
entering the Medicare generation, there will not be a substantial
increase in its relative size until the year 2011.[11,12] However, the
influence of several factors upon the health and well-being of the
elderly must be critically examined now and in the future.

Housing

The housing of older Americans is generally older and less adequate than the balance of the nation's housing.[6] Older women living
alone are more likely to be poor than older men or older couples.[10]
About 45% of those living in the community require assistance with
everyday activities, and 48% live alone.[12]

Insurance or Lack Thereof

It has been noted that one of the greatest threats to the economic
security of the elderly is the high out-of-pocket cost of health
care.[13] The major responsibility for payment for long-term care
services falls for the most part on the elderly themselves.[13] Thus,
the concerns of the elderly relate to payment not only for acute and
chronic health care services but also for the specter of payment for
long-term care services that may easily lead to destitution. Recently,
actuaries have suggested that it is not how long people live but how
many people live long that will stress the financial solvency of the
Medicare program.

Home Care, Senior Housing, and Assisted Living

The affluent elderly have created a market for retirement housing and assisted living environments.[14] Many nursing home facility chains have diversified into some type of retirement housing or assisted living. The elderly poor do not have the luxury of choosing such housing. Studies have shown that the elderly who are in poor health, older, and in need of assistance with the activities of daily living are supportive of accommodating housing (that which includes rehabilitation and disability services).[15] Residential homes provide an enhanced environment when compared with traditional nursing homes.[16] It has been shown that psychologically pleasing atmospheres, high levels of participation, and increased social contact are all enhanced in residential homes.[16]

Five million of the chronically ill elderly living in the community are in need of minimal assistance to perform activities of daily living according to the U.S. General Accounting Office.[17] By the year 2000, more than 15 million persons will suffer at least one chronic illness that will limit their functioning.[18] The demand for formal home care services will increase because of:

- The dramatic increase in single, childless adults.
- The aging of family members along with frail relatives.
- The shrinkage in the number of women caregivers.[17]

Despite trends to increase services, the delivery of home care is a fragmented system with different funding streams and mutually independent agencies offering services with differing standards. Service gaps, cumbersome distinctions among worker titles, lack of service continuity, stress, and poor working conditions result from this fragmentation.[17] Medicare coverage of home care accounts for less than 5% of expenditures, yet this is the largest source of public funding for home care.

SOCIAL AND BEHAVIORAL CONCERNS

Several variables influence social and behavioral concerns of the elderly. How pharmacists and others deal with these variables will influence the health and well-being of the elderly in the near and long term.

Ageism

The depiction of the elderly as a deprived, disadvantaged, disabled group of institutionalized patients is inappropriate. We must realize that old age is not necessarily a disabling period of life. The process of aging is subtle, gradual, and a lifetime process. The health not only of the elderly but also of the various professions depends upon how the elderly are viewed, with particular emphasis on the avoidance of any aspects of ageism.

Neglect and Abuse

Abuse and neglect of the elderly are cause for concern. Neglect can be unintentional or intentional. Dychtwald has noted that passive neglect, usually manifested as abusive inattention or isolation, is the most common form of abuse, followed by verbal, physical, and financial.[19]

Social Drug Abuse

The elderly are not immune to the misuse of social drugs. The effects of these agents cause irrevocable harm to the elderly. Often the elderly are not considered as potential abusers of alcohol, tobacco, or other agents. There is the oft-quoted phrase: "Let them do as they wish, they do not have long to live." However, the link between elderly social drug use and increased morbidity has been demonstrated in the literature.

Alcohol

Studies of the elderly show that the incidence of alcohol abuse ranges from 2% to 25% of the elderly.[20] The incidence of new alcoholism cases decreases with age but remains appreciable into the late sixties.[21] Other researchers have noted that because of a diminished capacity for metabolism of alcohol, the elderly should limit intake of alcohol to one drink per day.[22] The influence of alcohol misuse upon disease morbidity in the elderly should always be kept in mind. A history of ever being a heavy drinker is predictive of widespread morbidity and physical, social, and psychological dysfunction.[23]

Tobacco

The pervasively negative effects of smoking afflict all smokers, regardless of age.[24] However, the elderly smoker accumulates a tremendous potential for morbidity and mortality due to long-term smoking. Cardiovascular disease, cancer, endocrine effects, gastrointestinal disease, and respiratory morbidity are all compounded and/or precipitated by smoking. The occurrence of burns, accidents, and drug interactions also synergistically places the elderly smoker at risk. Smoking cessation and prevention activities need to be implemented for the elderly by all health care providers.[24] Smoking has been shown to be a cofactor in subsequent hospitalizations in elderly populations.[25]

Other Drugs

In the elderly, the abuse of legal substances is much more common than the abuse of illicit compounds.[20] But abuse of illicit drugs when it occurs in the elderly is just as devastating as it is for younger patients.

Falls

The elderly are susceptible to serious injuries due to falling. Fall management is much underestimated and misunderstood.[26] The mainstay of fall management has been restriction of activity through physical restraints. These actions may prevent falls, but they diminish the quality of life. A total 30 to 50% of older adults report recent falls. Falls increase linearly, but the prevalence is greater in females. Falls are the single most common cause of accident mortality, and mortality from fall-related injury is greater than that of pneumonia or diabetes.[26] Hip fractures are the most frequent fall-related injury to result in hospitalization. The major causes of falling for those less than 75 years of age are normal aging changes, decline in visual acuity, postural balance, foot steppage height, environmental hazards, low light, sliding rugs, and door thresholds.[26] For those older than 75 years of age, the major factors are neurological, musculoskeletal, cardiac disease, effects of medications, and usually a multiplicity of factors.[26]

GENERAL HEALTH CONCERNS

Older people account for 35% of hospital stays and 47% of all days of care in hospitals.[6] The elderly account for 40% of expenditures for health and health-related payments. Despite the obvious signs of morbidity in the elderly, only 29% have assessed their health as poor. This compares with 7% of those under age 65. This indicates that although the elderly experience significant morbidity often to multiple conditions, they perceive their health as good within the general constraints of illness and resultant morbidity.

The specter of occurrence of Alzheimer's disease and related dementias is a cause of fear and concern for many elderly and their families.[27] Alzheimer's disease is a leading cause of dementia in older adults and brings with it many difficult management problems. At least 6 million Americans suffer from some type of dementia. More than half of these individuals are afflicted with Alzheimer's disease.[28] Compounding the effects of Alzheimer's disease is the difficulty in obtaining an accurate diagnosis.[28] The ramifications of Alzheimer's disease are not difficult to ascertain: Alzheimer's disease is the fourth leading cause of elderly death in the U.S. It steals minds, destroys the character of individuals, and fills long-term care facilities.[29]

As previously noted, cardiovascular disease ranks high among the health problems facing the elderly. Contributing to the increase in cardiovascular problems are several factors: an increased sedentary life-style, increased anxiety, decreased social support, and other losses (emotional, financial, family, friends, etc.).[30] The prevalence of suspected coronary artery disease is estimated to increase from 10% at ages 51-60, to 20% at ages 70-90.[30]

An additional problem experienced by the elderly is urinary

incontinence. Prevalence rates vary from 8% to 51% in community dwelling populations.[31] Urinary incontinence is a facilitating and precipitating factor in the admission of patients into long-term care facilities.

The treatment of medical and social problems in the elderly can be accomplished by varying methods.[32] Too often a drug is "thrown" at a problem that may be amenable to treatment through other means and methods. The discontinuity of care provided for the elderly is problematic and leads to further problems, including hospital readmissions, repeated and unnecessary visits to health care providers, and consumption of unnecessary health resources.[33] One method of dealing with the ambulatory elderly in a coordinated fashion is the provision of care through geriatric assessment or evaluation units.[34] The provision of care through such units appears to be a useful model for providing care to medically frail elderly patients.[35] Potential targeted patients might be those with impaired functional status who are discharged from hospitals or emergency rooms.[36]

Eight of the ten leading causes of death in people age 65 and older are now related to chronic diseases. They are:

- Heart
- Malignant neoplasms
- Cerebrovascular disease
- Arteriosclerosis
- Diabetes
- Employment
- Cirrhosis of the liver
- Nephritis and nephrosis[37]

One of the top ten causes of mortality is accidents; another is infectious disease. It is important to note that health outcomes affect not only disease or mortality, but also quality of life, institutionalization, hospitalization, or iatrogenic complications.[37]

According to the American Association of Retired Persons (AARP), most older persons have at least one chronic condition and many have multiple conditions.[6] The most frequently occurring conditions per 100 elderly are:

- Arthritis (48)
- Hypertension (37)
- Hearing impairments (32)
- Heart disease (30)
- Orthopedic impairments (18)
- Cataracts (17)
- Sinusitis (14)
- Diabetes (10)
- Tinnitus and varicose veins (8 each)

MEDICATION-SPECIFIC CONCERNS

There is an increasing focus on the provision of ambulatory health care to the elderly. It is less expensive than similar care offered to the elderly as inpatients or residents of long-term care facilities. There is also an increased focus placed upon the quality of care provided and devising means to evaluate its effectiveness.[38] The use of drugs in the ambulatory setting is an essential component to examine when considering the use of ambulatory health resources by the elderly. This use of drugs can be positive or negative depending upon the appropriateness of the drug use process.

Prescription Medication Use

Pinpointing the exact rate of prescription drug use in the U.S. is an imprecise science for any population, but especially so for the elderly. A national survey estimate places the average rate of prescription consumption for the elderly at 15 prescriptions per year.[39] High-end users (> 20 prescriptions per year) were for the most part over the age of 80. A not surprising finding in a study of the elderly with a prescription drug benefit was the higher rates of utilization occurring in such programs. The patterns of drug use also persist and maintain levels of use from year to year.[40] Certain subunits of the elderly are high users of prescription drugs. It has been estimated that 75% of all therapies for home care patients are expenditures for drug treatments.[41]

It has been suggested that the beneficial attributes of drug therapies prolong the time patients can remain at home and avoid hospitalization or institutionalization.[42] As such, therapies for help in the avoidance of hip fractures, urinary incontinence, Alzheimer's disease, and hypertension and other cardiovascular disorders are continuously sought and eagerly awaited.

Providing for consistent and appropriate use of drugs is exceedingly important for the ambulatory elderly. Studies have shown that when it does not occur, hospitalizations occur due to noncompliance and to the occurrence of avoidable adverse drug reactions.[43] Significant predictors of preventable hospital readmission for the elderly include the occurrence of preventable adverse drug reactions, noncompliance, overdose, lack of a necessary drug therapy, and underdose.[44] Others have noted that 50% of drug-induced illnesses that require hospitalization could have been avoided.[45] Elsewhere, researchers have estimated that 75% of medications are misprescribed for the elderly, with overuse and underuse rampant.[46] There must be increasing efforts to ensure continuity of care for the ambulatory elderly to avoid these and other drug-related problems.[47] Because drug use in the elderly is dynamic and increases with proximity to death, pharmacists are key players to help the elderly avoid these drug-related problems.[48]

Therapeutic Noncompliance

The elderly should not be unduly singled out as noncompliant patients. Noncompliance with medication regimens is a pervasive problem in the health care system. Noncompliance has the potential to affect all patients with the same degree of probability. What makes the elderly unique with regard to noncompliance is not age per se but the chronicity, severity, and number of ailments that befall them as a group.[49] Poor compliance with needed chronic care medications has been shown to be the most common drug-related problem influencing the need for elderly hospitalizations.

Compliance can take various forms, as well as fluctuate over time. A review of the literature pertaining to elderly patient noncompliance pinpoints five major categories of factors negatively impinging upon patient compliance.[49] These factors include:

- Multiple drug regimens
- Duration of the regimen

- Social isolation
- Poor health status
- Lack of knowledge of the drug regimen

These factors are not mutually exclusive. When they are combined, they indeed become interactive and synergistic. When these factors are coupled with other impediments to proper drug use, such as misinformation, lack of information, or packaging that prevents easy and informed access to medication therapies, inappropriate use and noncompliance with regimens increases.[50]

Possible solutions to elderly drug-taking problems are listed in Table 20.1. Each of the suggested solutions may be tailored to any elderly patient to eliminate, or at least reduce, medication-related problems. Interventions must be tailored to the individual requirements of each patient.

Medication Packaging and the Elderly

When considering the elderly and methods to improve the drug-taking process, several strategies are important. Individualized interventions may be necessary for the optimum influence on elderly patient noncompliance. These interventions may include dosing interval adjustment, unit-of-use packaging, unit-dose packaging, blister packaging, or other options to be tried to enable the elderly patient to become optimally compliant. There can be no doubt that the standard and traditional packaging of medications for elderly consumption have been problematic and definite factors in the increasing rate of elderly patient noncompliance.

Dosing Intervals

The use of tailored dosing intervals for the enhancement of elderly patient compliance has been suggested to alleviate drug-related problems for cardiovascular patients.[51] Elsewhere, it has been proposed that drug regimen simplification, use of appropriate drug containers, labeling, information leaflets, and patient education and counseling will aid the elderly patient noncompliance problem.[52]

TABLE 20.1. Possible Solutions to Elderly Compliance Problems

Problem	Proposed Solution
Hearing Loss	Speech Adjustment
Lack of Information	Explicit Instructions
Memory Loss	Written Information, Telephone Patient Counseling
Noncompliance (access)	Non-Child-Resistant Closures, Compliance Packaging
Noncompliance (confusion)	Compliance Packaging, Calendars, Diaries, Verbal and Written Counseling
Regimen Complexity	Compliance Packaging, Simplify When and If Possible, Prioritize Regimens
Regimen Confusion	Compliance Packaging, Color Coding
Visual Loss	Large Letters, Boldface Type, Large Labels

Traditional Packaging as an Impediment to Elderly Patient Compliance

In a recently published study, Atkin and colleagues found that in a sample of geriatric patients the type of container influenced the geriatric patient's ability to access medication.[53] The following percentages indicate the proportion of subjects who *could not* open the specific container type and retrieve a tablet:

- Screw top (8.3%)
- Flip top (14.2%)

- Blister pack (20.8%)
- Dosette (24.2%)
- Foil wrap (30.0%)
- Child proof (56.6%)

These results indicate that the elderly must be provided medications in accessible packaging with directions explicitly indicating how to access and properly consume medications.

Unit of Use Packaging

In a study of elderly outpatients who were taking three or more medications, unit-of-use packaging and twice daily dosing improved medication compliance when compared to conventional packaging.[54] Others have suggested that various solutions exist for elderly patient noncompliance, such as tailoring of medication regimens, instruction/reminder sheets, calendar blister packaging, and using a cap on the medication that indicates the next dosing time.[55]

Compliance Packaging

Various studies and numerous researchers have noted the importance of compliance packaging in reducing patient noncompliance and potentially enhancing patient compliance in the elderly. Researchers in New Zealand have suggested that unit dose packaging is a cost-effective compliance enhancement in the elderly and should be reimbursable under health insurance plans.[56] In the United States, researchers have noted that compliance packaging and aids should be part of an overall plan to enhance patient compliance with regimens.[57] Others have noted that compliance aids (packaging, gadgets to improve physical dexterity and aid in administering metered dose inhalers, etc.) work and should be part of an overall compliance enhancement strategy for patients.[51,58] The importance of supplying educational materials along with specialized packaging has also been shown to be necessary.[59] In this instance, the educational materials also enhanced the ability of patients to access varying types of medication packaging (foil, blister packaging).[59]

Using a Global Strategy to Enhance Elderly Patient Compliance

The elderly patient population with noncompliance problems must be assessed and continuously encouraged to become more compliant and to remain as compliant as possible. This must be accompanied by a determined effort on the part of health professionals to do all that is possible to allow the elderly noncompliant patient to become compliant. Such an effort includes providing compliance-enhancing medication packaging for patients and better follow-up to ensure that packaging dispensed to the elderly is in fact what it is supposed to be. The literature is replete with studies indicating the elderly do not always receive medications in the appropriate packaging.[60] Others point to the need to incorporate varying strategies to enhance elderly patient compliance.[61] In a study of drug misadventures, a total of 28.5% of interventions over a month-long period involved the elderly, and a total of 21% of the interventions in the elderly involved patient noncompliance.[62] The noncompliance problems uncovered revolved around dosing intervals, packaging, and lack of proper instruction.[62] Finally, archaic and outdated state and federal regulatory requirements must be examined to determine regulations that impede patient compliance and that need to be eliminated. For example, restrictive requirements that prohibit multiple medication packaging or use of other compliance aids should be removed.[63]

Elderly patient noncompliance is one of the most vexing and difficult aspects of the geriatric drug use process. Patient noncompliance in the elderly is a cause for concern that is present and growing. Attempts to improve elderly patient compliance are complex as well and will become more complex as the number and percentage of the elderly increase dramatically in the years to come. Individualized therapies, the provision of enhanced verbal and written patient counseling, and use of unit-dose, blister packaged, and specialized compliance packaging are proven means of improving elderly patient compliance. All involved in the delivery of health care services to the elderly must realize that to improve elderly patient compliance, these and other techniques that allow patients to become and remain compliant must be continuously utilized in the growing geriatric population.

Nonprescription Medication Use

The use of over-the-counter (OTC) drug products in the elderly has not been quantified. It has been suggested that the elderly spend approximately one-half as much for OTC medications as they do for prescription drugs.[64] Various authors have noted the potential impact of OTC use on the health of the elderly. The vast majority of studies have simply enumerated the number of drugs taken, without assessing whether proper or informed use of the OTCs was occurring. Problematic indicators of OTC misuse have been identified and include:

- Rate and extent of use
- Inappropriate self-medication
- Misuse and inappropriate use
- Prolonged use
- Drug interactions
- Improper labeling
- Inappropriate advertising and promotion[65]

The elderly can reap benefits from self-care and OTC use. However, proper utilization is dependent upon proper information and properly informed elderly patients. Pharmacists must continue to do what they can to enhance OTC use by elderly patients.

Adverse Drug Effects

The elderly have been suggested to be vulnerable to the untoward effects of adverse drug reactions from both the standpoint of number of drugs consumed and the impact of drug therapy on the physiology of elderly patients. It becomes crucial to continuously monitor the appropriateness of therapeutic interventions—both type (class of drug) and intensity (dosage strength).[66]

Low Rate of Literacy

The elderly have special requirements for successful learning.[67] When normal physiological changes are coupled with low literacy, transfer of information from pharmacist to patient becomes problematic.[67] Tailoring educational interventions to the reading level of the patient is crucial for successful transfer of information.[68] Written information must be readable, which means the font size, kern, background, and color of printing must be considered. A multidisciplinary approach must be undertaken to have a successful impact upon the elderly patient.[69] It does little, if any, good for interventions to be done in isolation, without contact with and cooperation from other health care providers.

CONCLUSION AND COMMENTARY

Although it is easy to stereotype the elderly, it is inappropriate to do so. The elderly are a very heterogeneous population,[70] and because of this, special efforts must be undertaken to individualize treatment to every extent possible. This segment of our population is certain to increase in the years to come. With this increase will come a broader base of differing ethnic, social, health, language, and morbidity patterns for the elderly. It is crucial for us as pharmacists to examine our attitudes toward the elderly.[71] Treating the elderly as individuals with differing attributes, attitudes, and beliefs will enable the profession to better serve this segment of the population, which will, no doubt, make up an ever-increasing patient base for delivery and provision of pharmaceutical care.

REFERENCES

1. Randall T. Demographers ponder the aging of the aged and await unprecedented looming elder boom. JAMA 1993;269:2331-2.

2. U.S. Bureau of the Census. Center for International Research. International Data Base on Aging. 1993.

3. Healthy people: the Surgeon General's report on health promotion and disease prevention. DHEW (PHS) Publication #79-05571.

4. Fincham JE. The aging of America–how to deal with the geriatric patient in the community pharmacy. J Geriatr Drug Ther 1989;4(1):33-49.

5. Pucino F, Beck CL, Seifert RL, et al. Pharmacogeriatrics. Pharmacother 1985;5:314-26.

6. Anon. A profile of older Americans: 1993. Washington, DC: AARP, 1993.

7. Van Nostrand JF. Common beliefs about the rural elderly: what do national data tell us? Vital and Health Statistics. Washington, DC: National Center for Health Statistics, 3(28), 1993.

8. Public Health Service. Agency for Health Care Policy and Research. Health status and access to care of rural and urban populations. Braden J, Beauregard K. (AHCPR Publication Number 94-0031). National Medical Expenditure Survey Research Findings 18. Rockville, MD: Public Health Service, 1994.

9. Torres-Gil FM. The new aging, politics and change in America. New York: Auburn House, 1992.

10. Austin CD. Aging well: what are the odds? Generations 1991;15:73-5.

11. Avorn J. Patient compliance: evidence from health services research and the bedside. US Pharm 1987;12(Oct Suppl):24+.

12. Campion EW. The oldest old. JAMA 1994;300:1819-20.

13. Jennings MC, Porter M. The changing elderly market. Top Health Care Finan 1991;17(4):1-8.

14. Monroe SM. Retirement housing takes off. Provider 1994;20(10):89-91.

15. MacDonald M, Remus G, Laing G. Research considerations: the link between housing and health in the elderly. J Gerontol Nurs 1994;20(7):5-10.

16. Knapp M, Cambridge P, Thomason C, et al. Residential care as an alternative to long-stay hospitals: a cost-effectiveness evaluation of two pilot projects. Int J Geriatr Psychiatry 1994;9:297-304.

17. Monk A, Cox C. Home care for the elderly, an international perspective. Westport, CT: Auburn House, 1991.

18. Spiegal A. Home care. 2nd ed. Owings Mill, MD: Rynd Communications, 1987.

19. Dychtwald K, Flower J. Age wave. Los Angeles, CA: Jeremy P. Archer, Inc., 1989.

20. Barone JA, Holland M. Drug use and abuse in the elderly. U.S. Pharm 1987;12:82+.

21. Atkinson R. Late onset problem drinking in older adults. Int J Geriatr Psychiatry 1994;9:321-6.

22. Dufour MC, Archer L, Gordis E. Alcohol and the elderly. Clin Geriatr Med 1992;8(1):127-41.

23. Colsher PL, Wallace RB. Elderly men with histories of heavy drinking: correlates and consequences. J Stud Alcohol 1990;51(6):528-35.

24. Fincham JE. The etiology and pathogenesis of detrimental health effects in elderly smokers. J Geriatr Drug Ther 1992;7(1):5-21.

25. Lubben JE, Weiler PG, Chi I. Health practices of the elderly poor. Am J Public Health 1989;79:731-4.

26. Tideiksaar R. Falling in old age: its prevention and treatment. New York: Springer Publishing Company, 1989.

27. Mace NL, Rabins PV. The 36-hour day. Baltimore, MD: Johns Hopkins University Press, 1991.

28. Golden R. Dementia and Alzheimer's disease: indications, diagnosis, and treatment. Minn Med 1995;78(1):25-9.

29. Cleary BL. Alzheimer's disease: stressors and strategies associated with caregiving. In: Funk SG, Tornquist EM, Champagne MT, et al., eds. Key aspects of elder care: managing falls, incontinence, and cognitive impairment. New York: Springer, 1992:320-7.

30. Lamy PP. Elderly. US Pharm 1991;16:17-23.

31. Herzog AG, Fultz NH, Normolle DP, et al. Methods used to manage urinary incontinence by older adults in the community. J Am Geriatr Soc 1989;37:339-47.

32. Kailis SG. Drug usage in the elderly: role of the pharmacist. Aust J Pharm 1989;70:764-5.

33. Burns J, Sneddon I, Lovell M, et al. Elderly patients and their medication: a post-discharge follow-up study. Age Ageing 1992;21:178-81.

34. Kramer A, Deyo R, Applegate W, et al. Research strategies for geriatric evaluation and management: conference summary and recommendations. J Am Geriatr Soc 1991;39:53S-57S.

35. Rubin CD, Sizemore MT, Loftis PA, et al. Randomized, controlled trial of outpatient geriatric evaluation and management in a large public hospital. J Am Geriatr Soc 1993;41:1023-8.

36. Denman SJ, Ettinger WH, Zarkin BA, et al. Short-term outcomes of elderly patients discharged from an emergency department. J Am Geriatr Soc 1989;37:937-43.

37. Fried LP, Wallace RB. The complexity of chronic illness in the elderly: from clinic to community. In: Wallace RB, Woolson RF, eds. The epidemiologic study of the elderly. New York: Oxford University Press, 1992:10-9.

38. Ferris AK, Wyszewianski L. Quality of ambulatory care for the elderly: formulating evaluation criteria. Health Care Finan Rev 1990;12:31-8.

39. Public Health Service. National Center for Health Services Research and Health Care Technology Assessment. Prescribed medicines: a summary of use and expenditures by Medicare beneficiaries. Moeller J, Mathiowetz N. DHHS Publication 89-3448. National Medical Expenditure Survey Research Findings 3. Rockville, MD: Public Health Service, 1989.

40. Stuart B, Coulson NE. Dynamic aspects of prescription drug use in an elderly population. Health Serv Res 1993;28:237-64.

41. Koch KE, Wiser TH, Kleinman LS. Why pharmacists serve as a key component in compliance re home health care patients. Pharm Times 1986;52:67+.

42. Levy RA, Smith DL. Keeping the elderly patient at home. Am Pharm 1988;NS28:41-4.

43. Col N, Fanale JE, Kronholm P. Role of medication noncompliance and adverse drug reactions in hospitalizations of the elderly. Arch Intern Med 1990;150:841-5.

44. Bero LA, Lipton HL, Bird JA. Characterization of geriatric drug-related hospital readmissions. Med Care 1991;29:989-1003.

45. Choi T, Thomas NA, Nyugen M, et al. Effects of drug audit information to providers of high-risk elderly patients in a pre-paid group practice. Ann Rev Gerontol Geriatr 1992;12:76-94.

46. Brook RH, Kamberg CJ, Mayer-Oakes A, et al. Appropriateness of acute medical care for the elderly: an analysis of the literature. Santa Monica, CA: Rand Corporation, 1989.

47. Beland F. Continuity of ambulatory medical care: an exploratory study. J Aging Stud 1989;3:341-54.

48. Coulson NE, Stuart B. Persistence in the use of pharmaceuticals by the elderly: evidence from annual claims. J Health Econ 1992;11:315-28.

49. Fincham JE. Patient compliance in the ambulatory elderly: a review of the literature. J Geriatr Drug Ther 1988;2(4):31-52.

50. Fincham JE, Wertheimer AI. Elderly patient initial noncompliance: the drugs and the reasons. J Geriatr Drug Ther 1988;2(4):53-62.

51. Urquhart J. Partial compliance in cardiovascular disease: risk implications. Br J Clin Pract Symp Suppl 1994;73:2-12.

52. Wade B, Bowling A. Appropriate use of drugs by elderly people. J Adv Nurs 1986;11(1):47-55.

53. Atkin PA, et al. Functional ability of patients to manage medication packaging: a survey of geriatric inpatients. Age Ageing 1994;23(2):113-6.

54. Murray MD, et al. Medication compliance in elderly outpatients using twice-daily dosing and unit-of-use packaging. Ann Pharmacother 1993;27:616-621.

55. Kluckowski JC. Solving medication noncompliance in home care. Caring 1992;11(11):34-41.

56. Ware GJ, et al. Unit dose calendar packaging and elderly patient compliance. N Z Med J 1991;104(924):495-7.

57. Bond WS, Hussar DA. Detection methods and strategies for improving medication compliance. Am J Hosp Pharm 1991;48(9):1978-88.

58. Rivers PH. Compliance aids—do they work? Drugs-Aging 1992;2(2):103-11.

59. Whyte LA. Medication cards for elderly people: a study. Nurs Stand 1994;8(48):25-8.

60. Burns JM, et al. Elderly patients and their medication: a post-discharge follow-up study. Age Ageing 1992;21(3):178-81.

61. Botelho RJ, Dudrak R. Home assessment of adherence to long-term medication in the elderly. J Fam Pract 1992;35(1):61-5.

62. Fincham JE, et al. The true value of pharmacist care. NARD J 1995;17(3):29-31.

63. Anon. Standards for multiple medications packaging. Newsl National Assoc Boards Pharm 1995;24(1):2+.

64. Lamy P. Over-the-counter medications: the drug interactions we overlook. J Am Geriatr Soc 1982;30:S69-S72.

65. Fincham JE. Over-the-counter drug use and misuse by the ambulatory elderly: a review of the literature. J Geriatr Drug Ther 1986;1(2):3-21.

66. Fincham JE. Adverse drug reaction occurrence in elderly patients. J Geriatr Drug Ther 1991;5(4):39-50.

67. Hussey LC. Overcoming the clinical barriers of low literacy and medication noncompliance among the elderly. J Gerontol Nurs 1991;17:27-9.

68. Tymchuk AJ. Readability levels of over-the-counter medications commonly used by elderly people: a possible issue in compliance. Educ Gerontol 1990;16:491-6.

69. O'Connell MB, Johnson JF. Evaluation of medication knowledge in elderly patients. Ann Pharmacother 1992;26:919-21.

70. Lamy PP. Patient package inserts: voluntary approach. Drug Info J 1990; 24:615-20.

71. Croteau DR, Moore RJ, Blackburn JL. Attitudes of pharmacists toward the elderly. Am J Pharm Educ 1991;55:113-9.

ADDITIONAL READINGS

1. Public Health Service. National Center for Health Services Research and Health Care Technology Assessment. Prescribed medicines: a summary of use and expenditures by Medicare beneficiaries. Rockville, MD: Department of Health and Human Services, 1989.

2. Bazargan M, Barbre AR, Hamm V. Failure to have prescriptions filled among black elderly. J Aging Health 1993;5:264-82.

3. Bernstein LR, Folkman S, Lazarus RS. Characterization of the use and misuse of medications by an elderly, ambulatory population. Med Care 1989;27:654-63.

4. Chrischilles EA, Foley DJ, Wallace RB, et al. Use of medications by persons 65 and over: data from the Established Populations for Epidemiologic Studies of the Elderly. J Gerontol 1992;47:M137-M144.

5. Esposito L. Medication knowledge and compliance: home care vs. senior housing clients. Caring 1992;11:42-5.

6. Feinberg M. Polymedicine and the elderly: is it avoidable? Pride Institute J Long Term Home Health Care 1989;8:15-25.

7. Gehres RW. Medication compliance aids. Consultant Pharm 1986;1: 218-20.

8. Gehres RW. A medication monitoring service for elderly patients offered by the pharmacist on a fee-for-service basis. J Geriatr Drug Ther 1986;1(2):81-9.

9. Grymonpre R, Sabiston C, Johns B. Development of a medication reminder card for elderly persons. Can J Hosp Pharm 1991;44:55-62.

10. Johnson AG, Day RO. Problems and pitfalls of NSAID therapy in the elderly. Drugs Aging 1991;16:17-23.

11. Johnson RE, Vollmer WM. Comparing sources of drug data about the elderly. J Am Geriatr Soc 1991;39:1079-84.

12. Kail BL. Special problems on non-compliance among elderly women of color. Lewiston, NY: Edwin Mellen Press, 1992.

13. Kimberlin DL, Berardo DH, Pendergast JF, et al. Effects of an education program for community pharmacists on detecting drug-related problems in elderly patients. Med Care 1993;31:451-68.

14. Kutner NG, Ory MG, Baker DI, et al. Measuring the quality of life of the elderly in health promotion intervention clinical trials. Public Health Rep 1992;107:530-9.

15. Larson DB, Lyons JS, Hohmann AA, et al. Psychotropics prescribed to the U.S. elderly in the early and mid 1980s: prescribing patterns of primary care practitioners, psychiatrists, and other physicians. Int J Geriatr Psychiatry 1991; 6:63-70.

16. Leirer VA, Marrow DG, Tanke ED, et al. Elders' nonadherence: its assessment and medication reminding by voice mail. Gerontologist 1991;31:514-20.

17. Lipton HL, Bero LA, Bird JA, et al. Impact of clinical pharmacists' consultations on physicians' geriatric drug prescribing: a randomized controlled trial. Med Care 1992;30:646-55.

18. Margolis RE. Over the counter drugs and the elderly: safe remedies or potential menaces? HealthSpan 1992; 9:15-6.

19. Pesznecker BL, Patsdaughter C, Moody KA, et al. Medication regimens and the home care client: a challenge for health care providers. Home Health Care Serv Q 1990;11:9-68.

20. Potempa KM, Folta A. Drug use and effects in older adults in the United States. Int J Nurs Stud 1992;29:17-26.

21. Raetzman SO. Older Americans and prescription drugs: utilization, expenditures and coverage. AARP Public Policy Institute issue brief, 1991, 9:1-15.

22. Rantucci M. Health professionals' perceptions of drug use problems of the elderly. Can J Aging 1989;8:164-72.

23. Rowe JW, Ahronheim, JC, Lawton MP. Annual review of gerontology and geriatrics: focus on medications and the elderly. New York: Springer, 1993.

24. Ruben DH. Aging and drug effects: a planning manual for medication and alcohol abuse treatment of the elderly. Jefferson, NC: McFarland, 1990.

25. Stuart B, Lago D. Prescription drug coverage and medical indigence among the elderly. J Aging Health 1989;1:451-69.

26. Thomas C, Kelman HR. Health services use among the elderly under alternative health service delivery systems. J Community Health 1990;15:77-92.

27. Yurkow J. Abuse and neglect of the frail elderly. Pride Institute J Long Term Home Health Care 1991;10:36-9.

Chapter 21

Long-Term Pharmaceutical Care: Social and Professional Implications

Royce A. Burruss
Norman V. Carroll

INTRODUCTION

Long-term care has been defined as "the range of services that addresses the health, personal care, and social needs of individuals who need assistance in caring for themselves."[1] Where care is administered is determined by an individual's specific care needs. Providers are available in a spectrum of facilities–extended care, skilled nursing, intermediate care, hospice, residential care, assisted living, and the home.[2] The goals of long-term care are the same in each setting and involve the restoration and/or maintenance of health and function. The goals include preventive services as well as management of acute and chronic illnesses. Care is focused on maximizing autonomy and minimizing complications and dependency.[1] The scope of services required depends upon the extent of the patient's disability and underlying diseases, the availability of support (e.g., caregiving family members, transportation, nutrition, etc.), and the potential for recovery.[1]

Although long-term care services may be necessary at any time during the patient's life, the need generally increases as patients get older.[1] About one-third of people receiving long-term care live in an institutional setting (e.g., nursing home, subacute care, or rehabilitative care), while the remaining two-thirds are cared for in the community (usually living at home or with others).[1] Physical, psychological, emotional, and social support are some of the primary

537

determinants of the type of long-term care individuals need.[1] The older and more infirm people become, the more dependent they become on their families, their communities, and available health care resources. As an example, a patient who has family members willing and able to provide care may be a good candidate for home care; however, if such family care is unavailable or if the patient is so ill that licensed nurses are needed, then the patient may need the support of a nursing home or acute care hospital.

A significant change in the health status of an individual can substantially alter interpersonal relationships. Because relationships give a person's life meaning, changes in these relationships may result in that person developing a substantially different concept of life's meaning or purpose. This may, through self-referencing iterations, result in additional changes in relationships and, potentially, in health status.* The paradigm shown in Figure 21.1 depicts relationships among the variables of health status, interpersonal relationships, and life's meaning. A review of the major types of long-term care will provide a better understanding of each.

LONG-TERM CARE: THE NURSING HOME

The structure and function of modern nursing homes were influenced by the Social Security Act of 1935 and the Medicare and Medicaid programs enacted in the 1960s.[1] Nursing homes currently provide a traditional institutional health care environment which combines some of the characteristics of hospitals as well as those of residential care. Nursing homes now compete with acute care hospitals for the some of the less acutely ill patients by providing a less expensive, step-down level of care (i.e., subacute care) relative to that available in hospitals. This is particularly evident today, given the emergence of subacute care units in nursing homes that provide skilled care. As an example, subacute care units in nursing homes may have long-term ventilator patients and those requiring rela-

---

*I am appreciative of Margaret J. Wheatly who, through her book *Leadership and the New Science* (1992), evoked in me new ways to consider the concepts of life's meaning and self-referencing from the perspective of an individual as well as that of an organization.

FIGURE 21.1. A Paradigm of Life Changes

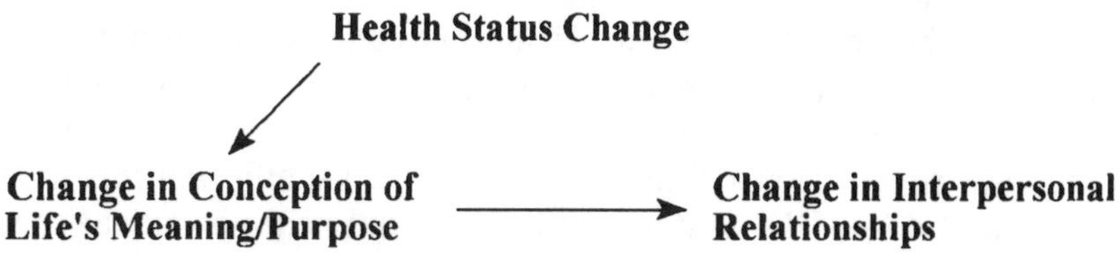

Health Status Change

Change in Conception of   **Change in Interpersonal**
Life's Meaning/Purpose   **Relationships**

tively complex intravenous pharmacotherapy (drug therapy). A decade ago, patients with these needs would have been cared for only in an acute care hospital.[2] The clinical staffing pattern in nursing homes is also different from that of hospitals. There are fewer licensed nurses per patient in nursing homes than there are in hospitals. A greater proportion of patient care in nursing homes consists of personal care (e.g., hygiene, assisted and personalized exercise, feeding, etc.). As a result, nursing homes employ more unlicensed technical/support staff than hospitals. Nursing homes do, however, offer a wide variety of professional health care services (e.g., administration, pharmacy, dietary, social workers, physical therapists, and physicians). Typically, physicians prescribe patient-specific care, and registered nurses (RNs) lead the nursing home's on-site health care team in providing that care.[1] Pharmacists participate as care team members by providing safe and effective drug distribution systems as well as sharing clinical responsibility with physicians for achieving desired pharmacotherapeutic outcomes.

Nursing home patients frequently have one or more significant clinical problems–both physical and mental. Examples include depression, confusion, dementia, malnutrition, pressure sores, coronary artery disease, hypertension, constipation, osteoarthritis, diabetes mellitus, sleep disorders, acute and chronic pain, renal dysfunction, pulmonary dysfunction, and cancer.[3] The number of significant clinical problems that coexist in an individual tends to increase with advancing age. Between the ages of 65 and 69, the average number of coexisting clinical conditions is 4.0; however, the average number of conditions increases to 5.1 (approximately a

25% increase) for the next decade.[3] A similar association between disability and aging has been reported.[4]

Nursing home patients typically are not capable of self-administering medication (with some exceptions) and usually have medications administered by a licensed practical nurse (LPN) functioning under the supervision of an RN. Drugs are usually administered from a unit dose (UD) system, with additional pharmaceutical care being provided by a consultant pharmacist. The unit dose system can be generally characterized as a 24-hour to 30-day supply of all regularly scheduled and most "as needed" medications prepared in single-dose or UD containers. UD containers are individually labeled with the drug name, dose, lot number, expiration date, and any special storage requirements (e.g., refrigerate) and are usually stored at the nursing home in lockable UD supply carts. These carts also contain medication administration adjuncts (e.g., needles, syringes, alcohol swabs, bandages, disposable cups, water carafe, etc.) and each patient's medication administration record (MAR). The MAR is a permanent part of the patient's medical record on which the nurse documents the administration of each dose. The MAR may be used as a billing document for drugs administered. Consultant pharmacist services may be numerous, consisting, in part, of pharmacokinetic dosing, patient and staff education, and drug regimen review (DRR).

DRR (including both prescription and over-the-counter medications) has been required for all nursing facility patients since 1974 and for all patients since the Omnibus Budget Reconciliation Act of 1987 (OBRA 87). It includes screening for potential drug interactions with other drugs and food; pharmacotherapy duplication; drug-disease contraindications; and incorrect or inappropriate drug, dose, frequency, and/or route. The pharmacist is required to offer the patient a discussion of pharmacotherapy issues deemed to be important, e.g., name of drug, route, dose, dosing frequency, duration of therapy, proper administration technique, special handling and precautions, signs and symptoms of side effects, adverse reactions, and interactions with other drugs and food, and the action to take in the event of a self-administration error. To accomplish this, the pharmacist should maintain (at a minimum) certain data in each patient's pharmacy record, e.g., patient-specific demographics such as name, address, birth date, age, race, and gender, relevant medical and medication histories, and pertinent clinical comments.

LONG-TERM CARE: ASSISTED LIVING

Generally, assisted living means living in a facility offering some combination of home care and nursing home facilities.[1] Individuals residing in assisted living facilities usually have a private bedroom and bathroom and share common dining and recreational areas. Such individuals generally can provide most of their own care, as mental and most physical capacities are functioning satisfactorily. They may, however, have physical limitations that result in the need for some assistance. For example, they may need help with bathing or getting from place to place. Most residents are capable of self-administering medication and are dispensed medications in multidose prescription containers; however, nurses or personal care aides may assist in medication administration by providing reminders of doses due to be taken and by counting remaining doses to make sure key medications are not inadvertently omitted by the patient. The pharmacist is frequently called upon by these patients for drug-related and other health care information. In the course of communicating with patients and their caregivers, the pharmacist may discover medication errors, medication mismanagement, and/or health status changes requiring pharmacotherapy changes. When this occurs, the pharmacist discusses these findings and pharmacotherapy recommendations with the patient and/or care team members to assure the best possible patient outcome related to drug therapy.

LONG-TERM CARE: CONTINUING CARE RETIREMENT COMMUNITY

The continuing care retirement community (CCRC), a relatively new living arrangement, provides individuals with a residential campus having easily accessible and attractive cultural and recreational activities as well as permanent housing for private dining, sleeping, and social interaction. This type of community also provides guaranteed coverage of the residents' health care needs.[1] Such living arrangements are typically quite expensive. The cost of living in a CCRC is probably comparable to purchasing a large home in an expensive neighborhood. Residents are capable of providing most of their own care and meeting their own personal needs; however, should a resident need assistance, it is readily available. The spectrum of available health care services ranges from assistance with personal hygiene to skilled nursing and subacute care. In this environment, therefore, one may expect to see the entire spectrum of pharmacy services, i.e., community practice to nursing home based practice, including unit dose drug distribution systems, parenteral therapies, and consultant pharmacy services.

LONG-TERM CARE: HOME HEALTH CARE

Home health care may be thought of as services provided to acutely or chronically ill patients in their residences as prescribed by a physician. Home health services may include nursing care, social services, personal care, home health aides, nutritional support, speech therapy, occupational therapy, and other specialized health and support services.[5] Support services are typically social and maintenance services that enable an incapacitated or otherwise disabled person to live at home. They include housekeeping, personal care, transportation, and shopping assistance.

There are a number of home care providers. These include both licensed providers, i.e., pharmacists, physicians, RNs, LPNs, and physical and occupational therapists, and unlicensed providers, i.e., certified nursing assistants (CNAs), personal care aides, social workers, respiratory therapists, delivery personnel, and family members. Unlicensed personnel are allowed to provide only personal or household assistance, i.e., personal hygiene, dressing, feeding, food preparation, and product or equipment delivery. Home health care products are typically administered by licensed care givers who may, in turn, instruct the patient and/or family in the safe and effective administration/application of those products or equipment.

Home health care today provides goods and services that may also be provided in hospitals, nursing homes, and assisted living arrangements. Rinehart reported that "the patterns of patient care in the long-term setting, as a whole, parallel closely the patterns of patient care in the acute care setting. Although various terms are used to differentiate between hospitals and long-term settings, the types of individuals cared for in the [respective settings] often have similar therapeutic needs."[2] Home health care patients can be acutely ill, chronically ill, or both. Chronically ill patients have long-term illnesses that typically require long-term treatment or palliation. Acutely ill patients have experienced recent or emergent health problems that may require intensive (often parenteral) pharmacotherapy and related equipment and supplies.

The Social Security Act of 1965 precipitated the evolution of home health care to what it is today. These changes established home health care as a reimbursable service by Medicare. However, the act did not include reimbursement for pharmaceuticals. Since 1965, most (but not all) pharmaceuticals and pharmacy services are reimbursable through other third-party payers, i.e., insurance companies and managed care organizations. There has been about a tenfold growth in home health care agencies (HHCA) in the U.S.A. HHCAs provide a spectrum of home health related goods and services, from personal assistance services to parenteral drugs and oxygen therapy. Numerous factors have been suggested to explain this growth:

- Aging of the population.
- Newer, more "user-friendly" medical devices and equipment.
- One wage earner in a single-parent family or two wage earners in a two-parent family (thus reducing the likelihood of there being at least one responsible family member at home to care for another ill, homebound family member).
- Cost reduction/containment initiatives (primarily by hospitals and insurance carriers) that have resulted in early hospital discharges of patients with care needs that are substantially higher than previously seen in discharged patients.
- An increase in patient preference to receive health care at home instead of in an institutional setting.

With the tremendous growth in HHCAs came a proportionate increase in demand for home care pharmacy (HCP) goods and services. HCPs provide durable medical equipment (DME), pharmaceuticals of all kinds (including intravenous and experimental medications), total parenteral and enteral nutrition (TPN and EN, respectively), all adjunct supplies, and the information required to use the drugs, supplies, and equipment safely and effectively. The home care pharmacist frequently communicates with home care providers and/or homebound patients to discuss medication history, home nutrition status, potential drug-drug and drug-food interactions, allergies, desired therapy outcomes, and the pharmaceutical goods and services required to achieve the outcomes. For a complete review of the pharmacist's role in home care the reader is referred to the American Society of Hospital Pharmacy's guidelines.[6]

PHARMACEUTICAL CARE: AN OVERVIEW

The profession of pharmacy continues to be challenged with tremendous economic, social, political, technological, and environmental changes. These changes, whether viewed from a macro or micro perspective, may be characterized as occurring simultaneously and to varying degrees of intensity; however, when viewed more critically over time, these numerous changes may be best described as part of the emergent evolution** of America's health care system. Pharmacy, being a highly adaptive and patient-focused profession as well as an integral part of this health care system, has undergone and will likely continue its own emergent evolution. Pharmacy's role in long-term care is, in large part, determined by the demands placed upon the pharmacist by long-term care patients and/or the long-term care facilities caring for them. Table 21.1 provides a brief summary by category of pharmaceutical goods and services typically provided to various categories of long-term care patients.[6]

Several decades ago, pharmacists had three primary career choices: community practice, hospital practice, and graduate school

---

**Emergent evolution–evolution characterized by the appearance at different levels of wholly new and unpredictable characters or qualities through a rearrangement of preexistent entities (*Webster's Seventh New Collegiate Dictionary*, 1965).

TABLE 21.1. Pharmaceutical Goods and Services Provided to Long-Term Care Patients

Location of Care	Initial Assess	Safety	ADRx	CQI	Drug Distrib	Drug Admin	DME	Supplies	Educ	P. Care Plan	Clin. Monitor	Comm. MD Pat.
NH	±	+	+	+	+	-	±	±	+	+	+	±
AL	+	+	+	+	+	-	±	±	+	+	+	+
CCRC	+	+	+	+	+	-	±	±	+	+	+	+
HC	+	+	+	+	+	+	+	+	+	+	+	+

NH = Nursing Home
AL = Assisted Living
CCRC = Continuous Care Retirement Community
HC = Home Care
ADRx = Adverse Drug Reaction Reporting
CQI = Quality Improvement
Drug Distrib. = Drug Distribution
Drug Admin. = Drug Administration
DME = Durable Medical Equipment
Educ. = Education of Patient and Caregivers
P. Care Plan = Pharmaceutical Care Plan
Clin. Monitor = Clinical Monitoring
Comm. MD, Pat. = Communication with Physician, Patient, et al.
+ = Typically provided by pharmacy
- = Typically not provided by pharmacy (usually provided by a different profession or by the family)
± = May or may not be provided by pharmacy; also, may or may not be provided by a different profession or by the family

and teaching.[7] Today, many other career choices are available: nursing home care, home health care, hospice care, subacute care, managed care, rehabilitation care, marketing, sales, consulting, and research. All of these careers share common bonds, but the one that exerts the most powerful influence is that of the concept and practice of pharmaceutical care.

Pharmaceutical care has been defined by Hepler and Strand as:

> . . . the responsible provision of drug therapy for the purpose of achieving definite outcomes that improve a patient's quality of life. These outcomes are (1) cure of a disease, (2) elimination or reduction of a patient's symptomatology, (3) arresting or slowing of a disease process, or (4) preventing a disease or symptomatology.
>
> Pharmaceutical care involves the process through which a pharmacist cooperates with a patient and other professionals in designing, implementing, and monitoring a therapeutic plan that will produce specific therapeutic outcomes for the patient. This, in turn, involves three major functions: (1) identifying potential and actual drug-related problems, (2) resolving actual drug-related problems, and (3) preventing potential drug-related problems.
>
> Pharmaceutical care is a necessary element of health care, and should be integrated with other elements. [It] is, however, provided for the direct benefit of the patient, and the pharmacist is responsible directly to the patient for the quality of that care. The fundamental relationship in pharmaceutical care is a mutually beneficial exchange in which the patient grants authority to the provider and the provider gives competence and commitment (accepts responsibility) to the patient.
>
> The fundamental goals, processes, and relationships of pharmaceutical care exist regardless of practice setting.[8]

Pharmaceutical care is that component of pharmacy practice that can be performed by no one other than a competent pharmacist committed to establishing a reciprocal relationship of truthfulness, confidentiality, and loyalty with patients whose pharmacotherapeutic needs are met through the application of knowledge and humanistic principles.[9] Strand and colleagues suggested that pharmacists perform the following process sequence in providing pharmaceutical care:

- Establish a pharmacist-patient relationship.
- Collect, synthesize, and interpret the relevant patient-related information.
- List and rank the patient's drug-related problems.
- Establish a desired pharmacotherapeutic outcome for each drug-related problem.
- Determine feasible pharmacotherapeutic alternatives.
- Select the ideal pharmacotherapeutic solution and individualize the therapeutic regimen.
- Design a pharmacotherapeutic monitoring plan.
- Implement the individualized regimen and monitoring plan.
- Follow up to measure success.[10]

Smith and colleagues proposed basic pharmacist functions common to all levels of pharmaceutical care:

- Develop and use a patient medication profile.
- Interpret, question, clarify, verify, and validate all drug-related orders.
- Provide a safe and efficient drug dispensing system.
- Monitor drug therapy for safety, efficacy, and desired clinical outcome.
- Screen for drug allergies, drug-drug interactions, drug-food interactions, and concomitant drug use.
- Detect and report drug allergies and adverse reactions.
- Recommend initial or alternative drug therapies.
- Respond to drug information requests from physicians, nurses, and patients.
- Teach health-care providers and patients about drug use.
- Obtain medication histories by interviewing patients.
- Assist in the selection of the drugs of choice and dosage forms.
- Conduct drug-use evaluations to gauge the appropriateness of drug use and achievement of desired therapeutic outcomes.
- Apply pharmaceutical principles for selected drug therapies.[11]

Smith and colleagues also suggested that there are four factors influencing the pharmaceutical care needs of the patient and the pharmacist-patient relationship:

- The patient's medical condition (e.g., diagnosis, acuity of illness, physiological status, response to drug use, the number of physicians providing care).
- The drug therapy the patient is receiving (e.g., the greater the number of drugs a patient is receiving, the greater the need for direct pharmacist involvement in pharmacotherapy).
- The degree of action required by the pharmacist.
- The interprofessional relationships between the pharmacist and other health-care providers (e.g., a collaborative relationship is ideal).[11]

Brodie suggested that health care is modeled similarly to the medical model and consists of three levels of care: primary, secondary, and tertiary.[12] Smith and colleagues elaborated on Brodie's model, as shown in Table 21.2.[11]

In providing pharmaceutical care, the pharmacist, in concert with other health care team members, determines on a patient-specific basis (1) the appropriate level of care, (2) the process to be followed, and (3) the clinical functions to be performed. The application of each of these principles is demonstrated in the case of AAA, a nursing home patient.

THE CASE OF AAA:
PHARMACEUTICAL CARE IN ACTION

AAA, a 100-year-old Caucasian female and a resident of XYZ long-term care facility for approximately five years, was admitted to that facility's subacute care unit after experiencing confusion, disorientation, nausea and vomiting, diarrhea, lower abdominal pain, and fever. One week prior to this event she had been lucid, writing letters to her family, ambulating with assistance, and had the following vital signs: blood pressure (BP) = 132/91, pulse (P) = 68/minute (1-2 skipped beats/minute), respiratory rate (RR) = 19, rectal temperature (RT) = 99.3°F. Her weight had been 102-105 lbs.

TABLE 21.2. Summary of the Three Levels of Pharmaceutical Care

Levels	Characteristics			
	Medical Condition	Drug Therapy	Degree of Action Required by R.Ph.	Interdisciplinary Interactions
Primary	Chronic, episodic	Not complex	Minor or occasional	Infrequent with MD or RN
Secondary	Acute	Multiple Lab monitor	Regular Lab monitor	Regular with MD and RN
Tertiary	Very acute	Same as secondary	Intensive Lot of monitoring	Very frequent with MD and RN

for the past three months, and her lung sounds were clear. Over the course of the week preceding admission to the subacute care unit, the aforementioned symptoms gradually emerged, increasing in severity during this time. She was on numerous prescription medications: digoxin 0.125mg QAM, hydrochlorothiazide 50mg QAM, KCl elixir 20mEq QAM, flurazepam 15mg QHS prn sleep, and ibuprofen 200mg Q6H prn joint pain.

Physician Exam in the Subacute Unit

AAA was noted to be thin, very pale, and disoriented. She had a history of hypertension (HTN) and congestive heart failure (CHF) associated with an infrequent atrial dysrhythmia. Her vital signs were: BP = 90/60, P = 55/minute with 2-3 skipped beats/minute, RR = 30/minute and shallow, RT = 102°F. Her weight was 95 lbs. Her abdomen was remarkably tender, particularly in the suprapubic area, and her bowel sounds were characterized as loud and frequent. During the exam, AAA complained of nausea and vomited approximately 200ml of dark green emesis followed by a loose, watery stool of approximately 100ml. It was noted that AAA's urine output over the past 24 hours was approximately 250ml and it had a dark yellow, cloudy appearance and a strong, foul odor. The stool was negative for occult blood. A clean catch urine specimen was sent to the laboratory for analysis, including culture and sensitivity. The only other relevant physician finding was 3+ bipedal edema where there had not been any edema two weeks before admission to the subacute care unit.

Radiographic and Laboratory Findings

Abdominal radiographic studies revealed no apparent disease. Paired blood cultures were drawn and were subsequently found to be negative. A urine culture and sensitivity test proved positive for *Proteus* species. The preliminary sensitivity results of the urine culture indicated the organism was sensitive to ciprofloxacin. A serum digoxin level was drawn and found to be 2.2ngm/ml. A standard battery of serum laboratory tests showed results within normal limits except for the following: BUN = 45mg/100ml, creatinine = 1.5mg/100ml, potassium = 6mMol/l, and WBC count = 14 $\times$ 10mm^3 (differential count desired but not done).

Subacute Unit Course

A peripherally inserted intravenous (IV) solution of D5%/0.45% saline was begun at 125ml/hr for 24 hours with the rate of infusion decreased to 75ml/hr thereafter for hydration and IV medication administration. Ciprofloxacin 250mg Q12H IV was initiated, with therapy prescribed for 14 days. The oral digoxin was not administered for 48 hours and was restarted at 0.125mg IV QAM; however, prior to the first dose of 0.125mg being administered, a repeat serum digoxin level was drawn. The repeat serum digoxin level was 1.2ngm/ml (within normal limits). Oral potassium chloride was also withheld for the first 48 hours postadmission to the subacute unit. On the third day in the subacute care unit, a serum potassium level was drawn and the level was 4.1mMol/l. With rehydration, the serum creatinine fell to 1.0mg/100ml, and the BUN fell to 19mg/100ml. After 72 hours of ciprofloxacin therapy, the patient became afebrile (RT = 99.2°F); her urine returned to a clear, pale yellow appearance without being malodorous; she no longer claimed to be nauseated; and she once again became lucid and coherent. Her vital signs returned to baseline: BP = 135/85, P = 70 (1-2 skipped beats/minute). Her weight had increased to 101 lbs. She began taking clear liquids by mouth which, within the next 24 hours, was followed by a return of appetite and resumption of her normal diet. At this time, her IV was discontinued, including IV ciprofloxacin. Oral ciprofloxacin 250mg Q12h was started to complete the 14-day course of therapy. Oral potassium supplementation was restarted at 10mEq PO BID. Oral digoxin 0.125mg QAM and hydrochlorothiazide 50mg QAM were also started. A repeat CBC revealed the WBC count was down to 8.5×10^3. The patient was transferred to her bed in the nursing home, where periodic serum digoxin, potassium, and creatinine levels were done with all levels within normal limits.

Pharmaceutical Care Process

The level of pharmaceutical care required by this centenarian can be characterized as a progression from primary to secondary and back to primary. The pharmacotherapeutic risk to her was moderate, considering the types of drug therapy, her remarkable constitution for a woman of such advanced years, and her response to pharmacotherapy.

First Pharmaceutical Care Event:
Establish a Pharmacist-Patient Relationship

The consulting pharmacist had known this patient for the previous eight years, which coincided with the time AAA had been in the nursing home. The relationship was one of mutual trust, respect, and cooperation.

Second Pharmaceutical Care Event:
Gather and Interpret Patient-Related Information

The pharmacist had a detailed database on AAA which was routinely updated when the pharmacist made monthly rounds to review pharmacotherapy and more frequently when AAA had changes in health status. This database consisted of objective patient-specific data, including relevant physical findings (e.g., vital signs and sensorium status), laboratory data (e.g., complete blood chemistries, serum digoxin levels, stool occult blood tests, urine and paired blood specimens for culture and sensitivity, and CBCs), radiographic studies (e.g., abdominal radiographs), other physical observations (e.g., appearance, smell and volume of urine, appearance and frequency of stools, body weight, and quality of the pulse upon palpation). The patient had been extensively interviewed prior to the latest episode for allergies to medications, and none were discovered.

The educational needs of the patient were assessed, and it was determined that all AAA needed was encouragement to comply with the medication administration nurses' efforts to administer oral medications to her as prescribed by her physician (i.e., AAA was incapable of self-medicating). The pharmacist documented all these findings in the pharmacy's record for AAA.

Third Pharmaceutical Care Event:
Determine Actual and Potential Drug-Related Problems

The pharmacist reviewed all medications the patient was taking from the perspective of actual or potential problems. All drug-related problems were ranked with regard to severity, risk to the patient, and patient-related factors as shown in Figure 21.2.

Fourth Pharmaceutical Care Event:
Establish a Desired Outcome for Each Drug-Related Problem

Each outcome should be as specific as possible and should clearly identify the desires of both the patient and the pharmacist. For AAA, the desired outcomes are shown in Figure 21.3.

Fifth Pharmaceutical Care Event:
Determine Possible Alternative Pharmacotherapies

Alternative(s)   **Resolve Drug-Related Problem(s)**
ACE inhibitors   Eliminate digoxin-, KCl-, and hydrochlorothiazide-related toxicity

Any alternative must be considered in the context of the benefits versus risks of changing pharmacotherapy.

FIGURE 21.2. Ranked Drug-Related Problems

Ranking	Drug	Problem (actual or potential)	Patient-Specific Factors for AAA: Age	Hydration	Renal	Infection
1	Cipro™	GU toxicity, crystalluria	++	++	++	+
2	Digoxin	Dig. toxicity	++	+	++	
3	KCl	Hyperkalemia, cardiac conduction block	+	+	++	
4	Hydrochloro-thiazide	Dehydration	+	++	++	
5	Ibuprofen	GU and GI toxicity		+	+	

Ranking: 1 = highest, 5 = lowest
+ = Significant factor
++ = Very significant factor

FIGURE 21.3. Desired Outcomes

Drug	Outcomes Desired by AAA and the Pharmacist
Ciprofloxacin	No crystalluria No interstitial nephritis No reduced renal function
Digoxin	No signs and symptoms of digoxin toxicity Return to/improvement in previous cardiovascular status Serum digoxin levels within normal limits for AAA
Potassium chloride	Serum potassium levels within normal limits No evidence of potassium-induced cardiac dysfunction
Hydrochlorothiazide	No edema No hypokalemia Satisfactory blood pressure for this patient, i.e., (120-140)/(75-90)
Ibuprofen	No GI toxicity, e.g., pain, nausea and vomiting, bleeding Satisfactory relief of joint pain

Sixth Pharmaceutical Care Event:
Select and Individualize the Preferred Pharmacotherapy(ies)

In this case, the alternative of changing to an ACE inhibitor may look attractive on the surface. However, AAA's congestive heart failure and hypertension had been treated to both the patient's and the physician's satisfaction for many years with digoxin, potassium supplementation, and hydrochlorothiazide. The patient responded well to a change in digoxin dose. In AAA's case, it made more sense to the team of caregivers (which included the pharmacist) to continue the original pharmacotherapy. Should AAA develop recalcitrant digoxin-related problems without the emergence of clinically significant atrial fibrillation, ACE inhibitors should be considered a viable alternative.

Seventh Pharmaceutical Care Event:
Develop the Plan to Monitor Drug Therapy

Additional patient-specific information or laboratory tests may be required to determine patient response to pharmacotherapy. Appropriate laboratory tests were ordered throughout AAA's nursing home stay (i.e., baseline data were available prior to AAA's acute episode, and new laboratory data were obtained during the acute episode).

Eighth Pharmaceutical Care Event:
Start the Drug Therapy and Begin Monitoring It

Communicating the pharmaceutical care plan to the patient and caregivers is essential if they are to accept and commit themselves to it. The pharmacotherapy plan was reviewed and agreed upon by the team of caregivers. AAA had informed the caregivers on numerous occasions that she liked her daily routines, including how and when she took her medications. AAA generally resisted changes to her routines, including pharmacotherapy, so it was agreed by all concerned that her pharmacotherapy would not be changed unless clinical circumstances required changes. It was also agreed that AAA would be advised of any necessary changes and would receive appropriate education about them. This approach left AAA feeling in control and having significant input into her care. This minimized her sense of helplessness in spite of AAA's realization that she was becoming increasingly dependent on others due to advanced longevity and slowly deteriorating condition.

Ninth Pharmaceutical Care Event:
Follow Up to Determine That the Desired Outcome(s)
Were Achieved

In this step, it is important for the pharmacist to determine whether the pharmaceutical care plan was successful and to identify any new drug-related problems. Following initiation of pharmacotherapeutic changes to address AAA's acute episode, appropriate laboratory and direct patient assessment data were obtained. The data supported the conclusion that the changes were achieving the desired outcomes (e.g., repeat urine culture, blood chemistries, serum digoxin and potassium levels, patient weight, and vital signs). The patient was kept informed of her improved condition. This improved her spirits and sense of security. This, in turn, contributed to a relatively rapid return to her baseline condition (i.e., reasonably healthy).

It should be emphasized that throughout all these pharmaceutical care processes and events, the pharmacist played an active role as an integral part of an interdisciplinary caregiving team. The pharmacist was consulted on numerous occasions and worked in concert with AAA's nurses on patient education issues as well as with AAA's physicians in selecting and implementing appropriate pharmacotherapy.

CONCLUSION

The entire spectrum of long-term care pharmacy practice will play an increasingly important role in our health care delivery system and, more specifically, in providing our sick and frail with alternatives to hospital-based services. Patients want to feel in charge of their lives, and if they are provided with alternatives to hospitalization (and the external controls typically exerted upon inpatients by hospital processes), patients may be afforded a sense of empowerment.[13] As our society ages, quality long-term care services will be critical to a larger segment of our population.[13] This will increase the demand for pharmaceutical care practice that integrates both the clinical and the drug distribution role of the pharmacist with the well-established models of medical practice. The ultimate beneficiaries of pharmacy's embracing and operationalizing this integrated model will be patients, caregivers, and pharmacy practitioners.

REFERENCES

1. Williams ME. The American Geriatric Society's complete guide to aging and health. New York: Harmony Books, 1995.

2. Rinehart J. Long-term care pharmacy services: a new dimension of pharmacy practice. Top Hosp Pharm Manage 1990;10(3):25-9.

3. Lamy PP. The needs of the elderly patient. In: Catania PN, Rosner MM, eds. Home care practice. Palo Alto, CA: Health Markets Research, 1994:53-66.

4. Menton KG, Hausner T. A multidimensional approach to case mix for home health services. Health Care Finan Rev 1986;7(4):33.

5. Catania PN. Introduction to health care. In: Home health care practice. 2nd ed. Palo Alto, CA: Health Markets Research, 1994:1-10.

6. Anon. ASHP guidelines on the pharmacist's role in home care. Am J Hosp Pharm 1993;50:1940-4.

7. Penna PP. Pharmaceutical care: pharmacy's mission for the 1990s. Am J Hosp Pharm 1990;47:543-9.

8. Hepler CD, Strand LM. Opportunities and responsibilities in pharmaceutical care. Am J Hosp Pharm 1990;47:533-43.

9. Strand LM, Cipolle RJ, Morley PC, et al. Levels of pharmaceutical care: a needs based approach. Am J Hosp Pharm 1991;48:547-50.

10. Strand LM, Cipolle RJ, Morley PC, et al. Drug related problems: their structure and function. Drug Intell Clin Pharm 1990;24:1093-7.

11. Smith WE, Benderev K. Levels of pharmaceutical care: a theoretical model. Am J Hosp Pharm 1991;48:540-6.

12. Brodie DC. Pharmaceutical education in perspective. Am J Hosp Pharm 1982;39:587-94.

13. Cleary A. The long view on long-term care. Hosp Health Networks 1995;69(6):61-4.

ADDITIONAL READINGS

1. Strand LM, Guerrero RM, Nickman NA et al. Integrated patient-specific model of pharmacy practice. Am J Hosp Pharm 1990;47:550-4.

2. Schraa CC, Small RE. Pharmaceutical care for home health care patients. In: Catania PN, Rosner MM, eds. Home health care practice. 2nd ed. Palo Alto, CA: Health Markets Research, 1994:89-96.

3. Sawyer WT, Eckel FM. Do we have to document pharmacotherapeutic interventions? Top Hosp Pharm Manage 1992;11(4):1-9.

4. Lepinski PW, Woller TW, Abramowitz PW. Implementation, justification, and expansion of ambulatory clinical pharmacy services. Top Hosp Pharm Manage 1992;11(4):86-92.

5. American Society of Hospital Pharmacists. ASHP statement on pharmaceutical care. Am J Hosp Pharm 1993;50:1720-3.

6. Mount JK. Contributions of the social sciences. In: Wertheimer AI, Smith MC, eds. Pharmacy practice: social and behavioral aspects. 3rd ed. Baltimore, MD: Williams & Wilkins, 1989:1-15.

7. Kozma CM. Outcomes research and pharmacy practice. Am Pharm 1995;NS35(7):35-41.

8. American Pharmaceutical Association. The third strategic planning conference for pharmacy practice. Am Pharm 1995;NS35(Suppl 7):1-51.

9. Kalies RF, Muhich M. Effectiveness of behavior modification medication monitoring system. Consultant Pharm 1990;5:531-4.

10. Laucka JG, Hoffman NB. Decreasing medication use in a nursing-home patient-care unit. Am J Hosp Pharm 1992;49:96-9.

11. Mahoney JM, Miller SW, Phillips CR. Impact of an adverse drug reaction reporting system in a geriatric health care center. Consultant Pharm 1991;6:907-14.

12. Lund MDE. The social organization of drug management in two nursing homes. University of Illinois Medical Center, 1979.

13. Cooper JW. Clinical outcomes research in pharmacy practice. Am Pharm 1993;NS33(Suppl 12):7-13.

14. Dole EJ. Beyond pharmaceutical care. Am J Hosp Pharm 1994;51:2183-4.

15. Campagna KD. Pharmacists' levels of performance in making drug therapy decisions. Am J Hosp Pharm 1995;52:640-5.

16. Grainger-Rouseau TJ, Segal R. Economic, clinical, and psychosocial outcomes of home infusion therapy: a review of published studies. Pharm Pract Manage Q 1995;15(1):57-65.

17. American Society of Hospital Pharmacists. ASHP guidelines for providing pediatric pharmaceutical services in organized health care systems. Am J Hosp Pharm 1994;51:1690-2.

18. American Society of Hospital Pharmacists. ASHP statement on the pharmacist's role in patient education programs. Am J Hosp Pharm 1991;48:1780.

19. American Society of Hospital Pharmacists. ASHP guidelines on pharmacist-conducted patient counseling. Am J Hosp Pharm 1993;50:505-6.

20. Joint Commission on the Accreditation of Healthcare Organizations. Accreditation manual for healthcare organizations. Oakbrook Terrace, IL: Joint Commission on the Accreditation for Healthcare Organizations, 1993.

21. American Society of Hospital Pharmacists. ASHP technical assistance bulletin on handling cytotoxic and hazardous drugs. Am J Hosp Pharm 1990;47:1033-49.

22. Schaffner A. Safety precautions in home chemotherapy. Am J Nurs 1984;84:346-7.

23. American Society of Hospital Pharmacists. ASHP statement on the pharmacist's role in infection control. Am J Hosp Pharm 1986;43:2006-8.

24. American Society of Hospital Pharmacists. ASHP guidelines on adverse drug reaction monitoring and reporting. Am J Hosp Pharm 1989;46:336-7.

25. American Society of Hospital Pharmacists. ASHP guidelines for the use of investigational drugs in organized health-care settings. Am J Hosp Pharm 1991;48:315-9.

Chapter 22

Hospitalized Patients

Andrew M. Peterson
Paul G. Pierpaoli

INTRODUCTION

Over the last five decades, the role of the hospital in America's health care system has been the "invisible hand" shaping patient, provider, payer, and policy-maker behaviors. In the recent wake of unprecedented market restructuring, government attempts at health care reform, and technological innovation, the very meaning of being hospitalized as a patient has been recast both in form and substance. Developing a conceptual model for providing pharmaceutical care to the hospitalized patient must reflect such changes if the profession is to fulfill its social mandate.

ACUTE HEALTH CARE SERVICES REVISITED

From their simple and uncomplicated origins in antiquity, hospitals have been regarded as bastions of benevolence. In fact, the root of the word hospital is the Latin *hospitalis*–referring to a guest. Since the inception of hospitals in the Western World (A.D. 1100), their core purpose has been charitable–literally serving as refuges for the needy, aged, and infirm. Some have suggested that during this period hospitals, most of which were owned by private non-profit organizations (usually religiously affiliated), fulfilled a public health service role by isolating the infirm from the rest of society. It was not until the early twentieth century that hospitals provided

559

anything but basic custodial care and comfort nursing care services to their patients.[1] The seeds for change were sown in the late nineteenth century with the great scientific advances in the germ theory and asepsis developed by Pasteur and Lister.[2] From this period on, scientific and technological advances have greatly shaped and influenced the leading edge of change in U.S. hospitals.

The Hospital: A Changing Milieu and Function Profile

Not unlike the other components of modern medicine, the hospital is essentially a product of the twentieth century, having been shaped by the social, cultural, and technological forces that have ultimately defined the nature and character of all the health professions.[3]

It is noteworthy that it was not until the turn of this century that mortality statistics decreased to less than 50% for the first time since their inception.[4] By the end of World War II, scientific and technological advances in medicine and their attendant costs, and resultant concentration in hospitals subsequently redefined the hospital as the "physician's workshop." By the 1970s, the explosion of diagnostic and treatment services, as well as medical practice specialization, had irrevocably converted the hospital into a community-wide resource for diagnosis and treatment. By the 1980s, the hospital had, in addition, assumed a prominent role as the focus for the entire span of health resources and services in health promotion, disease prevention, and rehabilitation. In short, the hospital had–for better or worse–become the dominant and central focus of America's health care delivery system.

With approximately 6,300 acute care hospitals in the U.S., there is ample evidence of continuing consolidation and an accelerating reduction in the number of acute care beds, while clinical case acuity continues to rise. This is speculated to be the result of a combination of excess acute care bed capacity, uncontrolled hospital costs, and unprecedented structural change in the health care sector. The rapid growth of managed care and private-sector employer and government initiatives has served as a major catalyst for such changes as well.

Economic competition is now a constant in the hospital's operating environment. This is a sharp contrast from the 1970s and 1980s, when hospitals competed for patients by attempting to improve patients' perception of quality and by expanding the range of services offered to attract physicians. Physicians, as a result of fee-for-service reimbursement arrangements, were provided incentives to admit more patients to the hospital. Hospitals subsequently provided the latest in technology, adequate and quality nursing services, and other professional and management support staff to meet these needs. Given a scenario that allowed for adequate reimbursement for virtually all patient services from both insurance carriers and other third-party payers, cost was of no significant concern to either patients or physicians.

Third-party payers, large employers, and managed care entities have revolutionized the concept of competition for hospitals as they negotiate with groups of physicians and hospitals for services, usually committing a predetermined volume of services and care for a guaranteed payment at a predetermined and discounted price. In short, hospitals now assume significant financial risk by competing on a newly defined economic playing field.[5]

Hospital Industry Consolidation and Integration

Consolidation and integration appear to be immutable trends in the acute care sector for the present and foreseeable future. Pressures from business coalitions organized to control and drive costs lower, competing hospitals, physician-owned organizations, and other large multihospital systems have forced many smaller hospitals to either close or be absorbed by larger hospitals as parts of multihospital systems. At least one prominent spokesman has predicted that only one-half of U.S. hospitals in existence in 1988 will be needed by the year 2010.[6]

Vertical and horizontal integration of different kinds of health care service units has also accompanied the recent hospital industry consolidation described above. A growing trend is for hospitals to integrate vertically (i.e., ambulatory care entities, long-term care, home care, alternate site care) into a comprehensive "seamless" system. Such systems are thought to be advantageous in that they provide buyers of health care services with a comprehensive and bundled approach to a complete continuum of services.

In horizontal integration of hospitals and health care delivery entities, hospitals and/or health care units such as those providing long-term care are legally and/or administratively combined to varying degrees to share common business and medical technology ends. Often these combinations include community hospitals integrating with academic medical centers to maximize the use of high-cost medical technology and increase access to capital markets while providing an expanded referral base for tertiary care. Such multihospital systems have become a mainstream phenomenon, with well over one-half of U.S. community hospitals belonging to such multihospital systems.

Horizontal and vertical consolidations have been viewed with varying advantages and disadvantages, depending on the evaluator's perspective (the physician, the hospital executive, the community at-large, or the patient). There does appear to be agreement, however, that such consolidations do offer the advantages of improved access to capital markets for facilities improvement and expansion as well as economies of scale for improvement of technical services. The ultimate net benefit of such trends in consolidation and integration are, as of this writing, not entirely clear or final by any means.[3]

In light of these changes as well as the emergence and growing influence of managed care, the patient referral base for hospitals has been changing with increasing speed. Managed care initiatives and their penetration into the hospital market have had a profound influence in decreasing patient admissions, length of stay, and patient days in hospitals. An overarching theme of any managed care initiative is to promote and maintain health while controlling access, quality, and cost of health care services—especially acute care hospital stays, which consume the largest share of health care expenditures.

At a heightened scale and pace, large multispecialty or single-specialty physician group practices in varying degrees of affiliation with hospitals and/or managed care entities (HMOs, PPOs, IPAs, and other networks) provide an ever-increasing base for patient referrals to hospitals. Although the hospital still has a significant role as the "physician's technological workshop," the era of the solo practitioner of medicine as the dominant source of patient referrals to the hospital has all but vanished—a sequelae of the continual aggregation of buyers and providers in America's health care markets over the last decade.

THE HOSPITALIZED PATIENT:
A CONCEPTUAL BASIS OF CARE

As is the case in any service organization, there are always at least two perspectives that are critical to an understanding of how best to meet the organization's mission (purpose)—namely, those of the staff and those of the customers. The sociology of care in the hospitalized patient is complex and interesting but well beyond the scope of this discussion. Nonetheless, it is important to bear in mind that in the hospitalized patient the sick role is greatly accentuated, with the patient having a major obligation to comply with the physicians's prescribed regimen. In short, the hospitalized patient is much more dependent on the physician and other hospital staff, to whom the physician has delegated authority, to assume a "caring" role. In large measure, the sick role that the patient assumes while in the hospital subordinates him to cooperation and compliance with general hospital routine and the sick role. What is described as "the one-on-one intimacy and permissiveness of the traditional patient-practitioner relationship present for the outpatient relationship is absent for the inpatient."[3]

A published personal account by a physician who was hospitalized presents an insightful and revealing commentary on the acute and unique nature of the sick role in the hospitalized patient:

> I had to be hospitalized, suddenly and urgently. In the space of only an hour or two, I went from . . . staff to inmate in a total institution. At one moment, I was a physician: elite, technically skilled, vested with authority, wielding power over others, affectively neutral. The next moment I was a patient: dependent, anxious, sanctioned in illness only if I was cooperative. A protected dependency and the promise of effective technical help were mine—if I accepted a considerable degree of psychological and social servitude.[3]

Once a patient is hospitalized, he or she is, in effect, a "guest" in a hotel (a hospital) which also brings to bear all of the cutting-edge knowledge, skill, and technology for the effective diagnosis and treatment of the illness as well as (both inside and outside the hospital) care. Literally dozens of groups are involved in the care of a hospitalized patient. Inside the hospital the patient may interact with a complete range of professionals, from physicians and nurses to pharmacists, physical therapists, social workers, rehabilitation specialists, technicians, and countless others. Moreover, housekeeping personnel, dietary workers, administrative personnel, and scores of other support staff provide a range of services available to a "guest" at a typical hotel. Hospitals are, in one sense, microcosms of daily life–highly differentiated and complex.

From an external perspective, the hospitalized patient is involved to varying extents with relatives, friends, and third-party payers and is directly and indirectly affected by voluntary accrediting agencies, state and federal regulations, and the community at large.

Figures 22.1 and 22.2 provide a systems view of the internal and external relationships that affect patients and hospitals and their mutually reinforcing nature. The authors suggest an analogy of concentric circles moving inwardly and outwardly as if a handful of pebbles were thrown into the water.[1]

Marshaling and integrating these elements of care is only part of the contemporary hospital's challenge in the 1990s. Providing quality care and value to patients and society is now a theme that reverberates through the entire health care community. It is clear that our health care efforts are now finally becoming reconciled with a public and professional mandate for assessing the quality of health care services in terms of "outcomes," outcomes being defined as the effects of care on the health status of patients and populations.[7]

Outcomes management has become a preoccupation of all health care organizations. Whether outcomes are described in clinical (morbidity and mortality), functional (quality of life), economic, or patient-satisfaction terms, they are the new yardstick for the professional community, the public, government, and payers.

Drug therapy for the hospitalized patient is now considered one of the major determinants of hospital care outcomes. On a conceptual level, optimal, safe, and efficient drug therapy outcomes in the hospital setting can be considered the ultimate output of a "drug use influence system" that has four major operant forces (inputs) affecting it.[8]

FIGURE 22.1

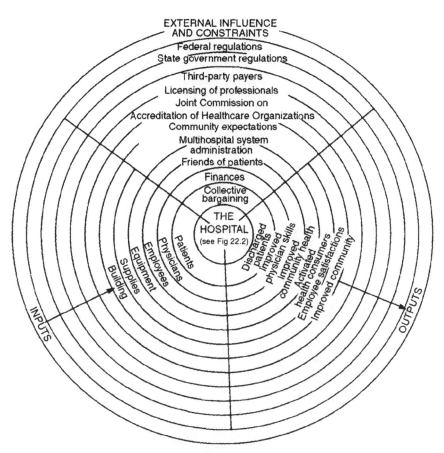

Source: Reprinted with permission from Schulz R, Johnson AC. "Organizational Arrangements of Health Services Delivery Systems," Chapter 8, *Management of Hospitals and Health Services—Strategies, Issues, and Performance, 3rd Edition*. St. Louis, MO: The C. V. Mosby Company, 1991, p. 131.

Providing pharmaceutical care within the context of this "influence system" represents the very essence of the hospital-based pharmacist's role. The pharmacy service "influence" input must be built on a conceptual foundation of providing pharmaceutical care to the hospitalized patient (Figure 22.3).

FIGURE 22.2

EXTERNAL SYSTEM (see Fig.22.1)

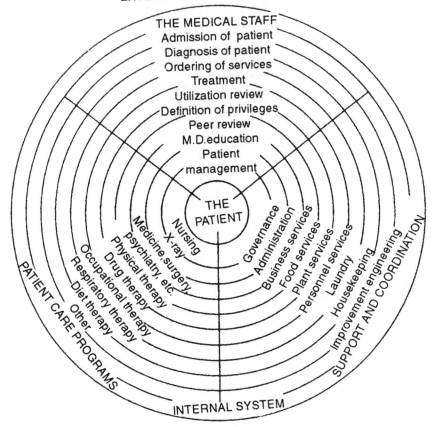

Source: Reprinted with permission from Schulz R, Johnson AC. "Organizational Arrangements of Health Services Delivery Systems," Chapter 8, *Management of Hospitals and Health Services–Strategies, Issues, and Performance, 3rd Edition.* St. Louis, MO: The C. V. Mosby Company, 1991, p. 132.

Framework for Delineating the Role of the Pharmacist

Pharmaceutical care is the "responsible provision of drug therapy for the purpose of achieving definite outcomes that improve a patient's quality of life."[9] This definition embraces the broader

FIGURE 22.3. Inputs to Optimal, Safe, and Efficient Drug Therapy Outcomes

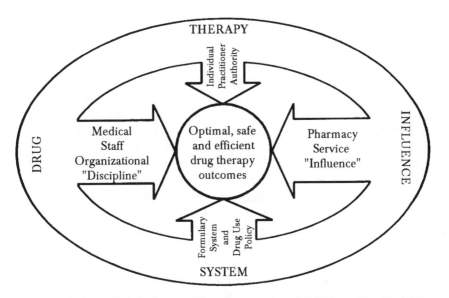

Source: Reproduced with permission from Pierpaoli PG, Lee SD. *Evolving Concepts in Hospital Pharmacy Management: Strategic Methods for Using the Drug Therapy Influence System in a Hospital.* Springfield, NY: Scientific Therapeutics Information, Inc., 1990:15.

societal mission of pharmacy to "help people make the best use of their medications"[10] regardless of practice setting. The philosophy of practice implicit in pharmaceutical care requires the pharmacist to assume personal and direct responsibility for the quality of care he or she provides. This framework–helping people through direct and personal responsibility–obligates the pharmacist to use his or her knowledge of drug therapy to ensure that the patient receives optimal drug therapy. Hepler and Strand further define the pharmacist's functions in pharmaceutical care as: (1) identifying potential and actual drug-related problems, (2) resolving actual drug-related problems, and (3) preventing drug-related problems.[9] Although by definition pharmaceutical care is provided directly to a patient by a pharmacist, the pharmacist must do so in collaboration with other health professionals. This integrated relationship is a key element of pharmaceutical care, especially in the complex world of today's hospitals.

Making the Transition from Outpatient to Inpatient

In the hospital environment, the patient too often remains a faceless entity, a medical record number or a chart outside a room. Strand and colleagues[11] suggest that pharmaceutical care is a component of pharmacy practice, residing at the level between the pharmacist and the patient or family (Figure 22.4). Medication histories are one mechanism for pharmacists to assist the patient in making the transition from home to hospital. While the purposes of a medication history are manyfold, often this is a first step to establishing the pharmacist-patient relationship.[12-14] Interviewing a patient and reviewing a patient's medical record are one means of providing pharmaceutical care. Although a medication history may provide important information about the patient and his or her drug therapy, time and manpower limitations may well preclude the pharmacist from interviewing each patient. Several authors have described specific criteria for screening those patients most likely to benefit from pharmacist-conducted medication histories.[15,16] Examples of these criteria are seen in Table 22.1. Other authors suggest the use of computer technology in obtaining medication histories.[17] Although this may help in collecting the necessary information, an in-person follow-up by the pharmacist is critical to establishing the pharmacist-patient relationship. Professional judgment is critical in deciding which patient or patients will benefit most from such pharmacist interaction.

Pharmacists caring for patients with complex medication histories or patients with limited recall capacity might well benefit from personally contacting the patient's own local or primary pharmacist. This professional-to-professional contact by the hospital pharmacist to the patient's primary pharmacist can lead to a clearer understanding of a patient and his or her medication regimen and assure a smoother transition from outpatient setting to inpatient setting.

In all instances of patient interaction, especially medication histories, the pharmacist should discuss the findings of the interview with the other health care professionals responsible for the patient's care. This includes, but is not limited to, the medical, nursing, and dietary staff caring for the patient. Practitioners should document

FIGURE 22.4. Patient-Specific Components of Pharmacy Practice

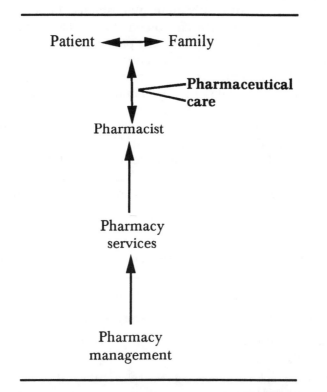

Source: Reprinted with permission from Strand LM, Cipolle RJ, Morley PC. The Pharmacist's Responsibilities, Current Concepts. *Pharmaceutical Care: An Introduction*: 12.

the medication history in the patient's permanent medical record to serve as a legal record for the interaction and as an educational reminder for the other health care professionals caring for the patient. Other authors advocate the documentation of all interventions performed by a pharmacist.[18]

The Hospital Stay

Whenever possible, it is important that the pharmacist-patient relationship established at the time of a medication history be maintained throughout the hospital stay. This requires continued contact with the patient and/or family. In many hospitals, if not most, it is not feasible for the same pharmacist to continue the care, and colleagues are relied upon to follow-up with the patient. This limitation can be effectively overcome by a system of documentation that is consistent and thorough. One framework, outlined by Strand and colleagues[19] is the Pharmacist Workup of Drug Therapy (PWDT). This system of documentation assists the pharmacist in resolving and preventing drug therapy related problems and documenting the decisions and interventions made by the pharmacist. The six steps of the PWDT are shown in Table 22.2. This process facilitates a step-by-step problem-solving method implicit in pharmaceutical care.

A key element of pharmaceutical care is the follow-up monitor-

TABLE 22.1. Criteria for Selecting Patients Requiring Medication Histories

1. Patient receiving more than five medications upon admission

2. Patient admitted for a suspected adverse drug reaction

3. Patient admitted for suspected drug noncompliance (including accidental overdose)

4. Patient is a high-risk group (e.g., transplant patient, diabetic patient, anticoagulated patient)

5. Patient requires assistance in identifying medications taken at home

TABLE 22.2. Six Steps in the Pharmacist's Workup of Drug Therapy (PWDT)

1. Identify issues associated with the patient's drug therapy

2. Identify patient-specific, drug-related problems

3. Describe desired therapeutic outcome

4. List all therapeutic alternatives to achieve desired outcome

5. Recommend individualized drug therapy

6. Establish a therapeutic drug monitoring plan

ing of the patient's drug therapy. As part of the shared responsibility, the pharmacist is obliged to assess whether the drug therapy regimen designed for the patient achieved the desired outcome through follow-up with the patient and the medical team. If the desired outcome is not achieved, alternative treatment must be considered and implemented and the follow-up process repeated as necessary. Using the PWDT method facilitates this dynamic process and enables the pharmacist to assume an active role as the pharmacotherapeutic "conscience" of the medical team in addition to the role of "patient advocate."

During the course of a patient's hospital stay, the pharmacist must be aware of his or her role as a general caregiver. Maintaining patient autonomy and dignity during hospitalization is a fundamental tenet of pharmaceutical care. Research suggests that coping mechanisms play an important role in recovering from illness.[20] In addition, a caregiver's response to a patient's coping mechanism can also influence the outcome. As an example, Dahlem and colleagues[21] studied the effects of PRN (as needed) medication use in "panicky" patients. Their results demonstrated that patients experiencing anxiety about their asthma requested and used more corticosteroids than other patients. Such findings suggest, at least in part, that a patient's use of medication is not only dependent on the underlying medical condition but also is a function of other sociobehavioral factors. Factors such as age, ethnicity, and gender have been explicitly identified as influencing factors affecting patient medication use behavior.[22-24] As a provider of pharmaceutical care, the pharmacist must understand patients in a broader contextual sense which may include these sociobehavioral factors.

Mumford and colleagues[25] reviewed evidence backing the role of psychosocial support in recovery from illness and decreased use of pain medications, tranquilizers, and hypnotics. This type of support can facilitate the healing process while the patient is hospitalized. Conversely, staff attitudes and perceptions can be counterproductive to the healing process. Perry and Wu[26] noted that patients in a general hospital received an ordered PRN hypnotic agent often not by their own request, but at the behest of the nurse. Furthermore, the researchers noted that patients on one medical team received more hypnotics than similar patients receiving care through another medical service. The authors evaluated the demographics and severity of illness of the patients and found no significant difference between the groups, leading to their conclusion that the observed disparity was due, in large part, to staff attitudes and not specific patient variables per se.

As noted earlier, hospitalized patients often receive medications on a PRN, or as needed, basis. Analgesics represent one of the most frequently prescribed and administered classes of PRN medications in the hospital setting. Managing pain in the hospitalized patient requires both a sound comprehension of the clinical pharmacology of analgesics and a clear understanding of the patient's perception of pain. Pharmacists caring for chronically ill patients not only find themselves assisting the patient in dealing with the physical component of pain but also are often confronted by the patient's psychological, spiritual, and social perceptions of pain medications. Age is often a major consideration in achieving pain relief. For example, the elderly are less likely to complain about pain and therefore request fewer analgesics to alleviate the pain.[22,27] This was also the case in pediatric practice where, primarily due to children's inability to express their suffering adequately, it was commonly believed that children could not feel pain. Clinicians now realize, however, that pediatric patients experience as much pain as adults, and due to the identified communication barrier, clinicians need special techniques to assess and understand the extent of children's pain.

An additional consideration in the case of the hospitalized patient with pain is the potential for suboptimal use of PRN analgesics. Studies have shown that physicians are reluctant to prescribe the full amount of a narcotic analgesic to patients due to inherent fears of overdose, addiction, and personal disbelief of the extent of the patient's true level of pain.[28] Nurses have been identified as having similar misconceptions regarding patients' need for pain medications.[29] Moreover, patients may be reluctant to request medication from the nurse out of concern for disrupting a "busy nurse" or appearing to be a "troublesome patient."[30] As a provider of pharmaceutical care, the pharmacist's role is to participate proactively in a multidisciplinary approach to pain management. An equally important obligation of the pharmacist is educating the medical and nursing staff about the need to use appropriate doses of analgesics and to understand the staff-patient relationship and its effect on their willingness to request analgesic medications.

Although not the primary impetus for doing so, initiating a patient self-medication program could often reduce or eliminate some of the barriers cited above.[31] Many hospitals allow patients to take medication, under supervision of the staff, as they would at home. Such self-medication programs have been shown to lessen patient dependence on professional staff and to provide a greater degree of patient autonomy for improved control over the patient's health. Of course, in such programs patient education on the proper use of medication occurs virtually throughout the hospital stay, not just at the time of discharge. Studies have shown that such programs, in select patient populations, may lead to patient empowerment and thereby make the transition from the hospital to home easier.[29]

Similarly, patient-controlled analgesia (PCA) has had a significant impact on obviating some of the limitations and factors mentioned above. While a full discussion of PCA is beyond the scope of this chapter, PCA therapy essentially allows the patient to receive, via a controlled infusion pump, a prescribed dose of an analgesic on demand. The patient autonomy and control inherent in PCA therapy greatly facilitate the pharmacist's provision of pharmaceutical care. Some studies have shown that this type of therapy improves a patient's quality of life through reduced pain[32] and reduced length of hospital stay.[33] There is conflicting evidence suggesting a wide variation in a patient's personal view of pain, which may contribute significantly to the perception of efficacy of this type of therapy.[34] Nonetheless, in almost all cases, patients are more satisfied with PCA than traditional analgesic administration, despite the failure to demonstrate a decrease in pain perception consistently.

In many institutions, pharmacists care directly for the patients receiving PCA therapy. This management includes evaluating the patient as a candidate for therapy, educating the patient on the use of the PCA pump, and converting the patient to a more suitable therapy upon discontinuation of PCA therapy and/or discharge. The number of pharmacist-initiated and -managed PCA programs is growing throughout the United States and will continue to grow as patients assume a greater responsibility for their own health care outcomes.

The aforementioned discussion of pain therapy should serve as an example of the dynamics implicit in a drug use influence system. Using this conceptual model, it becomes obvious that the individual nurse or physician practitioner, the hospital's drug use policies, and the level of intensity of pharmacy services all represent major inputs into the achievement of optimal drug therapy outcomes in pain management.

Life-Style Alterations for the Hospitalized Patient

As a "guest" in the hospital, the patient is the recipient of an entire spectrum of comfort services, such as housekeeping, laundry, and meal service, as well as professional nursing care. At the same time, however, patients frequently complain that life-style changes inherent in hospitalization affect their "comfort." These changes include, but are not limited to, noise in the hospital, difficulty in falling asleep, and a generalized increase in anxiety.[35-37] Although not fully elucidated, some of the reasons for these alterations in sleep habits and anxietal changes can be attributed to the environmental changes (noise, new bed, ambient temperature differences) as well as underlying medical conditions.

Life-style changes may also affect a patient's medication-taking behavior. For example, hypnotics are widely used in the hospital setting, even on patients who may never have received one previously. Transient insomnia, while a natural consequence of the minor situational stress induced by the transition to the hospital environment, may or may not warrant drug treatment. Careful evaluation is essential in choosing the appropriate treatment for the patient because use of sedative/hypnotic medications to maintain a comfortable life-style transition is not without its consequences. Virtually all sedative/hypnotics predictably alter the architecture of a patient's sleep pattern (e.g., changes in REM), but the unpredictable effect of disturbed sleep patterns varies from patient to patient. Most commonly seen is the phenomenon of diminished daytime performance (i.e., the hangover effect) resulting from long-acting sleep agents. Additional adverse effects include an increased risk of patient falls, occasionally leading to hip or other bone fractures. For example,

Sobel and McCart[38] studied the relationship between falls caused by nonenvironmental factors (e.g., decreased vision or muscular rigidity) and medication use. In the 45 patients who had fallen, diuretic and sedative/hypnotic agent use was significantly greater than in a matched control population who had not fallen. In addition, dizziness, confusion, and other central motor dysfunctions were observed more often in the group of patients who had fallen. Similarly, Trewin and colleagues[39] observed a correlation between specific hypnotic-benzodiazepines and falls and recorded a 7.5% incidence of fractures in the elderly patients who had fallen. In the context of pharmaceutical care, it is the pharmacist's responsibility to continually evaluate a patient's use of medications to assure that he or she is receiving them for the appropriate indication and at the appropriate dose.

Another life-style alteration particularly disturbing to patients in a hospital is a change in bowel habits. The sedentary nature of hospitalization, coupled with dietary changes to accommodate medical needs, may leave a patient feeling irregular. Moreover, medications are frequently cited as reasons for diarrhea, constipation, nausea/vomiting, and general stomach upset.[40-42] The pharmacist can assist the medical, nursing, and nutrition staff in understanding patient reactions to medications, can recommend changes in the use patterns of drugs, and can educate patients on how to prevent or resolve these iatrogenic effects. Timely education and proactive attention to these lifestyle changes may be conducive to allaying the anxiety commonly seen in patients. When providing pharmaceutical care to hospitalized patients, pharmacists must be ever vigilant of operant life-style changes affecting patients.

Patient and Family Education in the Hospital

Pharmaceutical care involves the process through which a pharmacist cooperates with a patient and other professionals in designing, implementing, and monitoring a therapeutic plan that will produce specific therapeutic outcomes for the patient.[9]

The collaborative nature of achieving a common therapeutic goal is explicitly stated in the definition of pharmaceutical care and serves as a foundation for multidisciplinary patient and family education.

In this vein, many hospitals have patient education committees designed to create and implement programs to inform patients on a variety of health-related issues. Many of these issues include the use of medications, both inside and outside the hospital. One such program, at Thomas Jefferson University Hospital in Philadelphia, is centered around the use of nicotine transdermal delivery systems (i.e., nicotine patches) to curtail the use of inhaled tobacco products. This smoking cessation program incorporates patient education material given to the patient the first time the patient receives a nicotine patch, follow-up education by the nurse or pharmacist, and a video program on the hospital's educational television channel. The information packet includes not only instruction on the use and disposal of the patch but also the names and numbers of participating pharmacies in the patient's home community who can help support the patient's smoking cessation efforts after discharge.

Diabetic teaching programs are not unlike smoking cessation programs, but they are more extensively and widely employed.[43] Such programs are usually designed for the newly diagnosed diabetic or the patient and family needing additional attention and education. The purpose of such programs is to educate the patient about the natural history of the disease, the proper technique for insulin storage and administration, and identifying and compensating for aberrations in blood glucose levels. Yarborough and Campbell[44] have suggested three levels of pharmacist-directed diabetes education. Level 1 (Survival Skills) provides basic education about the disease and gives the newly diagnosed diabetic the technical skills necessary to "survive" (e.g., how to draw up insulin). The next level, Home Management Skills, focuses on the behavioral aspects of the patient's self-management. Aspects routinely addressed at this level of education include dietary and exercise management as well as recognizing or anticipating changes in insulin requirements. The final and most comprehensive level (Level 3) includes patient comprehension of specific life-style changes aimed at minimizing the consequences of diabetes or related diseases (e.g., stop smoking, stress-coping programs, hypertension control).

Whether such programs are mediated solely through pharmacist involvement or represent a team effort, they are highly effective and provide a valuable service to the patient and family. Hospital pharmacists may incorporate Levels 1 and 2 into a predischarge patient and family education program. Because continued behavioral modification is often employed at Level 3, it may not be practical for all hospital pharmacists to maintain such a contiguous relationship with the patient. Most hospital pharmacists can assist in enrolling a patient in a program (e.g., stress management) and then transfer responsibility to a diabetes outpatient clinic pharmacist or the patient's primary pharmacist for follow-up care. To achieve a more "seamless" transition, some hospitals have developed their own outpatient diabetes clinics, which are staffed by a pharmacist, nurse, or other health care professional.[45] In any case, the pharmacist referring or enrolling a patient into a program should discuss the patient's level of understanding of the disease and its management with the health care professional accepting responsibility for the patient.

Anticoagulant teaching and monitoring[46] programs are yet another form of pharmacy-managed patient care services. These services, for inpatients and outpatients, include the management of patients on oral anticoagulants (e.g., warfarin). Ellis[47] and colleagues have demonstrated in a before-and-after trial that inpatient warfarin therapy managed by pharmacists resulted in improved warfarin dosing. Those patients managed by inpatient pharmacists and referred to an outpatient clinic were 5.4 times more likely to have a therapeutic prothrombin time (PT) at the initial follow-up clinic visit than those not managed by pharmacists. In addition, the research also showed an increased compliance with clinic visits among a subset of the patients studied. Moreover, there was an increase in referrals to the anticoagulation clinic. However, the study did not show an effect on key outcome indicators such as decreased hospital readmission rate related to toxicity or recurrence of thrombosis.

In contrast to disease-focused programs such as smoking cessation or diabetes education, anticoagulant monitoring programs are medication-focused programs. These examples of highly specialized approaches to pharmaceutical care provide a valuable service to patients by helping them make the best use of their medications. Other pharmacist-managed patient education initiatives include organ transplantation program, asthma clinics, and antihypertensive clinics.

Continuity of Care: The Nonhospitalized Patient

Given the newly defined economic imperatives for shortening a patient's length of hospital stay and transition to vertically integrated health care services, discharge planning and alternate site care referral have become dominant themes in hospitals. Discharge planning programs employ social workers and/or utilization case managers and nurses working with the medical team to facilitate a safe, efficient, and effective transition of patient care to the home or alternate subacute care setting. The pharmacist's role is pivotal in such programs.

Medication counseling, including appropriate drug use, potential drug-drug and drug-nutrient interactions, and anticipated adverse effects, is an essential component to discharge planning. It is noteworthy that patient noncompliance is responsible for nearly 5% of hospital admissions.[48] Equally important, patient drug education has been shown to improve patients' understanding of drug therapy regimens significantly and to promote compliance.[49,50] Fundamental to effective discharge counseling is the delivery of the necessary information in the context of language and meaning that is understandable by the patient. One study, conducted by Bourhis and colleagues,[51] demonstrated that nurses and physicians frequently talked to patients in medical language (ML) versus everyday language (EL). However, all three groups (physicians, nurses, and patients), felt that use of ML had frequently led to miscommunication. Conversely, patients felt that the use of EL did not lead to communication problems nearly as much. This divergence of communication styles is an important consideration for the pharmacist during any interaction with the patient. It assumes even more critical proportions at the actual time of patient discharge in light of any number of clinical and social interests faced by the patient and his family.

The importance of providing effective patient-specific written information cannot be understated.[52] When a patient is leaving the hospital, he or she is deluged with after-care instructions, including wound management, valet parking instructions, billing issues, etc. At these times, it is difficult to sustain the attention of a departing patient to ascertain if he or she has understood the medication use instructions delivered by the health care provider. In this vein, it is important to leave the patient with written information regarding the appropriate use, side effects, and administration of the agent for later referral.[53] Although not commonplace, a written summary of the medication regimen changes that occurred during the patient's hospitalization and recommendations to the primary pharmacist for continued follow-up and care should be considered a fundamental service found under the rubric of pharmaceutical care.

Pharmacist discharge planning also includes assisting the patient in locating sources for unique or hard-to-get medications after the patient leaves the hospital. Compounded medication formulas, unique to the hospital, should be shared with the patient's primary pharmacist or family before the patient attempts to fill the discharge prescription.

Home care and other forms of alternate site therapy are now key components in the health care system. Patients can be discharged from the hospital to a variety of medically supported, subacute settings: home with self-care, home with assisted care (commonly called "home health care"), nursing home, skilled nursing facility, or hospice. Each of these levels of intensity of care requires different degrees of discharge planning and coordination. The role of the pharmacist for each of these levels varies depending on the status of the patient, the destination, and the amount of coordination provided by other health care providers (nurses, social workers). In addition to assisting the patient in obtaining hard-to-get medications, home health care pharmacists play a key role in coordinating a patient's intravenous medication use outside the hospital. Many hospital pharmacies are becoming involved in home health care or home infusion services.[54]

CONCLUSION

The hospital is no longer the all-encompassing focal point of health care delivery in the United States. In fact, as we approach the beginning of the new millennium, there will be fewer hospitals with fewer beds and even fewer patients to fill those beds. This loss of centrality, however, does not diminish the need to provide responsible care to patients. There is every indication that drug therapy will continue to become more intensive and complex with advances in science and technology in the foreseeable future. Translating the philosophy of pharmaceutical care into practice through constantly identifying, resolving, and preventing drug-related problems is the essence of the pharmacist's existence. This includes assisting the hospitalized patient in the transition into the hospital, effectively using his or her medication while hospitalized, and then ultimately coordinating the transition to an outpatient or home setting.

REFERENCES

1. Schulz R, Johnson AC. Organizational arrangements of health services delivery systems. In: Management of hospitals and health services: strategic issues and performance. 3rd ed. St. Louis, MO: C. V. Mosby Company, 1990:129-41.

2. Rakich JS, Longest BB, Darr K. The health care delivery system. In: Managing health services organizations. 2nd ed. Philadelphia, PA: W. B. Saunders Company, 1985:27-79.

3. Wolinsky FD. Hospitals. In: The sociology of health principles, practitioners and issues. 2nd ed. Belmont, CA: Wadsworth Publishing Company, 1988:288-314.

4. Johnson EA, Johnson RL. Shifting roles in a competitive environment. In: Hospitals under fire–strategies for survival. Rockville, MD: Aspen Publishers Inc., 1986:11-2.

5. Goldsmith J. Half of today's hospitals won't be needed two decades from now. Am Hosp Assoc News 1988;24(August 15).

6. Schulz R, Johnson AC. Environmental pressures from cost controls and consolidation. In: Management of hospitals and health services: strategic issues and performance. 3rd ed. St. Louis, MO: C. V. Mosby Company, 1990:35-56.

7. Donabedian A. The quality of care–how can it be assessed? JAMA 1988;260:1743-8.

8. Pierpaoli PG, Lee SD. Evolving concepts in hospital pharmacy management–strategic methods for using the drug therapy influence system in a hospital. Springfield, NJ: Scientific Therapeutics Information Inc., 1990.

9. Hepler CD, Strand LM. Opportunities and responsibilities in pharmaceutical care. Am J Hosp Pharm 1990;47:550-4.

10. Zellmer WJ. Expressing the mission of pharmacy practice. Am J Hosp Pharm 1991;48:1195.

11. Strand LM, Cippole RJ, Morley PC, Perrier DG. Levels of pharmaceutical care: a needs-based approach. Am J Hosp Pharm 1991;48:547-50.

12. Badowski SA, Rosenbloom D, Dawson PH. Clinical importance of pharmacist obtained medication histories using a validated questionnaire. Am J Hosp Pharm 1984;41:731-2.

13. Wilson RS, Kabet HF. Pharmacist initiated patient drug histories. Am J Hosp Pharm 1971;28:49-53.

14. Trudeau T, Oleen M. How drug histories obtained by pharmacists help the hospitalized patient. Pharm Times 1978;44:65-7.

15. Frisk PA, Cooper JW, Campbell NA. Community-hospital pharmacist detection of drug related problems upon admission to small hospitals. Am J Hosp Pharm 1977;34:738-42.

16. Opdycke RA, Ascione FJ, Shimp LA, Rosen RI. A systematic approach to educating elderly about their medications. Patient Educ Counsel 1992;19:43-60.

17. Deleo JM, Pucino F, Calis KA, Crawford KW, Dorworth T, Gallelli JF. Patient-interactive computer system for obtaining medication histories. Am J Hosp Pharm 1993;50:2348-52.

18. Guerrero RM, Tyler LS, Nickman NA. Documenting the provision of pharmaceutical care. Top Hosp Pharm Manage 1992;(Jan):16-29.

19. Strand LM, Cippole RJ, Morley PC. Documenting the clinical pharmacist's activities: back to basics. Drug Intell Clin Pharm 1988;22:63-7.

20. Cohen F, Lazarus RS. Coping with the stresses of illness. In: Stone GC, Chen F, Adler NE, et al., eds. Health psychology–a handbook. San Francisco, CA: Jossey-Bass, 1979:217-54.

21. Dahlem NW, Kinsman RA. Panic-fear in asthma: requests for as-needed medications in relation to pulmonary function measurements. J Allergy Clin Immunol 1977;60:295-300.

22. Mechanic E, Angel R. Some factors associated with the report and evaluation of back pain. J Health Soc Behav 1987;28:131-9.

23. Zborowski M. Cultural components in response to pain. J Soc Issues 1952;8:16-30.

24. Verbrugge LM. Sex differences in legal drug use. J Soc Issues 1982;59-76.

25. Mumford E, Schlesinger H, Glass G. The effects of psychological intervention on recovery from surgery and heart attacks: an analysis of the literature. Am J Public Health 1982;72:141-51.

26. Perry SW, Wu A. Rationale for the use of hypnotic agents in a general hospital. Ann Intern Med 1984;100:441-6.

27. Ferrell BA, Ferrell BR. Assessment of pain in the elderly. Geriatr Med Today 1989;8:123-34.

28. Marks RM, Sachar EJ. Undertreatment of medical inpatients with narcotic analgesics. Ann Intern Med 1973;78:173.

29. Bird C, Hassall J. The benefits. In: Self-administration of drugs: guide to implementation. London: Scutari Press, 1993:56-67.

30. Ferrell BA, Ferrell BR, Osterweil D. Pain in the nursing home. J Am Geriatr Soc 1990;38:409-14.

31. Nelson WJ, Edwards SA, Roberts AW, Keller RJ. Comprehensive self-medication program for epileptic patients. Am J Hosp Pharm 1978;35:798-801.

32. Hecker BR, Albert L. Patient-controlled analgesia: a randomized, prospective comparison between two commercially available PCA pumps and conventional analgesic therapy for postoperative pain. Pain 1988;35:115-20.

33. Wasylak TJ, Abbott FV, English MJM, Jeans ME. Reduction of postoperative morbidity following patient-controlled morphine sulfate. Can J Anaesth 1990;37:726-31.

34. Knudsen WP, Boettcher R, Vollmer WM, Griggs DK. A comparison of patient-controlled and intramuscular morphine in patients after abdominal surgery. Hosp Pharm 1993;28:117+.

35. Fife D, Rappaport E. Noise and hospital stay. Am J Public Health 1976;66:680-1.

36. Brumet GW. Pandemonium in the modern hospital. N Engl J Med 1993;328:433-7.

37. Parker WA. Effects of hospitalization on patient use of hypnotics. Am J Hosp Pharm 1983;40:446-7.

38. Sobel KB, McCart GM. Drug use and accidental falls in an intermediate care facility. Drug Intell Clin Pharm 1983;17:539-42.

39. Trewin VF, Lawrence CJ, Veitch GBA. An investigation of the association of benzodiazepines and other hypnotics with the incidence of falls in the elderly. J Clin Pharm Ther 1992;17:129-33.

40. Rowland MA. When drug therapy causes diarrhea. RN 1989;52:32-5.

41. Lisi DM. Drug-induced constipation. Arch Intern Med 1994;154:461-8.

42. Bramble MG, Record CO. Drug-induced gastrointestinal disease. Drugs 1978;15:451-63.

43. Huff PS, Ives TJ, Almond SN, Griffin NW. Pharmacist managed diabetes education service. Am J Hosp Pharm 1983;40:991-4.

44. Yarborough C, Campbell RK. Developing a diabetes program for your pharmacy. In: Campbell RK, ed. Pharmaceutical services for patients with diabetes: a 4-part, 6 hour continuing education self-study program for pharmacists. Indianapolis, IN: Eli Lilly and Co., 1989:53-60.

45. Kanitz J, Birken B, Ward V. Pharmacist involvement in a diabetic education centre. Can J Hosp Pharm 1982;35:114-5.

46. Scalley RD, Kearney E, Jakobs E. Interdisciplinary warfarin education program. Am J Hosp Pharm 1979;36:219-20.

47. Ellis RF, Stephens MA, Sharp GB. Evaluation of a pharmacy-managed warfarin monitoring service to coordinate inpatient and outpatient therapy. Am J Hosp Pharm 1992;49:387-94.

48. Sullivan SD, Kreling DH, Hazlet TK. Noncompliance with medication regimens and subsequent hospitalizations: a literature analysis and cost of hospitalization estimate. J Res Pharm Econ 1990;2(2):19-33.

49. Baker DM. A study contrasting different modalities of medication discharge counseling. Hosp Pharm 1984;19:545-54.

50. Hawe P, Higgins G. Can medication education improve the drug compliance of the elderly? Evaluation of an in hospital program. Patient Educ Counsel 1990;16:151-60.

51. Bourhis R, Roth S, MacQueen G. Communication in the hospital setting: a survey of medical and everyday language use amongst patients, nurses and doctors. Soc Sci Med 1989;28:339-46.

52. Kessler DA. Communicating with patients about their medications. N Engl J Med 1991;325:1650-2.

53. McGinty MK, Chase SL, Mercer ME. Pharmacy-nursing discharge counseling program for cardiac patients. Am J Hosp Pharm 1988;45:1545-8.

54. Rich DS. Home care standards for hospitals. Hosp Pharm 1993;28:85-6.

ADDITIONAL READINGS

1. Bird C, Hassall J. Self-administration of drugs: guide to implementation. London: Scutari Press, 1993.

2. Bourhis R, Roth S, MacQueen G. Communication in the hospital setting: a survey of medical and everyday language use amongst patients, nurses and doctors. Soc Sci Med 1989;28:339-46.

3. Brumet GW. Pandemonium in the modern hospital. N Engl J Med 1993;328:433-7.

4. Donabedian A. The quality of care–how can it be assessed? JAMA 1988;260:1743-8.

5. Ellis RF, Stephens MA, Sharp GB. Evaluation of a pharmacy-managed warfarin monitoring service to coordinate inpatient and outpatient therapy. Am J Hosp Pharm 1992;49:387-94.

6. Fife D, Rappaport E. Noise and hospital stay. Am J Public Health 1976;66:680-1.

7. Guerrero RM, Tyler LS, Nickman NA. Documenting the provision of pharmaceutical care. Top Hosp Pharm Manage 1992;(Jan):16-29.

8. Hawe P, Higgins G. Can medication education improve the drug compliance of the elderly? Evaluation of an in hospital program. Patient Educ Counsel 1990;16:151-60.

9. Hepler CD, Strand LM. Opportunities and responsibilities in pharmaceutical care. Am J Hosp Pharm 1990;47:550-4.

10. Kessler DA. Communicating with patients about their medications. N Engl J Med 1991;325:1650-2.

11. Marks RM, Sachar EJ. Undertreatment of medical inpatients with narcotic analgesics. Ann Intern Med 1973;78:173.

12. McGinty MK, Chase SL, Mercer ME. Pharmacy-nursing discharge counseling program for cardiac patients. Am J Hosp Pharm 1988;45:1545-8.

13. Opdycke RA, Ascione FJ, Shimp LA, Rosen RI. A systematic approach to educating elderly about their medications. Patient Educ Counsel 1992;19:43-60.

14. Parker WA. Effects of hospitalization on patient use of hypnotics. Am J Hosp Pharm 1983;40:446-7.

15. Perry SW, Wu A. Rationale for the use of hypnotic agents in a general hospital. Ann Intern Med 1984;100:441-6.

16. Pierpaoli PG, Lee SD. Evolving concepts in hospital pharmacy management–strategic methods for using the drug therapy influence system in a hospital. Springfield, NJ: Scientific Therapeutics Information Inc., 1990.

17. Strand LM, Cippole RJ, Morley PC, Perrier DG. Levels of pharmaceutical care: a needs-based approach. Am J Hosp Pharm 1991;48:547-50.

18. Strand LM, Cippole RJ, Morley PC. Documenting the clinical pharmacist's activities: back to basics. Drug Intell Clin Pharm 1988;22:63-7.

19. Sullivan SD, Kreling DH, Hazlet TK. Noncompliance with medication regimens and subsequent hospitalizations: a literature analysis and cost of hospitalization estimate. J Res Pharm Econ 1990;2(2):19-33.

20. Verbrugge LM. Sex differences in legal drug use. J Soc Issues 1982;59-76.

21. Wolinsky FD. Hospitals. In: The sociology of health principles, practitioners and issues. 2nd ed. Belmont, CA: Wadsworth Publishing Company, 1988:288-314.

22. Zborowski M. Cultural components in response to pain. J Soc Issues 1952;8:16-23.

Chapter 23

Pharmaceutical Care
of Terminally Ill Patients

Arthur G. Lipman
Joni I. Berry

INTRODUCTION

Until just a few years ago, Americans commonly whispered the word "cancer" and the names of other fatal diseases when those dread diseases affected family members or friends. Families rarely discussed death and dying until forced to do so by tragic circumstances, and even then, euphemisms and avoidance language were more common than open discussions of these normal life-cycle events. Health professionals often participated in this conspiracy of silence by honoring a family member's wish not to tell a loved one that his or her prognosis was terminal. As a result, most individuals and families were ill prepared to deal with this normal but extremely difficult life event.

Fortunately, this attitude has now changed for many people in American society. In the 1970s, the modern hospice movement came to America.[1] This movement introduced the concept of palliative care to many health professionals. Palliative care is the provision of symptom control when cure is no longer a reasonable expectation. When health professionals enter their profession, they commonly take an oath. Physicians recite the Hippocratic Oath, nurses the Florence Nightingale Pledge, and pharmacists the Apothecary Oath. None of these pledges suggests that our efforts always lead to cure. That is not always possible. But we must always care for our patients. There is never a situation in which there is "nothing we can do." When cure is no longer possible, we have a moral and professional obligation to continue to provide care–in this case, palliative care.

The dramatic improvements in drug therapy and medical technology that have occurred in recent years have led to survival following many previously fatal traumatic events, infectious diseases, myocardial infarctions, and strokes. As a result, people live longer. But because of that longevity, more of us will experience cancer or other degenerative diseases that lead to end of life. For increasing numbers of people, life will end in a gradual process that will cause pain and other symptoms for weeks or months. We have the knowledge to manage these symptoms effectively in most cases. Unfortunately, because of deeply ingrained misconceptions in the use of analgesics and other symptom control measures, the symptoms often are not managed optimally, resulting in avoidable suffering.

Pharmaceutical expertise in drug therapy, dosage form compounding, drug information, and patient counseling have become integral elements in the care of terminally ill patients. Palliative care can be professionally challenging and rewarding. Because pharmacists are the health professionals who are most accessible and most often asked for advice by the general public, we are in an excellent position to participate in the palliative care and to make referrals to other care providers. It is therefore important that pharmacists be aware of the agencies and services in their communities that provide terminal care to patients and their families.

Palliative care is now recognized as a medical specialty in England, Canada, and several other countries through the establishment of professorships of palliative medicine. In the United States and Canada, some medical centers have established divisions of palliative medicine, primarily in conjunction with oncology (cancer) services. More important, physicians, nurses, pharmacists, and the general public are increasingly accepting palliative care as an appropriate alternative to aggressively seeking cure when the disease is advanced and irreversible. Palliative care is provided in hospitals, in long-term care facilities, and in patients' homes. By far the most common providers of palliative care are hospice programs, which provide care in conjunction with the patients' personal physicians, families, and other caregivers.

HOSPICE CARE

The hospice movement has formalized palliative care within American society. Hospice is a program of care for patients with advanced, irreversible disease and a life expectancy that is more reasonably measured in weeks or months than in years.[2] The term hospice is derived from a medieval French word used to describe resting places for Crusaders on their journeys to and from Europe and the Holy Land. The term was adopted at the end of the last century by a Catholic order which developed facilities for physical and spiritual care of dying patients. In the 1950s, St. Joseph's Hospice in the East End of London, England, began to incorporate modern psychosocial and medical modalities into that physical and spiritual care. The first truly modern hospice, St. Christopher's Hospice in southeast London (Sydenham), England, opened in 1967. Since then, dozens of additional hospice facilities have been established in the United Kingdom.

In the late 1960s, as a part of a research project on symptom control in advanced breast cancer at Yale University, an informal group of health professionals met regularly to discuss the study patients' progress. The interdisciplinary discussion group, which included nurses, physicians, and pharmacists, evolved into the first hospice in the United States. That program was formally established in the early 1970s as Hospice, Inc. (now the Connecticut Hospice) and received support from the National Cancer Institute (of the National Institutes of Health) as a hospice demonstration project from 1974 through 1977. Hospice care has evolved from that single program 25 years ago to over 2,000 hospice programs throughout the United States today.

Hospice care has grown so rapidly because of real patient needs and social phenomena. In the past century, many important events in people's lives have become institutionalized. In the past, births and deaths commonly occurred in the home, where an extended kinship group encompassing several generations was available for support. With mobility as the norm in our society, more often than not only the nuclear family members—or possibly friends—are available for support. More sophisticated medical technology also has favored institutionalization of care to provide efficiency and increased access to new technology. As a result, most patients with advanced disease spend their last days in hospitals.

St. Christopher's Hospice was developed as an inpatient facility for patients with a life expectancy of only a few months who required intensive symptom control. Many patients who were admitted for terminal care were subsequently discharged home in relative comfort for days, weeks, or months before they were readmitted for terminal care. In that original model, home care was a relatively small part of the program. In this country, hospice care started as a home care program, and most hospice care is still provided in the patients' homes. This is because home is the most natural and comfortable setting for most patients and their caregivers.

In recent years, many people have concluded that high-technology interventions often are not indicated or desired in the last days of life, especially when cure is not a reasonable expectation. Patients whose lives are coming to an end nearly always have unfinished business. There are goodbyes to be said, plans to be made, and other issues to be brought to closure. The home is a far more comfortable and appropriate environment for this important work than an institutional setting.

Concurrent with this recognition has been the discovery that we can usually control pain and other symptoms associated with advanced disease with medications administered by oral, rectal, and other noninvasive routes in combination with excellent psychosocial support. Hospitals and skilled nursing facilities are needed for the intermittent or permanent care of some patients. But the vast majority of patients can spend their last weeks at home in reasonable comfort if they receive excellent palliative care.

Hospices do just that. Today, the majority of American hospices are independent home care programs or services of comprehensive home health agencies. Other hospices are affiliated with hospitals or skilled nursing facilities. Most hospices that do not have their own inpatient facilities have contractual agreements with inpatient providers to be used as necessary for short symptom control or terminal admissions. A relatively small number of inpatient hospices exist either as stand-alone facilities or as dedicated parts of hospitals or skilled nursing facilities. In those cases, home care programs are nearly always offered as well.

Just a few years ago, hospice was seen as a small, alternative type of care for terminally ill patients. Hospitalization and the use of technology such as blood transfusions, intravenous fluids, and suction machines were shunned by many early hospice caregivers as inappropriate and interfering with the patient's natural and comfortable demise. Seldom was hospice even mentioned in health professional education and training. It was seen by some as questionable alternative care. However, hospice care has evolved greatly in the past two decades. Unnecessary interventions are still avoided due to their disruption of patients' lives. But interventions that improve the patients' quality of life are employed regularly.

The focus of hospice care is control of physical and psychological symptoms to allow patients to live until they die. Hospice care is not intended to extend life, per se, although some patients do seem to live longer than expected once their symptoms are controlled. Hospice care focuses on improving the quality of remaining life. Today, the philosophy and technology espoused in hospice care is being increasingly integrated into mainstream health care education and practice.

Terminally ill patients often experience pain and suffering. Saunders described this as "total" pain consisting of physical, emotional, spiritual, and financial pain.[3] Foley has added to that definition the aspect of existential suffering.[4] Both the lay press and the professional literature have been replete with discussions of assisted suicide and euthanasia in the past few years. Ineffective symptom control may be a factor that leads many patients to take such drastic measures. We now have the knowledge, drugs, and technology to make the vast majority of terminally ill patients reasonably comfortable without excessive sedation. It is tragic that only about two-thirds of the patients who can achieve good symptom control receive the care needed for such control. Pharmacists can play an important role by advising patients of the types of care that are available and by participating in the provision of that care.

The Medicare Hospice Benefit allows Medicare beneficiaries to receive total care from a Medicare-certified hospice program at no charge to the patient. There are over 1,500 Medicare-certified hospices throughout the United States today. All of those programs must have formalized pharmaceutical services through an in-house pharmacy or, more commonly, through contracts with community pharmacies.

THE HOSPICE TEAM

Hospice care is truly interdisciplinary. This is different from multidisciplinary care, in which the patient is seen successively by a range of health care providers who refer the patient to others after their own interventions fail.[5] The lead person on the hospice team is usually the home health nurse who visits the patient regularly and consults with the other team members in planning and implementing care. Hospice social workers also work regularly with the patients, and other team members also interact as needed. The patient's personal physician normally continues to direct the care, with the hospice medical director serving as a consultant.

The focus of hospice care is the patient *and* the family. The term "family" is used in hospice care to mean those persons who are deeply involved in the patient's care and who may need support during the patient's illness and during their own grieving or bereavement for that patient. Family members of terminally ill patients need emotional support and need to learn how to care effectively for the patient's physical and emotional needs. A primary caregiver, most commonly the spouse, child, parent, or close friend of the patient, is supported by hospice personnel. Hospice programs involve the patient's support group in the patient's care as much as possible because such involvement leads to closer relationships and easier grieving after the patient dies.

To provide this support, an interdisciplinary team is needed. The typical hospice team consists of nurses, physicians, pharmacists, social workers, chaplains (clergy who attend to patients' spiritual needs in a nondenominational manner), and volunteers. These core team members are supported by a range of other health care and psychosocial support providers as illustrated in Figure 23.1. Note that pharmacists often participate on both direct patient care and support levels as discussed below.

THE PHARMACIST'S RESPONSIBILITY TO TERMINALLY ILL PATIENTS

Although many pharmacists are now working directly with hospice programs, all pharmacists who have any direct or indirect

FIGURE 23.1. The Hospice Interdisciplinary Team

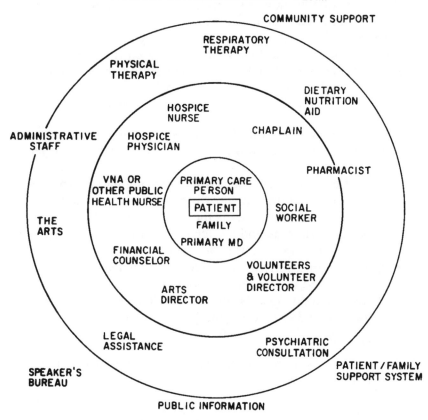

HOSPICE INTERDISCIPLINARY TEAM

The patient, primary caregiver and family are the focus of the hospice team's efforts in collaboration with the patient's primary physician. The core team is represented by the next circle away from the center. The support team is indicated by the outer circle. Pharmacists serve both on the core team by providing direct pharmaceutical care to the patient and family, and on the support level by providing professional and public education on drug therapy in the care of terminally ill patients.

contact with terminally ill patients or their caregivers have important professional responsibilities toward that patient and family system. It is crucial that the pharmacist be aware of the sensitive nature of this time in the patient/family's life and that the pharmacist facilitate—not impede—the patient care process.

Barriers to the effective management of symptoms in the terminal phase of a patient's illness have been identified by the World Health Organization (WHO) and various state Cancer Pain Initiatives. These include but are not limited to:[6]

- Lack of education of health professionals on effective modalities for treating the common symptoms found in this patient population.
- Fear of controlled substance (narcotic) regulations (both state and federal) that often do not differentiate legitimate from illegitimate use of opioids and other agents for symptom control.
- Lack of appropriate and sufficient information in well-known professional publications about symptomatic therapy.
- Promulgation of myths or misconceptions among health professionals and the lay public about the (unfounded) possibility of addiction to opioids and unfounded fears of respiratory depression and sedation in this patient population.

To remove these barriers, all pharmacists who provide care to terminally ill patients must:

- Understand the requirements of state and federal laws that pertain to the use of controlled substances and use this knowledge to contribute to—not impede—patient care.
- Be aware of the many symptoms that are experienced by terminally ill patients and understand the multifaceted treatment modalities that may be necessary to manage those symptoms.
- Ensure the availability of opioids and other medications needed in the treatment of these symptoms.
- Continually counsel and reassure patients and caregivers about the availability of effective treatment for distressing symptoms

- Understand the common side effects of these medications, such as constipation, nausea, and vomiting, and the appropriate preventative measures to decrease the occurrence and severity of these side effects.
- Monitor each patient's response to treatment and provide the health care team with information on therapeutic and adverse outcomes of drug therapy.

HOSPICE PHARMACEUTICAL CARE

Hospice pharmaceutical care (HPC) encompasses the delivery of administrative, clinical, educational, and dispensing services to a hospice by a pharmacist to improve patient drug therapy outcomes and to control drug therapy costs. Hospice is unique among health care systems in that many hospices' pharmacists have always participated as full members of the interdisciplinary team. This has allowed pharmacists to develop and deliver a more complete range of services to the hospice organization. Important also is the fact that the development of pharmaceutical services in hospices and the benefits of HPC to hospice patients have been documented.

In 1975, Lipman published the first formal paper on pharmaceutical care of terminally ill patients in the American pharmacy literature.[7] That paper proposed a different philosophical approach to care of patients with advanced, irreversible disease and described novel drug therapy for symptom control. In 1981, Berry and colleagues published the first description of pharmaceutical services in American hospices.[8] In 1993, Berry and Arter published the results of a national survey on pharmaceutical care to hospice patients.[9] The numbers in all of these surveys were small, but they do document changes over more than a decade. The differences in pharmacists' activities among the three surveys are illustrated in Table 23.1. Marked increases in clinical and direct patient care services have occurred.

The 1993 survey provides the broadest description of pharmaceutical services in hospice care. Respondents to that survey reported 1-15 years of experience in hospice pharmaceutical practice. About one-half were paid for their services, one-third were

TABLE 23.1. Changes in Pharmaceutical Services to Hospices from 1979 to 1991

Pharmacist activities	1979[a] n = 37		1981[b] n = 30		1991[a] n = 52	
	#	%	#	%	#	%
Member of interdisciplinary health care team	17	46	25	83	39	75
Monitors patients' drug regimen	–	–	20	67	40	77
Provides drug information to hospice staff	35	95	29	97	49	94
Provides drug information to patient/family	19	51	16	53	39	75
Makes home visits in order to provide recommendations, i.e., symptom management	2	5	5	17	19	39
Provides inservice education to hospice staff	26	70	–	–	41	78

a: n = all pharmacists surveyed
b: n = those pharmacists with an expanded role with the hospice termed "Hospice Consultant Pharmacists" or HCPs

Source: Reprinted with permission from Arter SG, Berry JI. The provision of pharmaceutical care to hospice patients: results of the national hospice pharmacist survey. *J Pharm Care Pain Sympt Control,* 1993;1(1):25-42.

volunteers, and the balance were paid for part of their time. Nearly one-half of the pharmaceutical service payments were based on dispensing fees, but 30% were explicitly for cognitive services. The pharmacists practiced in hospices that were fairly evenly divided among home care hospice programs, free-standing hospice facilities, hospital-based programs, and home health agencies. A few were in nursing home affiliated hospice programs. Ninety percent of the hospices with which the responding pharmacists were affiliated were Medicare certified.

Over a dozen different pharmaceutical services were provided by most or many of the respondents. The percentages of various services provided are listed in Table 23.2. Technology-based services such as intravenous infusion services were provided by two-thirds of the respondents. Those services are listed in Table 23.3. The components of HPC are listed in Table 23.4. As in any setting, the specific services provided depend upon the needs of the individual hospice and the expertise of the pharmacist providing the services. In some hospices, compounding, dispensing services, and delivery

TABLE 23.2. Pharmacy Services Provided to Hospices

Service	#	%
Dispensing of medications	38	73
Participating member of health care team	39	75
Attending regular team meetings	31	60
Monitoring patient drug regimens	40	77
Providing drug information to staff	49	94
Providing drug information to patients and families	39	75
Making home visits/providing recommendations for symptom management	19	39
Providing hospital/nursing home recommendations for symptom management	22	42
Maintaining medication profiles	39	75
Education of hospice staff on symptom management	41	79
Education of other health professionals on symptom management	33	63
Being available by phone or beeper during normal working hours for consultation	42	81
Participation in various other hospice activities such as member of the Board of Directors, member of the Quality Assurance Committee and the professional advisory committee.	19	37

Source: Reprinted with permission from Arter SG, Berry JI. The provision of pharmaceutical care to hospice patients: results of the national hospice pharmacist survey. *J Pharm Care Pain Sympt Control,* 1993;1(1):25-42.

TABLE 23.3. Responses to the 1993 National Hospice Pharmacist Survey

Do you provide high-technology services?[b]

	#	%
Yes	26	66.7
No	13	33.3
Total responses	39	100

What services are provided?

	#	%
TPN	20	63
IV Hydration	26	81
Epidural catheter	23	72
Chemotherapy	17	53
IV Opioids	30	94
Blood transfusions	11	34
Other[c]	3	9
Total responses	32	

Do you plan to initiate high-technology services?

	#	%
Yes	2	18
No	9	82
Total responses	11	

a = Percentages calculated on total responses to each question.
b = More than one response possible to each question.
c = Other high-technology: durable medical equipment, suction, enteral feedings, tracheotomy care, oxygen.

Source: Reprinted with permission from Arter SG, Berry JI. The provision of pharmaceutical care to hospice patients: results of the national hospice pharmacist survey. *J Pharm Care Pain Sympt Control*, 1993;1(1):25-42.

TABLE 23.4. Components of Hospice Pharmaceutical Care

Administrative

• managing in-house pharmacy (if applicable)
• negotiating contracts with provider pharmacies
• reviewing state and federal laws and regulations as they relate to the provision of hospice pharmaceutical care to hospice patients
• developing drug-related policies and procedures
• participating in quality assurance activities including drug use evaluations, cost avoidance and cost effectiveness studies
• procuring medications for indigent patients through pharmaceutical industry indigent patient programs
• managing the hospice formulary

Clinical

• participate in hospice interdisciplinary team rounds
• perform drug regimen reviews
• provide pain and symptom management consultations
• prepare routine admission orders
• prepare drug use protocols
• be available by pager or telephone for consultation

Educational

• provide staff education in drug therapy for symptom control and other indications
• provide education to patients and their families on medication use
• provide physician education to hospice patients' primary physicians
• provide public education on drug use in terminal care

Provision of Drugs

• dispense medications
• deliver medications to patients' homes
• compound needed dosage forms
• provide home infusion service
• maintain patient medication profiles

of medications are the responsibility of a community pharmacist, while other administrative, clinical, and educational services are the responsibility of a consultant pharmacist.

Several models of HPC have been developed to deliver services to hospice programs. The model used by a particular hospice will depend upon the needs and resources of that program and community. HPC may be:

- Community pharmacy based
- Hospital based
- Independent consultant based
- Academically based
- Health maintenance organization (HMO) based
- Hospice pharmacy based

In all of these models, one or more interested pharmacists contract with the hospice to provide care. Recent experience has shown that the types of services provided are similar despite the type of pharmacy practice through which the services are obtained.

HOSPICE PHARMACIST JOB DESCRIPTION

Pharmacist Stephen G. Arter practices as a hospice pharmacist at Parkview Memorial Hospital in Fort Wayne, Indiana. Arter, who has long been a leader in the National Hospice Organization, developed the following job description for pharmacists practicing in hospice care over a decade ago when he was one of only a few pharmacists nationwide providing comprehensive pharmaceutical services to a hospice program. Today, many pharmacists provide some or all of the functions in this description, and additional opportunities for hospice pharmacy practice increase regularly as new hospice programs open and present programs expand.

The hospice pharmacist will be appointed to the hospice care team as an equal partner with the various other disciplines. The pharmacist will attend all hospice team meetings to advise the team on matters concerning pharmacy. Educating the hospice team and hospice patients on matters of pharmacology and compliance with dosage regimens is a function of the hospice pharmacist. Also, the procuring of drugs, compounding of prescriptions and maintaining of legal aspects of pharmacy are functions of the hospice pharmacist. The pharmacist will communicate with the patient, either through the team or in person, concerning compliance with the prescribed medications regimen. The care giver to the patient, through instructions by the pharmacist, must be made to understand and follow the directions provided with the medication.

The pharmacist will advise the team concerning the actions of drugs in the body including warnings, contraindications, side effects and incompatibilities. The pharmacist will communicate with the patient, either through the team or in person, concerning compliance with the prescribed medication regimen. The care giver to the patient, through instructions by the pharmacist, must be made to understand and follow the directions provided with the medication. When indicated, the pharmacist will visit the patients' homes to communicate directly with the patients and their care providers and to make needed assessments.

The pharmacist will be involved in locating medication that is difficult to obtain and will prepare dosage forms that are not available commercially. The pharmacist will compound medication, changing the flavor, eliminating or adjusting the ingredients, or changing strengths as needed. The medical director will be advised of different dosage forms that may be more easily used by the patient. The pharmacist will be responsible for communicating with pharmaceutical manufacturing firms in regard to searches for new and better drugs.

Proper devices for the measurement of liquid medication will be provided by the pharmacist. Any device that might hinder the patient or the patient's care provider from easily obtaining the medication is to be eliminated by the pharmacist. The pharmacist will assure that all auxiliary labeling is attached to each prescription container.

The pharmacist will maintain a patient medication profile that allows monitoring of medication prescribed by the patient's physician. These medications will be palliative for the most part, controlling such symptoms as pain, nausea, vomiting and constipation. The pharmacist will monitor patient progress reports, giving advice concerning the various medications as they fit into the total clinical picture of the patient. The pharmacist is responsible for disposal of medication left after the patient dies if the family chooses to give the medication to the hospice team.

Communicating with all official boards and agencies regarding hospice pharmacy matters is also the responsibility of the pharmacist.

PRINCIPLES OF PALLIATIVE CARE

A sixteenth century adage states that doctors cure rarely, relieve often, and care always. Health professionals practicing palliative care should always attempt cure when that outcome is a reasonable possibility. The most desirable outcome of a palliative care workup is the discovery that the patient has a condition for which cure is feasible. When cure is no longer feasible, relief of symptoms, providing psychosocial support, and simply "being there" for the patient and family become the goals of therapy. Some of the symptoms commonly seen in palliative care for which drug therapy is important are listed in Table 23.5. Major advances in drug therapy for symptom control have occurred in the past 25 years. These include many new drugs, but more important, we have learned how to use older drugs more effectively.

Pain is the symptom most feared by patients with cancer and many other advanced, irreversible diseases. Opioid analgesics are the mainstay of cancer pain management. Many myths about

TABLE 23.5. Symptoms Commonly Seen in Palliative Care for Which Drug Therapy Is Important

Pain
Weakness
Nausea
Vomiting
Anorexia
Constipation
Anxiety
Depression
Insomnia

opioids have developed over the years.[10] Some of these myths and the truth about them are listed in Table 23.6.

WHO published an Expert Committee Report in 1986 entitled *Cancer Pain Relief* and an expanded report in 1990 entitled *Cancer Pain Relief and Palliative Care*.[11,12] In those reports, WHO described five simple principles that can be applied in undeveloped, developing, and developed nations alike. These principles have been applied and studied in over 10,000 patients in countries ranging from the African continent to Southeast Asia and Western Europe to North America.[13-20] The results were patients' own

TABLE 23.6. Myths About Opioids

MYTH	FACT
Addiction in inevitable.	Addiction is exquisitely rare if opioids are used appropriately.
Tolerance is a problem.	Tolerance may occur in the early phase of opioid therapy, but is not a problem if doses are titrated to response.
Respiratory depression limits opioid usefulness.	Respiratory depression is not a problem for most patients.
Injections are more effective than oral doses.	Regularly scheduled oral doses are as effective as parenteral administration.
Opioids should be used in the lowest possible dose with the longest possible interval between doses.	Patients should receive the dose needed for comfort. Each dose should be administered before the previous dose has fully lost its effect.
Patients who demand more of the drug are probably becoming tolerant or addicted.	Most patients who ask for more drug are in pain and should get more medication.
There is a maximum safe dose of morphine.	There is a huge interpatient variance in morphine doses that are effective. The dose is limited only by side effects when they occur.
Morphine is the strongest pain reliever available. If patients don't respond to it, there is nothing else that will work.	Some types of pain are opioid-resistant. Adjuvant medications may increase the effectiveness of opioids. Non-drug therapy may be indicated.

reports of good to excellent pain relief in 80-85% of the patients using only these five simple principles. In developed countries such as the United States, more sophisticated pain interventions are available for use when the simple approaches fail. These more sophisticated approaches include epidural (spinal) analgesia, patient-controlled analgesia (PCA) pumps, nerve blocks, and surgical interventions.[21] It is important to note that those more invasive and more expensive treatments normally are not indicated unless the less invasive and less expensive approaches described in the WHO principles are used optimally and still are not effective.

The WHO principles are summarized in Table 23.7. These principles have been carefully reviewed and strongly endorsed in this country by the American Pain Society (APS), which has included very similar recommendations in the APS monograph entitled *Prin-*

TABLE 23.7. The World Health Organization Principles of Treating Cancer Pain

Cancer pain should be treated:

1.	By mouth	Use oral medications. Rectal and other noninvasive routes should be considered when patients are unable to take drugs orally. There is no reason to use parenteral medications in patients able to take drugs orally.
2.	For the individual	Each patient is an individual who will have different medication requirements and responses to drugs. Titrate doses to response.
3.	By the clock	Medications should be given on a regular schedule so that another dose is given before the previous dose loses its effect. Additional "as needed" doses should be available to manage breakthrough pain.
4.	By the ladder	The analgesic ladder illustrated in Figure 23.2 should be used.
5.	With adjuvants	Indicated adjuvants medication should be given in addition to analgesics. These might include laxatives (to prevent opioid-induced constipation), antianxiety agents, antidepressants, and other symptomatic medications.

ciples of Analgesic Use in the Treatment of Acute Pain and Cancer Pain.[22] In 1992, the Agency for Health Care Policy and Research (AHCPR), an agency of the U.S. Public Health Service, Department of Health and Human Services, published a landmark Clinical Practice Guideline entitled *Acute Pain Management: Operative and Medical Procedures and Trauma*, and in 1994, a Clinical Practice Guideline entitled *Cancer Pain Management.*[23,24] A pharmacist served on the interdisciplinary panel that developed the first guideline, and two pharmacists served on the panel that developed the second. Both of these guidelines incorporate the WHO principles as well.

The first of the WHO principles is to give medication by mouth. Oral administration is as effective as parenteral injections as long as the drug level in the body is maintained. Injections are more expensive, require specialized expertise, are uncomfortable, and often lead to noncompliance.

The second of the principles is to individualize the dose regimen for the patient. Great interpatient variance occurs in opioid dose requirements for cancer pain. While one patient may need only 5mg of morphine every four hours, another patient with the same diagnosis may require 500mg of the same drug. Titrating the dose to the patient response is the only way to determine the optimal dose. There is no predetermined maximum safe dose of morphine.

The third principle is to give medication "by the clock," i.e., on a regular schedule. If the pain is permitted to become firmly reestablished, the patient suffers unnecessarily and the dose of drug needed actually increases. In addition to the regularly scheduled doses, as needed (prn) doses should be available for breakthrough pain that may occur between the scheduled doses. While it is appropriate to give analgesics prn for pain that is rapidly resolving, that approach is wrong for an ongoing pain insult such as often occurs in cancer.

The fourth principle is to give medications by the analgesic ladder illustrated in Figure 23.2. Mild to moderate pain often responds to acetaminophen or a nonsteroidal anti-inflammatory agent such as aspirin or ibuprofen. If the pain is too severe for the level of analgesia or if that analgesic is no longer effective, a less potent opioid such as codeine or a low dose of a drug such as morphine is added.

FIGURE 23.2. The World Health Organization Analgesic Ladder

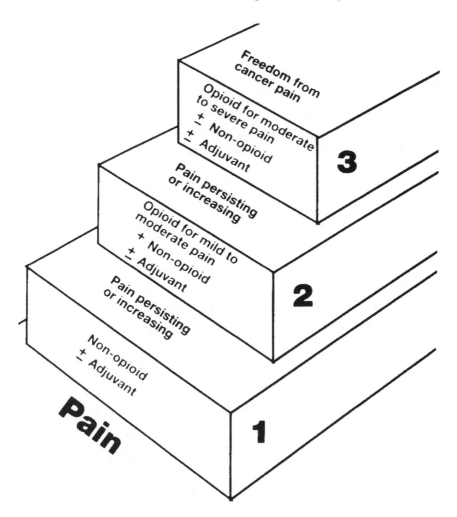

Drug therapy should proceed up the ladder as needed. If the pain is appropriate for a higher step, therapy should not be started on a lower step.

The third step of the ladder is a potent opioid such as morphine added to the NSAID.

The fifth principle is to use indicated adjunct medications. These include laxatives to prevent opioid-induced constipation, antianxiety agents, and antidepressants. Adjuncts are important because other symptoms often exacerbate pain.

A sixth principle that was not included by WHO should be considered by all pharmacists: to ensure that cost-effective medications are used. A recent survey of the cost to the patient for comparable NSAIDs demonstrated nearly a tenfold variance in cost for equianalgesic doses of medications.[25]

Pharmacists are in a good position to help ensure that these principles are considered in the planning and implementation of patient care plans.

CASE STUDIES IN PHARMACEUTICAL CARE OF TERMINALLY ILL PATIENTS

Case 1

GL was a 48-year-old female with a ten-year history of colon cancer. She was initially treated with surgery and radiation therapy. Subsequent recurrences were treated with various combinations of chemotherapy, radiation, and surgery. She had undergone a pelvic exoneration and a colostomy. Despite all this, she was able to maintain her home, raise her children, and live as normal a life as possible under these circumstances. Within the last year, GL had a massive recurrence of tumor in the sacral area resulting in intense vertebral involvement and sciatica. She had been hospitalized multiple times for pain crisis without obtaining relief. At the time of admission to hospice, GL was bedridden, oversedated, receiving epidural morphine, and still in extreme pain. A consult was called to the hospice pharmacist for pain control.

Recognizing that this type of pain was not opioid responsive, the pharmacist recommended discontinuing the epidural infusion and starting the following medications:

- Ibuprofen 400-600mg po qid
- Dexamethasone 1-4mg po qid

- Sustained acting 15mg morphine tablets (MS Contin® 15mg); 30-45mg every 12 hours, increase to every 8 hours if needed
- Standardized senna concentrate with stool softener (Senokot-S®) 2 po bid

In addition, a massage therapist and a physical therapist were consulted and a transcutaneous electrical nerve stimulation (TENS) unit was added to the regimen.

With this treatment, GL was able to be out of bed again, maintain her home, buy groceries, and cook for her family. This continued for four months, during which the hospice pharmacist and nurse made medication adjustments as needed. After four months, GL began to decline, and two weeks before her death, she became bed bound. Her pain was still controlled on an oral regimen until eight hours before her death, when IV morphine was started. She died quietly in her husband's arms.

Case 2

RA was an 80-year-old female diagnosed with breast cancer in 1975. She was treated by mastectomy and radiation at that time and remained disease free until bone metastases were discovered in 1990. In February of 1994, she was referred to hospice with the following problems:

1. Back pain secondary to bone metastasis
2. Nausea and vomiting
3. Constipation

Her medication regimen at that time was:

- Hydromorphone (Dilaudid®) 2mg q 4 h prn
- Promethazine syrup 12.5mg per 5mL; 5-10mL q 4-6 hrs prn
- Standardized senna concentrate (Senokot) liquid prn for constipation

Upon admission, the hospice nurse consulted the hospice pharmacist about an appropriate medication regimen. Recognizing that the patient's symptoms were not controlled by the above regimen and noting that the patient was having difficulty swallowing solid dosage forms, the pharmacist recommended the following regimen, which was accepted by the patient's primary physician:

- Morphine solution 10mg per 5mL, 5-10mL orally q 4 h and indomethacin 50mg rectally q 12 h for pain
- Prochlorperazine suppositories 5mg 1-2 rectally q 6 h for nausea
- Standardized senna concentrate (Senokot) liquid plus docusate (Colace®) liquid bid orally for constipation

Over the ensuing three months, the patient's morphine dose escalated to 500mg po (100mg per 5mL) q 3 h. The patient began having difficulty swallowing the volume of liquid involved, and her primary care physician decided to start an IV and deliver the morphine by infusion. Upon receiving that order, the hospice pharmacist called the physician and persuaded him to keep the patient on a compounded oral morphine preparation of 50mg per mL. The order was agreed upon. The solution consisted of morphine powder dissolved in sterile water; no flavoring was added. The solution had a two-week expiration date and was to be kept refrigerated.

The patient was then titrated to a dose of 800mg of morphine q 3 h while continuing her other medications. She began to complain of hallucinations. Haloperidol 0.5mg po bid was ordered but did not completely control the hallucinations. On her own, the patient began to decrease the morphine until she was at a dose of 200mg po q 3 h. She lived two more months and was able to keep her pain under control with escalating doses of liquid morphine. During the last two weeks of her life, she was unable to swallow any medication. The hospice pharmacist compounded 100mg suppositories, and these were used to keep her comfortable until her death.

CONCLUSION

Pharmaceutical care for patients with advanced, irreversible disease requires pharmacists skilled in a range of services. These include both traditional and clinical roles involving purchasing, formulating, dispensing, and controlling medications and related medical supplies. Many terminally ill patients have difficulty taking commercially available dosage forms, so their pharmacists need specialized expertise in extemporaneous compounding of oral liquid, buccal/sublingual, rectal, parenteral, topical, and other dosage forms. Many of these patients require high doses of controlled substance medications; therefore, meticulous drug control by their pharmacists is essential. Many pharmacists provide in-service education to staff and volunteers of hospice programs and counseling to terminally ill patients and their families. Because patients' primary care providers usually administer the medications in the home, those persons need to understand the purpose and use of each drug. Pharmacists are the best professionals to provide that education. Many hospice pharmacists participate in the development of standing medication orders and protocols for disposal of medications after patients die. The interdisciplinary orientation of hospice personnel creates opportunities for pharmacists to provide drug information to other professionals, to participate in planning drug therapy, and to provide drug therapy monitoring services.

In addition to analgesics and adjunct medications, many terminally ill patients receive palliative chemotherapy (for symptom control, not remission attempts), anti-infective therapy for infections secondary to impaired immune response, and other medications for specific problems. Drug interactions and adverse drug reactions can be serious problems for these very ill patients. Continual review of medication regimens and suggesting changes are common responsibilities of the pharmacists who care for these patients.

Providing pharmaceutical care to terminally ill patients can be challenging. It also can be very rewarding. As our population continues to age, there will be an increasing need for pharmacists skilled in the range of services necessary to provide good pharmaceutical care to terminally ill patients.

REFERENCES

1. Arter SG, Lipman AG. Hospice care: a new opportunity for pharmacists. J Pharm Pract 1990;3:28-33.

2. Lipman AG. Pharmacy services in hospice. Wellcome Trends Hosp Pharm 1990;(June).

3. Saunders C. The management of terminal illness. London: Edward Arnold, 1967.

4. Foley KM. Pain, physician assisted suicide and euthanasia. Am Pain Soc J 1994;3:(in press).

5. Russell SW. Interdisciplinary versus multidisciplinary pain management. J Pharm Care Pain Symptom Control 1993;2(2):89-94.

6. Cleeland CS. Barriers to the management of cancer pain. Oncology 1987;45(Suppl):19-26.

7. Lipman AG. Drug therapy in terminally ill patients. Am J Hosp Pharm 1975;31:270-6.

8. Berry JI, Pulliam CC, Caiola SM, Eckel FM. Pharmaceutical services in hospices. Am J Hosp Pharm 1981;30:1010-4.

9. Arter SG, Berry JI. The provision of pharmaceutical care to hospice patients: results of the national hospice pharmacist survey. J Pharm Care Pain Symptom Control 1993;1(1):25-42.

10. Lipman AG. Opioid use in the treatment of pain: refuting 10 common myths. Pain Manage 1991;4(4):13-7.

11. WHO Expert Committee. Cancer pain relief. Geneva: World Health Organization, 1986.

12. WHO Expert Committee. Cancer pain relief and palliative care. Technical Report Series 804. Geneva: World Health Organization, 1990.

13. Rappaz O, et al. Soins palliatif et traitment de la doleur cancereuse en geriatrie. Therapeutische Umschau 1985;42:843-8.

14. Takeda F. Results of field-testing in Japan of the WHO draft interim guidelines on relief of cancer pain. Pain Clin 1986;1:83-9.

15. Ventafridda V, et al. A validation study of the WHO method for cancer pain relief. Cancer 1987;59:851-6.

16. Walker VA, et al. Evaluation of the WHO analgesic guidelines for cancer pain in a hospital based palliative care unit. J Pain Symptom Manage 1988;3:145-9.

17. Vijayarm S, et al. Experience with oral morphine for cancer pain relief. J Pain Symptom Manage 1989;4:130-4.

18. Giosis A, Giorini M, Ratti R, Luliri P. Application of the WHO protocol on medical therapy for oncologic pain in an internal medicine hospital. Tumori 1989;75:470-2.

19. Schug SA, Zech D, Dorr U. Cancer pain management according to the WHO analgesic guidelines. J Pain Symptom Manage 1990;5:27-32.

20. Grond S, Zech D, Schug SA, Lynch J, Lehmann KA. Validation of the World Health Organization guidelines for cancer pain relief during the last days and hours of life. J Pain Symptom Manage 1991;6:411-22.

21. Ashburn MA, Lipman AG. Management of pain in the cancer patient. Anesth Analg 1993;76:402-16.

22. American Pain Society. Principles of analgesic use in the treatment of acute pain and cancer pain. 3rd ed. Oakbrook, IL: American Pain Society, 1992.

23. Acute Pain Management Guideline Panel. Acute pain management: operative or medical procedures and trauma. Clinical Practice Guideline. AHCPR Pub. No. 92-0032. Rockville, MD: U.S. Department of Health and Human Services, Public Health Service, Agency for Health Care Policy and Research, February 1992.

24. U.S. Management of Cancer Pain Guideline Panel. Management of cancer pain. Clinical Practice Guideline No. 9. AHCPR Pub. No. 94-0592. Rockville,

MD: U.S. Department of Health and Human Services, Public Health Service, Agency for Health Care Policy and Research, March 1994.

25. Lipman AG. A call for rational and cost-effective NSAID therapy. J Pharm Care Pain Symptom Control 1994;2(2):39-46.

ADDITIONAL READINGS

1. Krant MJ. Hospice philosophy in late-stage cancer care. JAMA 1981; 245:1061-2.

2. Lipman AG. Drug therapy in cancer pain. In: Corr CA, Corr DN, eds. Dying persons and hospice. New York: Springer Publishing Co., 1982.

3. Greene PE. America responds to cancer pain; a survey of state cancer pain initiatives. Cancer Pract 1993;11:65-71.

4. Lipman AG. Opioid analgesics in the management of cancer pain: pharmacokinetic considerations. Am J Hospice Care 1988;5(Sept/Oct):26-42.

5. Lipman AG. Pain management in the home care and hospice patient. J Pharm Pract 1990;3:1-11

6. Lipman AG. Clinically relevant differences among the opioid analgesics. Clin Pharm 1990;47(Suppl 1):S7-S13.

7. Bayer R, Callahan D, Fletcher J, et al. The care of the terminally ill: morality and economics. N Engl J Med 1983;309:1490-4.

8. Wilkins C. Pharmacy and hospice: a personal experience. Consult Pharm 1988;3:462-5.

9. Arter SG, DuBé JE, Mahoney J. Hospice care and the pharmacist. Am Pharm 1987;NS27:616-24.

10. Murphy DH. The delicate art of caring: treating the whole person. Am Pharm 1984;NS24:388-90.

11. Anon. The hospice connection. Am Pharm 1984;NS24:146-9.

12. Quigley JL, Quigley MA, Gumbhir AK. Pharmacy and hospice: partners in patient care. Am Pharm 1982;NS22:420-3.

13. Berry JI. Pharmacy services in hospice organizations. Hosp Form 1982;17:1333-8.

14. Mullan PA. Pharmacy and hospice: opportunities for service. Am Pharm 1982;NS22:424-6.

15. Oliver CH. Pharmacy and hospice: talk with the dying patient. Am Pharm 1982;NS22:429-33.

16. DuBé JE. Hospice care and the pharmacist. US Pharm 1981;(Apr):25+.

17. Wroblewski J. Pharmacists looking to define role in hospice care. Drug Top 1981;(Nov):48-50.

18. Gable FB. Death with dignity: the pharmacist's role in hospice care. Apothecary 1980;(Sept/Oct):62-7.

19. Hammel MI, Trinca CE. Patient needs come first at Hillhaven hospice: pharmacy services essential for pain control. Am Pharm 1978;NS18:655-7.

20. Manasse H, Garb E. The pharmacist and the dying patient. Pharm Int 1985;(Jan):9-14.

Chapter 24

Mental Disorders

Julie Magno Zito

INTRODUCTION

This chapter extends the pharmaceutical care model to the area of mental disorders. To achieve this purpose, drug therapy for mental disorders will be discussed in a biopsychosocial approach.[1] A bio-psychosocial treatment approach involves an integration of somatic treatment—typically medications—with supportive psychotherapy (talking therapy, individually or in groups) or behavioral treatment (e.g., behavior modification for phobias). From this underlying model, therapy for mental disorders takes account of the whole person being treated and the treatment setting as well as the medications used to control psychiatric symptoms.

The etiology of mental disorders deserves some comment before discussion of the behavioral and social aspects of drug therapy for such disorders. In recent years, there has been an emphasis in the lay press and among some researchers on genetic and biological factors as primary etiologic agents in the development of mental disorders. The research findings supporting a genetic etiology[2] were modest, at best, and have been subsequently revised,[3] although the media tends to ignore these revisions. Time and more rigorous studies will determine whether a biological basis is both necessary and sufficient for the etiology. Thus, the counseling pharmacist can remain tentative about such matters. Further, for the purpose of providing pharmaceutical care, it should be recognized that a treatment model can be completely *independent* of the etiology of the underlying disorder. This conclusion follows from the fact that the pharmacology of the drug explains its mechanism of action but need not be related to the etiology of the mental disorder.

In general, psychotherapeutic drug therapy relieves target symptoms of mental disorders and is not curative. The goal of drug therapy is to control target symptoms during the acute phase of a mental disorder to permit the next phase of treatment: psychosocial therapy that is aimed at social and vocational rehabilitation of the individual. Admittedly, many patients cannot be managed during the maintenance phase of treatment with psychotherapy alone, and in these instances, drug therapy is continued. The goals of the psychosocial treatment programs are to restore or improve the individual's ability to interact socially with family and neighbors (social functioning) and to perform appropriately, whether at school, as a homemaker, or at work (work functioning). These goals are in addition to activities of daily living such as grooming and personal hygiene.

The biopsychosocial model for treating mental disorders is most appropriately viewed as a longitudinal or long-term care approach. The ideal pharmacist to manage therapy in this approach is the pharmacist with access to several key resources: the medical record, the individual in treatment, and the providers of drug and psychosocial treatment for that individual. It is also desirable that the pharmacist be able to provide this monitoring for relatively large blocks of time. While some pharmacists do not have this expanded clinical role, this chapter is developed for the individual with an extended clinical database who wishes to provide the sophisticated monitoring necessary for the delivery of pharmaceutical care to those with mental disorders.

This chapter is necessarily too brief to be comprehensive. For more extensive information on the monitoring of psychotherapeutic agents, readers are referred to texts that emphasize a total longitudinal approach including efficacy, safety, and drug combinations.[4,5] There are numerous classes of drugs used extensively for the treatment of mental disorders: antipsychotics (neuroleptics, formerly called major tranquilizers) for psychotic disorders, neuroactive agents (e.g., anticholinergics) for unintended effects related to the use of antipsychotics, lithium for mania and other disorders, antidepressants for affective and other disorders, and benzodiazepines and other agents for anxiety and sleep disorder therapy. In recent years, there are trends for prescribing agents in addition to the primary drug class used to control symptoms of a major psychotic disorder. These treatments are called ancillary drug therapy and include scheduled benzodiazepines when antipsychotics alone are not sufficient[6] for controlling acute schizophrenia or bipolar disorder.[7] Also, anticonvulsants (e.g., carbamazepine and valproic acid) have been added to lithium for the control of mania.[8] Unfortunately, these approaches received limited study before being introduced, and there is controversy because some studies or experiences have failed to confirm or persuade clinicians of the effectiveness of these approaches.[9-11] Complicated regimens, exemplified by ancillary therapy, present a real challenge for the monitoring and evaluation of treatments for mental disorders. The range of drug classes involved in these treatments suggests that the amount of relevant pharmacological knowledge can be formidable; this pharmacology is an important primary information base upon which to build an understanding of the psychosocial aspects of the treatments. Because of space limitations, only antipsychotics, the most prominent drug class for the treatment of mental disorders, will be examined in regard to the psychosocial dimensions of their clinical use.

This chapter is organized into the following sections to introduce the pharmaceutical care reader to major psychosocial aspects of the issues relevant to a better understanding of the use of psychotherapeutic agents. The topics are: mental disorders, their classification and evaluation; epidemiology of mental disorders; selected psychosocial principles of mental disorder therapy; antipsychotics for psychosis and other disorders; and principles of longitudinal monitoring. For the pharmacological class of antipsychotics, a drug class used frequently in the treatment of major mental disorders, the section is organized in the following way: general indications, efficacy versus effectiveness, side effect monitoring, and behavioral toxicity monitoring. The goal of this chapter is to review areas of mental disorder therapy with an emphasis on the psychosocial and behavioral aspects of pharmaceutical care so that an integrated monitoring system develops with both drug and nondrug treatment factors. From this description of the issues, future research goals should crystallize.

MENTAL DISORDERS:
THEIR CLASSIFICATION AND EVALUATION

An introduction to the highly developed field of mental diagnosis begins by distinguishing between psychotic and nonpsychotic disorders. Psychosis involves emotional, cognitive, and behavioral abnormalities of such proportion that the individual is unable to function at a level appropriate to his or her age and level of experience. There is usually a severity range associated with it. Psychosis associated with disorders (e.g., schizophrenia or mania) typically manifests itself as a lifetime disorder with acute episodes interspersed with periods of remission. The diagnosis of mental disorders typically requires a judgment on the presence of symptoms that reflect the internal state of the individual and are elicited by interviewing the individual and then rating the pattern of the symptoms. This situation represents a critical departure from the more objective, externally validated (gold standard) criteria available for determining most medical diagnoses. As a consequence, there is a significant body of work devoted to issues of the reliability and validity of psychiatric diagnosis.[12] To increase reliability (and, hopefully, validity as well), there has been a succession of American diagnostic instruments, called the *Diagnostic and Statistical Manual*,[13] which was released in its fourth and latest version (*DSM-IV*) in 1994.

A brief review of the psychotic disorders of *DSM-IV* for adults includes schizophrenia, delusional (paranoid) disorder, psychotic mood disorders, and organic mental disorders. Many of this last are associated with substance abuse. Mood disorders are divided into bipolar disorder (mania) and depression. While some depression may be of a psychotic type, most depression does not involve psychotic symptoms. Nonpsychotic disorders are divided into anxiety disorders including phobias, somatoform, dissociative, sexual, sleep, factitious, and adjustment disorders. Axis I of the *DSM-IV* system codes the major mental disorders. Axis II codes any associated personality disorders which may be present. Axis III codes medical disorders. Axis II and Axis III may be less completely coded than the major mental disorders, depending on the source of the recorded diagnosis. The international counterpart of the *DSM* is the *International Classification of Disease* which is in its tenth revision (*ICD-10*).

As the *DSM* versions have evolved, there has been an emphasis on operational criteria, generally a good way to increase reliability. But users may have been too easily led into an overemphasis on specific symptoms and their treatment as opposed to viewing a prominent symptom in the context of other symptoms and, most importantly, in terms of the overall level of impairment or dysfunction. It is not known yet whether *DSM-IV* will be successful in overcoming this imbalance. Perhaps a corrective measure for this problem lies in the methods we select for evaluating improvement during the course of treatment. This subject is described below.

Improvement during the course of a chronic mental disorder is best described by a longitudinal monitoring model involving symptoms and functioning. In many instances, drug therapy will be a significant part of the therapeutic management. There are numerous rating scales available for measuring the improvement in mental disorder symptoms based on interviewing the patient and his or her caretakers. Originating in the research arena, several of these scales have achieved widespread recognition for their reliability. For example, the Brief Psychiatric Rating Scale (BPRS)[14,15] consists of 18 (sometimes 19) items which characterize the major positive and negative symptoms of schizophrenia. The items are scored on a seven-point severity scale from not present to extremely severe. Instruments have been specifically designed for the evaluation of other disorders such as depression[16] and obsessive compulsive disorder.[17] But these instruments are not complete measures of the degree of improvement because they account only for the change in symptoms and do not measure the change in functioning.

To fully address questions of improvement, social and vocational functioning must be included. A convenient, simple instrument for this purpose is the Clinical Global Impression[16] (CGI), although this is not a sensitive instrument from a psychometric standpoint. A more recently developed scale for the evaluation of hospitalized individuals with chronic mental disorders is the Strauss-Carpenter Scale,[18] which measures outcome in terms of symptoms and hospitalized days, as well as social and vocational functioning, thus providing a measure that fits the underlying biopsychosocial approach to the treatment of mental disorders.

The implications of this review of outcome measures for mental disorders for pharmaceutical care are straightforward. The clinician who evaluates the outcome of drug therapy should begin by knowing the clinical baseline in terms of diagnosis, target symptoms and functioning, and the severity of the symptoms. At subsequent treatment evaluation points, measures of improvement should include not only interviewing the individual about the target symptoms but also asking about the course of social interactions and work or school performance. Taken together, these four dimensions will reflect overall improvement and avoid situations that overvalue short-term gains in one area of need–namely, symptoms–at the eventual expense of overall global improvement.

EPIDEMIOLOGY OF MENTAL DISORDERS

Rigorously derived information on the epidemiology of mental disorders in the United States was not available until federal initiatives in the early 1980s developed the Epidemiologic Catchment Area (ECA) studies.[19] These large-scale, regional studies consisted of household surveys and were population-based. They have been completed and allow reliable U.S. estimates of the prevalence of major mental disorders[20,21] upon which estimates of mental health services utilization and expenditures[19] have been based. Unfortunately, little drug treatment data were collected, and no data have been published yet.

The lack of reliable drug treatment prevalence for mental disorders suggests that the epidemiology of drug therapy (i.e., pharmacoepidemiology of mental disorders) is an area greatly in need of further development. This topic has been elaborated upon in several previous reviews targeted at adult[22] and child/adolescent[23] patient population needs. Major mental disorders lend themselves to pharmacoepidemiological study because of the episodic nature of the disorders, the strong reliance on drug therapy, and the tendency of many patients with severe mental disorders to be available in the same setting (e.g., at a community mental health center or psychiatric facility) for considerable periods of time. Naturalistic drug information related to cohorts of individuals with major mental disorders, if examined in the context of sufficient clinical (both medical and psychosocial) and sociodemographic information to characterize the individuals, would be helpful in learning how effective treatments are in the usual practice setting.

Several examples of information gained from pharmacoepidemiological studies illustrate its usefulness for the study of mental disorders. In a study of newly admitted patients with a diagnosis of schizophrenia, no association was found between the intensity of daily dose exposure [low (≤ 700mg), moderate (701-1400mg), high (≥ 1400mg)] expressed on the relative scale of chlorpromazine equivalents and the length of hospitalization during a 90-day follow-up.[24] Evaluative data such as this, largely unavailable from drug trials, suggest the possibility of dose-lowering protocols without fear of delaying discharge. A second example of pharmacoepidemiological data illustrates the largely empirical use of medications for children[25] hospitalized with major mental disorders, a situation that is not consistent with recommendations based on the limited clinical trial data available. Greater hospitalization leading to greater complexity of therapy is also illustrated for young adult patients with schizophrenia.[26] Patients whose hospitalization exceeded 180 days were more likely to receive greater antipsychotic dosage and more ancillary agents than those discharged in less than 180 days.

Treatment of special populations of individuals with mental disorders requires knowledge of the particular disorders, the natural course of the disorders, unique patient characteristics, and treatments available. Age groups requiring special attention include children and adolescents, adults, and the elderly. In addition, gender issues in psychopharmacology are receiving more attention from the FDA and medical practitioners as a result of greater emphasis on women's issues, including health. Research studies on psychotherapeutic drug exposure[27] by gender are beginning to receive attention. Traditionally, psychiatric care has been provided to those with mental deficiency, developmental disabilities, and increasingly, to those with problems associated with alcohol and substance abuse. The overlap (comorbidity) of substance abuse with major mental disorders is a significant challenge to successful psychopharmacological management and warrants pharmacoepidemiological evaluation.

The environment in which mental disorders are treated varies a great deal: from the most secure lockup unit of the state psychiatric facility to the short-stay, day care, outpatient clinic of the modern community mental health center or private practitioner's office. The philosophy of care has changed greatly during the past 50 years in the direction of greater deinstitutionalization. This trend is partly explained by the advent of more effective treatments, namely, chlorpromazine, an antipsychotic agent discovered in 1950 and made widely available by 1954 to the thousands of institutionalized mentally ill in the U.S. mental health system. Early expectations of curative powers for this and subsequent antipsychotics were not borne out. Consequently, numerous psychosocial issues (e.g., compliance,[28] refusal among those with a commitment order for hospitalization,[29] and quality of life[30]) have emerged as serious aspects of the management of chronic mental disorders. In addition, a review of the history of modern mental health therapies illustrates the importance of an epidemiological approach, not just for the detection of low-frequency adverse drug effects but also for drug therapy evaluation—specifically for long-term effectiveness and for more recently identified problems of ineffectiveness described as treatment resistance.[31]

This section reviewed several aspects of the epidemiology of mental disorders to establish the need for longitudinal monitoring of drug and psychosocial therapy for these typically recurring, episodic disorders. These circumstances may result in continuous management with drug therapy over the lifetime of the individual. Pharmaceutical care specialists in mental health should be particularly cognizant of the need for intensive drug therapy monitoring in these settings, especially those providing a *least restrictive environment* because this may be one that is less well supervised in terms of substantial risks of medication side effects. Among these risks are falls[32] due to a more unstructured, active, and unsupervised living arrangement. A least restrictive environment is advocated often by legal theorists and consumer advocates to maximize the individual's personal freedom during treatment. Sophisticated monitoring and evaluation programs for drug therapy within safe, least restrictive environments is a useful goal of pharmaceutical care.

SELECTED PSYCHOSOCIAL PRINCIPLES OF MENTAL DISORDER THERAPY

Numerous psychosocial principles related to the care of those with severe mental disorders can be found within the viewpoints of each of the various disciplines of psychiatry, psychology, sociology, and social work. For this discussion, the following topics will be addressed as far as they relate to the pharmaceutical care of those with mental disorders: labeling theory, compliance, patient satisfaction, drug refusal, and treatment biases.

Labeling Theory

Labeling theory comes from sociology[33] and defines the individual whose behavior deviates from the social norm as one who is labeled as mentally ill. Upon receiving this label, the society of nondeviants then establishes a system for controlling the personal freedom of the labeled individual. Szasz[34] elaborated on the conflicts created by public psychiatrists who try to serve both the client (patient) and the state. In serving the state, they provide a judgment of patient incompetency that results in a court order of commitment. The labeled individual is stigmatized by being less able to recover status in the community once he or she is discharged from the mental hospital. More recently, the anthropologist Estroff [35] produced an ethnography of life within a community mental health center setting among individuals with severe mental disorders. In this exemplary, engaging study, the use of antipsychotic agents, along with their undesirable physical effects on motor function, was seen as a newer, modern form of stigmatization. Furthermore, the frequent occurrence of these long-term, irreversible adverse motor effects (e.g., tardive dyskinesia), which are easily recognizable among the community of patients, was interpreted as a crucial aspect of why patients are noncompliant with drug therapy.

Labeling in its various manifestations relates to the pharmaceutical care of those "labeled" with mental disorders to the extent that caregivers recognize the various ways that bias can occur in the management and evaluation of drug therapy. For example, individuals with a diagnosis of a major psychotic disorder such as schizophrenia or bipolar disorder may be viewed as incapable of self-reporting the effects and adverse effects associated with drug therapy. To counteract this bias, it is important to recall the earlier discussion of psychosis as an episodic disorder and one in which certain mental domains may be affected while others remain intact. Thus, individual counseling to assess the degree of judgment or competence and ability to self-report effects and adverse effects should be the guiding principle. This approach avoids false assumptions and bias in caring for those with mental disorders.

Compliance

Compliance (adherence) with the prescribed treatment regimen, in general, is a topic of wide interest among pharmaceutical caregivers.[36] In the area of psychiatric treatment, discussions of compliance as a probable explanation for treatment failure receive wide attention.[37,38] Kane reported the widely held belief that as many as 50% of outpatients with schizophrenia or another psychotic disorder are noncompliant with psychotherapeutic agents. It is easy to understand why the problem is important based on reasons already described, such as the need for drug therapy during acute exacerbation of symptoms when violent behaviors, if they will occur, would be most likely to occur. Also, noncompliance leading to abrupt withdrawal would be important for those who are continuously medicated after the acute exacerbation is resolved. Along with these clinical concerns, there is a clinical research literature[39] pointing to better outcomes for those who are compliant with therapy, although more rigorous evaluation is desirable because some critics are unpersuaded.[40] These criticisms stem from uncertainty as to whether the reasons for better outcomes among those with better compliance relate to patient characteristics. For example, those who are less severely ill may find it easier to comply, or they may be more able to embrace social and vocational rehabilitation.

Despite the attention compliance receives as a clinical issue, studies of compliance among those with severe mental disorders are quite limited. In a study of Veterans Administration patients with schizophrenia, a modest increase in improvement was observed for those in a medication education group when discharge from the hospital was used as the outcome measure.[38] Limited success, as was observed in this study, may point beyond cooperation with treatment to the limits of the effectiveness of drug therapy for the long-term medicated chronic patient population.

Patient Satisfaction

Patient satisfaction and quality of life is a topic that has gained increased attention as recent years have brought consumer advocacy to issues of medical and mental health. Lehman[30] established that quality of life could be reported by those with severe mental disorders, which gives empirical support for engaging those with chronic mental disorders in longitudinal monitoring of their drug treatment.

For pharmaceutical caregivers, satisfaction with drug therapy might be determined by asking the patient to identify those specific aspects of treatment that are not satisfactory. Encouraging this type of reporting and then attempting to verify and correct such problems in collaboration with the prescribing physician, family members, or other caregivers, is a means of improving patient satisfaction and compliance with psychotherapeutic agents.

Drug Refusal

Since the late 1970s, American patients in psychiatric hospitals have been recognized by a series of court decisions to have at least a qualified right to refuse psychiatric drug therapy.[41] This right exists despite involuntary commitment for psychiatric care. This right and its effects have been the subjects of intense controversy, considerable discussion in law and psychiatry, and empirical research.[41-47] These decisions gave a legal precedent for supporting individual rights but, in practice, they have given psychiatric patients limited authority beyond the expression of their dissatisfaction. Study findings from New York State[46] corroborate those from other states[44] and show that court review of patients' refusal in state psychiatric facilities results in an overwhelming approval for the physician to medicate patients over their objection. In a one-year evaluation,[46] among the 1% of patients with applications for forced medication, more than 90% were approved by the court. Similar results were found in a prospective evaluation,[29] suggesting that court review reaches a very narrow subgroup of noncompliers. Thus, the right is affirmed in theory but, apparently, without the kind of clinical, educational, and evaluative follow-up needed to launch new programs to address the variety of patient dissatisfactions that contribute to drug refusal. Moreover, many clinicians and some research findings[44,45] support the notion that refusal is a manifestation of acute psychosis and not based on drug-related reasons such as neuroleptic dysphoria[48] or abnormal involuntary movements, as Estroff suggested.[35]

The implications of drug refusal for those engaged in pharmaceutical care are numerous. First, acknowledging the limitations of long-term treatment with psychotherapeutic agents such as antipsychotics, lithium, or antidepressants lends itself to searching for optimal therapy and to an open, respectful therapeutic alliance[49] that has been recognized as the core of successful psychotherapy for mental disorders. In this way, putative drug-related problems can be identified, be monitored, and become part of a systematic search for that regimen offering the motivated individual the most satisfactory balance of benefits and risks. Second, drug refusal should trigger the reader of clinical pharmacology to recognize distinct problems with the major drug therapies in psychiatry. As an example, a comparison among the various classes is instructive. Whereas relatively few people report dissatisfaction with benzodiazepines–a drug class with significant liability for addiction–there is no such liability associated with antipsychotics, lithium, or the tricyclic antidepressants. It is too early to predict whether the patient satisfaction reported by some who take selective serotonin reuptake inhibitors (SSRIs) (e.g., fluoxetine) will lead to dependency or whether putative adverse effects will emerge to limit use.[50,51] For successful long-term monitoring of psychotherapeutic agents, patients should be given the opportunity (and training, if necessary) to report drug effects, unintended effects, and satisfaction.

Treatment Biases

Several biases which may suffuse the treatment process are important for understanding why a longitudinal, patient-specific monitoring program is ideal for psychotherapeutic agents. These are biological model bias, expectancy effects,[52] and negative placebo effects. First, as mentioned earlier, the current emphasis on the biological model for the etiology of mental disorders has probably resulted in an overemphasis on the drug portion of therapy, with a consequent increased drug prevalence.[53-55] Second, there is the potential for expectancy effects by the prescribing physician, the patient or surrogate decision maker, and those in a position to influence decisions about the individual's mental health therapy (e.g., school teachers, ward staff of an institution, or foster caregiver) to overestimate early treatment effects with a new drug. Whether these early (honeymoon) effects persist is a most critical question which remains unanswered by clinical trial data. As illustrated by one study of clozapine, questions about persistence of efficacy can most appropriately be answered by pharmacoepidemiological studies of follow-up after six months, or one or more years of treatment.[56] Third, and most difficult to demonstrate empirically, is the possibility of a negative placebo effect. This would occur when the patient believes he or she is better off with a nondrug treatment and so would show improvement when drugs are removed or, conversely, would show a worsening of the overall treatment effect when treated with a drug he or she believes is inappropriate treatment. The impact of each of these biases should be examined by research models to determine the extent of their practical (clinical) effect on therapy.

ANTIPSYCHOTICS FOR PSYCHOSIS AND OTHER DISORDERS

General Indications

There has been widespread use of antipsychotics for acute or chronic treatment of psychotic disorders for more than 40 years.

Symptomatic use for behavioral control of those with nonpsychotic diagnoses has occurred in the past but is no longer recommended because the risks of frequently occurring irreversible effects (e.g., tardive dyskinesia[57]) or rarely observed but life-threatening events (e.g., neuroleptic malignant syndrome[58,59]) outweigh the benefits. For the elderly, small doses of antipsychotics are used to treat symptoms associated with organic mental disorders (e.g., senile dementia).[60] For children and adolescents, antipsychotic indications are largely based on anecdotal rather than clinical trial experience, but there are some specific indications for children, such as haloperidol for the treatment of Tourette's Disorder. Behavioral control of those with mental deficiency[61] or developmental disabilities[62] is used when self-injurious behaviors or acting-out behaviors justify it.

Efficacy vs. Effectiveness

For the treatment of schizophrenia, antipsychotics are a mainstay of treatment and among the least expensive and most efficacious treatment modalities, according to leading psychiatric experts.[63] Numerous critical reviews of the large body of clinical trial data extant support the efficacy of antipsychotics under the following conditions: for short-term use (four to six week study duration); for symptom control (outcome measure is the change in "target," or key, symptoms from baseline); for selected subjects, i.e., those in a homogeneously defined diagnostic group (e.g., schizophrenia classified by research diagnostic criteria); and for those giving informed consent. Each of these factors may produce a barrier to generalizing from the clinical trial efficacy outcome to the long-term effectiveness outcome. For example, do antipsychotics continue to work as effectively after several years as they did when first selected?

In retrospect, early efficacy studies[64,65] showed impressive efficacy findings for antipsychotics in which 75% of those medicated with active drug were improved compared with 25% of those treated with placebo. Despite these early findings, more recent work on the atypical antipsychotic clozapine found that the drug improved only 30% of a study population with "treatment-resistant schizophrenia," defined as having failed to respond adequately to standard antipsychotic treatment.[66] The emergence of treatment-resistant schizophrenia is a telling indication of treatment failure in the 1980s after decades of antipsychotic treatment. This fact lends support to the conclusion that long-term effectiveness cannot be generalized from short-term clinical trial data. It should be expected that those in the clinical trial had fewer problems in regard to their diagnosis. This cleaner classification of subjects leads to greater homogeneity and, therefore, the outcome may not be reflective of clinical outcome for the usual practice population, which is complicated by the presence of atypical symptoms and comorbidities. Informed consent bias,[67] a form of volunteer bias, provides another reason why those who respond to treatment during the trial may be different from those who are given the treatment in the usual practice setting.

The pharmaceutical caregiver can respond best to issues of antipsychotic efficacy by acknowledging how limited the model is in regard to effectiveness. Pharmacoepidemiological data to characterize long-term outcome provide one solution to the gap between clinical trials and usual practice use. A second, perhaps greater, challenge involves a better understanding of how well antipsychotics work for maintenance treatment. The work of Hogarty[68] is methodologically sophisticated and supports continuing treatment for chronic psychotic disorders. While some research leaders see the maintenance treatment question as resolved,[39] others question the validity of the maintenance treatment study designs in which a randomized withdrawal from active drug to placebo was used.[40] These studies are quite dated now, but in light of greater awareness of physiological problems associated with drug withdrawal–specifically, anticholinergic rebound effects[69]–as well as controversial new dilemmas (e.g., tardive or supersensitivity psychosis[70]), there is a need to revisit this issue. Creative study designs that attempt to avoid the major pitfalls of early maintenance studies for schizophrenia would be useful in this regard. Alternative approaches to continuous maintenance treatment include the prodromal symptom monitoring of Herz,[71] the noncontinuous therapy of Carpenter,[72] and other low-dose strategies.[73]

Side Effect Monitoring

Because the antipsychotics have a range of central nervous system actions primarily related to the blockade of receptors in the dopaminergic, cholinergic, and alpha-adrenergic systems, each action can contribute unintended effects as well as the intended effect. The result is the frequent occurrence of unintended, relatively minor anticholinergic effects such as dry mouth, constipation, and blurred vision. Less frequently there are more serious effects, e.g., a loss of the gag reflex leading to the risk of asphyxia.[74] There are also alpha-adrenergic blocking effects such as orthostatic hypotension. Other serious cardiovascular effects include conduction delays and heart block as well as sudden unexplained death in young persons. The blockade of D-1 receptors in the basal ganglia is reported to be the cause of drug-induced parkinsonism, a syndrome of drooling, shuffling gait, tremor, and rigidity. Other conditions related to the extrapyramidal system result in a range of abnormal involuntary movements, e.g., dyskinesia or akinesia (difficult or reduced movements), dystonia (unusual wrenching muscle spasms of the neck or torso), and akathisia (usually seen as restlessness, pacing, and an inability to sit still). In addition, there are general central depressant effects commonly observed such as lethargy and depressed mood.

Behavioral Toxicity Monitoring

Two of the major antipsychotic side effects deserve further comment because they present a challenge in terms of the psychosocial aspects of the treatment of mental disorders described as behavioral toxicity.[75,76] First, akinesia shows up as poverty of movement and slowness of speech and can be confused with the underlying psychotic disorder (e.g., schizophrenia). Among the negative (passive type) of symptoms of schizophrenia are poverty of speech, poverty of thought, and social withdrawal. Confusion between drug-induced akinesia and the underlying disorder or de novo depression has been noted in the literature,[77] but the extent to which it is avoided in general practice is unclear. Second, akathisia, the drug-induced effect that produces a syndrome similar to anxiety, can be confused with an increase in acting out behavior or dyscontrol.[78] In each situation, without knowledge of the baseline condition of the individual being treated, familiarity with the dosing/response profile of the specific antipsychotic agent, and specific monitoring to rule out these drug-induced effects, it is likely that the effects will be misunderstood.

These and other psychological or behavioral effects are prominent in the treatment of mental disorders and present a real opportunity for pharmaceutical care to produce long-term care monitoring practices that will identify and correct such problems.

PRINCIPLES OF LONGITUDINAL MONITORING

Effectiveness Monitoring

This review of the past and current literature on the use of drug therapy for the treatment of mental disorders suggests that there is a need to clarify effectiveness for long-term use. This is particularly true in conjunction with measures of patient outcome that take account of the individual's functioning as well as symptom level. For example, in an evaluation of clozapine's effectiveness among patients with treatment-resistant schizophrenia hospitalized at state psychiatric facilities, Zito and colleagues[56] found a proportion of patients with *symptom improvement* at one year similar to the proportion reported by Kane and colleagues[66] (30%) after six weeks of a randomized trial. However, when another measure of outcome was applied to the former study population–namely, hospital discharge–less than 10% of the cohort had been discharged by the one-year follow-up, a considerably smaller number than those with symptom improvement. Thus, long-term effectiveness studies can provide a type of postmarketing surveillance[22] which helps with the policy and cost issues associated with new therapies and generally has been missing from drug evaluation. Pharmaceutical care should encompass multidimensional criteria for improvement (e.g., the Strauss-Carpenter scale), whether conducting the evaluation for clinical purposes or more extensively for the purpose of research study. A suggested approach for systematic clinical practice monitoring is provided as an appendix in a recent textbook.[5]

Research methodologies focusing on the individual patient response are available. For example, Guyatt and colleagues[79] described a clinical pharmacology unit at a major teaching hospital where systematic evaluations of treatments for individuals have been conducted. Key aspects of such systematic evaluations include some general rules of good clinical practice. For example, parsimonious drug regimens should be used. This rule is particularly useful in psychiatry where behavior and internal psychological states are the symptoms being monitored. Changes in therapy, wherever possible, should be made one drug at a time. Furthermore, dosing changes should not exceed 25% of the total daily dose[5] and should not be repeated sooner than one week. Such explicit protocols would help to eliminate transient pharmacological withdrawal states (rebound effects) from being misinterpreted as clinical decompensation. Because dose-response relationships are lacking for the antipsychotics,[80] the lowest effective dose is recommended. This approach reduces the risk of unnecessary long-term adverse events (e.g., tardive dyskinesia) and increases the likelihood of patient satisfaction.

Behavioral toxicity monitoring should be emphasized in the long-term management of psychotherapeutic agents. To facilitate the process of regular assessment of side effects for the drugs in question, there should be a working relationship with the individual being treated so issues related to dissatisfaction with treatment are likely to be shared. Sometimes the treated individual cannot express in words the distress being experienced (e.g., neuroleptic dysphoria[48]), but good interviewing can help the individual to articulate the problem.

Comorbidity monitoring demands attention to drug-drug and drug-disease interactions so medical and mental disorder therapies are integrated. In recent years, there is increasing overlap among cardiovascular medicines (e.g., propranolol, nifedipine, and clonidine as ancillary treatments for mental disorders) and mental disorder treatments. This additive drug burden greatly increases the likelihood of adverse events or reduced functioning which may lead to decreased satisfaction as well as the possibility of increased morbidity[32] or mortality. Unfortunately, epidemiological studies exploring the benefits versus risks of drug therapy in complex, multidrug regimens are not common. The very complexity of the situation scares off clinical researchers, who tend to be trained in the highly controlled clinical trial model. Future psychotherapeutic research should endeavor to provide information in a biopsychosocial framework and should take up the challenge of the multidrug, multiproblem individual because it is likely to shed light on the appropriate management of our most nettlesome and costly clinical problems.

CONCLUSION

Pharmaceutical care encompasses the care of those with mental disorders, a patient population ideally suited for long-term care management and individualization of treatment. The psychosocial aspects of this care are inherent in the way diagnosis is made and validated and the way treatment outcome is evaluated and are explicit in regard to patient characteristics such as compliance, refusal, and expectancy effects. Psychosocial factors also relate to the treatment model, and prescriber expectancy and societal expectancy regarding medicating individuals over their objection. This review has attempted to familiarize the pharmaceutical care reader with the nature of these issues. The goal of future work in this area should include a clinical practice model that integrates the psychosocial aspects of care with the biological aspects of care. Also, the research model should incorporate the study of effectiveness and behavioral toxicity into an agenda for the pharmacoepidemiology of mental disorders.

REFERENCES

1. Blackwell B. Schizophrenia and neuroleptic drugs: a biopsychosocial perspective. In: Doudera AE, Swazey JP, eds. Refusing treatment in mental health institutions–values in conflict: proceedings of a conference. Ann Arbor, MI: AUPHA Press, 1982:3-18.

2. Sherrington R, Brynjolfsson J, Petursson H, et al. Localization of a susceptibility locus for schizophrenia on chromosome 5. Nature 1988;336:164-7.

3. Detera-Wadleigh SD, Goldin LR, Sherrington R, et al. Exclusion of linkage to 5q11-13 in families with schizophrenia and other psychiatric disorders. Nature 1989;340:391-3.

4. Baldessarini RJ. Chemotherapy in psychiatry: principles and practice. Cambridge, MA: Harvard University Press, 1985.

5. Zito JM, ed. Psychotherapeutic drug manual. New York: Wiley and Sons, 1994.

6. Wolkowitz OM, Pickar D, Doran AR, Breier A, Tarell J, Paul SM. Combination alprazolam-neuroleptic treatment of the positive and negative symptoms of schizophrenia. Am J Psychiatry 1986;143:85-7.

7. Arana GW, Ornsteen ML, Kanter F, Friedman HL, Greenblatt DJ, Shader RI. The use of benzodiazepines for psychotic disorders: a literature review and preliminary clinical findings. Psychopharmacol Bull 1986;22:77-87.

8. Chou JC. Recent advances in treatment of acute mania. J Clin Psychopharmacol 1991;11:3-21.

9. Karson CN, Weinberger DR, Bigelow L, Wyatt RJ. Clonazepam treatment of chronic schizophrenia: negative results in a double-blind, placebo-controlled trial. Am J Psychiatry 1982;139:1627-8.

10. Lerer B, Moore N, Meyendorff E, Cho SR, Gershon S. Carbamazepine versus lithium in mania: a double-blind study. J Clin Psychiatry 1987;48:89-93.

11. Ruskin P, Averbukh I, Belmaker RH, Dasberg H. Benzodiazepines in chronic schizophrenia. Biol Psychiatry 1979;14:557-8.

12. Spitzer RL, Endicott J, Robins E. Research diagnostic criteria: rationale and reliability. Arch Gen Psychiatry 1978;35:773-82.

13. American Psychiatric Association. Diagnostic and statistical manual of mental disorders. Washington, DC: American Psychiatric Association, 1994.

14. Overall JE, Gorham DR. Brief Psychiatric Rating Scale. Psychol Rep 1962;10:799-812.

15. Rhoades HM, Overall JE. The semistructured BPRS interview and rating guide. Psychopharmacol Bull 1988;24:101-4.

16. Guy W. ECDEU assessment manual for psychopharmacology. Rockville, MD: National Institute of Mental Health, 1976.

17. Leonard HL, Swedo SE, Lenane MC, et al. A double-blind desipramine substitution during long-term clomipramine treatment in children and adolescents with obsessive-compulsive disorder. Arch Gen Psychiatry 1991;48:922-7.

18. Strauss JS, Carpenter WT Jr. The prediction of outcome in schizophrenia. I. Characteristics of outcome. Arch Gen Psychiatry 1972;27:739-46.

19. Shapiro S, Skinner EA, Kessler LG, et al. Utilization of health and mental health services: three Epidemiologic Catchment Area sites. Arch Gen Psychiatry 1984;41:971-8.

20. Regier DA, Farmer ME, Rae DS, et al. One-month prevalence of mental disorders in the United States and sociodemographic characteristics: the Epidemiologic Catchment Area Study. Acta Psychiatr Scand 1993;88:35-47.

21. Regier DA, Narrow WE, Rae DS, Manderscheid RW, Locke BZ, Goodwin FK. The de facto U.S. mental and addictive disorders service system: Epidemiologic Catchment Area prospective 1-year prevalence rates of disorders and services. Arch Gen Psychiatry 1993;50:85-94.

22. Zito JM, Craig TJ. Pharmacoepidemiology of psychiatric disorders. In: Hartzema AG, Porta MS, Tilson HH, eds. Pharmacoepidemiology: an introduction. Cincinnati, OH: Harvey Whitney Books, 1994.

23. Zito JM, Riddle MA. Psychiatric pharmacoepidemiology for children. Child Adolesc Psychiatr Clin North Am 1995;4.

24. Zito JM, Craig TJ, Wanderling J, Siegel C. Pharmacoepidemiology in 136 hospitalized schizophrenic patients. Am J Psychiatry 1987;144:778-82.

25. Zito JM, Craig TJ, Wanderling J. Pharmacoepidemiology of 330 child/adolescent psychiatric patients. J Pharmacoepidemiol 1994;3(1):47-62.

26. Zito JM, Craig TJ, Wanderling J, Siegel C, Green M. Pharmacotherapy of the hospitalized young adult schizophrenic patient. Compr Psychiatry 1988;29: 379-86.

27. Hohmann AA. Gender bias in psychotropic drug prescribing in primary care. Med Care 1989;27:478-90.

28. Blackwell B. Treatment adherence: a contemporary overview. Psychosomatics 1979;20:27-35.

29. Zito JM, Craig TJ, Vitrai J. Toward a therapeutic jurisprudence analysis of medication refusal in the court review model. Behav Sci Law 1993;11:151-63.

30. Lehman AF. A quality of life interview for the chronically mentally ill. Eval Program Plan 1988;11:51-62.

31. Davis JM, Schaffer CB, Killian GA, Kinard C, Chan C. Important issues in the drug treatment of schizophrenia. Schizophr Bull 1980;6:70-87.

32. Ray WA, Griffin MR, Schaffner W, Baugh DK, Melton LJ. Psychotropic drug use and the risk of hip fracture. N Engl J Med 1987;316:363-9.

33. Scheff TJ. Being mentally ill: a sociological theory. Chicago: Aldine, 1966.

34. Szasz T. The myth of mental illness: foundations of a theory of personal conduct. Rev. ed. New York: Harper and Row, 1974.

35. Estroff SE. Making it crazy: an ethnography of psychiatric clients in an American community. Berkeley, CA: University of California Press, 1981.

36. Haynes RB. A critical review of the "determinants" of patient compliance with therapeutic regimens. In: Sackett DL, Haynes RB, eds. Compliance with therapeutic regimens. Baltimore, MD: Johns Hopkins University Press, 1976:26-39.

37. Kane JM. Compliance issues in outpatient treatment. J Clin Psychopharmacol 1985;5:22S-27S.

38. Kelly GR, Scott JE, Mamon J. Medication compliance and health education among outpatients with chronic mental disorders (published erratum appears in Med Care 1991;29:889). Med Care 1990;28:1181-97.

39. Davis JM. Overview: maintenance therapy in psychiatry. I. Schizophrenia. Am J Psychiatry 1975;132:1237-45.

40. Tobias LL, MacDonald ML. Withdrawal of maintenance drugs with long-term hospitalized mental patients: a critical review. Psychol Bull 1974;81:107-25.

41. Appelbaum PS. The right to refuse treatment with antipsychotic medications: retrospect and prospect. Am J Psychiatry 1988;145:413-9.

42. Rachlin S. One right too many. Bull Am Acad Psychiatry Law 1975;3:99-102.

43. Brooks AD. The right to refuse antipsychotic medications: law and policy. Rutgers Law Rev 1987;39:339-76.

44. Hoge SK, Appelbaum PS, Lawlor T, et al. A prospective, multicenter study of patients' refusal of antipsychotic medication. Arch Gen Psychiatry 1990;47: 949-56.

45. Levin S, Brekke JS, Thomas P. A controlled comparison of involuntarily hospitalized medication refusers and acceptors. Bull Am Acad Psychiatry Law 1991;19:161-71.

46. Zito JM, Craig TJ, Wanderling J. New York under the Rivers decision: an epidemiologic study of drug treatment refusal. Am J Psychiatry 1991;148:904-9.

47. Zito JM, Hendel DD, Mitchell JE, Routt WW. Drug treatment refusal, diagnosis, and length of hospitalization in involuntary psychiatric patients. Behav Sci Law 1986;4:327-37.

48. Weiden PJ, Mann JJ, Haas G, Mattson M, Frances A. Clinical nonrecognition of neuroleptic-induced movement disorders: a cautionary study. Am J Psychiatry 1987;144:1148-53.

49. Gutheil TG. Drug therapy: alliance and compliance. Psychosomatics 1978;19:219-25.

50. Kramer PD. Listening to Prozac. New York: Viking, 1993.

51. Teicher MH, Glod C, Cole JO. Emergence of intense suicidal preoccupation during fluoxetine treatment. Am J Psychiatry 1990;147:207-10.

52. Rosenthal R. Experimental effects in behavior research. New York: Appleton-Century-Crofts, 1966.

53. Breggin PR, Ross Breggin G. Talking back to Prozac: what doctors won't tell you about today's most controversial drug. New York: St. Martin's Press, 1994.

54. Ruel JM, Hickey CP. Are too many children being treated with methylphenidate? Can J Psychiatry 1992;37:570-2.

55. Safer D, Krager JM. The increased rate of stimulant treatment for hyperactive/inattentive students in secondary schools. Pediatrics 1994;94:462-4.

56. Zito JM, Volavka J, Craig TJ, Czobor P, Banks S, Vitrai J. Pharmacoepidemiology of clozapine in 202 inpatients with schizophrenia. Ann Pharmacother 1993;27:1262-9.

57. Tardive dyskinesia: a task force report of the American Psychiatric Association. Washington, DC: American Psychiatric Association, 1992.

58. Pope HG Jr, Keck PE Jr, McElroy SL. Frequency and presentation of neuroleptic malignant syndrome in a large psychiatric hospital. Am J Psychiatry 1986;143:1227-33.

59. Sternberg DE. Neuroleptic malignant syndrome: the pendulum swings. Am J Psychiatry 1986;143:1273-5.

60. Raskind MA, Risse SC, Lampe TH. Dementia and antipsychotic drugs. J Clin Psychiatry 1987;48(5S):16-8.

61. Burd L, Fisher W, Vesely BN, Williams M, Kerbeshian J, Leech C. Prevalence of psychoactive drug use among North Dakota group home residents. Am J Ment Retard 1991;96:119-26.

62. Jacobson JW. Problem behavior and psychiatric impairment within a developmentally disabled population. III. Psychotropic medication. Res Dev Disabil 1988;9:23-38.

63. Kane JM. Treatment of schizophrenia. Schizophr Bull 1987;13:133-56.

64. Casey JF, Bennett IF, Lindley CJ. Drug therapy of schizophrenia. Arch Gen Psychiatry 1960;2:210-19.

65. Cole JO. Phenothiazine treatment in acute schizophrenia. Arch Gen Psychiatry 1964;10:246-61.

66. Kane J, Honigfeld G, Singer J, Meltzer H, Clozaril Collaborative Study Group. Clozapine for the treatment-resistant schizophrenic. Arch Gen Psychiatry 1988;45:789-96.

67. Edlund MJ, Craig TJ, Richardson MA. Informed consent as a form of volunteer bias. Am J Psychiatry 1985;142:624-7.

68. Hogarty GE, Schooler N, Ulrich RF, Mussare F, Ferro P, Herron E. Fluphenazine and social therapy in the aftercare of schizophrenic patients. Relapse analyses of a two-year controlled study of fluphenazine decanoate and fluphenazine hydrochloride. Arch Gen Psychiatry 1979;36:1283-94.

69. Lieberman J. Cholinergic rebound in neuroleptic withdrawal syndromes. Psychosomatics 1981;22:253-54.

70. Chouinard G, Jones BD. Neuroleptic-induced supersensitivity psychosis: clinical and pharmacologic characteristics. Am J Psychiatry 1980;137:16-21.

71. Herz MI, Melville C. Relapse in schizophrenia. Am J Psychiatry 1980;137:801-5.

72. Carpenter WT, Heinrichs DW. Intermittent pharmacotherapy of schizophrenia. In: Kane J, ed. Drug maintenance strategies in schizophrenia. Washington, DC: American Psychiatric Press, 1984:69-82.

73. Kane JM. Dosage reduction strategies in the long-term treatment of schizophrenia. In: Kane JM, ed. Drug maintenance strategies in schizophrenia. Washington, DC: APA Press, 1984:1-12.

74. Craig TJ. Medication use and deaths attributed to asphyxia among psychiatric patients. Am J Psychiatry 1980;137:1366-73.

75. DiMascio A. Behavioral toxicity. In: DiMascio A, Shader RI, eds. Clinical handbook of psychopharmacology. New York: Science House, 1970.

76. Van Putten T, Marder SR. Behavioral toxicity of antipsychotic drugs. J Clin Psychiatry 1987;48(Suppl):13-9.

77. Rifkin A, Quitkin F, Klein DF. Akinesia. Arch Gen Psychiatry 1975;32:672-4.

78. Van Putten T. The many faces of akathisia. Compr Psychiatry 1975;16:43-7.

79. Guyatt G, Sackett D, Taylor DW, Chong J, Roberts R, Pugsley S. Determining optimal therapy—randomized trials in individual patients. N Engl J Med 1986;314:889-92.

80. Baldessarini RJ, Cohen BM, Teicher MH. Significance of neuroleptic dose and plasma level in the pharmacological treatment of psychoses. Arch Gen Psychiatry 1988;45:79-91.

Chapter 25

Cultural Issues
in the Practice of Pharmacy

Pedro J. Lecca

INTRODUCTION

The future of pharmacy as a health care profession lies in its ability to add to the rational use of medication in health care. If pharmacy is to contribute in a comprehensive way, social and behavioral aspects of our consumer/patient population must be understood in this period of diversity. As pharmacy enters the twenty-first century as drug counselors and outcome monitors for other health professionals and expands its clinical responsibilities in the health care delivery system, it becomes essential for pharmacists to understand and be sensitive to cultural issues.

The level of cultural assimilation and integration is different within each ethnic group depending on its attitudes and how long its members have resided in the United States. In providing services to the culturally different client, it is crucial that pharmacists be aware and take into consideration the cultural factors and level of integration that will affect the process of assessment, diagnosis, and treatment of the patient. Furthermore, it is important in the assessment process to determine what is normal in the context of the patient culture. Culture refers to a total way of living which includes one's values, beliefs, norms, and traditions. For example, when assessing the patient, does the patient belong to a nuclear family or an extended family? The importance of the extended family is more significant than the nuclear family, as valued by many individuals of the host culture. Extended family members include aunts and uncles, grandmothers and grandfathers, cousins, etc.

In the Hispanic/Latino culture, males are expected to be strong, dominant, and the providers of the household. Thus, some Hispanic/Latino males may have difficulty expressing their feelings and emotions to practitioners because this could be viewed as a sign of weakness. Hispanic/Latina females, on the other hand, take on the role of being submissive, self-sacrificing, and respectful.

Conversely, in African-American families, females may be heads of households, children may take on parental roles in caring for younger siblings, and aunts may be primary caregivers. Thus, the pharmacist will need to assess the patient from the patient's perceptions.

Language also plays a major role in the assessment process. Misinterpretations of language usage or communication difficulty due to language barriers may lead to inappropriate or inaccurate assessments. For example, slang terms used by the culturally different patient may not be understood by or may have a different meaning for the pharmacist.

Nonverbal messages from the patient must be taken into consideration, as the minority patient may be reserved or hesitant upon receiving advice. In the American Indian and Japanese cultures, lack of eye contact is a sign of respect. If the pharmacist is not culturally competent, this behavior may be interpreted as a lack of interest on the part of the patient.

Another example of inaccurate assessment is when pharmacists make judgments based on a lack of cultural understanding. A sick baby with small razor nicks and bruises on its body, a healthy adolescent who is getting thin, a woman who is beaten by her husband: these are all situations seen routinely at health and mental health agencies. Many health professionals would assess the clients as being victims of child abuse, anorexia, and wife battering, respectively. However, these assessments are inaccurate when the situations are complicated by the patients being refugees from Vietnam, Cambodia, and Laos.

In many cultures, bloodletting and coin rubbing are the healing techniques used to balance the fluids and air which would, in turn, keep a baby healthy. In respect to the wife battering, in many cultures corporal punishment has been a familiar and acceptable means of restoring harmony to the family when proper conduct and attitudes are violated. This parallels the hierarchical forms of authority and male domination that are part of many cultures. Many refugees who come to the United States experience digestive problems due to a new diet, which may lead to significant weight loss. From these examples, one can observe that it is imperative for pharmacists to incorporate cultural understanding when working with culturally different patients.

CULTURAL PRACTICE ISSUES

The health professionals of today are no longer faced only with patients of the dominant culture; they also serve patients who are of different ethnic backgrounds and therefore different needs. Because of their lack of knowledge of the diverse backgrounds of their patients, a majority of these professionals have failed in their attempts to assist such a diverse clientele. Their lack of cultural awareness has created barriers between them and many minority patients. The prejudice and discrimination that have been directed toward minority groups and the lack of sensitivity of many health professionals have contributed to the creation and strengthening of such barriers, which, in turn, has led to the lack of utilization of health facilities by minorities. These practitioners are coming to the hard realization that they lack the training and knowledge to assist their diverse patients.

Culture can be described as consisting of all those things that people have learned and believe, their values, their religion, and their joy in their historical roots. It is the ideals, beliefs, skills, tools, systems, and institutions into which each member of society is born. Therefore, cultural sensitivity could be described as an awareness and understanding of the ideals and characteristics of different cultures. It is cultural sensitivity that the health practitioner must work to enhance. To become culturally sensitive, the practitioner must address factors such as language, stereotypes, racism, and counseling.

Language is a major barrier between the practitioner and the minority patient. Often practitioners are faced with patients who do not speak fluent English, if any at all. Such a barrier, if not addressed correctly, could greatly hinder the patient's understanding. For example, if a practitioner is questioning a client who cannot speak fluent

English, the responses to questions will depend not on what the practitioner asks but on how the patient interprets the question.

Because communication is an important aspect of practice, language can place the patient at a great disadvantage if the patient cannot interpret instructions correctly. A minority patient's responses could cause practitioners to misinterpret the patient's comments. The practitioner may misunderstand the patient as sullen or nonresponsive due to the language barrier.

One way to combat such misunderstandings and miscommunications is to employ well-trained translators who can communicate the practitioner's questions as well as the patient's responses to the other party. An even more efficient way of bridging the language gap between practitioner and client is for the practitioner to take crash programs for language acquisition. Monolingual practitioners should be able to communicate with the patient without the use of a third party.

But most important, practitioners should understand the culture of the client so as not to offend him or her with the words they use. For example, phrases such as "Indian giver" and "third-world people" can be considered insulting by various minority groups. Indian giver connotes someone who gives a gift and then takes it back. The term third-world people brings up the question of what is a first-world person and whether a first-world person is considered to be better than a third-world person. Practitioners should consider what it is they are saying before they say it.

Stereotypes and racism have also been sensitive issues for minorities. The opinions practitioners may have are sometimes unknowingly exhibited to the patient which, in turn, can cause distrust between the practitioner and the patient. Practice involves a certain level of trust between the practitioner and the patient. If such a trust is not found, patients will be less willing to self-explore and to self-disclose important information to the pharmacist.

Ethnic stereotyping may produce untoward reactions in either or both parties, including fear, contempt, resentment, pity, aggressiveness, mistrust, and guardedness. It is very important that practitioners not base their assessment on stereotypes they have formed about different cultural groups. Instead, they should realize that people cannot be placed in categories according to their cultural background.

Language could consciously or subconsciously be used as a tool of racism by the practitioner. Because Western society places such a high premium on a person's use of English, it is a short step to conclude that minorities are inferior, lack awareness, or lack conceptual thinking powers. Such misinterpretations are most clearly seen in the use and interpretation of psychological tests. IQ and achievement tests are notorious for their language bias.

Whatever the cause of the discrimination, practitioners should realize that by discriminating against the patient they are ignoring the cultural strengths of the minority group and unfairly blaming the victim for his or her problems. They fail to understand and appreciate the diversity of the minority patient.

The pharmacist should understand that ethnic clients usually bring their own misunderstandings and resentments about the dominant culture with them into their relationship with the staff. These misunderstandings between the practitioner and the patient prevent communication and the establishment of an individualized relationship. Therefore, it is very important that the pharmacist understand that both the pharmacist and the client will enter the situation with certain misunderstandings and that the pharmacist be willing to correct those misunderstandings. One method to ease such tensions is for practitioners to always be in pursuit of any information that can assist them in their goal of competency. Their own patients/customers are a good source of such information.

Pharmacists should not be afraid to ask questions pertaining to the patient's culture. The family's self-identification and self-description are sometimes the most important clinical data for the practitioner. By asking questions, the pharmacist also acknowledges that the patient knows more about himself than the pharmacist does. Asking questions allows both the pharmacist and the patient to acknowledge differences between their cultures and to establish common points of contact. It also allows the practitioner to identify and empathize with the minority patient.

LACK OF CULTURAL UNDERSTANDING

Another major problem that arises due to cultural ignorance is the fact that culturally ignorant practitioners tend to stereotype the different ethnic groups and think that a single method can be used to assist all people of all ethnic backgrounds. In fact, the method used to help minorities is the one that was actually created to help the dominant group. In other words, practitioners treat all patients the same, regardless of their color or ethnicity.

Considering the fact that the dominant culture does not contain the same value system as many minority cultures, the dominant method of practice cannot adequately deal with the problems of minority families in America. By ignoring the racial differences, pharmacists tend to categorize any differences that they do view during practice as forms of resistance by the patients. Hence, the patients do not receive the adequate counseling they deserve from their pharmacist, and issues about their own culture and identity are left unresolved. To ensure that appropriate practice is provided for the minority patient, pharmacists should learn about the client's culture. Pharmacists must understand and learn to relate to the cultures of the patients they are counseling.

In some cultures, such as the Hispanic/Latino culture, body language is just as important as—and sometimes even more important than—verbal conversation. Therefore, pharmacists must pay close attention to how the patient behaves. Characteristics such as body posture, voice tone, and facial expressions carry greater importance in cross-cultural understanding. By paying attention to the patient's body language, pharmacists will be able to gain much valuable information about the patient's views about certain issues.

HOLISTIC ASSESSMENT

The pharmacist should understand that the community and the family are intricate parts of the patient's life in some cultures. Therefore, the family unit should be incorporated in aspects of practice. In other words, the practitioner should ensure that the family plays a major role in the treatment of the patient. The pharmacist, when providing good solutions for patients of different cultures, should have an understanding of the communities in which the patients reside. By understanding the community, the pharmacist can make sure the solutions validate the patient's life strategy.

The pharmacist must also be able to identify and empathize with minority patients and make an effort to combine the patients' own cultural values and ideals into the treatment advice given by the practitioner. By making such an effort, the pharmacist is building cultural competency and telling the patient that he or she understands and values the patient's beliefs, thereby building a foundation of trust. To do this, the pharmacist should put aside his or her own culture temporarily and play by the patient's cultural rules. By seeing elements from the patient's own culture, the pharmacist will be able to incorporate the characteristics of the patient and tailor the treatment to fit those characteristics. Hence, a pharmacist will be able to find a method of care that best fits the minority group. The core focus of this method would be the needs and values of the minority group.

A good example of such a methodology would be *cuento* treatment. *Cuento* treatment is a culturally sensitive treatment modality for Puerto Rican children, in which Puerto Rican folktales about strength and perseverance are told to Puerto Rican children by their mothers. It is a culturally sensitive form of treatment because it is very accessible to the patients as it is the mothers who are providing the treatment to the children. Furthermore, it is administered by the culture's prime agent of acceptance, thereby rendering the treatment highly accessible and acceptable.

CULTURALLY SENSITIVE PRACTICES

The practices by which pharmacists are employed must also become culturally sensitive to their minority clientele. Pharmacy practices must have a thorough knowledge of the cultural factors that are particular to the target cultures. The practices should then incorporate the cultural knowledge into the framework and adapt to the cultural diversities of their patients. For example, they could create specific minority programs that would accommodate the needs of different minority groups. By infusing the cultural characteristics of their target populations into the assessment process, pharmacies can provide more accessible treatment to minority groups. For instance, because the family is a very important part of the Hispanic/Latino population, a pharmacy that has targeted such a population would provide family assessment as well as at-home assessments for its Hispanic/Latino patients. Therefore, the pharmacy would not only incorporate the family variable into its assessment but would also make such assessment more accessible for family members by providing the assessment at the home.

Pharmacies should also create special advisory groups composed of minority members to address the issues of that ethnic group and create evaluation committees to assess the cross-cultural performance of the group. These members could then assist the pharmacy in developing a methodology and an implementation plan that outlines goals they wish to accomplish within a set time period with the target populations. By doing this, standards could be set for the practice to follow, and cultural diversity could be incorporated into the service delivery system.

The owner of the pharmacy must also assist the practice in its drive for cultural competency. The owner should use some of the practice's financial resources to build cultural competency in the staff and service delivery by setting strict requirements in the hiring and training of employees. Furthermore, the owner should increase the community involvement in the practice so that pharmacies can become more accepted and more visible in minority communities.

CULTURE AND PHARMACY MANAGEMENT PRACTICE

The pharmacy management should emphasize diversity in much the same way as the policy makers of large organizations lead their organizations toward cultural competency by expressing their goals to their staff. They need to assess the predominant minority groups in their sector and prepare themselves accordingly. They should then provide special training courses to give their staff knowledge about different cultures and use up-to-date statistical information about different minority groups so that they may serve each patient accordingly. Furthermore, pharmacy managers should use opportunities such as training seminars to inform their staff that their ethnic patients usually bring their own misunderstandings and resentments about the dominant culture with them into their relationship with the staff. Those misunderstandings should be addressed, and key solutions should be pinpointed. Pharmacy managers should also periodically evaluate their staff on their level of understanding and competence.

The managers should try to recruit minority individuals who fit the needs of the predominant groups of their sector and ensure that all of their staff are culturally competent. For example, the staff should consist of bilingual and bicultural personnel. If well-trained personnel who are bilingual as well as bicultural are employed, these staff members will not only be able to provide appropriate information to the pharmacist and the patient but will also be able to act as cultural consultants who can explain the key cultural beliefs and values of the patient to the pharmacist.

Regardless of whether it is the practitioner, pharmacy, manager, or the owner, the purpose of the treatment remains the same; to treat and help patients. In order to treat and help the minority patients effectively, the needs and values of the minority group should be understood and be at the core of the treatment process. In pharmacies that are successful, you will find that efforts have been made to have the assessment process fit the client rather than to make the patient fit the treatment.

THE FUTURE OF CULTURE IN YOUR PRACTICE

The disproportionate increase among the minority population has had, and will continue to have, an impact upon the workforce in the United States. According to the Census Bureau Report (1996) the United States will undergo a profound demographic shift, and by the middle of the next century only about half of the population will be non-Hispanic whites. By 2050, the bureau indicates that immigration patterns and differences in birth rates, combined with an overall slow-down in growth of the country's population, will produce a United States in which 53% of the people will be non-Hispanic whites, down from 74% today.

As this trend continues, it will come with more diversity, a phenomenon that will present a host of new economic and social issues, particularly for the health professions.

With respect to this impact on pharmacy and pharmacy practice, by the year 2000, minority groups will make up 40 to 50% of the population in the health delivery system. As America becomes ethnically diverse, it is vital that pharmacists engage in culturally competent behavior in the delivery of pharmacy services to the myriad of ethnic groups.

CULTURAL TRAINING

Training for the pharmacist is vital to increase the effectiveness of the nonminority staff. Dr. Chau, in a recent work, proposed a model for teaching cross-cultural practice. Chau uses a value continuum of cultural pluralism and ethnocentrism (the view that the mainstream culture is superior to ethnic cultures) as a basis for teaching cross-cultural practice. The teaching of culture-sensitive practice uses appropriate instructional approaches.

Appropriate instructional approaches are needed to train health service staff. These teaching approaches aim to go beyond cognitive learning, including approaches that are focused on affective processes and skill development, because these are useful in helping staff to deal with the diversities and biases encountered in cross-cultural situations. Another training approach emphasizes minority field experiences to produce affective and skills development needed for minority practice. Additional approaches include creative literature, culture immersion, and the special impact of field learning.

Techniques and theories such as total quality management and Theory Y can also aid practitioners in their development of culture-related training programs. Total quality management (TQM) has recently been implemented in some health service organizations and, like the cultural competency model, it emphasizes self-improvement; learning; and a flexible, innovative management style and technique.

There is a strong need for culturally competent programs to be implemented in both undergraduate and graduate pharmacy programs. In addition, training programs should be extended to existing pharmacies, in both the private and the public sectors.

CONCLUSION

The current literature stresses the importance of cultural competency, urging practitioners to use skill and knowledge that are culturally sensitive to prevent the underutilization and premature termination of pharmacy services by minority individuals. If pharmacists do not incorporate culturally competent skills and knowledge in their assessment, diagnosis, and treatment, many minority patients will fall victim to the "culturally encapsulated" practitioner. Moreover, cultural competency in assessment, diagnosis, and treatment is crucial to meet the needs of the growing minority population in a country where the paradigm of a bicultural individual is becoming a reality.

Integration of cultural competence into a pharmacy is very beneficial to the staff as well as to the patients we serve. The competency of the pharmacy to provide culturally sensitive services to its patients will lessen stress upon the service practitioner, and the patient will be more receptive to pharmaceutical care. When patients are receptive to services they will be able to integrate themselves more appropriately into the new paradigm of pharmacy care. The road to cultural competency is identifying the barriers, understanding the problems arising from the barriers, and implementing the solutions (Table 25.1).

TABLE 25.1. The Road to Cultural Competency

Cultural Barriers	Problems Arising from Barriers	Solution
Language	• Minority patient does not speak fluent English and does not understand the practitioner. • Misunderstanding of pharmacist comments and misunderstanding of patient comments may result. • Certain statements made by pharmacist may unknowingly insult patient (e.g., Indian giver).	• Employ well-trained staff. • Learn the language. • Think before you speak.
Stereotypes and Racism	Practitioner's opinions of patient may be apparent to patient . . . no trust in patient-pharmacist relationship . . . no individualized relationship is established.	Learn about the patient's culture by asking the patient.
Assessment	When standard assessment is used regardless of ethnic difference, the patient does not receive appropriate assessment.	• Acknowledge that the standard method does not adequately treat all patients. • Incorporate beliefs and values of the client into assessment and treatment (e.g., family assessment and home-based assessment).

The most important ethical standard in any pharmacy is to ensure that the helping profession is actually being helpful. The way to maintain this standard when working with minority groups is to be *culturally competent.*

READINGS

1. Abel EM. Working more effectively with minority clients: training model for social workers. In: King SW, Jarrett AA, eds. World view of ethnic minorities and gender: problems and prospects of the '90s. Arlington, TX: University of Texas at Arlington Campus Printing Service, 1992:44-57.

2. Barringer F. Asian population in U.S. grew 70% in the '80s. New York Times 1990, Mar 2:A14.

3. Blank R, Slipp S. Voices of diversity. New York: Amacom, 1994.

4. Brocka B, Brocka MD. Quality management: implementing the ideas of the masters. Homewood, IL: Richard D. Irwin, Inc., 1992.

5. Bureau of the Census: Washington, DC, 1996.

6. Chau K. A model for teaching cross-cultural practice in social work. J Soc Work Educ 1990;2:124-32.

7. Costantino G, Malgady RG, Rogler LH. Cuento therapy. Maplewood: Waterfront Press, 1985.

8. Culturally competent system of care. Vol. I. Washington, DC: CASSP, Georgetown University, 1987.

9. Dean RG. Understanding health beliefs and behaviors: some theoretical principles of practice. In: Watkins EL, Johnson AE, eds. In: Removing cultural and ethnic barriers to health care. Chapel Hill, NC: University of North Carolina, 1979:49-67.

10. Fong R, Mokuau N. Not simply "Asian American," periodical literature review on Asian and Pacific Islanders. Soc Work 1994;39:298-305.

11. Gardenswartz L, Rowe A. Managing diversity. New York: Pfeiffer and Co., 1993.

12. Issacs MR, Benjamin MP. Towards a culturally competent system of care. Vol. II. Washington, DC: CASSP, Georgetown University, 1991.

13. Jackson SE. Diversity in the workplace: human resource initiatives. New York: Guilford Press, 1992.

14. Lappin J. On becoming a culturally conscious family therapist. In: Hanson JC, Falicov CJ, eds. Cultural perspectives in family therapy. Aspen Publications, 1983:122-35.

15. McGill D. Cultural concepts for family therapy. In: Hanson JC, Falicov CJ, eds. Cultural perspectives in family therapy. Aspen Publications, 1983:108-19.

16. Smith WD, Burlew AK, Mosles MH, Whitney WM. Minority issues in mental health. Reading, MA: Addison-Wesley Publishing Company, 1978.

17. Sue DW, Sue D. Barriers to effective cross cultural training. Personnel Guidance J 1976;51:386-9.

18. Sue DW, Sue D. Counseling the culturally different. New York: John Wiley and Sons, 1990.

19. Thomas RR. Beyond race and gender: unleashing the power of your total workforce by managing. New York: American Management Association, 1991.

20. Wright RW Jr, Saleeby D, Watts TD, Lecca PJ. Transcultural perspectives in the human services: organizational issues and trends. Springfield, IL: Charles C. Thomas, 1983.

PART III:
REVIEW OF USEFUL
CONCEPTS/MODELS

Chapter 26

A Macro View: Public Policy

Lon N. Larson

BACKGROUND AND OBJECTIVES

Are medical care and health services acts of political philosophy? Many health professionals–and students–may initially respond to this question in the negative. Instead, they prefer to regard health services as acts of science or rationality. At a minimum, they view health services as above politics. Yet, upon reflection, medical care is indeed an act of political philosophy.[1] The health care system is a social system. As such, it reflects the social and political values of society. The organization of health services–the manner in which services are delivered and paid for–is determined by the preferences, beliefs, and biases of society. For instance, some of the factors shaping health service delivery in the U.S. include professional dominance in organization and financing issues as well as clinical decisions, scientific medicine and a "faith" in technology, and a reliance on the private sector to deliver services.

Other developed countries comparable to the U.S. in wealth and knowledge have developed their own unique health systems that reflect their beliefs and values. To illustrate the differences, the goals stated in the constitutions of the U.S. and Canada are revealing. In the U.S., the goals are "life, liberty, and pursuit of happiness," while in Canada, they are "peace, order, and good govern-

Portions of this chapter appeared in a prior work by the author and colleagues: Larson LN, Bentley JP, Brenton MA. Values, participatory democracy, and health care reform. Am J Pharm Educ 1994;58:417-21.

ment."[2] Not surprisingly, the health care system in the U.S. is marked by individualism and a distrust of government. In Canada, more emphasis is placed on the collective good, and government is viewed more positively. Neither system is right or wrong; each simply reflects differing values and political philosophies.

In a democracy, health policy expresses society's view of what roles are appropriate for government to play with respect to two fundamental issues. These are the government's role in promoting the health status of its citizens and in its role providing health care services. These two issues are not the same, and they are treated separately in this chapter. Much of this chapter is devoted to the values that underlie the responses to these two issues in the U.S. First, two other topics are discussed: types or categories of policies and a model of interest-group politics. These are instructive in understanding the substance and intent of policies and the process of policy making, respectively. The chapter concludes with a discussion of public participation in policy making. This deals with another facet of the policy-making process and emphasizes the importance of effective participation.

The purpose of this chapter is to present a sociological perspective of health policy, focusing on the process and values that shape government activities in the health sector. The chapter does not give a history of health policy in the U.S., nor does it present a "how-to" on making government work for you. Specifically, the objectives of this chapter are to enable the reader to:

1. Describe the mechanisms of action of public policies.
2. Describe the supply and demand model of interest-group politics as a process for policy formulation and enactment.
3. List and describe three approaches to health promotion and explain which one(s) dominate public policy.
4. Describe the values that underlie the delivery of health care services and related public policies and compare individual-centered and community-centered values.
5. Describe the role of the citizenry and public participation in developing health policy.

THE MECHANISMS OF ACTION
OF PUBLIC POLICIES

Government entities have several means of affecting the delivery and financing of health care services. The mechanism of action of a policy is the relationship(s) between the policy or program and its target clients. Three types of policy based on these relationships are regulatory, allocative, and inducement.[3] (A fourth type of policy—budgetary policy—is discussed later.)

Regulatory Policies

Regulatory policies seek to govern the behavior of individuals or corporations to protect or promote the public welfare. Some examples are:

- The Occupational Safety and Health Administration (OSHA) regulates the workplace for safety and health threats.
- State pharmacy boards regulate admission into the profession, and they regulate the quality of practice.
- The Joint Commission on the Accreditation of Healthcare Organizations (JCAHO) regulates the behavior of hospitals and other organizations.
- The Food and Drug Administration (FDA) regulates the marketing of pharmaceutical products.
- Several states have laws requiring the use of seat belts, limiting smoking in public areas, and placing other restrictions on personal behaviors.

A variation on the theme of regulation, in terms of influencing behaviors, is tax policy. Taxing the behavior increases the cost of the behavior and thereby discourages its performance. For instance, cigarette and alcohol taxes discourage the use of these products.

Regulatory polices are common and affect many areas of everyday life. Regulations can be enforced by varying types of organizations. Some are purely public agencies (the FDA, OSHA); others are private (JCAHO); and finally, some are public agencies with independent boards (e.g., pharmacy boards are governed by members of the profession who are not public employees). Generally, violations of the regulations result in some form of punishment. The incentive for obeying the regulation is to avoid negative consequences.

Allocative Policies

A second type of policy is allocative. Allocative policies involve the delivery of services or income to designated individuals or communities. The potential recipient(s) must meet specified criteria to be eligible for receiving services, and often they must apply for services. Examples of these policies are the Medicare and Medicaid programs, in which government buys medical services from the private sector for the elderly and indigent, respectively. The Hill-Burton program, which built hospitals after World War II, is an example of an allocative policy in which communities were the recipients. Another allocative policy is the system of Veterans Affairs Medical Centers, which provide health services to former members of the armed services. Like Medicare, VA services are made available to a designated group, but unlike Medicare, services are provided directly by government entities rather than purchased through the private sector. In sum, allocative policies provide services to designated recipients. The services may be provided directly by a government agency, or the recipient may receive purchasing power to buy the services from the private sector.

Other allocative policies are less direct, but they allocate resources nonetheless. For instance, health care is the subject of many special provisions (i.e., tax breaks) in the tax code. Most notably, employer contributions to employee health benefits are tax-free income for the workers. The resulting incentive is to seek more comprehensive insurance coverage than if premiums were taxable income. This is a major subsidy on the part of the government to middle and upper income persons.[4] In 1991, the total tax revenue lost from special tax treatment of health care amounted to $53 billion.[5] As a means of comparison, in the same year, the amount spent on care for the poor through Medicaid was $90 billion.[6]

Medicare and Medicaid also exemplify the third type of policy—inducement policies. This type of policy provides incentives for private firms or individuals to help implement the policy. For instance, Medicare and Medicaid must be sufficiently attractive—

financially and otherwise–for providers to participate in them. The policy induces or encourages the private-sector providers to participate and provide services to eligible patients. Similarly, the orphan drug provisions are an incentive (inducement) for drug manufacturers to develop products that may be used for only a small number of persons. A final example is the former manpower policy of paying health professions schools to increase their class sizes (thus, receiving their cooperation in accomplishing the policy objective of increasing the supply of health manpower).

Multiple Policies

A single policy may include more than one of these mechanisms. For instance, Medicare has aspects of all three: participating providers must abide by the program's rules (regulatory), purchasing power is distributed to eligible persons (allocative), and the program's provisions must be sufficiently attractive for providers to induce their participation. Tax policy can also be used to discourage or encourage behaviors, and it has allocation implications (as described above).

Budgetary Policies

A fourth type of policy is budgetary policy. Budgets are not really part of the classification scheme mentioned above because they do not represent a unique mechanism of action. Rather, a budget provides the resources or fuel through which the other types of policy can become reality. Regulation cannot be strong if the regulatory agency's budget is low, and an allocative policy is of dubious significance if it does not have many dollars to allocate.

A budget reveals the relative importance of the competing objectives that government seeks to accomplish. For instance, in 1993, government spending (federal, state, and local) on personal health services amounted to $337 billion, while government spending on public health activities was about $25 billion.[6] Given these figures, which do you suspect has a higher priority in U.S. health policy: treating illness (via personal health services through Medicare or Medicaid) or preventing illness (via community health and public education programs)? As we will see later in the chapter, this difference reflects the relative priority given to the germ theory and scientific medicine versus the priority placed on the life-style and environmental theories as means of promoting the public's health.

Between 1980 and 1990, total spending per welfare recipient (including cash assistance, housing assistance, food benefits, jobs training, medical care, etc.) increased $616 (in constant dollars), but spending on medical care increased $682. In other words, even though overall welfare spending increased, spending for nonmedical assistance fell.[5] Further, during this period, cash aid through Supplemental Security Income (for the elderly and the disabled) increased $304 per recipient, and cash aid for Aid to Families with Dependent Children fell $119. From these data, what is the relative priority of medical care vis-à-vis other human services? Who has a higher priority in health policy: the elderly or children?

From a budgetary perspective, distinguishing between two types of programs is important. One type consists of programs that receive an annual appropriation, which is the amount they are allowed to spend during the year. They have a fixed budget. The Veterans Affairs medical system is such a program. The other type is entitlement programs. In these programs, persons are entitled to services if they meet the criteria for eligibility. Entitlements are exemplified by Medicare and Medicaid. Spending for entitlement programs is more difficult to control because the number of eligibles may change, the quantity of services used by each eligible person may change, or the prices of the services used may change. All three of these variables are potentially uncontrollable by the government (unless the law establishing the program is changed). As a result, controlling the expenditures of entitlement programs may be difficult, as evidenced by the cost problems faced by Medicaid and Medicare over the years.

Summary

In summary, there are four types of policy:

- Regulatory policies, which prescribe or proscribe behaviors.
- Allocative policies, which distribute services, income, or purchasing power to eligible recipients.

- Inducement policies, which encourage behaviors.
- Budgetary policies, which establish priorities among competing programs and services.

The first three are the tools available to governments to achieve their objectives of improving health status and improving the delivery of health services. Budgets reveal the relative commitment to accomplishing those objectives.

THE DEMAND FOR
AND SUPPLY OF PUBLIC POLICY

This section focuses on the process of making policy decisions, with special emphasis on the role of interest groups in that process. The model presented here was developed by a health economist, Paul Feldstein.[4,7,8] Feldstein views the policy process in a supply-and-demand context. Members of the legislative and executive branches of government are the suppliers of policies. Policies consist of legislation or laws enacted by the legislative branch, as well as the rules for implementation which are written and enforced by members of the executive branch. These individuals write legislation and related rules; they design and implement policies.

On the other side, interest groups are the demanders or purchasers of policies. Legislation redistributes wealth, so legislative decisions are important. Feldstein claims that interest groups define the public interest in terms of the economic interests of their members, despite their public pronouncements of promoting the public interest. In other words, interest groups support policies that promote the economic well-being of their members and oppose policies that impair those interests. For instance, one might expect the American Association of Retired Persons (AARP) to oppose any reductions in Social Security or Medicare benefits.

Similarly, hospital associations will support legislation that promotes the economic interests (i.e., current incomes) of hospitals. Medical associations will do the same for physicians, pharmacy associations for pharmacists, and so on. More specific policy goals of a health association in supporting legislation are to: increase the demand for its members' services, reduce price competition faced by its members, secure the highest reimbursement for its members, and increase the prices of competitors or negate their competitive advantages.

What is the medium of exchange between the demanders and the suppliers of policy? In economic markets, money serves this purpose. In the supply and demand of policy, payment is in the form of political support. Policy makers want to acquire political support. For elected officials, this may be votes, campaign contributions, or endorsements. For the regulator or bureaucrat, the rewards may be support for expanding (or maintaining) the agency, pleasing key legislators with oversight of the agency, or avoiding the loss of political support through bad press. For example, the Food and Drug Administration (FDA) has been criticized for unnecessarily delaying the marketing of effective drugs. The agency risks a greater loss of political support if it allows a dangerous drug on the market than if it slows the marketing of an effective drug. In both cases, people are harmed, but there is less publicity associated with the second. Given this situation, the agency understandably has been cautious in its review process. Yet agency behavior has changed with drugs for HIV infection, which have been backed by a politically strong and highly visible constituency.

In many cases, the only groups working to influence a particular policy are the groups that are directly affected by the policy, that is, the groups that stand to gain or loose economically from a policy. Interest groups with economic self-interests in a policy area will likely have the most clout in shaping those policy decisions. This is because they are willing to provide political support for decisions that are favorable to their economic interests. Further, the level of interest (and level of political support) is directly proportional to the amount of gain or loss facing the interest group. For instance, changes in Social Security benefits directly affect the elderly, but the effect on younger adults is less noticeable. As a result, nonelders will provide or withhold less political support over the issue than the elders. Similarly, a state pharmacy association may be keenly aware of Medicaid pharmacy reimbursement. However, this is hardly a hot topic on main street, even though all citizens of the state are paying the bill. As a result, the pharmacy association will provide more political support on this issue than will other groups.

Why, then, can't an interest group simply dictate policy in its area? AARP cannot necessarily increase benefits for the elderly, nor can pharmacy associations set reimbursement at the level they desire. One reason is that the association may not be able to provide enough political support to justify the time and effort to legislate and implement its wishes. The costs for the legislator in terms of time and effort are perceived to be greater than the benefits in political support. A second reason is that other interest groups may be directly opposed to the legislation. Third, other interest groups, capable of providing more political support, may have other agenda items that are attracting the attention of the suppliers. For instance, even if no interest group is opposed to an increase in pharmacy reimbursement, legislators may gain more political support by using the funds in other ways (e.g., tax cuts, other services) or by addressing an entirely different policy issue.

Finally, Feldstein suggests that in certain cases an interest group may support legislation that does *not* seem to be in the best economic interests of its members. This occurs when the cost of supporting a policy is perceived to exceed the benefits of the policy. A policy, although apparently not in the best interests of the members, may be supported because it is better than the alternative. Feldstein cites a professional association supporting mandatory continuing education because it is preferable to reexamination for continued licensure. In other cases, an interest group may want to avoid the negative image that may arise from opposition to a particular policy. In short, an interest group taking a position that appears inimical to its self-interests does not necessarily refute Feldstein's model.

In summary, in Feldstein's model of interest group politics, policy decisions are the result of interest groups providing political support to policy makers in exchange for decisions that enhance the economic well-being of the group's members. Further, a group's involvement (and disbursement of political support) in an issue is directly proportional to the impact of the issue on the group's economic status.

APPROACHES TO IMPROVING HEALTH STATUS

One of the overall goals of public policy is to increase the healthiness or health status of the population. To accomplish this goal, policies must address or alleviate the causes of ill health. Thus, the perceived cause(s) of ill health become critically important from a policy perspective because they provide the intellectual framework within which policy is developed. The relative priorities assigned to alternative causes determine the priorities and scope of health policy. Three theories are available to explain the cause of disease. These are germ theory, life-style theory, and environmental theory.[9]

Germ Theory

In germ theory, disease is caused by identifiable events within the body, including chemical or hormonal imbalances, as well as by germs or infectious agents. In this context, the germ theory is not limited to infections but encompasses any physiological and biochemical process that may lead to disease. The germ theory goes hand-in-hand with scientific medicine. Policies based on the germ theory emphasize personal health services (i.e., services to the individual rather than to the community or society), scientific research, and medical technology. With the germ theory, an individual is not responsible for his or her disease because the disease is beyond the individual's control. The germ theory leads a health system focused on the individual rather than the community; each body/mind suffers its own events. Preventive medicine focuses on screening, early detection, and vaccinations. Again, these are services provided to individuals.

Life-Style Theory

Life-style theory posits that ill health is caused by one's personal habits and life-style. Diet, exercise, smoking, drug consumption, and sexual behavior are life-style characteristics that affect health status. Unlike germ theory, life-style theory places responsibility on the individual person for health status. This can lead to "punishing" certain patients (e.g., if alcoholics are denied liver transplants). Like germ theory, life-style theory leads to an individual-oriented health care system: the life-style and behaviors of the individual are the cause of disease. Public health policy (e.g., programs handled by the Public Health Service) is largely based on the life-style theory. Health education programs and regulations to restrict behaviors (e.g., required seat belt use, no smoking areas) are examples of life-style-oriented policies.

Environmental Theory

The third theory of disease causation is the environmental theory. This theory focuses on environments–physical, occupational, social, economic–as the causes of disease. The physical environment includes clean air and water, as well as a safe food supply. However, social and economic environments also influence health status. Even in countries where there is free care, lower socioeconomic classes have lower health status than higher socioeconomic classes.[1] Mechanic emphasized that health education programs (designed to change life-styles) are often unsuccessful because they are not reinforced by social norms.[10] As an example, programs to reduce and relieve stress are not likely to affect behavior without changes in work habits, social definitions of success, and the quest for financial gain.

The environment is a major factor in health status. Some authors suggest that gains in health status over this century are due more to improvements in the physical environment (e.g., water and sewer systems), improved diet and personal hygiene, and increasing economic status than to advances in medical care.[1] Some have suggested that a critical element in improving health status is to enhance education (not health education, but education in general) because it is so closely linked to psychological modernity and health status.[10]

The environmental theory emphasizes communities rather than individuals. Health status, according to this theory, is dependent on socioeconomic status (i.e., income, education, type of occupation) and the physical environment. Improving health status requires changes at the community or societal level. Unlike the previous theories, the environmental theory places the community rather than the individual at the center of the health care system. Policies based on the environmental theory include public health programs related to safe drinking water and waste disposal, pesticide control, occupational health programs, antipoverty programs, aid to education, and community development programs.

Summary

To summarize this section, our views as to the cause of disease establish the framework for health policy. To alleviate a problem, policies must address its causes. All three of the theories discussed above are legitimate and play a role in promoting health status. In assessing policy, the issue is the relative emphasis or priority given to each. In the U.S., health policy has been dominated by the germ theory. Major policy initiatives have focused on developing and delivering medical services (curative and preventive) delivered to individuals. Policies associated with the germ theory include National Institutes of Health (research), Hill-Burton (hospital construction), manpower legislation, and Medicare and Medicaid.

Life-style theory is the basis for much of current public health policy in the U.S. Life-style theory suggests that an individual's living habits underlie health status. Policies based on this theory include smoking cessation and education programs, drug abuse education, sex education, and behavior regulations (e.g., mandatory seat belt use, no smoking areas).

Environmental theory is the theory that is possibly least readily accepted by health care professionals (including pharmacy students). We have difficulty accepting the idea that social structures—rather than our therapeutic prowess—are the major determinant of health status. Further, a community-oriented approach to improving health status is incongruous with the basic beliefs in an individualistic society such as ours. Yet the relationship between medical services and health status is increasingly dubious.[11] Ultimately, giving this theory a higher priority may be the most effective (but also the most difficult) of policy initiatives. Policies based on the environmental theory include clean water and air legislation, the Environmental Protection Agency, the Occupational Safety and Health Administration, community development programs, and antipoverty programs.

VALUES IN HEALTH CARE SERVICES DELIVERY

The previous section dealt with the policy issue of how to improve the health status of the population. As discussed, health services are not the sole determinant of health status. Nevertheless, the delivery of health services is an important topic in public policy. Health services (like food, housing, education) are viewed differently from other goods and services because of their impact on the quality of life and the fulfillment of human capacity.

This section focuses on the delivery of health services and, more specifically, on the values that underlie its related health policy. Values are the basic characteristics or attributes that we desire in our health care system; values lie at the core of public policy. Public policies attempt to correct or modify situations so that our values are more fully realized. For instance, one value underlying health policy is fair access. The priority given to this value by society is sufficient to enact Medicare and Medicaid but not sufficient to enact national health insurance. Public policy reflects society's priorities among values. With the recent public discussion of health care reform, attention has been given to stating the values that underlie the provision of health services.[12-15]

To illustrate the role of values in shaping policy with respect to health services delivery, two sets of health care values are compared here: individual-centered values which underlie our current system and community-centered values. With the former, the individual patient and provider reign supreme. They make decisions in their best interests with little or no regard for the interests of the community or society. The community-centered system views health care services as a community resource, with everyone sharing its benefits and burdens.

Individual-Centered Values

Our current system, with its emphasis on individual values, is marked by five priority values.[14,16] One of these is provider autonomy. Provider autonomy allows practitioners the freedom to practice their profession as they see fit, the opportunity for just compensation, and the right to refuse patients. It also allows the profession to regulate its members. Over the past century, the organization and financing of services in the U.S. have been greatly influenced by professional dominance.[17] Providers, especially physicians, have been free to practice as they see fit with little outside oversight and interference. In addition to this clinical independence of the individual practitioner, the health professions have been allowed to regulate themselves by setting entrance requirements and monitoring their members.

A second priority value in our current system is consumer sovereignty. This is the right of patients to select their providers, either individual providers or insurance plans. (A related point of the patient's right to decide treatment is included in the value of respect for patients.)

A third priority of the current system is high quality care: the right procedures are done in the right way, and customer (patient) expectations are met. In our current system, high quality is frequently equated with aggressive, intensive interventions involving sophisticated technology. Thus, we have the technologic imperative, that is, the belief that good medicine is using every technology available. Related to quality, the search for new procedures and technologies (i.e., research, development, and innovation) is also a priority in our current system.

The fourth priority is patient advocacy. This is the obligation of providers to be advocates for their patients and to promote the best interests of each patient. Providers are expected to act in the best interests of their patients. However, often this has been interpreted as doing everything medically possible for patients.

The fifth value of high priority in an individual-oriented system is respect for patients. This value involves several ethical aspects of the provider-patient relationship, including recognizing the patient's right and responsibility to make informed, voluntary decisions about care (patient autonomy and informed consent), maintaining the confidentiality of information, treating patients with respect and dignity, and promoting caring relationships between providers and patients. The effectiveness of the current system in realizing these attributes is debatable. Informed consent and confidentiality are common; patient decision making (autonomy) and caring relationships may be more suspect.

Community-Centered Values

A system based on community-centered values offers a contrast to the current system.[14-16] The key or priority values in a community-centered system include fair access, efficiency and wise allocation, fair burdens, respect for patients, quality care, and patient advocacy. While the last three are priorities of both sets of values, they are viewed somewhat differently in each.

The value of fair access includes two concepts. One is that everyone has access to essential or basic health care services without financial or other barriers. The second concept is that persons with similar medical needs are treated similarly without regard to socioeconomic status. In other words, there is not a multitiered system or separate-but-equal systems. The value of fair access recognizes the fundamental importance of health services. Health care is fundamentally important because it allows individuals to realize their potential as human beings, relieves pain and suffering, prolongs life, and gives information to plan life.[12,13]

In Priester's description of a community-centered system, fair access is the preeminent value.[14] In a conflict with any other essential value, fair access should be given priority. For Priester, fair access means that everyone can obtain a basic level of care. This is a minimum level of care below which no one falls. It does not imply universal access to all services, but universal access to basic services. His formulation of this value does not prohibit individuals from purchasing nonbasic services with their own financial resources.

Another high-priority value in a community-centered system is efficiency and wise allocation. In an efficient system, resources are used wisely, and they generate good value. Efficiency involves providing only necessary care (allocative efficiency) as well as producing necessary services in the least costly manner (technical efficiency). Efficiency is a high-priority value because resources are scarce and they have an opportunity cost. A dollar spent on a particular health service cannot be spent on another health service or on nonhealth services. Thus, any unproductive use of resources or inefficiency is to be avoided because such waste impedes the ability to accomplish other goals, including fair access. Wasting limited resources is economically and ethically undesirable. In addition, an efficient system is organized simply and is easy to use.

This value also requires that health care spending be balanced with other community needs (a facet of allocative efficiency). Limits on health care spending are set in a fair and open manner. Further, within the health care system, resources are allocated among the various services according to community needs. Finally, limits on the use of beneficial services are guided by publicly developed principles for rationing.

A third value of importance in a community-centered system is fair burdens. Fair burdens means that the costs of the health care system are spread across the entire community. Further, the costs are distributed according to ability to pay (i.e., through a mechanism like the progressive income tax). In contrast, our current system of monthly premiums that are virtually equal for everyone in an employee group is equivalent to a head tax and is not based on ability to pay.

As in our current system, the values of respect for patients, quality care, and patient advocacy are high priority values in a community-centered system. However, they are given different emphasis. Respect for patients in a community-oriented system involves making the patient the key decision maker (true patient autonomy and not merely obtaining informed consent). It also envisions providers as caregivers and not just curegivers. Patients are treated with respect and dignity. In essence, this value places the patient at the center of the health care system and makes the patient the key decision maker. Quality care in the community-centered system focuses on achieving patient-desired health status outcomes and satisfying patient expectations (in contrast to the technology-focused definition of quality in the current system). In a community-centered system, patient advocacy is done within previously established practice guidelines. Providers still serve as advocates for their patients and act in their patients' best interests, but providers are not compelled to do everything that medical science has to offer. Instead, they advocate for their patients within the approved boundaries of the system.

Two values of prime importance in the individual-centered system are absent from the list of high-priority values in a community-centered system: consumer sovereignty and provider autonomy. While not high-priority values, they are desirable in a community-centered system because they can help fulfill other, higher priority values. For instance, consumer sovereignty can improve quality and respect for patients by allowing patients to express their dissatisfaction through seeking another provider or plan. Similarly, provider autonomy can help realize other values of higher importance (such as quality care and patient advocacy). In sum, in a community-centered system, the values of consumer sovereignty and provider autonomy are desired as means to realize other values; they are not desired as ends unto themselves.

Use and Distribution of Services

The differences between these two sets of values may be most dramatic in how they allocate resources, the manner in which each system deals with such issues as the level of resources devoted to the health care system, the proliferation of new technologies, priorities among already available services, and who receives those services. The first four values described in the community-centered system (fair access, efficiency, wise allocation and fair rationing, fair burdens) deal with the use and distribution of resources.

Resource allocation is synonymous with rationing. Rationing occurs when "not all care expected to be beneficial is provided to all patients."[18] Medical technology has advanced to the point where–using this definition–rationing must occur in any system. We simply cannot afford to do everything that may be medically beneficial; rationing is imperative.[19] The question is not whether we are going to ration, but how. In the individual-centered system, rationing is based on ability to pay (or insurance coverage) and social class. And the rationing is done quietly. No formal, public decisions are made about the allocation of resources. In a community-centered system, the allocation issues and decisions are made explicitly and publicly, and they apply to everyone, regardless of socioeconomic status.

Two policy issues are closely tied to resource allocation.[11,20] One is the limits (if any) that are placed on the development and proliferation of technology in the medical care sector. Economists as well as ethicists argue persuasively that technology is the engine driving health care costs.[20-22] Currently, decisions to adopt or use technology are made quietly, without public debate. These decisions are made by individual institutions, organizations, professionals, and patients, guided by the incentives of insurance coverage. Hospitals invest in equipment, and physicians perform or prescribe the use of new procedures. Some patients demand to receive these services. Often the decision makers are immune from the cost considerations of their decisions. As a result, resources flow into the health sector, especially for personal health services, without restriction. This is the current method of allocating resources to the health care system and among services within the system.

Current efforts to control the proliferation and use of medical technologies focus on two areas: managed competition and outcomes research. The goal of managed competition is to enhance market forces and cost consciousness in health care utilization decisions. The assumption is that a price competitive marketplace will force managed care plans to use technology wisely, neither overusing it nor underusing it. The managed care plan that uses too much technology will have difficulty remaining price competitive, while the one that uses so little technology that outcomes suffer will be seen as a lower quality provider. Whether managed competition can effectively and efficiently employ technology is one of the key questions in health policy.

Outcomes research ties into the idea of developing practice guidelines to eliminate medically unnecessary procedures. Some believe that eliminating medically unnecessary procedures will solve the health care expenditure problem.[23] This idea is appealing. We all would like to hope that eliminating waste or unneeded services would be sufficient to solve our health care problems. This is a solution that harms no one. However, this may be wishful thinking. Callahan makes the argument that medical necessity is an elusive, ever-expanding concept that is ultimately ineffective as a basis for deciding what services should be made available.[22] Furthermore, even medically necessary services may not be affordable. For example, society does not provide children with every educationally necessary service or the defense department with every militarily necessary weapons system. Is society obliged to provide all medically necessary services?

The second issue relates to health services for the elderly. Is it in the community's best interests to have a medical care system (and a populace) whose motto is, "Never say die?" Or is it better to recognize and accept the human life span as limited? With programs like Medicare, health care services are requested by the individual but paid through community resources. Currently, Medicare recipients have a blank check to draw upon society's resources to whatever extent they want to satisfy their medical care demands. As the post World War II baby-boom generation reaches Medicare age, this issue raises very important ethical and economic issues. Callahan argues very persuasively that intensive lifesaving services should not be available to persons who have already lived an average life span, which he puts at 75 to 80 years of age.[22] Instead, he would put our emphasis on chronic care and palliative care, giving the elderly personal security and comfort, while accepting the reality of mortality and seeking to regain the meaning of growing old.

In sum, allocation decisions in our current system are made quietly and informally as individual providers and patients make investment and treatment decisions. In many markets, society is best served with individual buyers and sellers pursuing their best own interests. However, medical services are not the same as other goods and services. They affect human potential, and they are produced and used in the absence of economic incentives. A community-oriented system would include explicit decisions as to limits, priorities, and basic services; in other words, a system with explicit allocation decisions, a system that formally considers the needs and priorities of the community.

PUBLIC INPUT
AND PARTICIPATORY DEMOCRACY

As we have seen, health policy involves many difficult and emotional issues. How can such decisions best be made? Theoretically, cost-effectiveness analyses can tell us how best to spend a limited sum of money, but practically speaking, these analyses cannot give us the answers. While cost-effectiveness analyses are certainly useful, they do not offer us a complete way out of our dilemma. Can these decisions be left to panels of experts? Expert opinion is an important input, but such panels may not be effective in making allocation decisions. Many experts feel ill at ease offering opinions outside their area of expertise, and in their field, every procedure, screening test, and scientific advance is regarded as very significant. Having a group of experts allocate resources for the group's own area of expertise is not necessarily in the community's best interests.

Public Discussion and Input

An essential ingredient in making allocation and other policy decisions is public discussion and input: not just opinion polls, but discussion, deliberation, and debate. This would be participatory democracy. Jennings describes the need for and the needs of public discourse in resource allocation decisions.[24] He sees no single correct or best allocation pattern and suggests that allocation decisions would best be made in accordance with ethical criteria discussed at public forums. Jennings suggests that public forums can fruitfully discuss the importance or significance of such ethical criteria as the number of persons affected by a service, its effect on preventing disease, its effect on curing disease, its effect on prolonging life, its effect on enhancing quality of life, the worthiness of the typical patient receiving the service, and the social implications if the service becomes widely used. The results of these discussions could be considered in setting research priorities, setting priorities among existing services, and defining basic services (i.e., the services everyone is guaranteed as part of fair access).

Community discussions help make allocation decisions legitimate. They enhance procedural justice–the justice or fairness of the decision-making process. A legitimate decision is one that we respect and obey, even though we may disagree with it. With explicit allocation decisions, each individual will eventually encounter a situation in which some service with some potential to benefit a loved one is nonbasic and will not be covered by insurance. Such a circumstance is more likely to be accepted if the individual believes the decision-making process was fair and open.

Participatory Democracy vs. the Client State

Barber presents an excellent description of participatory democracy and the benefits that can be derived from an actively involved citizenry.[25] He contrasts a participatory democracy with another model of government which he labels the client state. These two models differ along three dimensions: the role of the individual in governance, the form of communication, and the role of leaders.

The client state has consumers (or clients) of government services but no citizens, while participatory democracy has actively involved citizens. The client state is marked by vertical communications from the leaders or elite to the masses, and occasionally vice versa. The participatory democracy also includes lateral communications, or citizens talking and listening to each other. In the client state, leaders develop policy initiatives, providing services for their clients or customers. In a participatory democracy, leaders facilitate citizen action and initiatives.

What are the consequences or results of a participatory democracy? "Blame-itis," which marks the client state, is diminished. This is the condition where all policy problems are blamed on the leaders and the leaders are looked to for solutions. In the participatory democracy, we take responsibility for our community's problems and their solutions. In short, we govern ourselves. Participatory democracy and community discussions cause us to listen more carefully and to understand better the positions of those with whom we disagree. The discussions help us to find common ground—not just a compromise between competing interests—and to identify our shared interests. We develop a sense of community. The values of social solidarity, social advocacy, personal responsibility, and service to the common good are realized.

Health Decisions Movement

One attempt to bring the virtues of participatory democracy into reality in the health care system is the Community Health Decisions movement.[26,27] Essentially, this is an attempt to develop grassroots bioethics networks whose purpose is to stimulate community discussion about values and about ethical issues facing the health care system. The first Community Health Decisions project originated in Oregon, where town meetings were held to discuss what the public viewed as important about health care services.[28] The results were one of several inputs in setting the priorities of treatment-condition pairs in Oregon's reformed Medicaid program.

American Health Decisions (AHD) is an umbrella organization of these local projects. The mission of AHD is to "use community education and discussion to promote and enhance understanding of ethical issues in health care, and direct involvement of citizens in personal, institutional and societal decisions about health care issues."[29] The Community Health Decisions projects often use town meetings as forums for public discussion. The meetings have thought-provoking and discussion-stimulating exercises; they are not simply public hearings. The Community Health Decisions projects promote community discussions and participatory democracy in health policy.

Many health professionals are skeptical of public participation in allocation decisions. The public appears to want everything medical science can provide without paying for it. The public wants high-tech services even if they may not be cost-effective and conversely, may not place a high priority on very cost-effective preventive services. The concept of a statistical life is not widely understood. Given this state of affairs, Jennings asks the key question: "Does health care allocation pose a dilemma that is irresolvable in a democratic society?"[24] Rather than bemoan the current limits of public participation in allocation decisions, maybe the time has come to begin building the skills of the citizenry.

REFERENCES

1. Cockerham WC. Medical sociology. 5th ed. Englewood Cliffs, NJ: Prentice Hall, 1992.

2. Graig LA. Health of nations: an international perspective of U.S. health care reform. Washington, DC: Congressional Quarterly, Inc., 1993.

3. Thompson FJ. The enduring challenge of health policy implementation. In: Littman TJ, Robins LS, eds. Health politics and policy. 2nd ed. Albany, NY: Delmar, 1991:43-56.

4. Feldstein PJ. Health care economics. 4th ed. Albany, NY: Delmar, 1993.

5. Center for Health Economics Research. The nation's health care bill: who bears the burden? Waltham, MA: Center for Health Economics Research, 1994.

6. Levit KR, et al. National health expenditures, 1993. Health Care Finan Rev 1994;16:247-94.

7. Feldstein PJ. The politics of health legislation: an economic perspective. Ann Arbor, MI: Health Administration Press, 1988.

8. Feldstein PJ. Health associations and the legislative process. In: Littman TJ, Robins LS, eds. Health politics and policy. 2nd ed. Albany, NY: Delmar, 1991:69-78.

9. Tesh SN. Hidden arguments: political ideology and disease prevention policy. In: Schwartz HD, ed. Dominant issues in medical sociology. 3rd ed. New York: McGraw-Hill, Inc., 1994:92-104.

10. Mechanic D. Promoting health. In: Schwartz HD, ed. Dominant issues in medical sociology. 3rd ed. New York: McGraw-Hill, Inc., 1994:87-92.

11. Fuchs VR. The Clinton plan: a researcher examines reform. Health Aff 1994;13:102-14.

12. Brock DW, Daniels N. Ethical foundations of the Clinton administration's proposed health care system. JAMA 1994;271:1189-96.

13. Dougherty CJ. Ethical values at stake in health care reform. JAMA 1992;268:2409-12.

14. Priester R. A values framework for health system reform. Health Aff 1992;11:84-107.

15. Daniels N. The articulation of values and principles involved in health care reform. J Med Philosophy 1994;19:425-33.

16. Larson LN, Bentley JP, Brenton MA. Values, participatory democracy, and health care reform. Am J Pharm Educ 1994;58:417-21.

17. Starr P. The social transformation of American medicine. New York: Basic Books, 1982.

18. Aaron HJ, Schwartz WB. The painful prescription: rationing hospital care. Washington, DC: Brookings Institution, 1984.

19. Eddy DM. Health system reform: will controlling costs require rationing services? JAMA 1994;272:324-8.

20. Fuchs VR. The future of health policy. Cambridge, MA: Harvard University Press, 1993.

21. Aaron HJ. Serious and unstable condition: financing America's health care. Washington, DC: Brookings Institution, 1991.

22. Callahan D. Setting limits: medical goals in an aging society. New York: Touchstone, 1987.

23. Hadorn DC, Brook RH. The health care resource allocation debate: defining our terms. JAMA 1991;266:3328-31.

24. Jennings B. Health policy in a new key: setting democratic priorities. J Soc Issues 1993;49:169-84.

25. Barber BR. Participatory democracy in health care: the role of the responsible citizen. Trends Health Care Law Ethics 1992;(Spr-Sum):9-13.

26. Jennings B. A grassroots movement in bioethics. Hastings Center Rep Special Supple 1988;(Jun-Jul):1-15.

27. Jennings B. Grassroots bioethics revisited: health care priorities and community values. Hastings Center Rep 1990;(Sept-Oct):16-23.

28. Crawshaw R, Garland MJ, Hines B, Lobitz C. Oregon health decisions: an experiment with informed community consent. JAMA 1985;254:3213-6.

29. American Health Decisions, Inc. By-Laws. Orange, CA: American Health Decisions, 1993.

ADDITIONAL READINGS

1. Littman TJ, Robins LS, eds. Health politics and policy. 2nd ed. Albany, NY: Delmar, 1991.

2. Feldstein PJ. The politics of health legislation: an economic perspective. Ann Arbor, MI: Health Administration Press, 1988.

3. Tesh SN. Hidden arguments: political ideology and disease prevention policy. In: Schwartz HD, ed. Dominant issues in medical sociology. 3rd ed. New York, NY: McGraw-Hill, Inc., 1994:92-104.

4. Priester R. A values framework for health system reform. Health Aff 1992;11:84-107.

5. Daniels N. The articulation of values and principles involved in health care reform. J Med Philosophy 1994;19:425-33.

6. Eddy DM. Health system reform: will controlling costs require rationing services? JAMA 1994;272:324-8.

7. Fuchs VR. The future of health policy. Cambridge, MA: Harvard University Press, 1993.

8. Jennings B. Health policy in a new key: setting democratic priorities. J Soc Issues 1993;49:169-84.

Chapter 27

The "Rebirth" of Cognitive Services

Kent H. Summers

INTRODUCTION

Modern pharmacy practice has evolved from a patient orientation in which the local druggist, "Old Doc," carefully evaluated the needs of individual patients and used professional judgment to decide which products would provide the best outcomes. However, pharmacy has strayed away from a patient orientation in recent years to a product orientation. The pharmaceutical care concept has been proposed to provide a framework for improving the quality of cognitive services. Hence, it may represent a "rebirth" of the cognitive services provided by pharmacists in past years. Reimbursement for these cognitive services is crucial to the future success of pharmacy.

THE EVOLUTION
OF MODERN PHARMACY PRACTICE

Pharmacy practice in the United States has witnessed dramatic change in this century. In the first half of this century, pharmacy was often practiced according to the scheme presented in Figure 27.1. "Druggists" of varying educational and experiential backgrounds positioned themselves to satisfy consumers' demands for

The author thanks Robert L. Wolf, RPh, MPA, Pharmacist Consultant (retired), Health Care Financing Administration, Region VII, for his helpful comments and suggestions on this chapter.

FIGURE 27.1. "Old-Fashioned" Pharmacy Practice

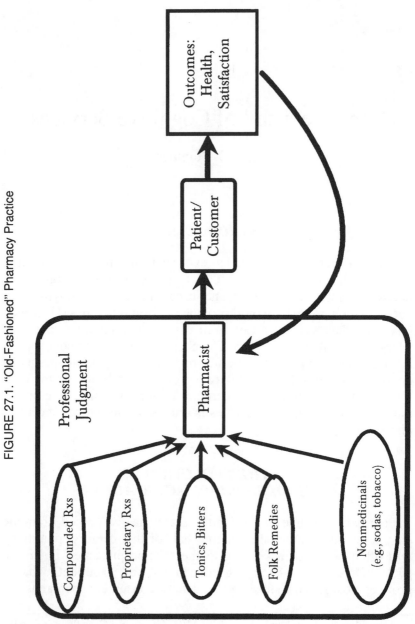

medicinal and nonmedicinal products. These early pharmacists established a source-of-supply for compounded and proprietary prescription products through independently owned pharmacies. At the same time, they satisfied customers' demands for tonics, bitters, and folk remedies. The sale of nonmedicinal products (such as sodas, health and beauty aids, as well as tobacco products) was inherent to the business of providing customers the products they demanded.

Pharmacists offered this wide array of products to customers using "professional" judgment to decide how to satisfy customers most effectively. In this environment, pharmacists' success was based on their willingness to *assume responsibility* for their patients' satisfactory outcomes. This assumption of responsibility was based on pharmacists' roles as independent business people. Early pharmacists adopted a customer-oriented focus in pharmacy practice because customer satisfaction was crucial to the financial success of these business owners. They diligently sought to supply customers with the products that would most satisfy their needs. Therefore, professional judgment and good business sense went hand in hand in this environment.

Variations in pharmacists' skills and practice of pharmacy, along with the variability of individually prepared products, combined to create an impetus for educational, ethical, and legal standardization of pharmacy and pharmacists. Local pharmacy associations sponsored the early pioneering efforts to provide pharmacist instruction.[1] Educational reform to raise the professional standard of pharmacy set a four-year course as a minimum standard in 1932, and the five-year course was adopted as the standard in 1960. The American Pharmaceutical Association has actively promulgated and revised pharmacists' codes of ethics. However, pharmacy leaders have had little success in enforcing compliance with codified ethos. Federal laws regulating the production of drug products and pharmacy practice were updated in 1938, 1952, and 1962.[2] The last set of amendments required the proof of drug products' safety and efficacy before they could enter the market.

The Era of "Count and Pour, Lick and Stick"

These legal and ethical controls in pharmacy combined with a more uniform education and competence among practitioners to support a professional maturing of pharmacy after World War II.

However, much of this uniformity came at the expense of autonomy in the expression of professional judgment in procuring the products patients wanted or needed.

Following World War II, the American pharmaceutical industry applied high technology to the production of medicines and rapidly became one of the most advanced industries in the world. This period also saw the introduction of effective antibiotics, corticosteroids, tranquilizers, antidepressants, antihypertensives, radioactive isotopes, and oral contraceptives. Pharmacies, which had served as a source of products for the relief of suffering and the treatment of minor ailments, became a source of preventatives and cures for serious diseases.

Pharmaceutical manufacturers supplied increasing numbers of new drugs, dosage forms, and marketing methods to shift physician prescribing away from complex mixtures of ingredients to ready-made, single-entity products. The need for consistency in these powerful medications encouraged their mass production in manufacturing plants, not individual pharmacies. Therefore, pharmaceutical manufacturers came to be viewed as the primary providers of high-quality medications, while pharmacists were relegated to dispensing them. Questions about quality subsided as government regulations and manufacturers' compliance with good manufacturing practices guidelines combined to provide consistently excellent pharmaceutical products.

Thus, the focus of pharmacy practice shifted from providing products tailor-made to the specific needs of individual patients to dispensing mass-produced products. With the introduction of effective new drug products came an increased number of prescriptions dispensed each year. Pharmacy income from the sale of prescription drugs increased faster than the sales of over-the-counter medicines, cosmetics, and other, "out-front" products that represent traditional drugstore fare.[2] This increase was based on the relatively simple function of "breaking bulk"–taking pills from large bottles and placing them in smaller, individually labeled prescription vials. Chain stores, grocery stores, other mass merchandisers, and mail-order dispensers rushed into the pharmacy business, displacing many independent "corner" drugstore operations. The success of larger pharmacies is based on their ability to deliver prescriptions more efficiently. This is accomplished through high-volume pre-scription dispensing. Consequently, the success of individual pharmacists is judged by their ability to dispense a large number of prescriptions per hour.

A change in the nature of pharmacist employment has also encouraged a product orientation in pharmacy practice. Pharmacy has become increasingly practiced by employee pharmacists rather than by owners. In many large organizations, employees tend to develop an inward focus where efforts are aimed at satisfying corporate managers. In such an environment, adherence to policies and procedures seemingly takes on more importance than altruistic demands of spending the time to deal with individual customer needs. Further, the success of high-volume prescription operations is often based on the organization's ability to concentrate pharmacists' efforts on dispensing the greatest number of prescriptions per hour. This environment does not encourage the time-consuming activity of patient discussions and counseling. Dissatisfied pharmacists who prefer to exercise professional judgment in patient interactions can be replaced by new pharmacists in need of the salaries offered by these large-volume pharmacies. Pharmacists quickly adapt to this environment and come to judge personal success by their adherence to corporate policies and the number of prescriptions dispensed each day. Customers' needs and wants often become secondary in this environment.

Many in pharmacy have come to accept the notion that the primary output of pharmacists' activities is the appropriate medication in a properly labeled vial. This is a convenient measure of output because it is tangible and easily measured. Pharmacy promotions are simplified when the basis of competition is reduced to the price per prescription dispensed. Such comparisons are simple and direct. Many third-party payers have used prescription prices to compare pharmacies and make decisions regarding the selection of providers.

Consequently, products have become the primary focus of pharmacy practice in many community pharmacies today. Figure 27.2 presents a scheme of pharmacy practice in this environment. Physicians, not pharmacists, are the center of most product information flows. In their promotional messages to physicians, pharmaceutical manufacturers assure product quality and patient satisfaction. Con-

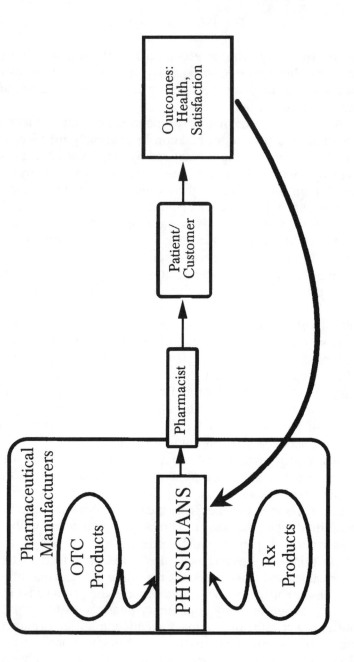

FIGURE 27.2. Product-Oriented Pharmacy Practice

firmation of product quality occurs in feedback to physicians from patients. Pharmacists in this environment play little role in product selection. Further, many pharmacists are not active in monitoring patient satisfaction. These pharmacists have avoided the responsibility for patients' satisfactory outcomes from the products they purchase. Consequently, many pharmacists play a minor role in the business of assuring patients' satisfaction with the outcomes of their pharmaceutical therapy.

The Development of Clinical Pharmacy and the Pharmaceutical Care System

Hospital settings provided the environment in which clinical pharmacy practice evolved. In the 1960s, many hospital pharmacies became decentralized, and pharmacists relocated from basement operations to medical units on the patient-care floors.[3] This enabled hospital pharmacists to better manage the drug distribution system. This relocation also exposed pharmacists to physicians and nurses who had many questions regarding patients' pharmaceutical therapy. Consequently, specialized clinical pharmacists evolved who focused their efforts not on the drug distribution system, but on enhancing their knowledge of drug therapy.

The notion that pharmacy must develop a greater patient orientation in community pharmacy is not new. In 1970, Mickey Smith suggested that pharmacy was developing an "ethical" marketing concept.[4] He thought that pharmacy education was presenting students with the view that patients' needs came before those of pharmacists. Eugene White is often cited as the founder of patient-oriented pharmacy.[5,6] He used a patient record system, as well as a pharmacy devoid of product displays, in a shift to greater patient orientation. Therefore, many in pharmacy have espoused a shift in community pharmacy from product orientation to patient orientation.

The concept of pharmaceutical care has developed to accommodate this shift. As defined by Hepler and Strand, "Pharmaceutical care is the responsible provision of drug therapy for the purpose of achieving definite outcomes that improve a patient's quality of life."[7] This concept involves the process through which pharmacists work with patients and other professionals to design, implement, and monitor a therapeutic plan that will produce beneficial outcomes for the patient. Pharmaceutical care has three major functions:

1. Identifying potential and actual drug-related problems.
2. Resolving actual drug-related problems.
3. Preventing drug-related problems.

Considerations of patient need drive decisions regarding the level of care required by and provided to a patient.[8] Thus, incorporation of pharmaceutical care into pharmacy practice necessitates a patient orientation and pharmacist acceptance of responsibility to the patient for achieving therapeutic effectiveness and improved quality of life.

The use of the pharmaceutical care concept in pharmacy practice permits a "rebirth" of cognitive services. The pharmaceutical care philosophy of practice is provided through an organizational structure called a pharmaceutical care system.[9] Figure 27.3 depicts pharmacy practice using pharmaceutical care as a guiding system. While pharmaceutical products play a role in achieving patient satisfaction, their selection and monitoring to achieve specific patient needs is more important. Further, the selection and application of appropriate pharmaceutical care services is crucial to gaining improvements in patients' health and satisfaction.

Pharmacists use management skills to assure the availability of pharmacy services and pharmaceutical products for patients. This managerial framework consists of five components:

1. Mission statement (documentation of management's agreement on the purpose of the business).
2. Organizational structure (availability of support staff, supervisors' support, and financial incentives to ensure pharmacist ability and motivation to deliver patient-focused care).
3. Practice standards (pharmacist productivity must be measured in a way that encourages continued patient focus).
4. Staff development (provision of educational opportunities to enable pharmacists to identify, resolve, and prevent patients' drug-related problems).
5. Documentation (formal written reports of pharmacist activity for communication with physicians, nurses, other health care providers, and payers).[3]

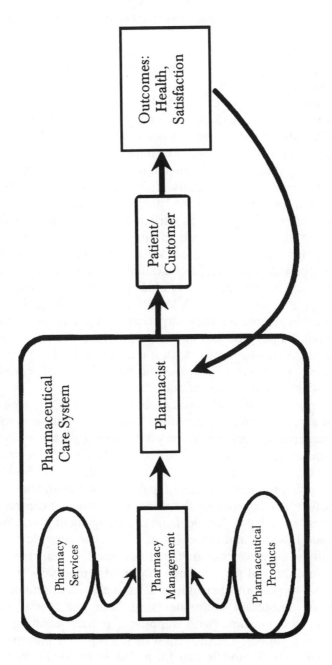

FIGURE 27.3. "Pharmaceutical Care" Pharmacy Practice

This managerial system provides the framework to enable the provision of pharmacy services and products as appropriate for specific patient needs.

A comparison of Figures 27.1 and 27.3 reveals many similarities. Success using either model of pharmacy practice is based on pharmacists' ability to improve health outcomes and patient satisfaction. The pharmaceutical care system provides a protocol to standardize decisions regarding the selection and implementation of appropriate pharmacy services and products. This replaces professional judgment used in the "old-fashioned" form of pharmacy practice. Professional/business judgment in earlier days would have permitted the stocking of tobacco products to satisfy customer wants. Nonmedicinal agents may not be excluded under a pharmaceutical care model of pharmacy practice. For example, pharmacists may choose to provide small doses of wine to elderly patients who would benefit from appetite stimulation. Alternatively, other patients suffering from mild hypertension may experience better health outcomes through nondrug approaches instead of pharmacotherapy. The pharmaceutical care concept provides a framework for the application of pharmacists' professional judgment in the delivery of services tailored to achieve maximum patient benefit.

The application of total quality management (TQM) techniques is touted as a solution to many of the woes facing business today.[10] The managerial framework of the pharmaceutical care concept has many similarities to the TQM approach. Table 27.1 provides an overview of these similarities. The mission statement in both managerial models provides guidance for decisions. Satisfaction of customer/patient needs provides the mission statement's central theme and serves as a basis for decisions. Management must provide appropriate rewards and empower employees to provide the scope and quality of services customers/patients need. Managers must use tangible measures of performance to provide employees the incentive needed to perform their duties effectively. Training and development are important components of both managerial models. Long-term relationships with suppliers, patients/customers, and their families support the provision of high-quality pharmacy services. Finally, both models call for a system to document the services provided and to provide information feed-back to managers,

TABLE 27.1. Similarities of the Important Components of Pharmaceutical Care Management (PCM) and Total Quality Management (TQM)

Documentation of Organizational Philosophy	Both depend on the development and use of a mission statement as a guiding principle for decisions.
Role of Customer/Patient	Customer/patient need is the basis of decisions.
Management	Management must empower employees and reward them for providing high quality services and products that satisfy customer/patient needs.
Productivity Measurements	Expectation standards must be measurable and objective. Employee advancement must be based on documentation of tangible activities that satisfy patient/customer needs.
Staff Development	Employees' access to training and development must be ongoing.
Relationships	Partnerships with suppliers, patients/customers and their families must be long-term.
Feedback	Services provided must be documented for feedback to managers (for assessment of process), referring providers, patients/customers, and payers.

Sources: Fisher DC, Horine JE, Carlisle TH, Williford SD. *Demystifying Baldridge.* New York: The Lincoln-Bradley Publishing Group, 1993. Strand LM, Buerrero RM, Nickman NA, Morely PC. Integrated patient-specific model of pharmacy practice. *Am J Hosp Pharm*, 1990;47:550-554. Summers KH, Monk-Tutor MR. Total quality management in health care organizations and consultant pharmacy. *Med Interface*, 1993;6(2):102-110.

referring providers, patients, and payers. Thus, pharmacists' adoption of the pharmaceutical care concept is consistent with the use of TQM efforts, which many organizations across the country have adopted to remain economically viable. Pharmacists' efforts to increase the quality of cognitive services they provide may be crucial to their success in our very competitive environment.

Competition and Its Impact on Pharmacy

Figure 27.4 depicts the competitive changes affecting pharmacy. In the post-World War II era, the number of prescriptions dispensed

FIGURE 27.4. Changes in the Nature of Competition

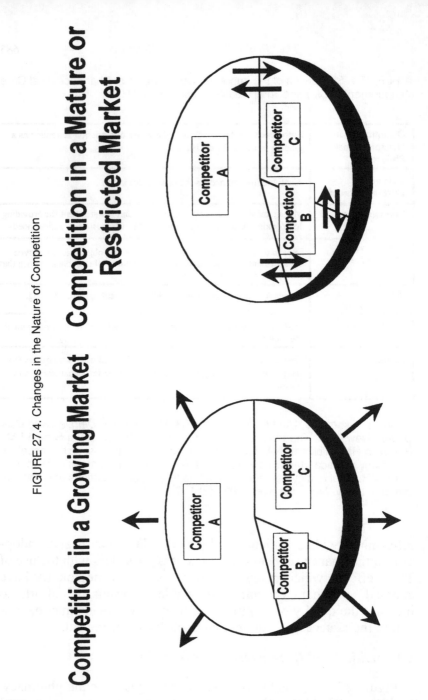

Competition in a Growing Market

Competition in a Mature or Restricted Market

was increasing, and competitors in this growing market exploited this expanding demand. New competitors in a given market area did not onerously affect existing pharmacies because the total market was growing. As the pharmacy market has matured, its growth has become restricted. Consequently, the growth of one competitor comes at the expense of others. Success in this market comes from effectively meeting the challenges of competitors and positioning the firm to have a sustainable competitive advantage over rivals.

The profitability of any firm is often directly related to the extent of competition in its market. Figure 27.5 shows the factors associated with competition in the community pharmacy market. Community pharmacies are subject to the threat of new pharmacies opening in their market area because pharmacies are relatively inexpensive to open and few legal restrictions present barriers to their entry. The bargaining power of buyers has increased, as third-party payers increasingly negotiate for prescription dispensing services on behalf of their patients. This serves to increase customer knowledge about prices and reduces pharmacists' ability to increase profitability through price manipulation. Government regulations affect competition by mandating uniformity of prescriptions. Finally, mail-order pharmacies represent the threat of alternative prescription delivery services. The lower prices offered by mail-order pharmacies may induce payers to view them as good substitutes for the more expensive, community pharmacy version of pharmacy practice. These factors have combined to increase the competitive pressures in community pharmacy practice.

A firm can gain a sustainable advantage over rivals in either of two ways: through a low-price leadership (emphasizing prescription dispensing) or through a differentiation approach (emphasizing pharmacy services).[11] Selection of a strategy is based on an assessment of expected future market opportunities and the pharmacist's abilities to compete for those opportunities.[12] Table 27.2 presents the factors likely to play a role in pharmacists' future success using either of these two strategies.

Price-leadership competitive strategy is exemplified by mail-order pharmacy outlets, in which the primary focus is the production of prescriptions for a lower price than that offered by the competition. Price-leadership competition is facilitated by government reg-

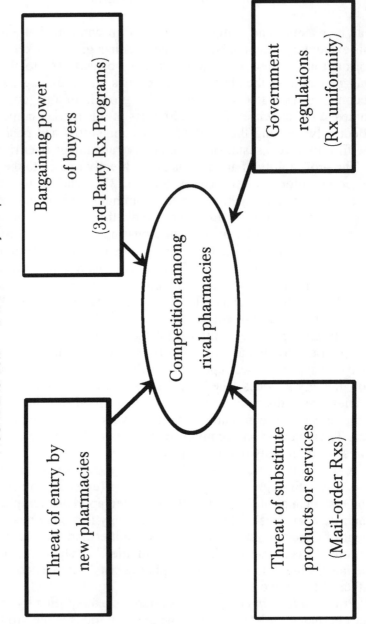

FIGURE 27.5. Determinants of Pharmacy Competition

TABLE 27.2. Factors Favoring the Two Strategies to Gain Sustainable Competitive Advantage

Price Leadership (e.g., mail-order)	Differentiation (e.g., cognitive services)
Access to new technologies that reduce the cost to dispense large numbers of prescriptions (economies of scale)	Access to personal computers and software that facilitate patient monitoring, services documentation, and billing
Access to low-cost labor, such as technicians	Access to highly motivated and competent pharmacists
Access to preferential pricing for prescription products	Reduced impact of preferential pricing on products
Payers' focus on the pharmaceutical budget (prescription costs) in "carve-out" benefits programs	Payers' focus on patients' total health care costs and patient productivity as a consequence of pharmaceutical therapy

ulations that require the consistency of the prescriptions pharmacists dispense. Thus, prescription products are a market commodity by legal mandate, making their dispensing vulnerable to price competition. Future success using this strategy will come from access to new technologies that reduce the cost to dispense prescriptions. The expense of this automation equipment hinders its widespread use in low-volume pharmacies. In addition, successful price-leadership competitors will make maximal use of pharmacy technicians to reduce the cost to dispense prescriptions further. Large volume operations will also have greater access to preferential prices for medications from manufacturers. Finally, price-leadership competitors will benefit from third-party payers' trend toward the provision of pharmaceutical "carve-out" benefits. In this situation, pharmaceutical benefits managers (PBMs) encourage competition on the cost of the pharmacy component alone. The impact of pharmaceutical therapy on other areas of health care costs is not a major concern. Therefore, the cost of the pharmacy budget is the primary concern to PBMs, and they will seek providers who can supply the lowest cost of medications per person.

Alternatively, a *differentiation* strategy focuses on the provision of pharmaceutical services. Many pharmacists now see the provision of cognitive services for ambulatory patients as a viable means of competing with the large-volume mail-order dispensers using a differentiation strategy. Future success using this strategy will involve access to personal computers and software that enable pharmacists to monitor patients, document services, and bill for provision of services. Many such tools are currently available, and more are likely to become available in the future. Unlike competition based on the dispensing of pharmaceutical products, the provision of pharmacy services is not subject to overhead costs such as the cost to carry inventory, theft, merchandising labor, and long hours of operation to provide retail customer convenience.[13] Further, success using a differentiation strategy involves access to highly qualified, motivated, competent pharmacists. Such pharmacists should become increasingly available as the large-volume pharmacies reduce their employment in dispensing positions. The delivery of effective pharmacy services is not directly affected by prescription product costs. Therefore, a focus on cognitive services as a differentiation strategy reduces the impact of preferential pricing on prescription products. Finally, service/differentiation competitors will benefit from third-party payers' focus on patients' total health care costs rather than the pharmacy budget alone. Pharmacists will likely receive reimbursement for cognitive services from payers who recognize the impact of such services on savings from reduced hospitalization, physician, and nursing home costs. Employers may increasingly focus on health care services' impact on patient productivity. Many employers may find it makes sense to pay for pharmacy services that help patients increase their work productivity by reducing absenteeism from illness.

Falling between these strategies are pharmacies attempting to offer both prescriptions *and* supporting (value-added) services. Such pharmacies provide no clear-cut focus of competitive advantage. A situation in which a competitive strategy has not been decided upon is termed a "muddled" approach.[11] Pharmacists' failure to decide may well lead to their pharmacies' failures, given increasingly intense competitive pressures in the community pharmacy market.

A review of Table 27.2 should facilitate a decision on how to compete most successfully in the future. Managers of mail-order operations have already recognized that their strength lies in using a price-leadership strategy. Many pharmacists may recognize that their greatest chance for success will come from using a differentiation strategy. Some chain drugstore organizations have already recognized these factors and are responding with enhanced pharmacy services offerings.[14] Therefore, depending on how a pharmacist prefers to compete, he or she should decide whether to concentrate on offering products (prescriptions) or to focus on providing pharmaceutical services.

GAINING REIMBURSEMENT FOR COGNITIVE SERVICES

Reimbursement for cognitive services enables community pharmacists to enhance the professionalism of their practice. When they are rewarded for patient-oriented activities, pharmacists can better satisfy the needs of their clients. Thus, cognitive services reimbursement permits a shift from product-oriented to patient-oriented pharmacy practice. For this to occur, pharmacists need a framework to gain reimbursement for their cognitive services.

Steps to Gain Reimbursement for Cognitive Services

Direct reimbursement of pharmacists' cognitive services is often a difficult proposition. Medicare permits direct billing only for physician services.[15] Major medical insurance programs take a similar approach. Some major medical plans may pay for pharmacy services if their provision and documentation complies with requirements. Health care payers that provide comprehensive services are most likely to appreciate cognitive services that reduce hospital and physician visit expenses. Therefore, group- and staff-model health maintenance organizations (HMOs) that integrate the pharmacy and medical benefits may represent organizations that could best appreciate the value of cognitive services. The keys to successful reimbursement are meeting with benefits managers to determine their needs and providing the services and documentation to satisfy those needs.

Many PBMs administer pharmacy "carve outs" of major medical programs. These carve-out programs provide stand-alone pharmacy benefit packages. These PBMs often focus their attention on the pharmacy budget, excluding concerns about benefits in other areas such as hospital and physician savings. Such PBMs may not appreciate the value of cognitive services that increase the pharmacy budget while decreasing other costs. Pharmacists' primary appeal to these administrators would include benefits that reduce drug costs. Therefore, generic and therapeutic substitution, as well as formulary enforcement, would be most attractive. In addition, drug utilization review programs that reduce the number of prescriptions dispensed to patients would also appeal to these PBMs.

Pharmacists' primary weakness in requesting payment for cognitive services is their failure to document thoroughly the services provided to each patient. If pharmacists do not document cognitive services, there is a question as to whether they actually occurred. Therefore, the most important steps in gaining reimbursement for cognitive services are: (1) documentation of the services provided, and (2) claims submissions to third-party payers for reimbursement. Each claim submission should contain a:

1. Cover letter
2. Statement of Medical Need
3. Pharmacotherapeutic Workup and Report
4. Invoice

Such documentation represents the missing link between providing cognitive services and getting paid for them.

The Cover Letter

A cover letter stating the submission's purpose should accompany the claim. This letter should indicate the patient's name and identification, as well as the plan's name and number. It should tell the administrator what benefits are likely to result from the cognitive services provided. The cover letter should include a list of the submission enclosures and complete identification of the pharmacist seeking reimbursement.

Statement of Medical Need

The Statement of Medical Need (SMN) tells the third-party administrator that a problem has been recognized by another health care expert and that the patient can benefit from a pharmacist's intervention. It represents a formal declaration of the patient's need for a pharmacist's cognitive services. The SMN, signed by the patient's physician, provides the authority many administrators need to provide reimbursement. The SMN also facilitates communication between the physician and pharmacist. This document may be viewed as the work order for cognitive services. Just as the prescription initiates dispensing activities in today's environment, the SMN may become the prescription's equivalent in the reimbursement of cognitive services.

Insurers should supply SMN forms, or pharmacists may develop their own based on the insurers' needs. Photocopies of these forms can be distributed to physicians who, after personal pharmacist visits, agree to refer patients having problems with their pharmaceutical therapy. Physicians who recognize the usefulness of cognitive services will likely refer patients to the pharmacist for care in the future. Pharmacists should make frequent visits to physicians' offices and check with the office staff to ensure that the SMN forms are readily available. Physicians need frequent reminders about the availability of cognitive services, their benefits, and the pharmacist's ability to provide them.

Pharmacotherapeutic Workup and Report

Target patient groups would include those most likely to benefit from a complete workup and report to the physician. The Pharmacotherapeutic Workup and Report (PWR) document is used to report the results of pharmacists' provision of cognitive services to physicians and third-party payers. With the information contained in the PWR, physicians can provide safer and more effective health care to their patients. This document also serves to prove to third-party plan administrators that the service was performed for the patient. Strand, Cipolle, and Morley, as well as Isetts, provide more complete descriptions of the development and use of the PWR document.[16,17]

Invoices

Invoices should include three components: (1) problem, (2) intervention (cognitive services), and (3) outcomes. The patient's problem is the reason for employing cognitive services. The intervention component is a comprehensive list of the activities provided in response to the patient's problem. Outcomes represent expected changes in drug therapy or patient behavior directly attributable to the intervention. Invoices should include patients' signatures verifying their receipt of cognitive services.

These components of claims submission are a synthesis of the available literature regarding the most effective means to gain reimbursement from a wide variety of third-party payers. However, the reimbursement of cognitive services has not gained general acceptance, and submission processes are not standardized. Each payer may have specific forms, certifications, or procedures that must be completed before reimbursement is considered. A key to getting reimbursed for cognitive services is to survey (personally, by telephone, or by mail) third-party payers to determine their requirements for documentation, payment authorization, and claims submissions. This approach to gaining reimbursement has been successfully employed by pharmacists.[18] Thus, pharmacists should consider the requirements of potential payers for documentation and/or verification of cognitive services before they develop their own submission forms.

Portner and Srnka have evaluated the use of the National Association of Retail Druggist (NARD) Pharmacist Care Claim Form (PCCF).[19] The PCCF represents an attempt to develop a universal claim form for cognitive services. Completion of the PCCF includes a combination of four data fields, taken from the two-digit codes associated with boxes checked in four sections of the form. Professional fee charges are based on codes for the cognitive service performed. Pharmacists must individually determine their professional service fees for cognitive services.

Establishing Professional Fees for Cognitive Services

The establishment of usual and customary professional fees is difficult when little preexisting charge experience exists. Pharmacists may develop fee-for-service (FFS) charges using cost-based (billable hours, prevailing fees) or market-based fees (customer willingness to pay or the savings expected from cognitive services) as the basis for pricing approaches. Alternatively, capitation (fixed, prospectively determined prices for a comprehensive set of services paid regardless of utilization) or a retainer fee may be used to bill for cognitive services.

Cost-Based Professional Fees

Billable Hours. The amount of time directly used in the provision of a cognitive service can provide a basis for determining professional fees. An example for one hour of services is below:[20]

$ 25.00	Direct labor (@ 2,000 hours, $50,000 annual salary)
+3.75	Benefits such as health insurance, Social Security, and unemployment insurance (15% of direct labor)
+ 6.25	Clerical support and facility overhead (25% of direct labor)
$ 35.00	Charge per billable hour

This approach covers costs but excludes a provision for profit. In addition, it does not take into account time in which pharmacists are not actively involved in providing cognitive services, such as time involved with paperwork and research. Consequently, this figure may represent an inadequate price, but it can serve as a minimum charge for pharmacists' usual and customary fees.

Prevailing Fees. Some authors have suggested charging from $40 to $85 per hour of pharmacist time.[21-23] A figure of $60 per hour is commonly cited.[13,14] Therefore, pharmacists might charge $1 per minute for the time spent providing cognitive activities.

However, hourly and prevailing fee reimbursement for cognitive services does not encourage pharmacists to use their time efficiently. From the perspective of many third-party payers, paying health care providers on an hourly basis is like providing an open checkbook. This approach, therefore, may not receive widespread support from third-party payers. Still, the time needed to perform cognitive services may provide a basis for calculations to determine the minimum cost of their provision.

Market-Based Professional Fees

Using market-based professional fees involves setting prices that the market will bear. This would require charging cash-paying customers the amount they would be willing to pay for cognitive services. Pricing for cognitive services should reflect the value of their benefits to the payer. Therefore, cash-paying patients will likely evaluate charges based on the personal benefits provided by cognitive services. Alternatively, third-party payers are more likely to base their valuation of cognitive services on the tangible benefits to their pharmaceutical benefits plan.

The benefits of pharmaceutical therapy may be categorized as shown in Table 27.3. Reducing or eliminating disease symptoms has an important impact on patients' satisfaction, comfort, and sense of well-being. These considerations are meaningful to patients, and they are important when patients pay for their own health care. However, administrators of health care organizations paying patients' bills may be more interested in how cognitive services reduce the utilization of hospital, physician, and pharmacy services. For example, the provision of comprehensive drug regimen reviews (DRRs) has been found to reduce the number and cost of medications taken by nursing home patients.[24-27] This directly benefits the third-party payer (Medicaid) through reduced drug costs. This experience has been an important reason for Medicaid's

TABLE 27.3. Pharmaceutical Therapy Benefits from Various Perspectives

Third-Party Payer	Employer	Patient
Decreased hospital, physician, nursing home costs; Decreased drug costs most important to managers of pharmacy "carve-out" benefits programs	Increased employee productivity; Decreased total health care costs	Reduced fear, pain, and suffering; Increased quality of life, satisfaction

Source: Summers KH. Pharmacists' guide for the reimbursement of competitive services for ambulatory patients. *Drug Topics*, Special Supplement, 1994.

paying pharmacists for DRR services in nursing homes. Therefore, pharmacists should adopt a payer's perspective when charging for cognitive services. Rather than focusing on the process of care, pharmacists should instead focus on the outcomes or benefits of cognitive services.

Consumers' Willingness to Pay. In 1985, Carroll reviewed the literature and found studies demonstrating that patients were willing to pay $1 to $5 for personalized counseling sessions.[28] Subsequently, Carroll and colleagues found that patients would be willing to pay an average of $5 per prescription for counseling provided in a private area.[29] Szeinbach reported a pilot study in which patients were willing to pay pharmacists from $2 to $8 for personal counseling.[30] Anecdotal reports suggest pharmacists have charged from $10 for cardiac risk consultations to $50 for home medication reviews.[31,32] Therefore, many patients are willing to pay for cognitive services. This occurs in spite of the fact that most pharmacy patrons are conditioned neither to expect cognitive services nor to pay for them.

Consumers usually purchase health care services without much knowledge about what they are buying. Consequently, patients often respond to their uncertainty by assuming high price is a good indicator of high quality.[33] Indeed, patients seeking psychological help may place a higher value on the services provided by psychologists with expensive fees.[34] Therefore, professional fees based on patients' willingness to pay would tend upward in a consumer market.

Professional Fees for Third-Party Payers. We might expect third-party payers would oppose this upward tendency of patients' willingness to pay professional fees. A better rationale for establishing professional fees would involve payer benefits as the basis for pricing. Stephens reported that between 8% and 13% of all hospitalizations result from inappropriate medication use.[35] Further, a pilot study by Hirsch, Gagnon, and Camp indicated that some third-party administrators would consider reimbursement for cognitive services that increase patient satisfaction.[36] Therefore, third-party payers would more likely consider paying for cognitive services that result in health care cost savings or patient benefits.

Capitated Professional Fees

Capitated reimbursement for professional fees would involve a fixed, prospectively determined charge for patients over a period of time (e.g., one year). This fee is fixed, regardless of patients' utilization of services. Capitation is appealing to many managed care administrators because it shifts the fiscal risk of service provision to the provider. Under capitation, third-party payers are not at financial risk if patient demand for cognitive services unexpectedly grows or pharmacists bill for unnecessary services.

A capitated professional fee might be appropriate when providing a comprehensive set of cognitive services to manage patients' pharmaceutical therapy. The term "pharmaceutical care" is often associated with the provision of a comprehensive set of pharmacist services that focus on satisfying the patient's needs associated with pharmacotherapy.[37] Third-party payers and patients would likely benefit from the outcomes of pharmaceutical care:

1. Curing the disease.
2. Reducing or eliminating the disease.
3. Arresting or slowing a disease process.
4. Preventing a disease or its symptoms.

The provision of cognitive services on a capitated basis is one way of getting patients the medical outcomes they need and third-party payers the payment mechanism they want. Providing comprehensive pharmacy services may represent a good basis to structure reimbursement on a prepaid basis.

This type of reimbursement would eliminate the administrative costs of individual claim submissions. However, this type of capitated reimbursement did not have great success when implemented in Iowa in 1981.[38] Perhaps the lack of a structured pharmaceutical care process hindered the program's success.

The Outcomes of Cognitive Services Provided by Community Pharmacists

Much literature is available describing pharmacists' provision of clinical services in hospital and nursing home institutions. A good review of studies documenting the economic value of clinical pharmacy services in the institutional setting was provided by Hatoum and colleagues in 1986.[39] Willett and colleagues provided a more recent review of the literature in 1989.[40] Thus, the body of literature documenting the economic value of clinical pharmacy services in the institutional setting is fairly substantial. However, the number of studies that formally evaluate outcomes of cognitive services provided by community pharmacists is relatively small.

Fee-for-Service Reimbursement

Most of the literature documenting the economic value of cognitive services is based on fee-for-service (FFS) reimbursement. The following discussion includes: self-care consultations, physician consultations, drug utilization review (DUR) and intervention, therapeutic drug monitoring, compliance consultations, and disease-specific consultations.

Self-Care Consultations. Patients' self-care is very important to those who pay for health care services. Without self-care and self-medication, the resulting increase in physician and hospital costs would collapse our health care system.[41] Components of self-care include the monitoring, diagnosis, prevention, symptomatic treatment, cure, and management of disease. Managed care organizations can save money when patients appropriately and effectively use self-care and over-the-counter (OTC) medications. Pharmacists are in a unique position to capitalize on this need of third-party payers.

Pharmacists' self-care consultations involve providing a combination of product and information. In addition to OTC drug products, pharmacists may also provide nondrug products such as self-testing and monitoring devices, wound treatment materials, braces, and supports.[42] The information component includes advice on product selection and use, nondrug self-care, referrals, and health assessment.

However, the literature documenting the value of pharmacists' self-care consultations is meager. A large, nationwide consumer survey supported the results of previous studies documenting patient acceptance of pharmacists' OTC product advice.[43] Sixty-six percent of the 1,356 respondents in a *Drug Topics* study indicated their extreme satisfaction with pharmacists' OTC advice. This was confirmed in Szeinbach and Banahan's study of 375 patients who received advice from specially trained pharmacists regarding the long-term use of antacid products.[44] Ninety-eight percent of these patients indicated their satisfaction with pharmacists' counseling, and 62% were willing to pay for this service. However, a dollar amount of payment was not established. In 1979, Pathak and Nold studied the value of pharmacist-conducted training programs for patients self-administering antihemophilic factor, calcitonin, cytarabine, injectable analgesics, and parenteral nutrients.[45] They estimated that for every dollar charged for the pharmacy training programs, 1.25 days of hospitalization and $321.90 in hospital charges were avoided.

Much more research needs to be done on the role of pharmacists in self-care consultations and the value of consultations to third-party payers. Training improves pharmacists' ability to recommend the appropriate self-care, self-medication, or referral to other health care providers.[44] However, the public and physicians may not view pharmacists' abilities to provide health information as highly as pharmacists think.[46] Therefore, pharmacists must establish their expertise in the minds of patients and physicians. The provision of self-care cognitive services may prove to be very valuable when widely provided and systematically studied.

Physician Consultations. Health care organizations often use generic and therapeutic substitution programs, as well as formularies, in an attempt to save money by limiting the availability of newer, more expensive medications. Pharmacists are used to enforce these cost-saving programs. A recent pilot study indicated that direct drug budget savings due to pharmacists' efforts averaged $14.66 per generic substitution and $37.37 per therapeutic substitution.[47] Physicians are often unaware of the individual restrictive formulary that applies to each patient because of the many formularies associated with the multitude of plans covering their patients. Consequently, physicians often prescribe medications that are not reimbursable under patients' health care plans. In 1975, Ryan and colleagues showed that hospital pharmacists' discharge consultations saved patients $9.13 per hour of consultation time when dealing with third-party payer issues.[48] No other studies document the value of pharmacists' consultations with physicians to enforce a plan's formulary. However, a reasonable approach might involve basing the professional fee on the cost of pharmacist time involved in such efforts, along with a reasonable profit.

An alternative to formularies might involve the use of pharmacists to provide educational consultations with physicians. A survey of 430 primary care physicians indicated that less than one-third personally searched the literature when information was needed about drugs, and most assessed the scientific value of literature primarily from personal experience.[49] Mailed continuing education programs have not proven very effective.[50] Therefore, we can infer that physicians receive much of their drug information from pharmaceutical sales representatives. McCombs and Nichol found pharmacy-enforced outpatient drug treatment protocols were a viable alternative to restrictive formularies and prior authorization.[51] Mead and McGhan studied the effect of continued physician consultations on the proper use of medications for peptic ulcer disease in an HMO.[52] Considering only drug budget savings, the authors found that these consultations produced a benefit:cost ratio of 4.3:1. Several researchers have evaluated the effect of pharmacists' provision of educational visits in physicians' offices (i.e., academic detailing).[53-55] One analysis of these visits demonstrated a benefit:cost ratio of 3:1 when high-volume prescribers were the targets of pharmacists' visits.[56] It appears that third-party payers are likely to experience $3.00 to $4.30 savings for every dollar they invest in paying pharmacists to provide educational consultations with physicians.

Drug Utilization Review and Intervention. Pharmacists' ongoing review of drug prescribing, dispensing, and administration has become an important means to control health care costs. DUR is defined as a structured, continuing program to review, analyze, and interpret patterns of drug use in a health care delivery system against predetermined standards, with the major focus being quantitative outcomes.[57] A closely related term, drug usage evaluation (DUE), places greater emphasis on the qualitative aspects of treatment outcomes.[58] Interest in DUR has grown due to the introduction of more costly drugs, increasingly complex medication regimens, questioning of medication outcomes, and computer systems that assist in data collection and analysis. This interest has coincided with the passage of the Omnibus Budget Reconciliation Act of 1990 (OBRA 90) which mandates that pharmacists provide prospective and retrospective DUR programs for their Medicaid patients.[59] DURs can occur before, during, or after prescription dispensing (i.e., prospective, concurrent, or retrospective DUR, respectively).

The benefits of DUR assessments and interventions are related to the identification and avoidance of medication regimens that may potentially expose patients to adverse drug experiences. Third-party payers would benefit from the avoidance of paying for these events. For example, geriatric Medicaid patients exposed to high doses of nonsteroidal anti-inflammatory drugs (NSAIDs) have a 10.1 times increased risk of developing peptic ulcers over nonusers.[60] The excess risk of ulcer disease associated with NSAID use in this study resulted in 17.4 more hospitalizations per 1,000 person-years of exposure. Therefore, pharmacists' DUR efforts to detect and prevent patient exposure to drug problems such as this would benefit health care payers.

Several researchers have documented the value of DUR services.[61-65] The direct benefit of DUR was documented in research conducted by Rupp and his associates.[61,63,64] In one study, 89 pharmacists dispensed 33,011 new prescriptions. Pharmacists identified 623 (1.9%) of these prescriptions as potentially associated with prescribing-related problems that required pharmacist intervention. An expert panel estimated savings of $76,615 would result from avoided emergency room visits, hospitalizations, and physician visits. Based on these savings, the estimated value of DUR and intervention was $2.32 per prescription screened. Dobie and Rascati used Rupp's methodology to prospectively evaluate 6,000 new prescriptions.[65] The provision of DUR and interventions by pharmacists was estimated to save about $3.50 in medical costs per prescription screened. This provides good evidence to support the notion that plan administrators should pay pharmacists to encourage provision of DUR screening and interventions when necessary. A reasonable professional fee would be in the range of $2.32 to $3.50 per prescription screened.

Therapeutic Drug Monitoring. Therapeutic drug monitoring

(TDM) may also represent a cognitive service that is valuable to third-party payers. Patients needing TDM include those with a history of poor compliance, drug overuse, toxicity, or poor therapeutic response.[66] Medications that sometimes require serum drug concentration measurements include theophylline, digoxin, quinidine, phenytoin, and phenobarbital. Benefits to third-party payers would include a reduction in emergency room visits, hospitalizations, and physician visits for inadequate or toxic responses to pharmaceutical therapy.

Some literature is available regarding patients' willingness to pay for TDM services. In 1988, Einarson and colleagues found that patients were willing to pay an average of $11.54 for serum determinations of potassium or cholesterol.[67] After using the service, these 443 patients were satisfied with it and expressed a willingness to pay an average of $14.47 per test in the future. The authors predicted that such a service could be financially feasible if a pharmacist performed eight tests daily at a charge of $15. However, maintenance of quality and accuracy may require a larger daily volume of tests.

Compliance Consultations. The DUR process may be used to identify patients who would benefit from compliance consultations regarding their pharmaceutical therapy. Pharmacists can prospectively anticipate noncompliance among elderly patients taking more than five prescriptions who have difficulty reading prescription instructions and opening containers.[68] These patients would exhibit overuse, underuse, or erratic use of their medications.

Noncompliance has important effects on health care costs. Col and colleagues reviewed the records of and interviewed elderly patients admitted to an acute care hospital.[69] Of the 315 consecutive admissions studied, 11.4% were directly related to some form of noncompliance, at a cost of $2,150 per hospital admission. Smith has estimated that total health care system costs of noncompliance exceed $100 billion each year.[70] If only 10% of hospital admissions could be traced to noncompliance, the annual cost would exceed $25 billion. This becomes particularly important with the availability of more expensive and effective medications to treat chronic diseases. Further, variability in patient compliance may have greater impact on treatment outcomes than the use of more effective medications. Therefore, less effective (and less expensive) medications in combination with pharmacist counseling may, in some cases, perform as well as more expensive alternatives.

Noncompliance has been identified as an important cause of elderly patients' nursing home admissions.[71] Noncompliance among noninstitutionalized elderly patients is estimated to range from 49% to 75%, with underutilization being the most common form of noncompliance.[72,73] Underutilization may result in increased hospital and physician costs to treat patients with symptoms of poorly controlled disease. Controlling for demographic factors and blood pressure, Maronde and colleagues convincingly demonstrated that the underutilization of antihypertensive medications is associated with significantly higher hospital readmissions for the treatment of hypertension.[74] In addition, a negative impact on a third-party payer's pharmacy budget may result from patients who receive, but do not consume, excessive quantities of medications (i.e., hoarding). Alternatively, patients consuming excessive quantities of medications may cost the system in terms of unnecessary hospitalizations and physician visits to treat drug toxicities.

Pharmacists can help health care payers reduce these costs by providing patients medication regimen compliance consultations. Jones and colleagues demonstrated the value of a pharmacist-operated refill clinic.[75] Compared to physician clinic visits, the pharmacist refill evaluation clinic benefited patients through reduced waiting times, physician fees, and drug costs. Additional benefits included more accurate documentation of medical records, drug abuse surveillance, and therapy monitoring. However, this study did not evaluate the impact of pharmacists' direct interventions to enhance medication compliance. Compliance-enhancing medication containers and computerized refill reminder systems have demonstrated their usefulness in improving patients' compliance.[76-79] In addition, a variety of pharmaceutical manufacturer-sponsored compliance programs are available to enhance patients' knowledge about their diseases and medications.[80] Sclar demonstrated that an educational program for hypertensive patients increased pharmacy costs while decreasing hospitalization and physician visit costs.[81] This program decreased total costs for all new and continuing hypertensive patients (n = 985) an average of $92.97 and $127.79

per year, respectively. Based on the savings demonstrated in this study, educational consultations for hypertensive patients may be worth $100 per patient per year. However, many researchers in this area agree that providing patients with information alone is necessary for their understanding, but not sufficient for their engaging in compliant behaviors.[82-84] The provision of drug information is twice as effective in improving patient knowledge as it is in improving compliance with the medication regimen.

Community pharmacists' ability to provide personal contact and follow-up with patients puts them in a unique position to effectively achieve patients' appropriate use of medications. A recent European, multicenter study evaluated the quality of life (QOL) benefits of antihypertensive medications.[85] The only consistent beneficial effects on QOL in the study occurred during the washout period in which patients received personal care from the researchers but no medication. The same improvements in QOL were not experienced during the treatment period, when patients were taking antihypertensive medications. Erfurt and colleagues demonstrated that follow-up to health education was vitally important to the success of a wellness program.[86] Programs that included follow-up counseling and health promotion were nine to ten times more cost-effective than programs using education alone. Thus, pharmacists may use their patient contact to provide the most effective combination of compliance-enhancing interventions, based on individual needs.

The literature, then, supports the notion that pharmacists can enhance patients' compliance with their medication regimens. Preliminary work documents the value of providing patients with information about their medications. Such drug information programs may be worth $100 per patient per year. Much more work is needed to document the additional benefit of pharmacists' provision of personal compliance consultations on patients' health outcomes and third-party payers' costs.

Disease-Specific Consultations. Pharmacists may increase their effectiveness if they focus efforts on specific patient groups. This way, they can develop the expertise needed to achieve optimal patient outcomes from their consultations. Pharmacists may elect to specialize in care for patients suffering from cardiovascular diseases, diabetes mellitus, respiratory diseases, cancer, and psychiatric disorders. While these specialized consultations appear promising, more studies are needed to document the economic outcomes of these cognitive services.

Capitated Reimbursement

The provision of a comprehensive set of cognitive services (pharmaceutical care) may be amenable to capitated reimbursement. However, the literature is lacking in studies documenting such reimbursement. Anecdotal reports indicate some pharmacists have charged from \$25 to \$30 per family per year to provide comprehensive cognitive services on a retainer basis.[31,87,88] The emergence of studies documenting the value of comprehensive pharmaceutical care services to third-party payers will help in establishing capitated professional fees.

CONCLUSION

The nature of pharmacy practice has evolved from a customer-orientation in which the corner druggist carefully evaluated the needs of each patient and used professional judgment to decide which products would best satisfy him or her. These pharmacists' efforts to satisfy patient needs were compensated in the product prices. However, the increase in prescription business brought large business and third-party payers into pharmacy competition. Market share changes in this competitive environment involved price reductions for pharmaceutical products. This decreased pharmacists' ability to get compensation for their cognitive services. Thus, pharmacy has strayed from a patient orientation to a product orientation. The pharmaceutical care concept provides a framework for returning to the earlier orientation of pharmacy practice. It helps improve the quality of cognitive services, thereby permitting individual pharmacists to better compete with large prescription-dispensing operations. Reimbursement for these cognitive services is crucial to this evolution of pharmacy.

Acquisition of such reimbursement is not easy. Rupp has suggested that seeking reimbursement for cognitive services represents an area of "uncharted territory."[89] Because no generally accepted policies regarding reimbursement exist, such payments are likely to be obtained one service and one payer at a time. Pharmacoeconomic studies documenting the value of pharmacists' cognitive services are not exhaustive and conclusive at this time. However, many studies are now underway that will serve to establish pharmacists' value. This documentation of pharmacists' value will facilitate reimbursement of cognitive services and enable pharmacists to return to the patient-orientation of earlier pharmacy practice.

REFERENCES

1. Sonnedecker G. Evolution of pharmacy. In: Osol A, Hoover JE, eds. Remington's pharmaceutical sciences. 15th ed. Easton, PA: Mack Publishing Company, 1975.

2. Higby GJ. Evolution of pharmacy. In: Gennaro AR, ed. Remington's pharmaceutical sciences. 18th ed. Easton, PA: Mack Publishing Company, 1990.

3. Strand LM, Guerrero RM, Nickman NA, Morley PC. Integrated patient-specific model of pharmacy practice. Am J Hosp Pharm 1990;47:550-4.

4. Smith MC. Pharmacy education? J Am Pharm Assoc 1970;NS10:548-51.

5. Catizone C. Office-based pharmacy in the United States: development of a practice alternative. Am Pharm 1984;NS24:24-32.

6. White EV. Development of the family prescription record system. J Am Pharm Assoc 1973;NS13:357-9.

7. Hepler CD, Strand LM. Opportunities and responsibilities in pharmaceutical care. Am J Hosp Pharm 1990;47:533-43.

8. Strand LM, Cipolle RJ, Morley PC, Perier DG. Levels of pharmaceutical care: a needs-based approach. Am J Hosp Pharm 1991;48:547-50.

9. Hepler CD. The future of pharmacy: pharmaceutical care. Am Pharm 1990;NS30:583-9.

10. Summers KH, Monk-Tutor MR. Total quality management in health care organizations and consultant pharmacy. Med Interface 1993;6(2):102-10.

11. Porter MF. Competitive advantage: creating and sustaining superior performance. New York: Collier Macmillan, 1985.

12. Schwartz A, Sogol EM. Part 2: Developing a marketing plan. Drug Top 1987;(Aug 3):63-71.

13. Anon. Richard Brychell's little revolution in pharmacy. NARD J 1992; 114(9):38-41.

14. Meade V. New services emerge in chain pharmacy. Am Pharm 1993; NS33(2):23-7.

15. Goode MA, Gums JG. Therapeutic drug monitoring in ambulatory care. Ann Pharmacother 1993;27:502-5.

16. Strand LS, Cipolle RJ, Morley PC. Documenting the clinical pharmacists' activities: back to basics. Drug Intell Clin Pharm 1988;22:63-6.

17. Isetts BJ. Monitoring and managing patient care. Am Pharm 1992; NS32(1):77-83.

18. Vaczek D. Pharmacists take action in getting paid for cognitive services. Pharm Times 1994;(Feb):27-39.

19. Portner T, Srnka Q. NARD's Pharmacist Care Claim Form. NARD J 1994;116(7):55-65.

20. Ukens C. Cognitive services: pharmacy's new hope. Drug Top 1991; 135(14):37-40.

21. Srnka QM. Implementing a self-care-consulting practice. Am Pharm 1993;NS33(1):61-71.

22. Bergin DM. Pharmacy provides health information—for a few. Drug Top 1990;134(3):88.

23. Martin S. Teaching pharmacists to provide and bill for clinical services. Am Pharm 1990;NS30(5):24-6.

24. Harrington C, Swan JH. The impact of state Medicaid nursing home policies on utilization and expenditures. Inquiry 1987;24:157-72.

25. Strandberg LR, et al. Effect of comprehensive pharmaceutical services on drug use in long term care facilities. Am J Hosp Pharm 1980;37:92-4.

26. Kidder S. The cost-benefit of drug reviews in long-term care facilities. Am Pharm 1982;NS22(7):63-6.

27. Cooper JW. Pharmacy service to long-term care patients. In: Brown TR, ed. Handbook of institutional pharmacy practice. 3rd ed. Bethesda, MD: American Society of Hospital Pharmacists, 1992.

28. Carroll NV. Consumer demand for patient oriented services in community pharmacies—a review and comment. J Soc Admin Pharm 1985;3:64-9.

29. Carroll NV, Perri M III, Eve EE, et al. Estimating demand for health information: pharmacy counseling services. J Health Care Market 1987;7(4):33-40.

30. Szeinbach SL. Placing a monetary value on consumers' willingness-to-pay for counseling services. Unpublished.

31. Robinson B. Cardiac risk screening gets Florida pharmacy test. Drug Top 1988;132(3):12.

32. Meade V. Conducting at-home medication reviews. Am Pharm 1992; NS32(6):37-9.

33. Robinson JC. Hospital quality competition and economics of imperfect competition. Milbank Q 1988;66:465-81.

34. Schneider LJ, Watkins CE Jr. Perceptions of therapists as a function of professional fees and treatment modalities. J Clin Psychol 1990;46:923-7.

35. Stephens D. Insurer looks at the pharmacist's cost-containment function. Am Druggist 1991;203(3):62.

36. Hirsch JD, Gagnon JP, Camp R. Value of pharmacy services: perceptions of consumers, physicians, and third party prescription plan administrators. Am Pharm 1990;NS30(3):20-25.

37. Hepler CD, Strand LM. Opportunities and responsibilities in pharmaceutical care. Am J Hosp Pharm 1990;47:533-43.

38. Yesalis CD, Lipson DP, Norwood GJ, et al. Capitation payment of pharmacy services. Med Care 1984;22:737-45.

39. Hatoum HT, Catizone C, Hutchinson RA, et al. An eleven-year review of the pharmacy literature: documentation of the value and acceptance of clinical pharmacy. Drug Intell Clin Pharm 1986;20:33-48.

40. Willett MS, Bertch KE, Rich DS, et al. Prospectus on the economic value of clinical pharmacy services: a position statement of the American College of Clinical Pharmacy. Pharmacotherapy 1989;9(1):45-56.

41. Cranz H. Health economics and self-medication. J Soc Admin Pharm 1990;7:184-9.

42. Srnka QM. Implementing a self-care-consulting practice. Am Pharm 1993;NS33(1):61-71.

43. Gannon K. Patients content with OTC information from RPhs. Drug Top 1990;(Mar 19):26-8.

44. Szeinbach SL, Banahan BF. The pharmacist as "gatekeeper." US Pharm 1993;(Dec):85-96.

45. Pathak DS, Nold EG. Cost-effectiveness of clinical pharmaceutical services: a follow-up report. Am J Hosp Pharm 1979;36:1527-9.

46. Stratton TP, Stewart EE. The role of the community pharmacist in providing drug and health information: a pilot survey among the public, physicians, and pharmacists. J Pharm Market Manage 1991;5(4):3-26.

47. Fincham JE, Karnik KA, Hospodka RJ, et al. Documenting the worth of pharmaceutical services: a pilot project. Final report submitted to the NARD Foundation, 1994.

48. Ryan PB, Johnson CA, Rapp RP. Economic justification of pharmacist involvement in patient medication consultation. Am J Hosp Pharm 1975;32: 389-92.

49. Williamson JW, German PS, Weiss R, et al. Health science information management and continuing education of physicians. Ann Intern Med 1989;110: 151-60.

50. Evans CE, Haynes RB, Birkett NJ, et al. Does a mailed continuing education program improve physician performance? Results of a randomized trial in antihypertensive care. JAMA 1986;255:501-4.

51. McCombs JS, Nichol MB. Pharmacy-enforced outpatient drug treatment protocols: a case study of Medi-Cal restrictions for cefaclor. Ann Pharmacother 1993;27:155-61.

52. Mead RA, McGhan WF. Use of H_2-receptor blocking agents and sucralfate in a health maintenance organization following continued clinical pharmacist intervention. Drug Intell Clin Pharm 1988;22:466-9.

53. Avorn J, Soumerai SB. Improving drug-therapy decisions through educational outreach: a randomized controlled trial of academically based "detailing." N Engl J Med 1983;308:1457-63.

54. Schaffner W, Ray WA, Federspiel CF, et al. Improving antibiotic prescribing in office practice: a controlled trial of three educational methods. JAMA 1983;250:1728-32.

55. Ray WA, Blazer DG II, Schaffner W, Federspiel CF, Fink R. Reducing long-term diazepam prescribing in office practice: a controlled trial of educational visits. JAMA 1986;256:2536-9.

56. Crowley S. Economic evaluation of academic detailing programs. Australian Prescriber 1993;16(Suppl 1):21-4.

57. Brodie DC, Smith WE Jr, Hylnka JN. Model for drug usage review in a hospital. Am J Hosp Pharm 1977;34:251-4.

58. Mehl B. Evolving concepts in hospital pharmacy management: drug usage evaluation. Springfield, NJ: Scientific Therapeutics Information, Inc., 1990.

59. Lipton HL. Drug utilization review in ambulatory settings: state of the science and directions for outcomes research. Med Care 1993;31:1069-82.

60. Griffin MR, Piper JM, Daugherty JR, Snowden M, Ray WA. Nonsteroidal anti-inflammatory drug use and increased risk for peptic ulcer disease in elderly persons. Ann Intern Med 1991;114:257-63.

61. Rupp MT. Evaluation of prescribing errors and pharmacist interventions in community practice: an estimate of "value added." Am Pharm 1988;NS28(12):22-6.

62. Britton ML, Lurvey PL. Impact of medication profile review on prescribing in a general medicine clinic. Am J Hosp Pharm 1991;48:265-70.

63. Rupp MT. Value of community pharmacists' interventions to correct prescribing errors. Ann Pharmacother 1992;26:1580-4.

64. Rupp MT, DeYoung M, Schondelmeyer SW. Prescribing problems and pharmacist interventions in community practice. Med Care 1992;30:926-40.

65. Dobie RL III, Rascati KL. Documenting the value of pharmacist interventions. Am Pharm 1994;NS34(5):50-4.

66. Gums JG, Robinson JD. Pharmacokinetic monitoring in the community health-care setting. Drug Intell Clin Pharm 1987;21:422-6.

67. Einarson TR, Bootman JL, Larson LN, McGhan WF. Blood level testing in a community pharmacy: consumer demand and financial feasibility. Am Pharm 1988;NS28(3):76-9.

68. Murray MD, Darnell J, Weinberger M, et al. Factors contributing to medication noncompliance in elderly public housing tenants. Drug Intell Clin Pharm 1986;20:146-52.

69. Col N, Fanale JE, Kronholm P. The role of medication noncompliance and adverse drug reactions in hospitalizations of the elderly. Arch Intern Med 1990;150:841-8.

70. Smith MC. Noncompliance with medication regimens: an economic tragedy. Emerg Issues Pharm Cost Containment 1992;2(2):10.

71. Strandberg LR, Dawson GW, Mathieson D, et al. Effect of comprehensive pharmaceutical services in long-term care facilities. Am J Hosp Pharm 1980;37:92.

72. Cooper JK, Love DW, Raffoul PR. Intentional prescription nonadherence by the elderly. J Am Geriatr Soc 1982;30:329.

73. Ostrom JR, Hammerlund ER, Christenson DB, et al. Medication usage in an elderly population. Med Care 1985;23:157.

74. Maronde RF, Chan LS, Larsen FJ, et al. Under-utilization of antihypertensive drugs and associated hospitalization. Med Care 1989;27:1159-66.

75. Jones RJ, Goldman MP, Rockwood RP, et al. Beneficial effect of a pharmacist refill evaluation clinic. Hosp Pharm 1987;22:166-8.

76. Sclar DA, Skaer TL, Chin A, et al. Effectiveness of the C Cap™ in promoting prescription refill compliance among patients with glaucoma. Clin Ther 1991;13:396-400.

77. Williams RF, Shepherd MD, Jowdy AW. Effect of a call-in prescription refill system on workload in an outpatient pharmacy. Am J Hosp Pharm 1983;40:1954-6.

78. Baird TK, Brooekemeier RL, Anderson MW. Effectiveness of a computer-supported refill reminder system. Am J Hosp Pharm 1984;41:2395-7.

79. Simkins CV, Wenzloff NJ. Evaluation of a computerized reminder system in the enhancement of patient medication refill compliance. Drug Intell Clin Pharm 1986;20:799-802.

80. Debrovner D. Did you take your pill today? Am Druggist 1992;206(6):60-6.

81. Sclar DA, Skaer TL, Chin A, et al. Effect of health education on the utilization of HMO services: a prospective trial among patients with hypertension. Prim Cardiol 1992;1(Suppl):30-5.

82. Dolinsky D. How do the elderly make decisions about taking medications? J Soc Admin Pharm 1989;6:127-37.

83. Reid LD, Hasuike BL. Comparing measure of beliefs about drug-taking compliance. J Soc Admin Pharm 1985;3:53-8.

84. Christensen DB. Understanding patient drug-taking compliance. J Soc Admin Pharm 1985;3:70-7.

85. Goggin T. Quality of life in hypertension—the effect of "care" versus "therapy." Drug Info J 1994;28:115-21.

86. Erfurt JC, Foote A, Heirich MA. The cost-effectiveness of work-site wellness programs for hypertension control, weight loss, and smoking cessation. J Occup Med 1991;33:962-70.

87. Martin S. Traditional values, innovative practices: pharmacists who provide cognitive services. Am Pharm 1990;NS30(4):22-7.

88. Penna RP. Compensation for pharmaceutical service: it's time for a change. Pharm Times 1990;56(10):43+.

89. Rupp MT. Strategic reimbursement: to secure third-party reimbursement for cognitive services, pharmacists need to answer four key questions. Am Pharm 1992;NS32(7):48-59.

ADDITIONAL READINGS

1. Strand LM, Guerrero RM, Nickman NA, Morley PC. Integrated patient-specific model of pharmacy practice. Am J Hosp Pharm 1990;47:550-4.

2. Catizone C. Office-based pharmacy in the United States: development of a practice alternative. Am Pharm 1984;NS24:24-32.

3. Hepler CD, Strand LM. Opportunities and responsibilities in pharmaceutical care. Am J Hosp Pharm 1990;47:533-43.

4. Strand LM, Cipolle RJ, Morley PC, Perier DG. Levels of pharmaceutical care: a needs-based approach. Am J Hosp Pharm 1991;48:547-50.

5. Hepler CD. The future of pharmacy: pharmaceutical care. Am Pharm 1990;NS30(10):23-9.

6. Goode MA, Gums JG. Therapeutic drug monitoring in ambulatory care. Ann Pharmacother 1993;27:502-5.

7. Strand LM, Cipolle RJ, Morley PC. Documenting the clinical pharmacists' activities: back to basics. Drug Intell Clin Pharm 1988;22:63-6.

8. Isetts BJ. Monitoring and managing patient care. Am Pharm 1992; NS32(1):77-83.

9. Srnka QM. Implementing a self-care-consulting practice. Am Pharm 1993;NS33(1):61-71.

10. Kidder S. The cost-benefit of drug reviews in long-term care facilities. Am Pharm 1982;NS22(7):63-6.

11. Hatoum HT, Catizone C, Hutchinson RA, et al. An eleven-year review of the pharmacy literature: documentation of the value and acceptance of clinical pharmacy. Drug Intell Clin Pharm 1986;20:33-48.

12. Willett MS, Bertch KE, Rich DS, et al. Prospectus on the economic value of clinical pharmacy services: a position statement of the American College of Clinical Pharmacy. Pharmacotherapy 1989;9(1):45-56.

13. Avorn J, Soumerai SB. Improving drug-therapy decisions through educational outreach: a randomized controlled trial of academically based "detailing." N Engl J Med 1983;308:1457-63.

14. Ray WA, Blazer DG II, Shaffner W, et al. Reducing long-term diazepam prescribing in office practice: a controlled trial of educational visits. JAMA 1986;256:2536-9.

15. Rupp MT. Value of community pharmacists' interventions to correct prescribing errors. Ann Pharmacother 1992;26:1580-4.

16. Col N, Fanale JE, Kronholm P. The role of medication noncompliance and adverse drug reactions in hospitalizations of the elderly. Arch Intern Med 1990;150:841-8.

17. Maronde RF, Chan LS, Larsen FJ, et al. Under-utilization of antihypertensive drugs and associated hospitalization. Med Care 1989;27:1159-66.

18. Sclar DA, Skaer TL, Chin A, et al. Effect of health education on the utilization of HMO services: a prospective trial among patients with hypertension. Primary Cardiol 1992;1(Suppl):30-5.

19. Dolinsky D. How do the elderly make decisions about taking medications? J Soc Admin Pharm 1989;6:127-37.

20. Goggin T. Quality of life in hypertension—the effect of "care" versus "therapy." Drug Info J 1994;28:115-21.

21. Rupp MT. Strategic reimbursement: to secure third-party reimbursement for cognitive services, pharmacists need to answer four key questions. Am Pharm 1992;NS32(7):48-59.

22. Norwood GJ, Yesalis C, Lipson D, Johnson N. Reimbursement by capitation: it's new, controversial, and strengthens incentive. Am Pharm 1979;NS19(1): 37-40.

23. Yesalis CE III, Norwood GJ, Lipson DP, et al. Capitation payment for pharmacy services: impact on generic substitution. Med Care 1980;18:816-28.

24. Lipson DP, Yesalis CE III, Norwood GJ. Capitation payment for pharmacy services: rationale, findings, and future plans. Med Market Media 1980;15(Oct):25-32.

25. Lipson DP, Yesalis CE III, Kohout FJ, Norwood GJ. Capitation payment for Medicaid pharmacy services: impact on non-Medicaid prescriptions. Med Care 1981;19:342-53.

26. Helling DK, Yesalis CE III, Norwood GJ, et al. Effects of capitation payment for pharmacy services on pharmacist dispensing and physician-prescribing behavior: I. Prescription quantity and dose analysis. Drug Intell Clin Pharm 1981;15:581-9.

Chapter 28

Recent Developments in Behavioral Medicine

Donna E. Dolinsky

WHAT IS BEHAVIORAL MEDICINE?

Attempts to frighten adolescents with negative consequences of cigarette smoking or recommendations to "just say no" to smoking and drugs have not been effective in deterring smoking in junior high students. Telling coronary heart disease patients to lose weight and to increase exercise, without helping them to change their behavior, has not been effective for most patients.

These common-sense, rational solutions to modify health-related behavior have not been effective because they do not reflect mechanisms or principles of human behavioral change. Interventions based upon research on human behavior have been effective in modifying behavior and subsequent physiological and behavioral health outcomes. Examples of these interventions are:

- Adolescents trained to teach their peers skills in choosing to smoke or not to smoke have been effective in reducing initiation of smoking.[1]
- Relaxation techniques involving muscle relaxation and breathing exercises can reduce chronic pain and decrease nocturnal enuresis.[2]
- Hypnosis and relaxation can increase percentages of T-lymphocytes in medical students during medical school examinations.[3]
- Hostility, the tendency to wish to inflict harm on others or the tendency to feel anger toward others, is correlated with coronary heart disease and other illnesses.[4,5]
- The quality of interpersonal relationships can influence immune function.[6]

Behavioral medicine studies these thoughts, feelings, and behaviors in health and illness and designs interventions based upon principles of human behavior resulting from research.

Behavioral medicine is an "interdisciplinary applied science concerned with the development and integration of behavioral and biomedical science, knowledge and techniques, related to health and illness, and the application of this knowledge and these techniques to prevention, diagnosis, treatment and rehabilitation."[7] The term "behavioral medicine" was first used in 1973, the first departments of behavioral medicine appeared in medical schools in late 1970, and the first review article on behavioral medicine appeared in 1982.[8,9]

Behavioral medicine is interdisciplinary. It integrates and applies knowledge and methodologies from behavioral and biomedical sciences to health and illness. The "parent" behavioral sciences are psychology, education, sociology, anthropology, and epidemiology. The "parent" biomedical sciences are anatomy, physiology, biochemistry, immunology, and neurology. Behavioral medicine is the joint application of these sciences to problems of health and illness. Cardiologists and psychologists, for example, have jointly studied the effect of hostility, a psychological variable, on the progression of coronary heart disease, a medical variable.

Behavioral medicine is primarily an applied science. The emphasis is upon patient care, as it is in medical and pharmaceutical care. When behavioral scientists conduct research, it is primarily applied, identifying behavioral treatments or interventions to affect health and illness. However, when theory, principles, or research on bidirectional links between behavior and health do not exist, researchers develop basic research to identify mechanisms that explain the effect of behavior on health and illness and the effect of health and illness on behavior.

One example of basic research is the work on peoples' mental representations of their illnesses.[10] Patients' representations, or common-sense beliefs about their illnesses and medications, influence their decisions about adhering to their medication regimens. If we can modify patients' representations about their illnesses, we can enhance medication adherence. However, we first need to conduct basic research to identify patients' representations of causes and consequences of their illnesses.

The early research and practice in behavioral medicine focused upon prevention: modifying behavior that contributed to the acquisition and modification of risk factors, conditions, and illness. Two risk factors studied were smoking and obesity. Insomnia was a health condition, and headache an illness that was studied. Recent research has addressed behavioral aspects in the prevention of AIDS, treatment of distress in cancer therapy, treatment and rehabilitation for coronary heart disease, and psychological treatment in cancer.

Scientists have also integrated and created new disciplines within behavioral medicine. One new discipline, psychoneuroimmunology, developed at the intersection of psychology, neurology, and immunology. Researchers in this discipline have examined relationships between behavioral variables and immune function, such as test anxiety and T-cell count. Biopsychology has investigated biological links to psychological behavior such as neurotransmitters and drugs in mental illness, biological models of eating disorders, and substance abuse as a function of reward systems in the brain.[11]

If the purpose of behavioral medicine is to modify human behavior related to illness, and if psychology is the science of human behavior, is behavioral medicine psychology? There is a branch of psychology called health psychology. Let us compare it to behavioral medicine.

WHAT IS HEALTH PSYCHOLOGY?

Health psychology is "the aggregate of the specific educational, scientific and professional contribution of the discipline of psychology to the promotion and maintenance of health prevention and treatment of illness and the identification of etiologic and diagnostic correlates of health, illness and related dysfunction."[12] While it is not an interdisciplinary behavioral science, but a division within the discipline of psychology, health psychology looks like behavioral medicine. Health psychology takes a biopsychosocial approach to human behavior in health and illness. This biopsychosocial approach combines biological, psychological, and sociological systems in describing and explaining physical illness.

Health psychologists originally focused upon causation and the process of illness, while behavioral medicine focused upon treatment and rehabilitation. Both behavioral medicine and health psychology have addressed prevention of illness. As we review research and applications in behavioral medicine, the overlap between behavioral medicine and health psychology may be more apparent than the differences. One author described behavioral medicine as a subspecialty of psychology.[12]

Behavioral medicine and health psychology clearly overlap in their definitions and in the topics discussed in two recent publications: a special issue of a psychological journal devoted to behavioral medicine and a review article on health psychology.[13,14] The issue on behavioral medicine presented research on behavioral and cognitive-behavioral treatments of obesity, smoking, chronic pain, headache, cancer, AIDS, insomnia, Type A behavior and coronary heart disease, gastrointestinal disorders, rheumatoid arthritis, diabetes, and asthma.* The health psychology article described variables related to health and illness such as hostility and inability to express emotion, cognitive factors such as patients' views of their illnesses, social variables such as social support, sociocultural variables such as ethnicity and poverty, coping behavior, adherence to a medication regimen, substance abuse, and exercise. The author discussed these ideas in relation to coronary heart disease and cancer.

It is not easy to differentiate between behavioral medicine and health psychology. This chapter's references could not be easily sorted into two piles labeled behavioral medicine and health psychology. The references that follow are taken from research published in journals in behavioral medicine, health psychology, and others.

WHAT BEHAVIORS ARE RELATED TO ILLNESS?

The matrix below presents behavioral variables that have been found to be related causally or correlationally to prevention of, treatment of, and rehabilitation following illness and disease. These terms are also called primary, secondary, and tertiary prevention.

Primary prevention includes promoting health behavior, such as exercise, good nutrition, and not smoking. This behavior can pre-

---

*The editor asked that the authors of the research articles focus upon treatment rather than prevention, diagnosis, or rehabilitation.

vent illness from occurring. Secondary prevention includes early detection of problem behavior and elimination of risk factors, such as stopping smoking once it has started. Tertiary prevention means treatment and rehabilitation.

Because some behaviors, like exercise, are implicated in more than one type of prevention, primary, secondary, and tertiary prevention will not be separated in the matrix below.

Behavioral aspects of diagnosis, another aspect of behavioral medicine, are beyond the scope of this chapter.[15]

In studying behavioral aspects of health and illness, behavioral scientists separate human behavior into four different constructs. Constructs are variables that are based in theory and whose existence is inferred from patterns of behavior. Receptors were considered to be constructs before they were isolated and identified.

These constructs used to describe human experience are: cognition (knowing and thinking), affect (feeling), behavior (doing), and interpersonal (thinking, feeling, and doing with someone else). Beliefs, defined as an assertion that something is true, will be categorized with cognitive variables, although they also contain affective characteristics. Think about the emotionality of people with extreme political beliefs. The belief is the assertion that there is a right way to believe and act, and the emotion is the feeling with which the individual pursues his or her political goal.

These categories of human experience were designed by behavioral scientists—primarily Western behavioral scientists—to study, describe, and explain aspects of human behavior. While they may not be directly measurable at a micro level (as are receptors) and thus may not be "real" in a biomedical sense, we can measure differences in physiological responses to differences in cognition, affect, behavior, and interpersonal relationships. Because these constructs are based upon interpretations of a unified human experience, descriptions of cognition, affect, behavior, and interpersonal interactions overlap. Anxiety (as in test anxiety), behavior bidirectionally related to physical illness, can have cognitive components (thinking one is not in control), affective components (fear), and behavioral components (trembling and sweaty palms).* Anxiety

---

*Anxiety can result in physical illness, and physical illness can result in anxiety.

can also be moderated by interpersonal relations. We may seek out other people when we are anxious.

At times, it may be difficult to separate cognition, affect, and behavior. As I walked outside after a discussion of these behavioral variables in a class, I overheard one student asking another, "Do I *think* it is raining, or do I *feel* it is raining?" While not distinct ideas, these categories of human experience will be used to help sort through and understand bidirectional relationships between human choice and action or experience as related to illness.

The cognitive, affective, behavioral, and interpersonal variables in Table 28.1 are some that have been found to be related to health and illness. The most easily understood relationships between behavioral variables and illness are those that directly influence physiological functioning and subsequent illness. These are listed under behavioral variables in the above matrix and described below.

Behavioral Variables

Alcohol consumption, calorie intake, drug consumption, nutrition and diet, physical activity, and smoking are observable variables. They do not have to be implied like cognition and affect. The leading chronic degenerative diseases—heart disease, cancer, and stroke—which account for 67% of fatal diseases, all have prevent-

TABLE 28.1. Behavioral Variables Related to Health and Illness

Cognitive	Affective	Behavioral	Interpersonal

Beliefs
Body image
Hardiness
Interpretation of pain
Learned helplessness
Locus of control
Self-efficacy
Self-esteem
Stage of willingness to
  change behavior
Stress
	Anxiety
	Depression
	Hostility
		Adherence to
		  treatment
		Alcohol consumption
		Drug consumption
		Nutrition and diet
		Physical activity
		Sex (safer)
		Smoking
		Stress-related
		  physiological arousal
		Sun (excessive)
			Loneliness
			Support & immune
			  function
			Interpersonal
			  conflict

able related behavioral risk factors.[16] Smoking, emotional stress, obesity, sedentary life-style, high-fat diet, and excessive exposure to sunlight are risks that lead to these and other chronic diseases. "Safer" sex reduces sexually transmitted diseases. Behavioral medicine practitioners have learned to modify these variables in health promotion and prevention, treatment, and rehabilitation.

An additional behavioral variable related to health and illness is adherence to treatment, or patient compliance. This behavioral variable, which has cognitive and affective precursors, is discussed in two other chapters in this text and will be addressed only briefly in this chapter. While adherence indirectly influences health outcome, it can influence it to a very great degree.

Patient adherence to medication regimen is astonishingly and reliably low. Patients are nonadherent about 20% of the time for short-term therapy, 50% of the time for longer-term symptomatic chronic conditions, and 70% of the time for long-term symptomatic conditions requiring a life-style change.[17] Nonadherence is very often under cognitive control of the patient, who is making the decision to modify the medication regimen. Patients' understanding, belief, affect, and feedback about their medication from past behavior influences their choices to follow their medication regimens. These representations of drugs' effects on the body that affect patients' adherence vary in degree of truth, logic, and emotion.[18]

Like most behavioral variables, adherence to medication is conscious, not automatic, and involves human choice. These behavioral variables function differently from biomedical variables like absorption, excretion, and receptor binding, in that while not 100% predictable, biomedical variables generally do not involve choice.*

Stress-related physiological arousal, listed here under "behav-

---

*This "choice" involved in behavioral variables may make pharmacy students uncomfortable. Sometimes the content of the applied pharmaceutical sciences is understood as "truth" to pharmacy students. It is believed to be either right or wrong, rather than an outcome of research with a probability of occurring by chance alone fewer than five times out of 100. The perceived certainty of applied pharmaceutical science course content, such as pharmaceutics and pharmacology, is easier to deal with than the perceived ambiguity of behavioral science variables, which almost always imply consciousness, choice, thus lack of certainty, ambiguity, and thus, discomfort.

ior," has been identified as a bidirectional component of several disorders: vascular and tension headache, essential hypertension, spasmodic dysmenorrhea, insomnia, gastrointestinal disorders, and asthma.[19] Relaxation is one nondrug therapy used in the treatment of these disorders.

Cognitive Variables

Beliefs about health and illness, here categorized under cognitive variables, influence behavior. Health behaviors are influenced by beliefs about the probability that an outcome will occur, the severity of an illness, the effectiveness of a particular health behavior in protecting against disease, the costs or barriers of engaging in a health behavior, the internal and external rewards of engaging in a behavior, the value of a behavioral outcome, and the belief that a person could successfully engage in a behavior.[20]

Here is an example of how a belief will influence behavior, which can lead to illness. Americans generally consider an attractive woman's body to be a very lean body. These beliefs about how one should look can influence eating behavior that can lead to a lean body that is not nourished which can lead to illness.[21] It is not "logical": it is psychological–and it is not uncommon.

Body image–one's perception of how attractive one's body is–is also related to eating behavior. As stated earlier, patients' representations of their conditions are not always true or logical. For example, one's own body image does not always correspond to reality as agreed to by others. A formerly obese woman was reported to have requested a seat-belt extension on an airplane. Her body image was of her previous body, not her present body.

Hardiness is a psychological construct thought to buffer the stress-illness connection. Commitment, control, and challenge are the three categories of beliefs that compose hardiness. Recent research suggests that hardy people are more likely to solve problems and to seek support when stressed. People who score lower on hardiness are more likely to engage in wishing and avoiding behavior, to avoid solving problems, and to become ill.[22]

Interpretation of pain, locus of control, and self-efficacy are three variables that have been linked to physiological responses.[2] Patients who interpreted their pain as not controllable; were passive in response to pain; or used hope, attention diversion, or praying to cope with pain had greater disability than patients who actively coped with their pain. Because the data were correlational, it is not possible to decide directionality of variables, i.e., did interpretation and subsequent coping influence disability, or did disability influence interpretation?

Locus of control is a cognitive variable in which the person locates blame as either internal (I did it to myself) or external (they did it to me). Patients scoring high on internal locus of control were less likely to report their pain as severe than patients scoring high on external locus of control.

Self-efficacy is a person's estimate of his or her ability to succeed. This cognitive variable has been linked to tolerance for pain and to carrying out health promotion behavior, such as stopping smoking and losing weight.[23] Self-efficacy is probably bidirectionally related to health and illness in that healthy, successful people may have higher beliefs in their self-efficacy.

Planning–thinking about contingencies of action–can affect health behavior. People who were asked how they might deal with problems of a sexual partner's reluctance to use a condom were found to take more free condoms, when offered, than people who were in traditional condom education programs that involved information giving and illustrating use.[24]

Self-esteem, the degree to which one holds oneself in regard, is bidirectionally related to desirable health behavior. People who engage in exercise programs have higher levels of self-esteem.[25]

Sometimes people with lower self-esteem will constantly find means to prove themselves in an effort to increase feelings of self-esteem. This behavior is one component of a complex of behavior, called Type A behavior, that has been linked to coronary heart disease. While there have been some problems in duplicating some of the research linking Type A behavior complex to coronary heart disease depending on the instrument used to measure Type A behavior, individual components of the complex, like self-esteem, have been identified as predictors of disease.[26]

Identifying the stage of willingness to change a behavior has been predictive of success in smoking cessation. Stopping smoking is a process that occurs in several identifiable stages. These are "pre-contemplation, not thinking about quitting smoking, contemplation, seriously considering quitting smoking in the next six months, action, quit or considering quitting within the next 30 days, and maintenance, quit for more than six months."[23] Interventions have been generally targeted at the action stage. Interventions may take different forms and have different degrees of success, depending on the stage targeted.[23] This staging idea may be useful in targeting interventions to modify other health behavior.

Affective Variables

Affective variables are those that address human emotion. Three emotions bidirectionally implicated in illness are anxiety, depression, and hostility.

Fifty-year-old to 65-year-old adults who exercised regularly over a 12-month period reported lower levels of anxiety and depression than adults who did not exercise.[27] Anxiety is a common precursor and reaction to physical illness. Anxiety correlated positively with denial of cardiac conditions and loss of self-esteem following coronary bypass surgery.[28]

Patients in chronic pain who were more depressed rated their pain as greater than those who were less depressed.[2]

Type A behavior, a behavioral complex consisting of time urgency, hostility, competitiveness, a strong drive to perform well, the need for recognition for performance, feelings of inadequacy, the need to constantly prove one's worth, acceleration of physical and mental activities, and intense concentration and alertness, is believed to be related to coronary heart disease.[26] Subsequent research has questioned the early findings because different results were obtained with different measures of the Type A behavior. But recent research has shown that hostility, "the tendency to react to unpleasant situations with responses that reflect anger, frustration, irritation, and disgust" may be one of the specific variables predicting coronary heart disease.[29] In retrospective research, Type A behavior was more likely to be found among middle-aged men with coronary heart disease than among men without the disease.[30]

Cynicism has been identified as a particularly virulent component of hostility. Cynicism is "an attitudinal set that stems from an inadequately developed sense of basic trust and centers around beliefs that other persons are generally mean, selfish and undependable."[31] Cynicism correlated with the incidence of coronary heart disease.

Intervention studies have attempted to reduce components of Type A behavior through cognitive behavioral interventions. The interventions were successful in altering Type A behavior in some men. These behavioral changes resulted in improved physical and psychosocial outcomes.[26]

Interpersonal Variables

Loneliness and lack of support have been found to be related to immune function in medical students and caregivers, to increased ability to stop smoking, and to increased pain for chronic back pain patients who had interpersonal conflict with their wives.[22,32,33]* In a review of research on social support and illness, Walker and Katon found that marriage, husbands' perception of love and support, absence of social isolation, and family support were correlated with health.[34]

Behavioral medicine has identified interventions to modify health-related cognition, affect, and behavioral and interpersonal interactions related to illness.

WHAT ARE BEHAVIORAL MEDICINE'S INTERVENTIONS?

The purpose of behavioral medicine interventions is to modify peoples' thoughts (cognition and beliefs), feelings (affect), and behavior related to illness and disease. Behavioral medicine interventions have been effective as primary (treatment of choice) and adjuvant treatment to drug therapy.[35] Several behavioral techniques have been used in the nonpharmacological treatment of hypertension as supplements to drug therapy.[36] They are used in primary prevention (stopping something before it starts by engaging in

---

*The study of the relation of behavior to immune function, psychoneuroimmunology, may be attractive to pharmacy students as it connects or grounds behavioral variables to measurable components of immune function.

healthy behaviors), secondary prevention (reducing risk factors), and tertiary prevention (treatment or rehabilitation).

The techniques reflect the matrix of behavioral variables: cognition, affective, behavioral, and interpersonal. Reduction of physiological arousal through relaxation training is added and precedes the other interventions.

Relaxation Training

Relaxation training involves reduction of physical and cognitive arousal that may lead to increased stress and subsequent related physiological symptoms. Relaxation techniques have been used in arthritis, asthma, bruxism, bulimia, cancer, smoking, dysmenorrhea, eczema, hemophilia, herpes, hypercholesterolemia, hypertension, insomnia, irritable colon, menopausal hot flashes, headache, obesity, pain, seizures, ulcers, and test anxiety.[37] There are several techniques used in relaxation training. These are progressive muscle relaxation, autogenic training, hypnosis, and biofeedback.

Progressive muscle relaxation involves tightening and releasing different muscles in sequence and focusing upon the relaxation in the muscles following tension release. A relaxation session may involve tightening and releasing hands, then arms, and then moving through other muscles throughout the body. The sessions may last as long as 45 minutes per session twice a day at the beginning of training and eventually result in a 5 to 15 minute session designed to reach the same goal much more efficiently by relaxing groups of muscles.[38]

A similar relaxation technique, autogenic training, involves suggestions and images of relaxation (e.g., feelings of heaviness in the limbs, warmth, easy heart beat, and an ordered and slow breathing).[37]

Hypnosis, a state of relaxation, distraction, and suggestion has been used to reduce physiological arousal in many illnesses.[16] A suggestion to relax the gastrointestinal tract along with hypnotherapy has been effective in reducing patients' gastrointestinal distress. Both group and individual hypnotherapy were found to be effective.[39]

Biofeedback means receiving direct, immediate, and understandable feedback on an aspect of physiological functioning that is usually not under voluntary control, such as heart rate, temperature, blood pressure, muscle tension, skin resistance, and brain waves that may signal relaxation.[40] Electronic equipment is designed to receive a signal from the physiological function and to send it to the person who wishes to monitor and learn to change the function. Biofeedback has been used in treatment of hypertension, migraine, irritable bowel syndrome, motor dysfunction with cerebral palsy and stroke, and to relax the pelvic floor in constipation and in chronic and recurrent rectal pain.[39,41]

An easy way to visualize biofeedback is to hold a mirror up to your mouth after aerobic exercise. If you have raised your heart rate to aerobic levels, your breath should steam up the mirror. Biofeedback is a more complicated, electronically mediated measure of physiological function.

Cognitive Behavioral Interventions

While we may see cognitive and behavioral interventions used separately, multiple interventions have been more successful. Cognitive interventions address relationships between what people think and believe about themselves and their illness and health outcomes. While success of cognitive interventions varies across disease states, there is strong evidence that the interventions were effective, and because of this, we can expect new treatment advances.[42] Behavioral interventions focus upon modifying specific behaviors. The techniques are based upon operant conditioning.[43]

Cognitive-behavioral interventions consist of packages of information listing and educating about behavior-illness connections, relaxation techniques, and coping with the stresses of an illness through modifying thoughts and illness related to behavior. The coping may include cognitive restructuring of beliefs and self-instructional training.

The rationale behind cognitive-behavioral interventions is that thoughts decide actions and that people can change their own behavior through naming and altering thoughts, beliefs, and images that define their perceptions of their behavior. These techniques have been successfully used in both physiological and mental disorders. Patients with irritable bowel syndrome were treated with education about the relationship between stress and bowel symptoms, thermal biofeedback (a technique to help relaxation), and cognitive stress-coping techniques. The experimental groups reported significantly greater reduction in bowel symptoms.[39]

A combination of stress management techniques has been used to treat patients rehabilitating from coronary bypass surgery. These have included addressing emotions of anxiety, depression, anger, and denial, and group and individual counseling to promote relaxation, time and personal organization, coping with life changes, resolution of interpersonal conflicts, preventing relapse, and modifying Type A behaviors. More than one-half of coronary bypass patients have been identified as Type A.[28]

The behavioral part of cognitive-behavioral interventions includes behavior modification techniques for specifying and measuring goals, and for monitoring progress (e.g., What am I doing now, what do I expect to do next week, in six weeks, and in the long term).[44]

Patients can modify cues that stimulate behavior. Insomnia patients, for example, can stop all activities in the bedroom that are incompatible with sleep, such as leaving the bed if lying awake for 20 minutes and not returning until sleepy.[45] Overweight people can remove potato chips and ice cream from the house.

Rewarding desirable behavior is an important part of behavior modification. We are likely to repeat rewarded behavior. We remember the reward, which can be internal (feeling good about succeeding) or external (lowered cholesterol), and it motivates behavior. Behavioral techniques have been used in treating abdominal pain, anxiety, alcoholism, assertive behavior, cancer, coronary prone behavior, dysmenorrhea, essential hypertension, hyperactivity, insomnia, irritable bowel syndrome, ischemic heart pain, lower back pain, migraine, physical fitness, sexual dysfunction, smoking, tension headache, and ulcer.[40]

Affective Interventions

These techniques address emotions implicated in health and illness. Psychotherapy is at one end of the continuum (e.g., people who used mental health services were less likely to use medical health services).[32] Discussing emotion is at the other end of the continuum. Students asked to keep a diary in which they wrote about troubling experiences had better immune function response than students who wrote about trivial experiences.[32]

Interpersonal

Patients recovered faster from illness when they could take part in and were comfortable with interactions with their physicians.[46] Spouse training and support groups have been effective in increasing health outcomes. Pharmacists in Copenhagen were trained to run support groups of smokers who were clients in their pharmacies. Their success rates were 60% at their last group meetings and 45% six months later.[47]

These are a few of the many studies describing how arousal reductions and cognitive, affective, behavioral, and interpersonal techniques have been effective in prevention, treatment, and rehabilitation of physical illness.

WHY IS BEHAVIORAL MEDICINE IMPORTANT FOR PATIENTS AND FOR THE HEALTH CARE SYSTEM?

Behavioral medicine, compared to biomedicine, may be less costly in the long term, as it may lead to changes in behavior that can decrease physical illness, it has fewer negative side effects than drug therapy, and it may be preferred by patients because it gives them greater autonomy and control. These reasons are briefly discussed below.

Behavioral interventions that promote prevention and coping with physical decease are less costly and less stressful to the patient than surgery, hospitalization, or long-term care.

Behavioral treatment may lead to behavioral change and the diminution or removal of risk factors, conditions, and diseases. Drugs may cure but are more likely to control the disease. They may have to be taken for very long periods and may have negative side effects. Side effects of behavioral treatments can be positive, e.g., skills in modifying smoking behavior may affect other systems of behavior, like eating behavior.

Patients may prefer nondrug therapy.* Behavioral medicine interventions may give them greater control over their diseases, greater autonomy, and less dependence upon the health care system.

The "challenge of the future lies more in chronic care than acute care and the most significant power of the physician or another health care provider may lie not in technological expertise but in his or her capacity to act as a behavioral change agent and a catalyst to get patients to modify a variety of self-injurious behavior."[48]

DO THE NONDRUG TREATMENTS OF BEHAVIORAL MEDICINE HAVE A PLACE IN PHARMACEUTICAL CARE?

Pharmacists delivering pharmaceutical care need to understand behavioral medicine and use the principles in making drug therapy decisions, in identifying drug-related problems, in interactions with patients, and in monitoring health outcomes for the following reasons.

Pharmaceutical care should lead to positive patient health outcomes, including quality of life. If a patient can achieve a comparable health outcome with a behavioral medicine treatment rather than a drug treatment, the behavioral treatment may be preferable in the long term for the reasons stated above: less costly, less stressful, fewer negative side effects, preferable to patients.

Behavioral treatments are also used as adjuvants to drug therapy, to reduce medication use, and to enhance outcomes.

Drug misadventuring, a general term describing drug-related problems, can lead to drug-related morbidity and mortality.[49] The responsibility of pharmacists to identify drug-related problems includes identifying when patients receive no drug when one is needed, receive the wrong drug, receive too little drug, receive too much drug, experience adverse drug reactions, experience drug interactions, or take a drug for a medically nonvalid indication.[50]

---

*Some patients may prefer drug therapy. The short-term benefits of medications may outweigh the long-term cost of modifying a behavior. Behavior modification takes time, commitment, and self-management. The short-term benefit may be preferable to many patients, to their physicians, and to their pharmacists.

An additional drug-related problem is taking a drug when nondrug therapy is indicated.[51] Pharmacy recognized this as a need when the American Council on Pharmaceutical Education included "understanding relevant non-drug therapy" as one of the competencies for the entry-level PharmD degree.[52] Nondrug therapy is part of pharmaceutical care.

The American Council on Pharmaceutical Education has also identified health promotion and disease prevention as part of pharmaceutical care.[52] Pharmacists, for example, have been sending letters to new parents, and including an immunization schedule and have been keeping immunization status as part of the patient's record.[53]

Many aspects of behavioral medicine overlap with pharmaceutical care.

WHAT DOES THIS MEAN FOR A PHARMACIST?

There is safety in drugs. Drugs do not make decisions about how they will work in the body. There is not much ambiguity. With humans and their behavior, there is ambiguity that goes along with consciousness and choice. A medical student discussed his preference for hospital care over home care. "In the hospital . . . you have social problems where you want them, you don't have to deal with them. There is no social. . . . Whereas at home, there are all kinds of interactions that you really have no control over."[54] Is this similar to a pharmacist's preference for hospital care over ambulatory care, where patients may not make many choices and pharmacists have more control?

Humans are not always rational. Biomedicine purports to have a rational base. "Biomedicine applies rational techniques to sickness, focussing on the body as a machine by distancing from the emotional significance of illness, ordering the body according to a physiological scheme, and finally, demanding that patients adopt the detached perspective of the scientist toward their own bodies."[55] Behavioral medicine applies behavioral models to sickness.

Pharmacy should make a decision whether to adopt aspects of behavioral medicine as part of pharmaceutical care or to let someone else address those needs. At some time during your professional development, you will choose, either actively or passively, to align yourself with a biomedical model of practice, a drug product model, or a pharmaceutical care model which incorporates biopsychosocial aspects of health and illness.

REFERENCES

1. Evans RI, Smith CK, Raines BE. Deterring cigarette smoking in adolescents: a psychosocial-behavioral analysis of an intervention strategy. In: Baum A, Taylor SE, Singer JE, eds. Handbook of psychology and health. Vol. 4: Social psychological aspects of health. Hillsdale, NJ: L. Erlbaum Associates, 1984:23-35.

2. Keefe FJ, Dunsmore J, Burnett R. Behavioral and cognitive-behavioral approaches to chronic pain: recent advances and future directions. J Consult Clin Psychol 1992;60:528-36.

3. Kiecolt-Glaser JK, Glaser R, Strain EC, Stout JC, et al. Modulation of cellular immunity in medical students. J Behav Med 1986;9:5-21.

4. Chaplin JP. Dictionary of psychology. Rev. ed. New York: Dell, 1982.

5. Smith TW. Hostility and health: current status of a psychosomatic hypothesis. Health Psychol 1992;11:139-50.

6. Kiecolt-Glaser JK, Glaser R. Interpersonal relationships and immune function. In: Carstenses LL, Neale JM, eds. Mechanisms of psychological influence on physical health. New York: Plenum Press, 1989:93-114.

7. Shunichi A, ed. An integrated behavioral approach to health and illness. Amsterdam: Elsevier, 1992.

8. Birk L. Biofeedback: behavioral medicine. New York: Grune and Stratton, 1973.

9. Agras WS. Some structural changes that might facilitate the development of behavioral medicine. J Consult Clin Psychol 1992;60:499-504

10. Skelton JA, Croyle RT, eds. Mental representation in health and illness. New York: Springer-Verlag, 1991.

11. Pinel JPJ. Biopsychology. Boston: Allyn and Bacon, 1990.

12. Matarazzo JD. Behavioral health and behavioral medicine: frontiers for a new health psychology. Am Psychol 1980;35:807-17.

13. Blanchard EB, ed. Special issue on behavioral medicine: an update for the 1990s. J Consult Clin Psychol 1992;60(4).

14. Rodin J, Salovey P. Health psychology. Annu Rev Psychol 1989;40: 533-79.

15. Tryon W. Behavioral assessment in behavioral medicine. New York: Springer Publishing Co., 1985.

16. DiMatteo MR. The psychology of health, illness, and medical care. Pacific Grove, CA: Brooks/Cole Publishing Company, 1991.

17. Sherbourne CD, Hays RD, Ordway L, DiMatteo MR, Kravitz RL. Antecedents of adherence to medical recommendations: results from the Medical Outcomes Study. J Behav Med 1992;15:447-68.

18. Croyle RT, Skelton JA, eds. Mental representation in health and illness. New York: Springer-Verlag, 1991.

19. Carlson CR, Hoyle RH. Efficacy of abbreviated progressive muscle relaxation training: a quantitative review of behavioral medicine research. J Consult Clin Psychol 1993;61:1059-1067.

20. Weinstein ND. Testing four competing theories of health-protective behavior. Health Psychol 1993;12:324-33.

21. Brownwell KD. Dieting and the search for the perfect body; where physiology and culture collide. Behav Ther 1991;22:1-12.

22. Williams PG, Wiebe DJ, Smith TW. Coping processes as mediators of the relationship between hardiness and health. J Behav Med 1992;15:237-55.

23. Lichtenstein E, Glasgow RE. Smoking cessation: what have we learned over the past decade? J Consult Clin Psychol 1992;60:518-27.

24. Winter L, Goldy AS. Effects of prebehavioral cognitive work on adolescents' acceptance of condoms. Health Psychol 1993;12:308-12.

25. Taylor SE. Health psychology. New York: McGraw-Hill, Inc., 1991.

26. Thorensen CE, Powell LH. Type A behavior patterns: new perspectives in theory, assessment and interventions. J Consult Clin Psychol 1992;60:595-604.

27. King AC, Taylor CB, Haskell WL. Effects of differing intensities and formats of 12 months of exercise training on psychological outcomes in older adults. Health Psychol 1993;12:292-300.

28. Anderson MP. Psychological issues in cardiovascular rehabilitation. In: Elias JW, ed. Cardiovascular disease and behavior. Washington, DC: Hemisphere Publications Corporation, 1987.

29. Lachar BL. Coronary-prone behavior: Type A behavior revisited. Texas Heart Inst J 1993;20:143-151.

30. Miller TQ, Turner CW, Tindale RS, et al. Reasons for the trend toward null findings in research on Type A behavior. Psychol Bull 1991;110:469-85.

31. Williams RB, Anderson N. Hostility and coronary prone heart disease. In: Elias JW, ed. Cardiovascular disease and behavior. Washington, DC: Hemisphere Publications Corporation, 1987:40-53.

32. Kiecolt-Glaser JK, Glaser R. Psychoneuroimmunology: can psychological interventions modulate immunity? J Consult Clin Psychol 1992;60:569-75.

33. Schwartz L, Slater MA, Birchler GR. Interpersonal stress and pain behaviors in patients with chronic pain. J Consult Clin Psychol 1992;62:861-4.

34. Walker EA, Katon WJ. Psychological factors affecting medical conditions and stress responses. In: Stoudemire A, ed. Human behavior: an introduction for medical students. Philadelphia, PA: J. B. Lippincott Company, 1994:115-127.

35. Blanchard EB. Introduction to the special issue on behavioral medicine: an update for the 1990s. J Consult Clin Psychol 1992;60:491-2.

36. Shapiro AP. The non-pharmacological treatment of hypertension. In: Drantz DS, Baum A, Singer JE. Cardiovascular disorders and behavior. New York: Lawrence Erlbaum Associates, 1983:47-58.

37. Smith JC. Relaxation. In: Ramachandran VS, ed. Encyclopedia of human behavior. San Diego, CA: Academic Press, 1994:102-103.

38. Rice PL. Stress and health. Monterey, CA: Brooks/Cole Publishing Company, 1987.

39. Whitehead WE. Behavioral medicine approaches to gastrointestinal disorders. J Consult Clin Psychol 1992;60:605-12.

40. Pinkerton SS, Hughes H, Wenrich WW. Behavioral medicine–clinical applications. New York: Wiley, 1982.

41. Luiselli JK. Behavioral medicine. In: Ramachandran VS, ed. Encyclopedia of human behavior. San Diego, CA: Academic Press, 1994:273-275.

42. Emmelkamp PM, van Oppen P. Cognitive interventions in behavioral medicine. Psychother Psychosom 1993;59:116-30.

43. Skinner BF. Science and human behavior. New York: Macmillan, 1953.

44. Tunks E, Bellissimo A. Behavioral medicine: concepts and procedures. New York: Pergamon Press, 1991.

45. Lacks P, Morin CM. Recent advances in the assessment and treatment of insomnia. J Consult Clin Psychol 1992;60:586-94.

46. DiMatteo MR. A social-psychoanalysis of physician-patient rapport: toward a science of the art of medicine. J Soc Issues 1979;35:12-33.

47. Moller L. Pharmacists helping smokers to quit: the role of the pharmacist in smoking cessation. Int J Smoking Cessation 1994;3(1):15-7.

48. Baschle T. Introduction. In: Wedding D, ed. Behavior and medicine. St. Louis, MO: Mosby-Yearbook, 1990:1-5.

49. Manasse HR Jr. Medication use in an imperfect world. Baltimore: ASHP Research and Education Foundation, 1989.

50. Strand LM, Morley PC, Cipolle RJ, et al. Drug-related problems: their structure and function. DICP Ann Pharmacother 1990;24:1093-7.

51. Dolinsky D, Webb CR. The need for pharmaceutical care. In: Knowlton CH, Penna RP, eds. Pharmaceutical care. New York: Chapman & Hall, 1996.

52. American Council on Pharmaceutical Education. The proposed revision of accreditation standards and guidelines for the professional program in pharmacy leading to the Doctor of Pharmacy degree. Chicago: ACPE, April 7, 1993.

53. Nahata MC. Pharmacist's role in improving immunization status among preschool children. Ann Pharmacother 1994;28:952-3.

54. Sankar A. Medical care in the home. In: Lock M, Gordon D, eds. Biomedicine examined. Dordrecht: Kluwer Academic Publisher, 1988:543-548.

55. Kirmayer LJ. Mind and body as metaphors: hidden values in biomedicine. In: Wedding D, ed. Behavior and medicine. St. Louis, MO: Mosby-Yearbook, 1990:312-323.

ADDITIONAL READINGS

1. Matarazzo JD. Behavioral health and behavioral medicine: frontiers for a new health psychology. Am Psychol 1980;35:807-17.

2. Blanchard EB, ed. Special issue on behavioral medicine: an update for the 1990s. J Consult Clin Psychol 1992;60(4).

3. Rodin J, Salovey P. Health psychology. Annu Rev Psychol 1989;40: 533-79.

4. Lichtenstein E, Glasgow RE. Smoking cessation: what have we learned over the past decade? J Consult Clin Psychol 1992;60:518-27.

5. Thorensen CE, Powell LH. Type A behavior patterns: new perspectives in theory, assessment and interventions. J Consult Clin Psychol 1992;60:595-604.

6. Emmelkamp PM, van Oppen P. Cognitive interventions in behavioral medicine. Psychother Psychosom 1993;59:116-30.

7. Baschle T. Introduction. In: Wedding D, ed. Behavior and medicine. St. Louis, MO: Mosby-Yearbook, 1990:1-5.

Chapter 29

Expectations, Education, and Technology

Sheldon Xiaodong Kong
Stephanie Y. Crawford

INTRODUCTION

Pharmacy has come to a critical stage in its professional development. The profession is currently facing many issues, such as the protracted debate on national health care reform, development and utilization of new technologies, proliferation of new specialties, increased incidence and prevalence of chronic diseases, and new social expectations. This chapter discusses the issue of pharmaceutical care from three perspectives: social and professional expectations, education, and technology. Three questions are answered: (1) Why should pharmaceutical care be the focus of pharmacy practice in the 1990s? (2) How will new technologies foster the provision of pharmaceutical care? (3) What type of educational processes are needed for pharmacists to provide pharmaceutical care? A model (Figure 29.1) was developed based on the premise that pharmacy practice in the United States has been undergoing a "paradigm shift" since the 1970s, a shift from the conventional product-oriented pharmacy practice to the patient-focused and outcome-related pharmaceutical care.[1-3] This shift has been driven by forces both within and outside the pharmacy profession. This chapter first identifies some of the major external forces that shape pharmacy practice and then discusses factors within the profession affecting the shift of pharmacy practice to pharmaceutical care. Next, the impact of technology on pharmaceutical care is examined. The chapter concludes with suggested educational strategies for the pharmacy profession to deal with these changes.

FIGURE 29.1. Factors that Affect Pharmaceutical Care

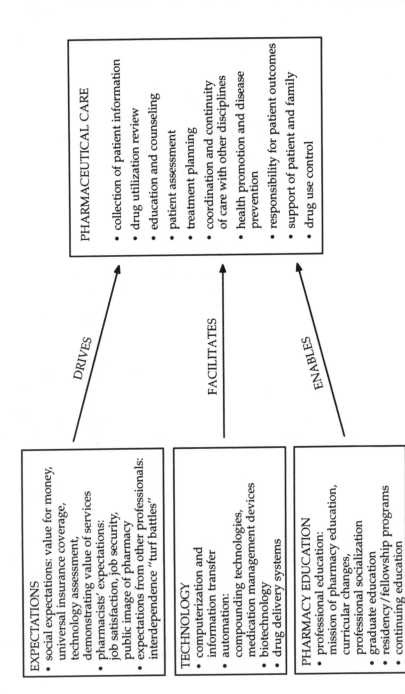

EXPECTATIONS

Many factors (e.g., social, economic, technological, and political) influence the future development of the U.S. health care system in general and the pharmacy profession in particular. This section discusses the expectations placed upon the pharmacy profession. These expectations come from society, pharmacists themselves, and other health care professionals.

First, the public is increasingly demanding high quality health care at an affordable price (value for its money). These demands are due to several critical factors, including the aging population, development and utilization of new medical and pharmaceutical technologies, and health care cost constraints.

Second, in the 1990s, pharmacists can no longer be limited to the dispensing role. Pharmaceutical care provides opportunities and challenges for pharmacists to assume more responsibilities related to patient care. These opportunities will bring an improved professional image, job satisfaction, and increased job/financial security for pharmacists.

Finally, new social and technological developments bring new challenges to all health professionals. On the one hand, new drugs, new knowledge, and new clinical frontiers (e.g., biotechnology) may encourage other health care professionals (e.g., physicians and dentists) to rely more on the pharmacist for drug-related decision making. On the other hand, these new responsibilities assumed by pharmacists may likely produce more "turf battles" among different professions in the cost-constrained health care environment.

Social Expectations: Value for Money

According to Relman,[4] the U.S. health care system is now on the threshold of its third revolution since World War II. The first revolution, which can be described as the Era of Expansion, began in the late 1940s and continued through the 1960s. The Era of Expansion was "characterized by rapid growth in hospital facilities and number of physicians, new developments in science and technology, and the extension of insurance coverage to the majority of the population."[4]

The open-ended system of health insurance payment and the

expansion of medical services of the first era led inevitably to the second revolution—the Era of Cost Containment.[4] To contain health care costs, third-party payers and government employed a variety of strategies, such as prospective payment systems (e.g., diagnosis-related groups, or DRGs, for hospital payments and certificates of need for hospital construction and medical equipment purchases) and managed competition (e.g., health maintenance organizations, formularies, utilization review, and capitation).

Despite efforts to contain health care expenditures in the Era of Cost Containment, the costs continued to escalate in the 1980s and 1990s. Compounding the frustration of third-party payers in their efforts to control costs was the growing concern about the unknown quality and outcomes of medical services.[4] New medical and pharmaceutical technologies were developed, dispensed, and utilized at an unprecedented rate. In the late 1980s, people began to realize that the time had come for a new national health care initiative focusing on the performance of health care institutions and practitioners. The emphasis has been shifted to the relative costs, safety, and effectiveness of all new technologies and health care interventions. The U.S. health care system has come to the third revolution—the Era of Assessment and Accountability.[4] The following are several major intended features of this new era for health care systems in the U.S.

Universal Coverage at an Affordable Price

American national health expenditures represent about 14% of the gross domestic product (GDP), and they are projected to reach 18% by the year 2000.[5] At the same time, roughly 35 to 40 million Americans have no health insurance coverage; millions more have inadequate health insurance that leaves them vulnerable to large out-of-pocket expenditures, excludes preexisting conditions, or may be lost if they become seriously ill.[6,7] The growing public concern about sharply escalating health care costs and the increasing uninsured and under-insured population have led the public, corporate executives, and union leaders to argue for some form of universal coverage. Although there is no universally accepted approach to these conditions, the agenda of health care reform is here to stay in the 1990s, despite President Clinton's initial setback by Congress.

Universal health care coverage may have conflicting effects on pharmacy practice. Added insurance coverage for medications will certainly increase access to pharmaceutical products and services for lower income and aged groups, who originally faced barriers to medication use due to lack of health insurance coverage. However, increased access to medical care and pharmaceutical products and services will inevitably balloon overall health care costs, which will force third-party payers to seek providers who can supply medications at the lowest per person cost. Pharmacists could employ two different strategies to gain advantages over rivals: the price leadership strategy and the differentiation approach.[8] Munroe and Rosenthal have described two similar strategies: (1) evolution–a gradual process of formation or growth, the immediate focus of which remains on the distribution aspects of pharmacy practice, and (2) revolution–a complete change in pharmacy practice, a proactive approach that focuses on patient-directed services (pharmaceutical care).[9]

The price leadership strategy focuses on the production of prescriptions for a lower price than that of the competition.[8] Pharmacies can use automation to increase dispensing volume and reduce costs; however, there are limits in efficiency that can be reached. The major focus of the differentiation strategy is to provide comprehensive pharmaceutical care. Unlike competition based on the dispensing of pharmaceutical products, provision of the service components of pharmaceutical care is not subject to overhead costs such as inventory, theft, merchandising, labor, or long hours of operation to provide retail customer convenience.[8] Provision of high quality, comprehensive pharmaceutical care requires highly qualified, motivated, and competent pharmacists who can demonstrate the value of their services based on patients' total health care outcomes (costs and quality of life) as a consequence of pharmacy services.

Technology Assessment

As discussed later in this chapter, the creation of sophisticated medical technologies and new drugs has been one of the most significant societal developments of the last two decades. The influence of high-cost medical and pharmaceutical technologies goes beyond the consequences for selected patient groups that may directly benefit from their application; the new technologies influence virtually everybody in society because of their high costs and potential impact on lives. Therefore, it is essential to perform rigorous evaluations of the safety, efficacy, and cost-effectiveness of these new technologies.[10] Since the 1980s, new research methodologies have been developed to address the question, How attractive does a new technology have to be to warrant adoption and utilization?[11-14] Assessments of some new technologies and drug therapies have already been performed, such as life-sustaining devices in home care,[15] pulse oximetry,[16] low-osmolar contrast media,[17] and drug treatment of cisplatin-induced nausea and vomiting.[18]

At the moment, two of the major producers/sponsors and consumers of technology and new drug assessment research are pharmaceutical manufacturers and managed care institutions. The major challenge to pharmacists is to understand and apply the research results from these studies. Substantial knowledge is required in the area of outcomes research to apply the results in drug formulary management and clinical decision making.[19-21]

Demonstrating the Value of Professional Services

In the Era of Assessment and Accountability, the public, health insurance companies, managed care organizations, and government are increasingly asking the question, Is society obtaining monetary value for its investment?[22,23] This question is directed not only toward new medical technologies and pharmaceuticals but also toward the performance of health care institutions (e.g., hospitals) and services provided by health care professionals (e.g., physicians and pharmacists). Length of stay (LOS) has been used by third-party payers as one of the key indicators for hospital (economic) efficiency.[24,25] Although health care professionals are increasingly asked to demonstrate the value of their professional services, there have been no universally accepted indicators available for assessment.[23,26]

Pharmacist payment has usually been linked to the sale of a product rather than to provision of services. To justify their services (or existence) economically, pharmacists must demonstrate the value of pharmaceutical services to the public, third-party payers, and other health care professionals.[27-30] The methodologies for evaluating the outcomes of professional services are similar to those used in assessing medical technologies and pharmaceutical products.[11-13] In general, the assessment requires detailed documentation of pharmacists' activities in specific settings, assigning monetary values to professional activities, and determining the costs and benefits/effectiveness of the services, all of which require advanced training in outcomes research.[31-34] We discuss the issue of pharmacy education later in this chapter.

Major Pressures Imposed on the Pharmacy Profession

Several major forces in the Era of Assessment and Accountability will shape the development of the pharmacy profession. The major aim of these forces is to control costs and ensure quality health care. Discussed below are some of the major activities from the federal government and managed care institutions that demand that pharmacists demonstrate the value of pharmaceutical products and services in health care.

The Government

The most obvious governmental solution to escalating health care costs is introduction of legislation and regulations. The Clinton Administration targeted health care reform as its main agenda. There seem to be two conflicting objectives of the current health care reform efforts: to provide high quality health care to everyone (universal coverage) and to contain the increase in health care costs (affordability).

The increased cost of pharmaceutical products has drawn significant attention from the federal government. In 1990, Congress passed the Medicaid drug rebate legislation (also known as OBRA 90) which requires pharmaceutical companies to offer rebates to state Medicaid programs.[35] Due, in part, to legislative changes, the annual increase in the producer price index for drugs and pharmaceuticals diminished from 6.9% in 1991 to 4.3% in the first half of 1993.[36] The declining rate of increase in the price of pharmaceuti-

cals will not only cut the profit margin of pharmaceutical manufacturers but may also influence the profitability of dispensing prescriptions in pharmacies. Therefore, pharmacists are motivated to
consider alternate sources of revenue, such as cognitive services,
both for the good of their patients and for the financial viability of
their pharmacies.[8]

The Omnibus Budget Reconciliation Act of 1990 (OBRA 90)
also mandates the implementation of prospective and retrospective
drug utilization review programs for Medicaid outpatient drug therapies.[29,37] This legislation is intended to save taxpayers money by
reducing the government's cost for pharmaceuticals and improving
patient care outcomes in the Medicaid program. The legislators
expected that pharmacists' interventions would lead to cost containment in health care. Pharmacy leaders and educators have seen the
passage of OBRA 90 as a golden opportunity for pharmacists to
demonstrate the value of their services. They believe that "pharmacists must rise to the occasion and show that the trust placed in them
by Congress has not been misplaced."[37]

Managed Care Programs

Within the last three decades, managed care has expanded from
the relatively simple health maintenance organization (HMO)
model to the diversified plans that are in existence.[38,39] Although
there is currently no universally accepted plan, managed care plays
an essential role in almost all health care reform proposals currently
under consideration in Congress. Miller and Luft[40] have summarized some of the key features in the current health care marketplace:

- The rapid growth of individual practice associations (IPAs),
  network and mixed-model HMOs, preferred provider organizations (PPOs), and point-of-service plans.
- The creation and transformation of many new networks and
  organizations, and increased market competition among health
  plans and among health care providers.
- Reductions in indemnity plan coverage.
- Changes in Medicare payment methods.
- A general shift from the hospital to ambulatory care settings.

As the market share of managed care programs increases in the U.S. health care marketplace, the impact of these plans on pharmacy will become more apparent. Managed care institutions view pharmacy as an area that can provide significant cost-control advantages,[9] and these institutions are more likely to contract with retail chains, use mail-order pharmacy services, set restricted formularies, and require generic substitution.[38,39,41] Some of the main challenges facing pharmacists in the managed care arena include increased competition within and outside pharmacy, cost management, drug utilization review, and demonstration of the economic and health care outcome value of pharmaceutical products and services.

Pharmacists' Expectations

The current social, economic, political, professional, and health care environments provide pharmacists with new challenges as well as opportunities. Wendell T. Hill Jr., in his 1989 Harvey A. K. Whitney Lecture, recommended that "since predicting the future of pharmacy is unreliable, it is necessary that pharmacists take another approach for securing an important role as health care providers."[42] He contended that pharmacists should take matters into their own hands and focus their attention on meeting the challenges of providing pharmaceutical services that are attractive to the public.[43]

Diversified Competition within Pharmacy

Only three decades ago, the term "pharmacy" denoted the hospital pharmacy and the independently owned and operated community pharmacy. Today, pharmacy comprises many types of competing businesses such as chain pharmacies, mass-merchandiser stores, food-drug combination stores, independent pharmacies, mail-order pharmacies, dispensing physicians, clinics, hospitals, home care providers, and managed care pharmacies.[9] The major payers for prescription medications have also shifted from solely the patient to the government, the employer (drug benefit managers), third party-payers, and the patient (out-of-pocket). In this environment, the pharmacist may not know how to respond to different audiences.

Pharmacy providers who practice in health care systems (e.g., hospitals, managed care settings, and chain pharmacies) may currently have an advantage over independent community pharmacy providers in responding to these divergent audiences unless independent pharmacists coalesce for better negotiating power and influence.

Outcomes of Pharmaceutical Care

Given the current health care environment in general and the increased competition in pharmacy in particular, it is important that pharmacists shift their practice toward patient-oriented pharmaceutical care. The potential outcomes/returns of cognitive services may be multifold for the pharmacist and the patient.

Improved Patient Health Care Outcomes and Reduced Health Care Costs. Because the pharmacist has the best training in pharmacotherapeutics among all health care professionals, one might speculate–and research has shown–that the pharmacist's involvement in drug-related patient health care reduces overall health care costs and improves the quality of care by reducing the number of subsequent clinic visits, preventing adverse drug events that require future interventions, and possibly preventing emergency-room visits and hospitalizations.[43] Drug-related illness accounts for 3 to 15% of all hospital admissions and visits, many of which might be prevented if pharmacists actively participated in the patient care process.[44-46]

Job Satisfaction. Many factors may influence job satisfaction among pharmacists, such as age, fringe benefits, work schedule, relationship with supervisor, practice location, and perceived opportunity for promotion.[47-50] Given the current rate of automation in pharmacies and increased financial pressure, pharmacists are increasingly asked to fill more prescriptions per unit of work time. Therefore, pharmacists have long been considered the overeducated and underutilized health care professionals. The career expectations of the new generation of pharmacists cannot be met merely by filling prescriptions. To reverse the trend of insufficient mental challenge, the role of pharmacists should be expanded beyond the traditional role of dispensing to pharmaceutical care.[49-51]

Increased Job Security. As discussed previously in this chapter, the increased cost-containment strategies employed by third-party payers, use of pharmacy technicians, and advanced automation in pharmacies will greatly diminish the pharmacist's dispensing function and reduce the manpower need for pharmacists. On the other hand, the increased public demand for the demonstration of "value for money" has created a unique opportunity for pharmacists to be involved in the patient care process. Pharmacists must resolve the fear that patient-focused care would eliminate their jobs, overcome other barriers for providing cognitive services, and grasp the opportunity to demonstrate the value of their services.[52-55] To accomplish this, pharmacy must go through a process of occupational reconstruction and self-renewal.[56]

Improved Professional Image. Hepler listed two negative perceptions of pharmacy's public image that affect the profession: (1) the pharmacist may be perceived as a "passive conduit" for drug technology where the pharmacist adds too few valued, complex, and scientific services, and (2) pharmacy practice may be perceived as a confusing hybrid of business and profession.[56] Confusion about pharmacy's professional standing may lead the patient to suspect that the professional (pharmacist) is not bringing sufficient competence to the exchange of trust between the profession and the public. In the Era of Assessment of Accountability, pharmacy's standing in the eyes of the public may indeed be more important than its prestige among health care managers.[56]

Confusion about pharmacy's purpose (business versus profession) is more troublesome because the business and professional systems are theoretically incompatible.[56] The issue for pharmacy is how to regain professional authority. In a professional covenant, the professional promises competence, including knowledge, skill, and appropriate attitudes of caring for the patient's welfare. In exchange, the patient grants the professional authority.[56]

Gallup polls have revealed that pharmacy has been ranked as one of the most respected and trustworthy (in terms of honesty and ethical standards) professions by the American public. Public trust and perception are key elements to the profession's authority. Pharmacists should capitalize on the public's high perception of the profession and show that the public's trust in them has not been misplaced.

Expectations from Other Health Care Professionals

In the 1990s, a cooperative relationship between pharmacists and other health care professionals, especially physicians and nurses, will be crucial. Two conflicting forces will shape the interprofessional relationships between pharmacists and other health professionals: competition for limited resources due to cost containment, and increased interdependence among health care workers due to proliferation of sophisticated new technologies and pharmaceuticals. Due to complicated medical technologies and new drugs, physicians and nurses will depend more on pharmacists for pharmacotherapeutics-related advice. On the other hand, the increased involvement by pharmacists in patient care will threaten the medical profession, which already has a surplus of personnel in certain specialties.

Most of the territorial disputes between health care professionals have been focused on the intrusion into the "physician's turf" by other health care professionals, such as nurses, midwives, social workers, occupational therapists, and pharmacists.[57-64] In his Morris Saffron Lecture presented to the Medical History Society of New Jersey, Cowen nicely summarized the "love and hate" relationship between pharmacists and physicians.[64] Throughout the history of American medicine and pharmacy, according to Cowen, the relationship between these two professions has been sometimes at ease and sometimes stressed. In the late nineteenth century, the pharmacy profession began to gain respect from the medical profession, partially because of the growth in both medical and pharmaceutical sciences and partially because of the physicians' ability to earn a living without having to sell pharmaceuticals. However, while they have recognized the key role of pharmacists in compounding and dispensing drugs, physicians have been very protective about their rights with respect to prescription and therapeutic decision making.[64]

Since the early twentieth century, the pharmacy profession has undergone substantial changes both in education and contemporary practice.[3,65] Pharmacist education has expanded from the two-year program (the beginning of this century), to a three-year program (1920s), four-year baccalaureate (1930s), and, finally, five-year baccalaureate and six-year doctorate programs (since 1960s). The advanced education of pharmacists, compounded by the increased potency and number of pharmaceuticals and the inadequate training of physicians in therapeutics, has increased the reliance of physicians on pharmacists for pharmacotherapeutic information and pharmaceutical services. Some of the pharmacist's services, such as pharmacokinetic monitoring of drugs[66] and nutrition support,[67] have gained physicians' acceptance in many institutions.

The medical establishment's reaction to increased pharmacist involvement in patient pharmaceutical care, however, has not been always positive.[64,68] Many physicians are increasingly concerned that the growing role of pharmacists in direct patient care, such as in drug therapy maintenance and therapeutic interchange, will threaten the medical authority of physicians. The reaction from some in the medical profession has been that, "if the pharmacist wants to get into the clinical management of patients, he/she ought to go to medical school."[69,70]

Hepler identified three major steps for pharmacy to deal effectively with interprofessional relationships between pharmacists and other health care providers:

> First, pharmacy should realistically assess the strength of its evidence for efficacy in drug-use control. Second, pharmacy should try to identify the new gatekeeper in medicine, for there is no reason to believe that medicine will be allowed to decide on everyone else's function. Third, pharmacy should seek alliances with other occupations, especially nursing, in pursuit of common long-term goals.[56]

TECHNOLOGY

Depending on one's perspective, technology has either forced or enabled the provision of pharmaceutical care and the paradigm shift in pharmacy. Since the first half of the twentieth century, thanks to the U.S. pharmaceutical industry, pharmaceuticals have been mass-produced in large quantities with high product quality. When the "fine art and science" of compounding and dispensing was replaced by high-precision machines, many pharmacists were left with the job description of "count, pour, lick, and stick." In the late 1960s, the profession of pharmacy began to realize that its societal need could no longer be fulfilled simply by distributing the product; hence, the clinical pharmacy movement was spawned. Today, the profession is advocating pharmaceutical care, which includes responsibility for providing all medication-related care (distributive, clinical, and interpersonal) to achieve positive patient outcomes.[54,71] How will new technologies foster pharmaceutical care?

Computerization and Information Transfer

Currently, computerized pharmacy systems (for maintenance of patient profiles and generation of filling or dispensing lists) are used in 75% of hospitals[72] and independent community pharmacies[73] and in 100% of chain pharmacies.[73] Additionally, 85% of hospitals have at least one microcomputer, and the most frequently used software applications are word processing, spreadsheet, and clinical information management.[72]

A plethora of technological developments has exploded and affected pharmacy practice. Point-of-sale electronic terminals are available in pharmacies to track product sales and utilization. Facsimile machines are utilized for prescription ordering. Synthesized voice systems can telephone patients with refill reminders. Electronic bulletin boards promote rapid information exchange.[73] Computers are used to document clinical interventions by pharmacists.[74-76] Other technological advances in information transfer that affect the delivery of health care include optical disc storage for medical records, smart cards containing patient medication histories, voice recognition technologies, and expert systems.[73] A new discipline, pharmacoinformatics, has emerged which deals with computer-supported storage, retrieval, and use of biomedical data for drug-related issues including drug therapy monitoring, drug information delivery, pharmacokinetics, technology assessment, and formulary decision making.[77] Some emerging hospital information systems have the functional capacity for decision support or artificial intelligence. These computerized expert systems assist in drug therapy decisions that are not completely objective, such as screening drug interactions and drug therapy monitoring.[77-79] Computerized systems in organized health care settings (primarily hospitals) have become more standardized, afford software portability to work in a variety of computers, and provide connectivity (linkage) to interface with different computer systems throughout the organization.[73,77,79] These systems facilitate the rapid transfer and sharing of clinical and other specific data on patient populations.[80] For example, the Department of Veterans Affairs (VA) has developed a computerized infrastructure for its hospital information systems, electronic mail systems, and wide-area network to connect its sites for the transfer and sharing of medical data through remote data exchange. This system has enabled the VA to establish a consolidated mail-out pharmacy program that is operational in several regions.[81]

Communications technology can eliminate information barriers that impede the provision of patient-oriented pharmaceutical services in both inpatient and ambulatory care settings. However, pharmacists must demonstrate that they can add value to the medical care process beyond drug dispensing.[56] The pharmacist's image will determine whether the profession will be included in the list of crucial health providers who must have access to critical diagnostic and other information.[82] If the pharmacist is seen only as a "drug merchant" or supervisor of robots, the public and policy makers will fail to see the need for pharmacists to access the information and will deny them the opportunity to affect patient outcomes.[56]

Pharmacists who practice in this environment of information explosion and rapidly developing technology must have a clear, basic understanding of computer operations. Additionally, health care administrators who consider longer-term measures to save costs and labor should understand the need to invest in technology that has the potential to reallocate staffing activities from a product focus to a patient focus. Managers should bear in mind that technological changes will have an important impact on the health care enterprise. As such, managers should prepare for an expansion of clinical decision-support systems; intense dependence on enhanced user interfaces and complex systems that encompass data, voice, and video;[83] and the necessity of educational support mechanisms for staff to keep abreast of the new technology.[84]

Automation

For the successful implementation of pharmaceutical care, pharmacists must ensure the existence of a safe, efficient drug distribution system. For example, the validity of the drug distribution conceived over 20 years ago in hospital settings (i.e., unit dose) has not been reevaluated. The impact on patient care of eliminating time-consuming manual steps in drug distribution is unknown.[85,86] Technological developments in computer-driven robotics and other automated devices for distributive medication management should enable pharmaceutical care by providing reliable support systems that free the pharmacist for more patient-oriented activities.[87] Automation should also help improve the quality of care. Despite numerous system innovations over the past decades, the manual medication dispensing process is fraught with opportunities for error, waste, and insufficient documentation. Initial studies have shown decreased error rates with use of automated medication management systems.[87,88] Automated aspects of the drug use process include compounding devices and medication distribution devices.[87,89-96]

Automated Compounding Technologies

New devices have been designed and introduced to automate the repetitive aspects of intravenous product compounding. These devices have been used primarily for compounding total parenteral nutrition (TPN) solutions. Devices are also being developed and used to automate the compounding of large- and small-volume parenterals (including syringes) for hospital-based practices. In cases where pharmacists were directly involved in compounding, these systems have shifted pharmacist time availability. Although more research is needed to document that the shifted activities are more clinically directed, some studies have shown increased accuracy, personnel time reduction, and labor reduction.[87,92-95,97]

Automated Medication Management Devices

There are several hospital-based automated medication distribution devices commercially available or currently under development. Essentially, these devices help with (1) distribution of medication to and from the patient care unit, (2) administration of medication directly to the patient, (3) inventory control, (4) management of controlled substances, and (5) documentation of medication dispensing and administration.[96] Some of the medication management systems are based in the patient care unit (e.g., Access, Argus, Lionville CDModule, Meditrol, MedStation, SelecTrac-Rx, Sure-Med Unit Dose Center, and Sure-Med Dispensing Center) and are designed to replace manual unit-dose cart filling or to increase control over floor stock medications and controlled substances.

Several medication management systems have been developed for central location in the pharmacy (e.g., ATC-212, Automated Pharmacy Station, and Medispense) and are designed to improve the manual system for filling unit-dose carts. Additionally, several point-of-care information systems (e.g., Automated Medication Administration Tracking, CliniCare, MedLynk, and MedTake) include medication management functions. These systems enable caregivers to enter and receive patient-specific data (based on bar-code technology) at the patient's bedside. The choice of which medication management system to use should be decided based on consideration of costs, regulatory requirements, environmental considerations (including space availability, staffing, existing information systems, and systems linkage capabilities), and improvements in patient care.[96,98] Similar advances are being developed for ambulatory care pharmacy settings. For example, the VA has developed an automated outpatient medication management system that can help dispense 10,000 prescriptions in a typical day shift.[96]

The high cost of such automated devices, however, poses a threat to their acquisition, use, and continued development. Well-designed assessment analyses are needed to evaluate the cost-effectiveness of these systems. Research on use of automated medication dispensing systems has shown decreased medication error rates,[88] increased time for interpersonal interactions between pharmacists and patients,[91] and reduced labor requirements.[89,90] More research is needed to examine the effect of automated medication management on the quality of patient care. In some cases, anecdotal evidence suggests that robotics and other automated systems are used merely as attempts to control and reduce costs without concomitant shift of personnel activities toward patient care functions (rather, the professionals may be expected to fill higher quotas in the distributive system).

Implications

Computerization and automation will largely displace technical aspects of pharmaceutical distribution (prescription and cart filling), reduce response time, decrease the error potential of manual systems, and facilitate documentation of services and interventions—all of which should improve the quality of patient care. Technology should be viewed as an adjunct enabling pharmacists to provide more patient-focused care while maintaining drug-use control. Use of these technologies should free up valuable pharmacist time for therapeutic assessments and professional judgment, creative problem solving, and professional and interpersonal communication with patients and other practitioners. Enright surmised that "critical to the overall success of the transition will be the ability to reallocate, retrain, and motivate pharmacists to assume patient-care roles."[87]

If pharmacists do indeed shift their activities to these cognitive areas, drug use control will remain a professional imperative for pharmacists for a long time. As Hepler stated, the provider who cannot control drug use may be unable to influence patient outcomes by preventing drug-related morbidity and may be unable to control health care costs.[56]

Biotechnology

Biotechnology products have revolutionized health care over the past decade. Commercially available biotechnology products (and their major indications) include:[99-102]

- Antihemophilic factor (Factor VIII deficiency)
- Coagulation factor IX (Factor IX deficiency)
- Erythropoietin (anemia)
- Granulocyte colony stimulating factor (chemotherapy-induced neutropenia)
- Granulocyte macrophage colony stimulating factor (myeloid recovery in bone marrow transplantation)
- Growth hormones (human growth hormone deficiency)
- Hemophilus B conjugate vaccine (*Haemophilus influenzae* type B)

- Hepatitis B vaccine (hepatitis B prevention)
- Human insulin (diabetes)
- Interferon-α (hairy cell leukemia, AIDS-related Kaposi's sarcoma, hepatitis C, genital warts)
- Interferon-γ (chronic granulomatous disease)
- Muromonab CD3 (renal transplant rejection)
- Tissue plasminogen activator (acute myocardial infarction)

Other biotechnology products in development include dismutases (reperfusion injury), growth factors, blood factors, interferons, interleukins (cancer, AIDS, bone marrow deficiencies), recombinant soluble CD4s (AIDS), tumor necrosis factors (cancer), and vaccines.[99,101,102]

Because of the high costs and range of unlabeled uses for biotechnology products, pharmacist interventions are needed to achieve positive patient outcomes in a cost-effective manner. Improved patient outcomes will be achieved through drug therapy monitoring, individualized dosing, and management of adverse events.[100] Pharmacists' responsibilities for biotechnology products include:[101,103]

- Clinical research.
- Development of prescribing protocols and guidelines.
- Technology assessments comparing alternative therapies.
- Inventory control in consideration of special storage conditions and short half-lives.
- Assessment of economic impact.
- Development of reimbursement strategies.
- Therapeutic monitoring and surveillance programs.
- Patient education and counseling.
- Education of physicians and other providers with respect to biotechnology product properties, dosage forms, indications, and cost considerations.

Reluctance among some pharmacists may be due to lack of familiarity with sources of information available on biotechnology products, reimbursement support, and other educational materials.[104] Pharmacists should cooperate with pharmaceutical manufacturers to manage costs, obtain reimbursements, and address educational issues.[105] Finally, pharmacists must exercise personal judgment, as well as professional judgment, in the resolution of ethical dilemmas with respect to biotechnology agents (e.g., rationing of therapy, control of agents, access to test results, and gene therapy to alter genetic makeup).

Drug Delivery Systems

Drug delivery systems are systems where a medication is integrated with another chemical, device, or process to control the rate or site of release. The pharmacist must ensure that new drug delivery systems for traditional and biotechnology products are used appropriately in patient care. The rational design of these drug delivery systems centers around relationships among physiological barriers; disease characteristics; and physicochemical, pharmacokinetic, and pharmacodynamic drug properties in addition to material science.[106] Some of the new drug delivery systems include:[102,106]

- Oral drug delivery systems that use polymers for targeted delivery to the gastrointestinal system, with minimized adverse events and increased efficacy.
- Rate-controlled drug delivery systems, including passive, preprogrammed systems (e.g., implants, transdermal patches); active, preprogrammed systems (e.g., infusion pumps, implantable pumps, implantable insulins under development); and active, self-programmed systems with biosensors to measure physical parameters such as temperature and glucose levels.
- Noninvasive drug delivery systems, such as transdermal, intranasal, ophthalmic, and buccal, that are used to achieve systemic (rather than local) effects.

It is questionable whether pharmacy will be identified as the most knowledgeable—and hence, responsible—profession for control of these agents and delivery systems. The precedent for non-pharmacy control of technological medication-related agents and devices has been established (e.g., radiopharmaceuticals are controlled by only 4% of pharmacy departments in community hospitals; similarly, only 14% control IV pumps).[72] The profession must decide whether pharmacy can increase its value by controlling the use of new technology or whether the opportunity costs of this control will be too high and will detract from cognitive professional responsibilities.[56] Although some might argue that pharmacy should rid itself of the distributive function, complete interpretation of pharmaceutical care provision would argue that pharmacy must be involved in the whole medication use process to ensure the rational use of drug therapy. Product sponsors largely direct the distribution channels of their products. Pharmacists must demonstrate and publicize the value of pharmaceutical care to the product originators (i.e., the pharmaceutical and biotechnology industries). A more collaborative relationship between the profession and industry during product development would help ensure the proper use of these agents.[56]

If pharmacy desires to control this technology, competition from other providers should be expected. The pharmacist's role should include therapeutic monitoring, compounding individualized dosage forms, recommendation of effective dosage, patient counseling and education, selection of drug delivery systems, and research (including controlled clinical trials).[102,106,107] It is especially critical that community pharmacists (e.g., home care pharmacists) and hospital pharmacists be knowledgeable about therapeutic applications and proper use of these systems.[108]

PHARMACY EDUCATION

Pharmacy has been challenged to demonstrate and communicate its value in health care. In this environment, pharmacy must revisit the question, What does it take to be a pharmacist today?[109] It must also reevaluate existing curricula and shift its instructional focus, if necessary. Because the focus of pharmacy practice is changing from primarily dispensing to patient-oriented pharmaceutical care, new topics are needed in the pharmacy curriculum. A needed addition in pharmacy education is the introduction of human values education so pharmacy graduates will be equipped with the ability to realize the opportunities and welcome the challenges in the new pharmacy paradigm.[2,3] In the 1990s and beyond, the U.S. population will be more diverse; therefore, pharmacy education must prepare graduates to meet the distinct needs and expectations of different population groups.

Entry-Level/Professional Education

After years of lingering debate about the entry-level Doctor of Pharmacy (PharmD) degree, in 1989 the American Council on Pharmaceutical Education (ACPE) issued a "Declaration of Intent" which indicated that future accreditation standards would focus on the PharmD degree as the sole professional degree. Seven national organizations have expressed support for a sole, uniform doctor of pharmacy degree.[110-112] The increased length of pharmacy education and the PharmD degree should enhance the academic stature for pharmacists among other health care professionals and the public.[113] The existing pharmacy curricula (both BS and PharmD) provide comprehensive instruction on basic pharmaceutical sciences, biomedical sciences, and pharmacy practice—all of which form the basis of contemporary pharmacy practice. However, the addition of new topics will better enable new graduates to provide pharmaceutical care.

Mission of Pharmaceutical Education

In 1993, the American Association of Colleges of Pharmacy Commission to Implement Change in Pharmaceutical Education (AACP Commission) suggested that the mission of pharmacy education and curricular content should be organized around pharmaceutical care.[114,115] The AACP Commission stated that:

- The mission of pharmaceutical education is to prepare students to enter pharmacy practice with the knowledge, skills, attitudes, and values needed to render pharmaceutical care to patients at a level defined by the evolving mission statement of pharmacy practice.
- Pharmaceutical education should be organized around four broad competencies: conceptual competence, technical competence, integrative competence, and career marketability.

The AACP Commission contended that each college or school of pharmacy should have a well-defined educational philosophy and curriculum so that pharmacy graduates will be equipped with thinking abilities, communication abilities, facility with values and ethical principles, personal awareness and social responsibility, self-learning ability and habits, and social interaction and citizenship.[114,115]

Curricular Changes

At present, there exist standard, well-developed curricula for professional pharmacy education.[116] While the focus on basic pharmaceutical sciences and biomedical/clinical sciences in traditional pharmacy curricula has provided the knowledge foundation for providing clinical pharmacy services, certain curricular changes and new instructions are needed for pharmacy education in the Era of Assessment and Accountability. These changes and new instructions should be based on the "progression from general ability-based outcome goals to professional ability-based outcome goals" based on integration of different pharmaceutical disciplines and outcomes assessment.[117]

The ACPE Proposed Revision of Accreditation Standards and Guidelines for the Professional Program in Pharmacy Leading to the Doctor of Pharmacy Degree specifies that the revised "curriculum should provide the student with a basic core of knowledge, skills, and attitudes, which in composite, relates to the expectation for the stated set of entry-level professional competencies necessary to become a generalist practitioner, who renders pharmaceutical care."[118] The curriculum should keep an appropriate balance among four areas: biomedical sciences; pharmaceutical sciences; behavioral, social, and administrative pharmacy sciences; and pharmacy practice.[118] New ideas and course modules have been proposed, such as research evaluation, therapeutics, pharmacoeconomics, and others.[117,119-122] The additional topics of research evaluation, outcomes research, health care systems, and technology should be incorporated into the new curricula.

Research Evaluation. Topics such as research design, statistics, and scientific writing should be included. The major objective for this type of instruction is to teach pharmacy students how to evaluate clinical, pharmaceutical, and outcomes studies in the primary literature. The emphasis should be on skills in reading, interpreting, analyzing, and critically evaluating research using a set of criteria.[119] Although sophisticated research skills are not as essential to practicing pharmacists as to research-oriented pharmacists with graduate degrees, without a good understanding of research methodologies, pharmacy practitioners may not be able to understand the research literature and may not be able to communicate effectively with third-party payers and others on research findings that demonstrate the value of pharmaceutical products and services in health care.

Outcomes Research. The new curricula should incorporate a comprehensive model of health care that includes clinical, economic, and humanistic (e.g., quality of life) outcomes so pharmacy graduates will be able to assume responsibilities for drug-related patient health care, assess patient outcomes, and demonstrate the value of pharmaceutical care.

Health Care Systems. As stated by the AACP Commission, "pharmacists must enter practice prepared to take active roles in shaping policies, practices and future directions of the profession."[114] Pharmacy graduates must understand that their actions may have broad environmental, social, and individual ramifications.[123] To communicate the value of pharmaceutical products/services to the public, policy makers, and other health care professionals, pharmacists should first comprehend the health care system. Changes may be needed in the existing instruction on health care systems.

Technology. Relevant emerging technology should be integrated into education and training programs with the goal of improving quality care, rational drug use, and cost effectiveness.[124] The professional curricula should include introduction to immunology, biotechnology, and compounding and dispensing of novel dosage forms.[56,101,125] Pharmacy education should also expose students to the use of automation in pharmacy practice. Introduction of the appropriate role of technology (e.g., robotics) in drug use control as an adjunct to pharmaceutical care fosters more receptivity toward its use and less threat among pharmacy students.[126]

Broad-Based Professional Socialization

Socialization is the process by which people acquire all skills, knowledge, and disposition that enable them to play a more or less active role in society. Professional socialization is the general process whereby students learn about the professional role of pharmacists and the expectations of performance in that role.[127,128] Pharmacy educators and practitioners must realize that "role models, namely faculty; institutional philosophy; student's values/reasons for selecting the profession; peers; previous and concurrent practice experiences and experiential programs all affect the (socialization) process."[128] Faculty at schools/colleges of pharmacy should form the base for fostering student academic learning so that pharmacy graduates can perform all competencies required for practice. However, the work environment (e.g., chain pharmacies, pharmacies located in mass-merchandiser stores, hospitals, mail-order pharmacies, independent pharmacies, etc.) in which pharmacists practice their profession and where pharmacy students are trained plays a key role in shaping professional philosophies, values, attitudes, and behaviors of pharmacy graduates. Pharmacy educators and practitioners should combine their efforts to educate and socialize pharmacy students on ideal professional responsibilities and constraints in the real world to avoid disillusionment as students progress through a pharmacy curriculum.[128]

Graduate Education

AACP defines graduate education as "post-baccalaureate and post-PharmD educational programs intended to produce scholars capable of conducting independent research in the pharmaceutical sciences."[129] The major focus of graduate education is "to teach students how to do research and how to use research tools in their disciplines to address important issues."[129] To conduct necessary research in new areas (e.g., pharmacoeconomics, biotechnology, health informatics, and communications), pharmacists may need graduate education.

Traditionally, most MS-degree pharmacy practitioners (as well as pharmacists with MBA degrees) have been in administrative practice affiliated with institutional pharmacies, whereas the majority of PhD degree holders, especially in chemistry and pharmaceutics, work in the pharmaceutical industry and academe. In the Era of Assessment and Accountability, new graduate programs should be established for new specialty areas, such as pharmacoeconomics, outcomes research, pharmacoepidemiology, and biotechnology.

Postgraduate Training: Residency and Fellowship Programs

The distinguishing feature of any clinical profession is the quality and scope of its postgraduate residency training program.[109] A consortium of national pharmacy organizations defined a pharmacy residency as "an organized, directed, postgraduate training program in a defined area of pharmacy practice" that is designed to train pharmacists in practice and management.[130] About 5% of hospitals offer accredited pharmacy residency programs, while an additional 8% offer residency programs that have not been accredited by any formal accreditation agency.[72] The effects of residency training extend beyond the individuals trained because residency-trained pharmacists provide the preceptorship needed for many of the other programs through which complex skills for providing pharmaceutical care can be attained.[131] These programs include:

- Residencies and fellowships
- Postbaccalaureate PharmD programs
- Continuing education (CE)
- On-the-job training by qualified preceptors
- Part-time residencies
- Short-term residencies or certificate programs

In 1992, the American Society of Hospital Pharmacists (ASHP, now the American Society of Health-System Pharmacists) revised its accreditation standards for residency training in pharmacy practice with an emphasis on pharmaceutical care.[132,133] ASHP currently is the only accrediting body for pharmacy residency programs, and most of the accredited pharmacy residency sites are in hospital-based settings. In the 1990s, the emphasis of health care is shifting to ambulatory and primary care. However, no primary care role for pharmacists has been recognized.[2,131,134] Special manpower training is needed for pharmacists' primary care and ambulatory care activities. Very few residency programs are available to pharmacy graduates who practice in community settings. Due to institutional structures, business environments, and other environmental constraints, it is still perceived as a challenge to practice pharmaceutical care in traditional community settings.

A pharmacy fellowship is defined as "a directed, highly individualized, postgraduate program designed to prepare the participant to become an independent researcher."[130] Fellowship programs are designed to develop pharmacists' research competencies and skills, and to foster investigation of pharmacy services and drug therapy needs. There are over 200 pharmacy fellowship positions currently offered.[135] Projected growth in pharmacy fellowship programs will foster research and new knowledge in defined areas of practice, including clinical outcomes management.

Continuing Education

Lifelong learning is necessary for pharmacy practice. Because pharmacists need many different types of knowledge, skills, and experiences (e.g., information-retrieval skills, knowledge about pathophysiology and immunology, experiences with patients with multiple diseases, and outcomes assessment), and most of these areas were not part of the curriculum at the time when they attended pharmacy school, special continuing education programs are needed to train, retrain, and continually develop pharmacists to provide pharmaceutical care. Pharmacists may learn new knowledge required to provide pharmaceutical care by reading scientific journals, attending scientific and professional meetings, and taking evening courses. However, certificate programs may be more effective in teaching pharmacists new knowledge and skills necessary to render pharmaceutical care. The unprecedented development and utilization of new technologies, increased economic pressure, and new clinical frontiers will encourage practitioners to accept new practice challenges. Certificate programs should provide pharmacists the opportunities to acquire new knowledge and skills to enhance their practices or to pursue alternate career paths.[136] The main objective of continuing education should be expanded from updating knowledge to both updating and retraining.

AACP defines certificate programs as "structured and systematic postgraduate education and training experiences for pharmacists that are generally small in magnitude and shorter in time than degree programs, and that impart knowledge, skills, attitudes and performance behaviors designed to meet specific pharmacy practice objectives."[136] Although at present the duration of certificate programs is potentially broad and may range from several days to weeks or years, AACP suggests that a certificate program should have competency-based objectives, measurable outcomes, didactic and experiential components, program and participant evaluations, and practitioner input to determine program objectives and content.[137] Schools/colleges of pharmacy should assume a greater responsibility for continuing education. Additionally, professional associations should assist in sponsoring programs and creating mechanisms to learn about new therapies and technologies. Information transfer via telecommunications and other technologies should be incorporated.[124]

In summary, pharmacy education (i.e., professional education, graduate education, postgraduate training, and continuing education) is the foundation that provides the necessary skills for pharmacists to meet new challenges in the new era of health care delivery.

CONCLUSION

In the 1990s, the U.S. health care system has come to an Era of Assessment and Accountability. Due to the escalating health care costs and the growing concern about the unknown quality and outcomes of medical and pharmaceutical services, the public, the government, and third-party payers are demanding value for their investments. Many forces will shape the future of pharmacy, such as expectations (society's, pharmacists', and other health care professionals'), technologies (new drugs, new medical technologies, and automation), and economics. The unprecedented development and use of new pharmaceuticals (such as biotechnology) and increased medical information will enhance the role of pharmacists in drug therapy related patient care.

On the other hand, because of automation, the use of pharmacy technicians, and managed care's perception of pharmacy as an area that can provide sufficient cost-control advantages, the importance of the dispensing function in pharmacy has been rapidly diminishing. Pharmacy practice and education should reflect the rapidly changing social environment. Pharmacy education (professional, graduate, residency, fellowship, and continuing education) should be organized to meet the new challenges in pharmacy practice and research. Pharmacy education should promote the knowledge, skills, values, and attitudes necessary to provide pharmaceutical care to enable the pharmacist to collaborate with other health professionals and share in responsibility for health care outcomes.

REFERENCES

1. Hepler CD. The third wave in pharmaceutical education: the clinical movement. Am J Pharm Educ 1987;51:369-85.

2. Plein JB. Pharmacy's paradigm: welcoming the challenges and realizing the opportunities. Am J Pharm Educ 1992;56:283-7.

3. Mrtek RG, Mrtek MB. Parsing the paradigms: the case for human values in the pharmacy curriculum. Am J Pharm Educ 1991;55:79-84.

4. Relman AS. Assessment and accountability: the third revolution in health care. N Engl J Med 1988;319:1220-2.

5. Burner ST, Waldo DR, McKusick DR et al. National health expenditure projects through 2030. Health Care Finan Rev 1992;14(1):1-29.

6. Friedman E. The uninsured. From dilemma to crisis. JAMA 1991;265:2491-5.

7. Wilensky GR. Filling the gaps in health insurance: impact on competition. Health Aff 1988;7:133-49.

8. Summers KH. Getting reimbursed for nondispensing services. Drug Top 1993;137(17):74-89.

9. Munroe WP, Rosenthal TG. Implementing pharmaceutical care: evolution vs. revolution. Am Pharm 1994;NS34(4):57-69.

10. Joint report of the Council on Scientific Affairs and the Council on Medical Service. Technology assessments in medicine. Arch Intern Med 1992;152:46-50.

11. Laupacis A, Feedy D, Desky A, et al. How attractive does a new technology have to be to warrant adoption and utilization? Tentative guidelines for using clinical and economic evaluations. Can Med Assoc J 1992;146:473-81.

12. Rutten FF, Bonsel GJ. High cost technology in health care: a benefit or a burden? Soc Sci Med 1992;35:576-7.

13. Deber RB. Translating technology assessment into policy. Conceptual issues and tough choices. Int J Technol Assess Health Care 1992;8:131-7.

14. Drummond M, Davies L. Economic evaluation of drugs in peripheral vascular disease and stroke. J Cardiovasc Pharmacol 1994;23(Suppl 3):S4-7.

15. Goldberg AI. Technology assessment and support of life-sustaining devices in home care. The home physician perspective. Chest 1994;105:1448-53.

16. Society of Critical Care Medicine. Technology Assessment Task Force. A model for technology assessment applied to pulse oximetry. Crit Care Med 1993;21:615-24.

17. Jacobson PD, Rosenquist CJ. The introduction of low-osmolar contrast agents in radiology: medical, economic, legal, and public policy issues. JAMA 1988;260:1586-92.

18. Zbrozek AS, Cantor SB, Cardenas MP, et al. Pharmacoeconomic analysis of ondansetron versus metoclopramide for cisplatin-induced nausea and vomiting. Am J Hosp Pharm 1994;51:1555-63.

19. Skaer TL. Applying pharmacoeconomic and quality of life measures to the formulary management process. Hosp Form 1993;28:577-84.

20. Drummond M, Torrence G, Mason J. Cost-effectiveness league tables: more harm than good? Soc Sci Med 1993;37:33-40.

21. Himmelstein DU, Woolhandler S, Bor DH. Will cost effectiveness analysis worsen the cost effectiveness of health care? Int J Health Serv 1988;18:1-9.

22. Anisfeld MH. Valuation: how much can the world afford? Are we getting value for money? J Pharm Sci Technol 1994;48(1):45-8.

23. Lightfoot J. Demonstrating the value of health visiting. Health Visit 1994;67(1):19-20.

24. Pedersen SH, Douville LM, Eberlein TJ. Accelerating surgical stay programs: a mechanism to reduce health care costs. Ann Surg 1994;219:374-81.

25. Lutjens LR. Determinants of hospital length of stay. J Nurs Adm 1993;23:14-8.

26. Marshall EC. Rationing the public's health and the optometric agenda. Philos Trans R Soc Lond B Biol Sci 1993;342:287-91.

27. Meade V. Getting paid for cognitive services. Am Pharm 1994;NS34(6):32-36.

28. Reeder CE. Economic outcomes and contemporary pharmacy practice. Am Pharm 1993;NS33(12 Suppl):S3-S6.

29. Lipton HL, Bird JA. Drug utilization review in ambulatory settings: state of the science and directions for outcome research. Med Care 1993;31:1069-82.

30. Hatoum HT, Akhras K. 1993 bibliography: a 32-year literature review on the value and acceptance of ambulatory care provided by pharmacists. Ann Pharmacother 1993;27:1106-19.

31. Ferris KB, Kirking DM. Assessing the quality of pharmaceutical care. II. Application of concepts of quality assessment from medical care. Ann Pharmacother 1993;27:215-23.

32. American Society of Hospital Pharmacists. ASHP technical assistance bulletin on assessing cost-containment strategies for pharmacies in organized health care settings. Am J Hosp Pharm 1992;49:155-60.

33. Dobie RL III, Rascati KL. Documenting the value of pharmacist interventions. Am Pharm 1994;NS34(5):50-4.

34. Chrischilles EA, Helling DK, Rowland CR, et al. Cost-benefit analysis of clinical pharmacy services in three Iowa family practice offices. J Clin Hosp Pharm 1985;10:59-66.

35. American Society of Hospital Pharmacists. Government Affairs Division. Summary of 1990 Medicaid drug rebate legislation. Am J Hosp Pharm 1991;48:114-7.

36. Santell JP. Projecting future drug expenditures–1994. Am J Hosp Pharm 1994;51:177-87.

37. Brushwood DB, Catizone CA, Coster JM. OBRA 90: what it means to your practice. US Pharm 1992;17(Oct):64-74.

38. Curtiss FR. Managed care: the second generation. Am J Hosp Pharm 1990;47:2047-52.

39. Curtiss FR. Managed health care. Am J Hosp Pharm 1989;46:742-63.

40. Miller RH, Luft HS. Managed care plan performance since 1980. JAMA 1994;271:1512-9.

41. Kreling DH, Mucha RE. Drug product management in health maintenance organizations. Am J Hosp Pharm 1992;49:374-81.

42. Hill WT Jr. Taking charge of the profession. Am J Hosp Pharm 1989;46: 1557-61.

43. Lobas NH, Lepiniski PW, Abramowitz PW. Effects of pharmaceutical care on medication cost and quality of patient care in an ambulatory-care clinic. Am J Hosp Pharm 1992;49:1681-8.

44. Miller RR. Hospital admissions due to adverse drug reactions: a report from the Post Collaborative Drug Surveillance Program. Arch Intern Med 1974;134:219-33.

45. Prince BS, Goetz CM, Rign TL, et al. Drug-related emergency room department visits and hospital admissions. Am J Hosp Pharm 1992;49:1696-700.

46. Colt HG, Shapiro AP. Drug-induced illness as cause for admission to a community hospital. J Am Geriatr Soc 1989;37:323-6.

47. Mason HA, Gaither CA, Hoffman EJ, et al. Benefits and work-schedule options for female hospital pharmacists. Am J Hosp Pharm 1994;51:790-7.

48. Lahoz MR, Mason HL. Burnout among pharmacists. J Fam Pract 1990;31:305-9.

49. Stewart JE, Smith SN. Work expectations and organizational attachment of hospital pharmacists. Am J Hosp Pharm 1987;44:1105-10.

50. Noel MW, Hammel RJ, Bootman JL, et al. Job satisfaction and the future of pharmacy. Am J Hosp Pharm 1982;39:649-51.

51. Rauch TM. Job satisfaction in the practice of clinical pharmacy. Am J Public Health 1981;71:527-9.

52. Vogel DP. Patient-focused care. Am J Hosp Pharm 1993;50:2321-9.

53. Raisch DW. Barriers to providing cognitive services. Am Pharm 1993;NS33(12):54-8.

54. Hepler CD, Strand LM. Opportunities and responsibilities in pharmaceutical care. Am J Hosp Pharm 1990;47:533-43.

55. Penna RP. Pharmaceutical care: pharmacy's mission for the 1990s. Am J Hosp Pharm 1990;47:543-9.

56. Hepler CD. Unresolved issues in the future of pharmacy. Am J Hosp Pharm 1988;45:1071-81.

57. Fagin CM. Collaboration between nurses and physicians: no longer a choice. Acad Med 1992;67:295-303.

58. Alpert HB, Goldman LD, Kilroy CM, et al. 7 Gryzmish: toward an understanding of collaboration. Nurs Clin North Am 1992;27:47-59.

59. Blais R, Meheux B, Lambert J, et al. Midwives defined by physicians, nurses and midwives: the birth of a consensus? Can Med Assoc J 1994;150:691-7.

60. Mizrahi T, Abramson J. Sources of strain between physicians and social workers: implications for social workers in health care settings. Soc Work Health Care 1985;10(3):33-51.

61. Gross AM, Gross J. Attitudes of physicians and nurses towards the role of social workers in primary health care: what promotes collaboration? Fam Pract 1987;4:266-70.

62. Roberts CS. Conflicting professional values in social work and medicine. Health Soc Work 1989;14:211-8.

63. Gilfoyle EM. Creative partnerships: the profession's plan. Am J Occup Ther 1987;41:779-81.

64. Cowen DL. Changing relationship between pharmacists and physicians. Am J Hosp Pharm 1992;49:2715-21.

65. Mrtek RG. Pharmaceutical education in these United States—an interpretive historical essay of the twentieth century. Am J Pharm Educ 1976;40:339-65.

66. Lewis RP. Clinical use of serum digoxin concentration. Am J Cardiol 1992;69(18):97G-106G.

67. Meade V. Nutrition support pharmacist helps physician practice. Am Pharm 1993;NS33(10):45-54.

68. Ballin JC. Therapeutic substitution: usurpation of the physician's prerogative. JAMA 1987;257:529.

69. Strickland DA. Pharmacists encroaching on care. Med World News 1991;32(Jan):42.

70. Williams CB. Everyone wants to play doctor. Med Econ 1992;69(Jan 20):37.

71. American Society of Hospital Pharmacists. ASHP statement on pharmaceutical care. Am J Hosp Pharm 1993;50:1720-3.

72. Crawford SY, Myers CE. ASHP national survey of hospital-based pharmaceutical services—1992. Am J Hosp Pharm 1993;50:1371-404.

73. Barker KN, Allan EL, Swensson ES. Effect of technological changes in information transfer on the delivery of pharmacy services. Am J Pharm Educ 1989;53(Suppl):27S-40S.

74. Smith SR, Utterback CM, Parr DD, et al. Pharmacist clinical intervention program. Top Hosp Pharm Manage 1993;13(2):1-15.

75. Schumock GT, Guenette AJ, Clark T, et al. Hospital mainframe computer documentation of pharmacist interventions. Top Hosp Pharm Manage 1993;13(2):16-24.

76. Bluml BM, Enlow M. Use of hand-held computers to record and analyze intervention data. Top Hosp Pharm Manage 1993;13(2):25-31.

77. Dasta JF, Greer ML, Speedie SM. Computers in healthcare: overview and bibliography. Ann Pharmacother 1992;26:109-17.

78. Morrell R, Wasilauskas B, Winslow R. Expert systems. Am J Hosp Pharm 1994;51:2022-30.

79. Hynniman CE. Drug product distribution systems and departmental operations. Am J Hosp Pharm 1991;48(Suppl 1):S24-35.

80. Grabenstein JD, Schroeder DL, Bjornson DC, et al. Pharmacoepidemiology and military medical automation: opportunity for excellence. Mil Med 1992;157:302-7.

81. Dayhoff RE, Maloney DL. Exchange of Veterans Affairs medical data using national and local networks. Ann NY Acad Sci 1992;670:50-66.

82. Beto JA, Geraci MC, Marshall PA, et al. Pharmacy computer prescription databases: methodologic issues of access and confidentiality. Ann Pharmacother 1992;26:686-91.

83. Kerr JK, Jelinek R. Impact of technology in health care and health administration: hospitals and alternative care delivery systems. J Health Admin Educ 1990;8:5-10.

84. Kalis DJ. New technology: a dilemma for the pharmacy manager. Top Hosp Pharm Manage 1989;8(4):11-25.

85. Shane RR. Prerequisites for pharmaceutical care. Am J Hosp Pharm 1992;49:2790-1.

86. Talley CR. Whither unit dose drug distribution? Am J Hosp Pharm 1994;51:1879.

87. Enright SM. Supporting pharmaceutical care through automation. Top Hosp Pharm Manage 1992;12(3):73-82.

88. Barker KN, Pearson RE, Hepler CD, et al. Effect of an automated bedside dispensing machine on medication errors. Am J Hosp Pharm 1984;41:1352-8.

89. Sham SM, Hollis RC. A cooperative approach to implementing automated medication distribution. Top Hosp Pharm Manage 1991;10(4):59-66.

90. Jones DG, Crane VS, Trussel RG. Automated medication dispensing: the ATC 212 System. Hosp Pharm 1989;24:604-10.

91. Lee LW, Wellman GS, Birdwell SW, et al. Use of an automated medication storage and distribution system. Am J Hosp Pharm 1992;49:851-5.

92. Dickson LB, Somani SM, Herrman G, et al. Automated compounder for adding ingredients to parenteral nutrient base solutions. Am J Hosp Pharm 1993;50:678-82.

93. Cote DD, Torchia MG. Robotic system for i.v. antineoplastic drug preparation: description and preliminary evaluation under simulated conditions. Am J Hosp Pharm 1989;46:2286-93.

94. Somani SM, Woller TW. Automating the drug distribution system. Top Hosp Pharm Manage 1989;9(1):19-34.

95. Thielke TS. Automation support of patient-focused care. Top Hosp Pharm Manage 1994;14(1):53-9.

96. Perini VJ, Vermeulen LC Jr. Comparison of automated medication-management systems. Am J Hosp Pharm 1994;51:1883-91.

97. Achusim LE, Woller TW, Somani SM, et al. Comparison of automated and manual methods of syringe filling. Am J Hosp Pharm 1990;47:2492-5.

98. Sterne J. Consultant offers sound advice on how to pick automation. Hosp Pharm Rep 1994;8(6):44.

99. Huber SL. Strategic planning for the colony-stimulating factors. Pharmacotherapy 1992;12(2 Pt 2):39S-43S.

100. Huber SL. Strategic management of biotechnology agents. Am J Hosp Pharm 1993;50(Suppl 3):S31-3.

101. Piascik MM. Research and development of drugs and biologic entities. Am J Hosp Pharm 1991;48(Suppl 1):S4-13.

102. MacKeigan LD, McGhan WF. Impact of technology on pharmacy manpower requirements. J Res Pharm Econ 1994;5(4):69-112.

103. Herfindal ET. Formulary management of biotechnological drugs. Am J Hosp Pharm 1989;46:2516-20.

104. Piascik P. Getting information about biotechnology products. Am Pharm 1993;NS33(4):18-9.

105. Bleecker GC. Reimbursement and pharmacoeconomic perspectives in biotechnology. Am J Hosp Pharm 1993;(Suppl 3):S27-30.

106. Robinson DH, Mauger JW. Drug delivery systems. Am J Hosp Pharm 1991;48(Suppl 1):S14-23.

107. Kwan JW. High-technology I.V. infusion devices. Am J Hosp Pharm 1991;48(Suppl 1):S36-51.

108. Kwan JW, Anderson RW. Pharmacists' knowledge of infusion devices. Am J Hosp Pharm 1991;48(Suppl 1):S52-3.

109. Anderson RW. Of perceived value. Am J Hosp Pharm 1992;49:1919-24.

110. Anon. ASHP, APhA, NARD issue joint statement on entry-level Doctor of Pharmacy degree. Am J Hosp Pharm 1992;49:244-56.

111. Anon. Five pharmacy groups issue commentary on joint statement on the entry-level degree in pharmacy. Am J Hosp Pharm 1992;49:2082-6.

112. Buttaro M. AACP house of delegates vote: colleges to move to sole entry-level Pharm.D. Am J Hosp Pharm 1992;49:2346-50.

113. Manasse HR Jr, Giblin PW. Commitments for the future of pharmacy: a review and opinion of the Pharm.D. curricular debate. Drug Intell Clin Pharm 1984;18:420-7.

114. American Association of Colleges of Pharmacy. Commission to Implement Change in Pharmaceutical Education. Background paper I: What is the mission of pharmaceutical education? Am J Pharm Educ 1993;57:374-6.

115. American Association of Colleges of Pharmacy. Commission to Implement Change in Pharmaceutical Education. Background paper II: Entry-level, curricular outcomes, curricular content and educational process. Am J Pharm Educ 1993;57:377-85.

116. American Council on Pharmaceutical Education. Standards and Guidelines for Accreditation of Professional Degree Programs in Pharmacy. Revised. July 1, 1984.

117. Chalmers RK, Grotpeter JJ, Hollenbeck RG, et al. Changing to an outcome-based, assessment-guided curriculum: a report of the Focus Group on Liberalization of the Professional Curriculum. Am J Pharm Educ 1994;58:108-16.

118. American Council on Pharmaceutical Education. The Proposed Revision on Accreditation Standards and Guidelines for the Professional Program in Pharmacy Leading to the Doctor of Pharmacy Degree. April 7, 1993.

119. Dolinsky D. Teaching skills of research evaluation. Am J Pharm Educ 1994;58:82-6.

120. Draugalis JR, Slack MK. A course model for teaching research evaluation in colleges of pharmacy. Am J Pharm Educ 1992;56:48-51.

121. Winslade N. Large group problem-based learning: a revision from traditional to pharmaceutical care-based therapeutics. Am J Pharm Educ 1994;58: 64-73.

122. Draugalis JR, Coons SJ. The role of colleges of pharmacy in meeting the pharmacoeconomic needs of the pharmaceutical industry: a conference report. Clin Ther 1994;16:523-37.

123. Coster JM. Pharmacy, politics, and public policy. Am J Hosp Pharm 1992;49:1759-62.

124. Brodie DC, Smith WE. Implications of the new technology for pharmacy education and practice. Am J Pharm Educ 1984;49:282-94.

125. Allen LV Jr, Bloss CS, Brazeau GA, et al. Strategies for education. Am J Pharm Educ 1993;57:265-8.

126. Grussing PG, Crawford SY, Rice JA, et al. Attitudes of pharmacy and engineering students toward robotics in pharmacy. American Association of Colleges of Pharmacy, 95th Annual Meeting, Albuquerque, NM, July 19, 1994.

127. Hatoum HT, Smith MC. Identifying patterns of professional socialization for pharmacists during pharmacy schooling and after one year in practice. Am J Pharm Educ 1987;51:7-17.

128. Chalmers R, Adler D, Haddad A, et al. The essential linkages of professional socialization and pharmaceutical care. Report to the COF Committee on Changing the Culture within Our Schools/Colleges of Pharmacy. July, 1994.

129. American Association of Colleges of Pharmacy. Commission to Implement Change in Pharmaceutical Education. The responsibility of pharmaceutical education for scholarship, graduate education, fellowships, and postgraduate processional education and training. Am J Pharm Educ 1993;57:377-85.

130. Anon. Definitions of pharmacy residencies and fellowships. Am J Hosp Pharm 1987;44:1142-4.

131. Knapp KK. Pharmacy manpower: implications for pharmaceutical care and health care reform. Am J Hosp Pharm 1994;51:1212-20.

132. American Society of Hospital Pharmacists. ASHP accreditation standard for residency in pharmacy practice (with an emphasis on pharmaceutical care). Am J Hosp Pharm 1992;49:146-53.

133. Ray MD. Proceedings of the 1992 National Residency Preceptors Conference: planning residencies for pharmaceutical care. Am J Hosp Pharm 1992; 49:2161-6.

134. Schwab PM, Paavola FG. Future paradigm: challenge for pharmaceutical education. Am J Pharm Educ 1992;56:279-82.

135. Knapp KK, Sorby DL. The impact of specialization on pharmacy manpower. In: Directions for specialization in pharmacy practice, Part II. Proceedings of an invitational conference sponsored by the American Association of Colleges of Pharmacy, the American College of Clinical Pharmacy, the American Pharmaceutical Association, and the American Society of Hospital Pharmacists. Am J Hosp Pharm 1991;48:691-700.

136. Popovich NG. Chair report for the professional affairs committee. Am J Pharm Educ 1988;52:415-7.

137. AACP Memorandum (Richard P. Penna) to Deans of Member Schools. AACP/ACPE Conference on Certificate Programs, November 19, 1989.

ADDITIONAL READINGS

For the Instructor

1. Starr P. The social transformation of American Medicine: the rise of a sovereign profession and the making of a vast industry. New York: Basic Books, Inc., 1982:3-30, 290-449.

2. Hepler CD. The third wave in pharmaceutical education: the clinical movement. Am J Pharm Educ 1987;51:369-85.

3. Burner ST, Waldo DR, McKusick DR, et al. National health expenditure projects through 2030. Health Care Finan Rev 1992;14(1):1-29.

4. American Council on Pharmaceutical Education. The Proposed Revision on Accreditation Standards and Guidelines for the Professional Program in Pharmacy Leading to the Doctor of Pharmacy Degree. April 7, 1993.

5. Mrtek RG. Pharmaceutical education in these United States–an interpretive historical essay of the twentieth century. Am J Pharm Educ 1976;40:339-65.

6. American Association of Colleges of Pharmacy. Commission to Implement Change in Pharmaceutical Education. Background paper I: What is the mission of pharmaceutical education? Am J Pharm Educ 1993;57:374-6.

7. American Association of Colleges of Pharmacy. Commission to Implement Change in Pharmaceutical Education. Background paper II: Entry-level, curricular outcomes, curricular content and educational process. Am J Pharm Educ 1993;57:377-85.

For the Student

1. Hepler CD. Unresolved issues in the future of pharmacy. Am J Hosp Pharm 1988;45:1071-81.

2. Knapp KK. Pharmacy manpower: implications for pharmaceutical care and health care reform. Am J Hosp Pharm 1994;51:1212-20.

3. Munroe WP, Rosenthal TG. Implementing pharmaceutical care: evolution vs. revolution. Am Pharm 1994;NS34(4):57-69.

4. Cowen DL. Changing relationship between pharmacists and physicians. Am J Hosp Pharm 1992;49:2715-21.

Chapter 30

Ethical Concerns in Drug Research

Kenneth A. Speranza Sr.
Amy M. Haddad

INTRODUCTION

Contemporary drug research and the need to ensure scientific and ethical integrity in drug development, clinical testing, human experimentation, and marketing had their beginnings in the late 1930s. This was a period of significant social and scientific change and the beginning of the therapeutic revolution in medicine and pharmacy. As the political and social stage was being set for U.S. involvement in World War II, the sulfa drugs, forerunners of the twentieth century's new "miracle drugs," were being introduced as highly effective tools in the treatment of infections, both at home and on the war front. Concern for drug safety at this time (intensified by the death of approximately 100 people due to a poisonous sulfanilamide elixir), however, resulted in the passage of the Federal Food, Drug, and Cosmetic Act in 1938.

During World War II, penicillin was developed. After the war, "penicillin was followed by other antibiotics, by anti-malarials, synthetic vitamins, new and improved vaccines, steroid hormone-like compounds, antihistamines, tranquilizers, and other potent therapeutic agents."[1] In 1962, the Federal Food, Drug, and Cosmetic Act was expanded by the Kefauver-Harris Amendment to ensure that the nation's rapidly developing and enormous drug supply was not only safe for use but also effective. Today the Federal Food, Drug, and Cosmetic Act governs much of the activity required for the development and marketing of safe and effective drugs. The research and approval process that evolved from this law is presented in Figure 30.1.[2]

Concomitant with the development of the scientific procedures and methods necessary for the conduct of valid drug research was the formulation of ethical principles and guidelines deemed imperative for socially responsible drug testing in humans. "The beginnings of modern concepts of ethics and regulation of clinical research began after World War II when the Nazi physicians' atrocities became known to the world."[3] The Nuremberg Code, written after the conclusion of the war trials, detailed ten principles of

FIGURE 30.1. Drug Research and Approval Process

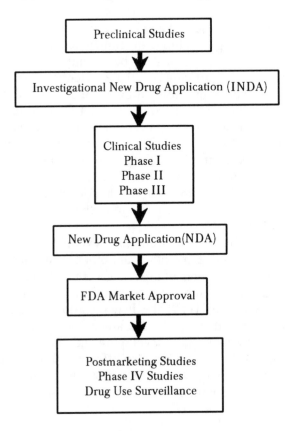

permissible medical experimentation and emphasized the ethical necessity of voluntary consent of the human subjects in clinical research (see Appendix A). The Declaration of Helsinki, published by the World Health Association in 1964 and revised and extended by the Twenty-Ninth World Medical Assembly in Tokyo in 1975, expanded ethical considerations regarding human experimentation by emphasizing that only qualified medical persons should conduct clinical research and that human beings must be free to give their informed consent before participating in clinical research studies (see Appendix B).

The Nuremberg Code and the Declaration of Helsinki established ethical norms for the conduct of research in humans. By the mid-1960s the U.S. government had begun to translate these ethical norms into federal regulations. In 1966, the U.S. Public Health Service issued a policy statement requiring committee review of research conducted under government financial support. The National Research Act, which was passed in 1974, established the National Commission for the Protection of Human Subjects of Biomedical and Behavioral Research. Institutional review boards (IRBs) were required in regulations promulgated by the Food and Drug Administration and the Department of Health and Human Services in 1981. In addition to specifying the composition and procedures of the IRBs, the regulations outlined elements of informed consent that must be provided to any research subject.

These codes, federal agencies, laws, and regulations have been and continue to be instrumental in establishing the myriad of guidelines intended to ensure the ethical conduct of drug research and the protection of human subjects. However, even the most detailed set of ethical guidelines cannot anticipate all of the moral problems that could arise during a clinical trial. Thus, it is incumbent upon researchers to consider more deeply the ethical principles and normative expectations that guide each phase of a research study involving human subjects.

ETHICAL PRINCIPLES UNDERLYING DRUG RESEARCH

Ethics is that branch of philosophical study concerned with the rational analysis and justification of moral principles that provides knowledge regarding what we ought to do to attain the most of what is best in human life.[4] Health care ethics addresses problems that occur within the context of the health care system. So, for the specific purposes of this chapter, we focus on those principles and virtues that govern the moral life in the area of drug research and human experimentation.

Ethical principles are action guides that help us assess what is morally right or wrong regarding human action in general. There are numerous ethical constructs and principles, but the most central concepts to a discussion of human experimentation are autonomy, beneficence, justice, and truth telling. Together, these principles provide a framework for the evaluation and analysis of the moral acceptability of particular decisions or acts.

Autonomy

The basic principle of respect for persons asserts that human beings are unconditionally worthy agents and, therefore, should be treated with respect and regard. The notion of respect for persons is understood at its most fundamental level as noninterference, i.e., the duty to leave others alone to pursue their own interests. The principle of respect for persons is most commonly perceived in light of individual autonomy. Autonomy has been variously defined and associated with the concepts of liberty, self-determination, freedom, and self-governance. The most general idea of autonomy is probably best understood as "personal rule of the self that is free from both controlling interferences by others and from personal limitations that prevent meaningful choice, such as inadequate understanding."[5]

Autonomy is an ethical principle of particular importance when one considers that human experimentation, by its nature, involves encroachment upon the privacy and physical well-being of subjects. Thus, an investigator must be certain that any interference with human subjects is in keeping with their autonomous wishes. The investigator assures that he or she is in accord with the subjects' choices by the informed consent process.

The criteria for and challenges of informed consent specific to research are discussed later in the chapter. However, one specific challenge of informed consent merits discussion here. It is clear that

some individuals lack the capacity to make a substantially autonomous decision, i.e., a decision based on adequate information and as
freely given as is possible within the constraints of the role of
patient in therapeutic research or as a healthy volunteer in nontherapeutic studies. Not all potential research subjects can appreciate the
information given to them about a research protocol or understand
the possible risks involved. Investigators must make a special effort
to protect the interests of inherently vulnerable subjects, particularly concerning the involvement of subjects in the decision-making
process–regardless of limitations–and the role of surrogate or proxy
decision makers.

Even competent adults are at a decided disadvantage in giving
valid consent when they are in the patient role. The patient role
implies a certain degree of dependence and passivity. After all, one
would not be a patient unless one needed some type of assistance or
help with a health problem. Patients are often frightened, in pain,
and in strange surroundings, all of which combine to reduce their
autonomy. Furthermore, health care professionals are at a decided
advantage in their relationship with patients. Health care professionals have considerable status and authority in the eyes of their
patients. This inequity in power certainly affects the informed consent process. As Beecher notes, "If suitably approached, patients
will accede, on the basis of trust, to about any request their physician may make."[6] The burden then falls on health professional
investigators to honor the trust patients place in them and to avoid
assiduously taking advantage of it.

Confidentiality and Privacy

Confidentiality is defined as the assurance of secrecy regarding
information imparted to another. Clearly, there is a need for personal, private information sharing in all types of relationships
between health care professionals and patients. This information
sharing is particularly critical in a research study involving human
subjects. Patients expect that the information they share will be held
in strictest confidence unless they have been informed differently.
The ethical questions involving confidentiality revolve around the
obligation the investigator has to the subjects and who should have
access to personal information. It is commonly held that an ethically

justifiable reason for breaking a confidence, let us say the identity of a specific subject, is to protect the well-being of said subject. Furthermore, it is ethically justifiable to breach confidence to protect oneself or an innocent other from harm or injury. Of course, the harm in question would have to be sufficient to outweigh the obligation to maintain confidentiality.

Confidentiality is related to privacy. "The claim to privacy is fragile but persistent; it is as subtle and powerful as the need for personal dignity; it is a fundamental aspect of individual freedom and worth."[7] Investigators must make certain that all subjects are accorded the right to privacy, not just those who express concerns about it in the informed consent process.

Beneficence

The principle of beneficence is best understood in light of the relationships that health care professionals hold with patients. Beneficence requires that at all times we should avoid harming the patient and, further, we should do good. A corollary principle, proportionality, requires not only that we meet our obligation to help those under our care (which clearly includes subjects in a research study), but also that we weigh the harms and benefits of our actions and choose those actions that maximize benefits and minimize harms. Simply put, the gains of an investigation must outweigh the risks or harms involved, whether the study is a therapeutic one that might have direct benefit for the subject or a nontherapeutic one in which the subject does not usually receive any direct benefit.

Justice

There are numerous types of justice, but those most germane to a discussion of human experimentation are distributive and compensatory. Distributive justice focuses on the fair distribution of burdens and benefits within a group or community. Generally, the distribution question centers on matters of scarcity and burden. There does not seem to be much debate over the distribution of surplus goods. In matters of human experimentation, justice enters the picture when we discuss who will be enrolled in a study and how subjects are selected for inclusion or exclusion. How do we fairly distribute the burden of being a research subject when we assume that most individuals would prefer not to volunteer? On a broader level, how do we fairly allocate research dollars that eventually result in research findings that further knowledge generally?

In answer to the former question, fairness requires that investigators not take advantage of vulnerable subjects who are often the target for research studies because of their availability, i.e., the institutionalized; those with little or no education; or the subjects from these groups who may not be capable of giving informed consent, not only because they do not fully understand the information given them, but because they are desperate, bored, or frightened. This makes a mockery of the informed consent process and is a form of exploitation.

On the other hand, one should not uniformly exclude vulnerable subjects from research studies without scientific justification. In a recent study by Larson in which 754 approved research protocols were reviewed from a two-year period, it was noted that the elderly, the poor, and ethnic minorities were found to be excluded without identifiable justification from research protocols.[8] Additionally, barriers to women enrolling in clinical trials have only recently been assailed. Women have been largely excluded from involvement in clinical research in two ways. "First women have been barred from participating as subjects in clinical trials, both *de jure* (explicit exclusion criterion in the protocol) and *de facto* ("inadvertent" failure to recruit women or to conduct the trial in a manner that realistically permits women to participate)."[9] Elimination of the disparity of involvement of vulnerable groups and the need to eliminate exploitation are not mutually exclusive goals. Steps should be taken to ensure informed consent, minimize coercion and unfair inducements, and fairly represent the groups affected by the research in question. A revision in the Food and Drug Administration's policy regarding the inclusion of fertile women in early clinical trials has not changed the FDA's commitment to the safe development of drugs, but gives more flexibility to IRBs, investigators, and patients in determining how best to ensure the safety of study subjects.[10]

Compensatory justice revolves around the moral question of the appropriateness of compensation for injuries sustained as a result of human experimentation. It appears that a consensus has been reached regarding the moral obligation to provide compensation, although it is not required. The general ethical argument that supports compensation for research-related injuries is as follows: When those who bear the risks of clinical research are not the direct beneficiaries of the research, it is felt that the scales of justice are out of balance; thus, "compensation may be regarded as a means of restoring the balance after the fact, when the residuum of risk not eliminated by the protective devices has eventuated in injury. Like the protective devices, compensation is a further means of limiting the burdens born by individual subjects in research."[11]

Truth Telling

Perhaps the most important ethical principle related to human experimentation is the principle of veracity, or truth telling. "The first and most basic ethical principle guiding research is to avoid dishonesty and deception, to conduct and report research with as much objectivity as possible with no deliberate bias or misinterpretation."[12] The requirements of objectivity go beyond honestly presenting the facts to subjects regarding research protocol and candidly reporting research findings. Objectivity also encompasses a number of additional responsibilities, such as acknowledging factors that could affect judgment and identifying possible conflicts of interest that could have an impact on the design and interpretation of findings.[13]

Previously in this chapter, we mentioned that one way of showing respect to individuals is to leave them alone to conduct their own affairs. Another way of showing respect is to be honest. Informed consent cannot exist without an honest presentation of the risks and benefits of a research protocol. Even if an investigator wants to be scrupulously honest, it may not be possible because of his or her perspective. It is questionable whether an investigator can ever truly appreciate the risks of a study as a subject would perceive them. This is where the involvement of an IRB is especially critical because the IRB does not have a vested interest in the outcomes of the protocol in question save to protect the well-being of subjects.

ETHICAL NORMS OF DRUG RESEARCH

The fundamental ethical principles embodied in the codes and regulations concerned with human experimentation have produced a set of "ethical norms" that serve as behavioral guideposts for scientists to follow when conducting drug research. These ethical norms prescribe what investigators should or should not do to ethically conduct experimental drug studies in human subjects. There are five general ethical norms derived from the ethical principles of drug research previously discussed. These norms are (1) good research design and procedures, (2) competent investigators, (3) informed consent, (4) a favorable balance of harm and benefit, and (5) equitable selection of subjects. A sixth norm that is gaining recognition is concerned with compensation for injury caused by the research.[3] These norms of drug research and the ethical principles from which they are derived are presented in Table 30.1.

Ethical codes and federal regulations describe procedures to be followed to ensure that investigators comply with the aforementioned norms of drug research. Documentation of informed consent and review by an IRB or an ethics committee are the most important procedural requirements relevant to all research involving human subjects. Documentation of informed consent shows evidence of compliance with the norm calling for informed consent. IRBs and ethics committees safeguard the interests of patient-subjects. An IRB or ethics committee may be composed of physicians, pharmacists, nurses, consumer organization representatives, clergy, and ethicists. The value of the IRB or ethics committee is to provide guidance to researchers when ethical dilemmas concerning the conduct of the research study are encountered.

The ethical behavior expectations associated with the norms, codes, and regulations guiding drug research are expressed in general rather than precise rules of behavior. Rigid rules of ethical behavior in drug research are not desirable because it is impossible to foresee all of the ethical ramifications of the research activity. Researchers must be granted sufficient latitude in ethical behavior

TABLE 30.1. Ethical Principles and Norms of Drug Research

Principle	Norm
Respect for Persons	Informed Consent
Autonomy	Confidentiality
Beneficence	Good Research Design and Procedures Competent Investigator/Clinician Favorable Balance of Harm vs. Benefit

Justice:
   Distributive
   Compensatory
	Equitable Selection of Subjects
	Compensation for Research-Induced Injury

Truth Telling	Harm vs. Benefit Objectivity

expectations to allow them to conduct scientifically valid and socially responsible drug research while at the same time honoring their ethical duties to human subjects. In the final analysis, therefore, "the investigator's personal sense of moral duty combined with review by an independent party to avoid abuse is probably a more reliable safeguard for the conduct of ethical drug research than even the best thought-out rules."[14]

Good Research Design and Procedures

This norm upholds the principle of beneficence. If the drug research study is not well designed or conducted appropriately, it will not benefit society. Badly designed research wastes money, exposes human subjects to needless risks and harm, and fails to contribute to the advancement of safe and effective drug therapy. The norm of good research design is also responsive to the principle of respect for persons. A poorly designed research study wastes the subjects' time, frustrates their desire to participate in a meaningful way, and negates their expectation that their participation in the study will result in something of value.

As depicted in Figure 30.1, there are three phases in the drug research process: preclinical investigations, clinical studies and postmarketing, and Phase IV clinical studies and drug surveillance. An Investigational New Drug Application (INDA) must be submitted to the FDA before an experimental drug can be tested in humans, and a New Drug Application (NDA) must be submitted to the FDA as the last step before the agency approves a drug's use in the marketplace as a recognized therapeutic agent. Preclinical investigations are not directly concerned with human experimentation, but are carried out to obtain data that justifies the testing of an experimental drug in human subjects. Major ethical issues to be considered in this phase of drug research are that all of the physical, chemical, toxicological, and biological prerequisite studies are conducted in a comprehensive and scientifically responsible manner, that required animal studies are conducted humanely and according to appropriate legal and ethical guidelines,* and that the data produced are analyzed in a sufficiently comprehensive, competent, and truthful manner to justify the testing of the drug in human subjects.

The clinical trial is at the heart of drug research. The testing of experimental drugs in humans cannot be challenged rationally unless society is willing to accept the alternatives of refusing the need for making therapeutic progress or is willing to adopt drug treatments that have not been adequately studied. However, "the fact that a clinical trial may lead to important therapeutic advances benefiting future patients in no way justifies depriving trial participants of existing therapy necessary for their condition or exposing them to an unreasonable risk."[15] Ethical issues of major concern when conducting clinical trials focus on the clinical trial techniques used, the selection of subjects, the use of healthy volunteers to demonstrate drug safety, informed consent, and the role of IRBs or ethics committees in the conduct of the trial.

The randomized controlled double-blind study is the scientific

---

*Some of the important laws, regulations, and documents that guide the use of animals by investigators are the Animal Welfare Act (Public Law 99-198), the Health Research Extension Act of 1985 (Public Law 99-158), the *Guide for the Care and Use of Laboratory Animals* (Department of Health and Human Services), and the *Public Health Service Policy on Humane Care and Use of Laboratory Animals by Awardee Institutions* (NIH, 1986).[15]

and legal experimental design of choice for the testing of new drugs. This approach randomly selects test subjects to receive the experimental drug or the control drug. The control drug is typically a placebo or a standard reference medication. Neither the subject (patient) nor investigator/physician knows which drug a subject receives. At the completion of the study, the effectiveness of the experimental drug is determined by statistically comparing its therapeutic outcomes to the therapeutic outcomes of the control drug.

Random assignment of subjects in a clinical trial is ethically justified only when it appears, a priori, that, given the current status of knowledge, the treatments to be compared are equal. The null hypothesis–there is no difference between the two drug treatments–is the scientific rational and ethical justification for random assignment of subjects in an experimental drug study. However, when there is no doubt which drug is best for a subject, it would be considered unethical, as well as scientifically irresponsible, to include that individual in the trial. Even with these constraints, some have argued that the randomized trial "requires doctors to act simultaneously as physicians and scientists. This puts them in a difficult and sometimes untenable ethical position. The conflicting moral demands arising from the use of the randomized, clinical trial reflect the classic conflict between rights-based moral theories and utilitarian ones."[16] This tension between the obligations one has as a health professional and those one has as a scientist exists in all facets of clinical research, but it does not have to result in exclusion of one set of obligations to satisfy the other.

A double-blind study is the experimental design of choice in drug research because it eliminates subject and physician bias and maximizes the statistical power necessary for treatment comparisons. If blinding entails risk to subjects, however, it would be considered ethically appropriate to use a less rigorous design, such as a "single-blind" or "open" (nonblind) design to conduct the drug study. Although these designs are statistically less powerful than the double-blind design, they minimize or eliminate risk while still allowing for randomization of subjects and the comparison of treatments. However, if a drug trial requires a double-blind design and there is a potential risk to a subject(s), the ethically acceptable approach for conducting this study is to establish procedures for breaking the blinding code if harm should occur to a subject.

It is ethically appropriate to use a placebo as the control drug when there is no convincing evidence that the advantages of the current standard drug treatments outweigh their disadvantages. If a standard reference medication is therapeutically effective and its advantages outweigh its risks, however, there is no ethical justification for using a placebo in place of that medication to conduct a comparison with the experimental drug. To do so would violate the ethical principle of do no harm to the patient. When this situation exists, special study designs may be employed that allow the use of a placebo along with the standard reference medication in the control group. Regardless of the situation under which a placebo is used, however, an ethical question to be considered is, How much information should be given a subject who participates in a study that uses a placebo? Research indicates that a subject's knowledge of placebo use may affect the study outcome. Not telling a subject, on the other hand, is a lie and violates the ethical principle of truth telling. Resolution of this ethical dilemma is complex and depends upon the goals of the clinical study, the investigator's personal sense of moral duty, and the perspective of the independent review agency responsible for ensuring that the subjects' interests are protected.

The use of healthy volunteers in a Phase I study to determine an experimental drug's safety is considered an ethically acceptable research procedure. For some disease conditions, however, the ethicalness of excluding sick individuals from the first phase of a clinical study has been questioned. It has been argued that it makes little sense to exclude terminally ill patients from the beginning stages of clinical investigations of these drugs because these patients are not healthy. This viewpoint has been expressed concerning drugs being tested for the treatment of cancer and AIDS. Fortunately, through the treatment IND and the parallel track research protocol, the FDA has developed procedures to legally address this critical issue.

Duplicating a trial when it has been demonstrated that one treatment is superior to another is also an ethically questionable drug research procedure. The only instance in which a trial should be duplicated is if the veracity (honesty) of the investigator(s) is questioned or the investigation is suspected of being a biased study. Otherwise, unnecessary duplication of valid research could be considered a waste of scarce resources, i.e., research funding.

There are instances when it may be ethically incumbent upon investigators to discontinue a clinical study. It is universally agreed if it becomes obvious that a drug trial is a serious hazard to the subjects, the study should be terminated. But sometimes the decision to terminate a trial is an ethical dilemma. Benefits perceived from such a decision are that human subjects are saved from further experimentation and a new therapeutic agent will be made available to society sooner than expected. The argument against terminating a trial in this situation, however, is that the investigators may not be able to draw a definite conclusion concerning the therapeutic superiority of the drug.

Competent Investigators

Like the norm for good research design and procedures, the norm for competent investigators is derived from the ethical principle of beneficence. Investigators are expected to be competent to produce valid research results that are of benefit to society while at the same time showing respect for research subjects. It is expected that all drug investigators will possess the appropriate scientific training and skills to achieve the purposes of the research study. They must be clinically qualified not only in the field under study but also in clinical trial methodology and ethics. To ensure that clinical investigators are competent to participate in drug research studies, the FDA requires all Investigational New Drug Applications to include a statement of every clinical investigator's education and experience (FDA Form 1573).

Because clinical investigators are in most cases clinicians, they are also expected to be competent to "take care" of the research subject. The dual roles of clinician and researcher present an investigator with an ethical dilemma. As a clinician, the investigator's allegiance is to the subject as a patient. As a researcher, the investigator's allegiance is to the research project and adherence to the research protocol to ensure valid results. The procedure typically used to address this ethical conflict is to assign one investigator/clinician member of a drug investigation team the responsibility for observing the subjects' welfare. Specific duties of this individual are assessment of subjects' symptoms, signs, and laboratory results; early detection of adverse effects of the drug and authority to withdraw or terminate the drug if necessary; and institution of appropriate intervention(s), i.e., an antidote, when necessary to minimize adverse effects of the trial and the subjects' discomfort.

Informed Consent

Informed consent is a normative expectation in drug research responsive to the ethical principle of respect for persons. Autonomy is a secondary principle addressed by investigators because of the need to determine the subjects' competence to give informed consent for their participation in a drug study.

Informed consent involves explaining the study's goals and methodology and its risks and benefits to patients so they may choose to participate in the study of their own free will. Department of Health and Human Services and FDA regulations outline the following eight basic elements of informed consent which must be provided to any research subject:[17]

- A statement that the study involves research, an explanation of the purposes of the research and the expected duration of the subject's participation, a description of the procedures to be followed, and identification of any procedures that are experimental;
- A description of any reasonably foreseeable risks or discomforts to the subject;
- A description of any benefits to the subject or to others that may reasonably be expected from the research;
- A disclosure of appropriate alternative procedures or courses of treatment, if any, that might be advantageous to the subject;
- A statement describing the extent, if any, to which confidentiality of records identifying the subject will be maintained;
- For research involving more than minimal risk, an explanation as to whether any compensation and an explanation as to whether any medical treatments are available if injury occurs, and if so, what they consist of or where further information may be obtained;

- An explanation of whom to contact for answers to pertinent questions about the research and research subjects' rights and whom to contact in the event of a research-related injury to the subject;
- A statement that participation is voluntary, refusal to participate will involve no penalty or loss of benefits to which the subject is otherwise entitled, and the subject may discontinue participation at any time without penalty or loss of benefits to which the subject is otherwise entitled.

Informed consent is not a one-time event, but an ongoing process that requires investigators to continually apprise subjects of information necessary for their continued informed participation in a study. Unfortunately, sometimes it is difficult to give full disclosure of information or to obtain a patient's consent. There are ethical arguments for and against telling patients with incurable illnesses that they may be receiving a placebo because there is no proven treatment for their illness. Some patients may not be competent to give their informed consent because they are incapable of full understanding, i.e., children, psychiatric patients, or comatose patients. In situations such as these, it is ethically permissible to obtain authorization from the patient's family or legally valid surrogate.

Typically, at the beginning of a study, the subject is asked to sign a consent form. The purpose of this form is to prove that the necessary information has been provided and that the subject agrees to participate in the study of his or her own free will. Some ethicists have argued that informed consent is never totally informed and free because of a patient's dependent relationship upon physicians and the fact that a patient may feel coerced to participate in a study. This latter argument is a major reason why prisoners are no longer included as subjects in experimental drug trials.

Favorable Balance of Harms and Benefits

The norm of favorable balance of harms and benefits is derived from the ethical principles of beneficence and respect for persons. The principle of respect for persons requires that investigators provide intended subjects with a complete assessment of risks and benefits when seeking the subjects' informed consent. From an ethical perspective, there is no justification for experimental drug research without a favorable balance between risk and benefit. Ideally, benefit is to be maximized and risk is to be minimized.

Risks associated with drug research are not as severe as typically perceived by the lay public. The majority of serious or life-threatening risks associated with an experimental drug usually have been determined before the drug is given to research subjects. Physical discomfort typically is the prevalent risk encountered by subjects during a clinical trial. The Department of Health and Human Services and the Food and Drug Administration have established a risk threshold criterion to assist investigators in determining whether the risk associated with a trial is worth the benefit derived from it. This criterion states "that the risks of harm anticipated in the proposed research are not greater, considering probability and magnitude, than those ordinarily encountered in daily life or during the performance of routine physical or psychological examinations or tests."[3]

In addition to physical or psychological injury, there are other forms of harm associated with drug research. If an experimental treatment is not superior to an old treatment, a patient participating in the study can be considered to have experienced harm. Additional cost to a patient participating in a trial–temporal, physical, or financial in nature–is also a harmful situation. If there is potential for harm due to risks associated with a drug investigation, the investigators are ethically bound to develop provisions for rectifying or preventing the harm, such as discontinuing the trial and providing various forms of compensation for injuries incurred. Some general ethical questions that focus on the subjects' participation and perception of risk are: Will there be a loss of independence? Will the subjects be asked personal information? What is the time and energy requirement of the subjects? Will there be any physical or mental discomfort? Will there be monetary costs to the subjects? Will there be a delay in recovery?[18]

Failure to maintain confidentiality is another harmful situation to be avoided in the conduct of a clinical trial. A subject's participation in an experimental drug study is a private matter. Because it may be necessary to identify subjects a posteriori, however, it is perfectly ethical to use identification techniques, such as codes, that identify subjects as subjects but not as persons.

Equitable Selection of Subjects

Equitable selection of subjects for participation in experimental drug trials is based on the ethical principle of distributive justice. Meeting the demands of distributive justice requires that there be fair distribution within subject populations of both the benefits and the burdens of research. Certain people, due to unique life situations, are more vulnerable to the burdens of research than others because they are incapable of protecting their own interests. Consequently, in carrying out drug research, investigators are duty bound to identify individuals who are unable to protect themselves from potential harm and should not be included in the study.

Population groups traditionally considered vulnerable and in need of protection in human experimentation are children, prisoners, the mentally infirm, the fetus, and women of childbearing age. Other individuals who should be considered as potentially vulnerable are people who are ill; uncomprehending individuals such as the senile, the unconscious, or the uneducated; impoverished persons; and individuals who have a dependent relationship with an investigator or institution. There is ethical justification for including a vulnerable population in a drug investigation, however, if it can be demonstrated that the condition to be studied is unique to the vulnerable population, that the risk or inconvenience associated with the trial is not an excessive burden, or that the knowledge to be gained is necessary for understanding the conditions contributing to the population's vulnerability.

Compensation for Research-Induced Injury

This norm is derived from the principle of compensatory justice. The ethical reasoning underlying this norm is that subjects injured by an experimental procedure are entitled to compensation. Most ethicists perceive compensation for research injury to be analogous to society's practice of compensating soldiers for war injuries received in the service of their country. Although this norm is not universally accepted, it is starting to appear in some ethical guidelines. Current federal regulations (DHHS) require investigators to advise research subjects of the potential for compensation due to injury, but they do not require institutions to compensate injured subjects. Although many institutions do not provide compensation for research-induced injury, most do provide free medical treatment to subjects for injuries received through participation in research studies.

ADDITIONAL CONSIDERATIONS
FOR THE PRACTICING PHARMACIST

Phase IV Clinical Studies/Postmarketing Surveillance

Federal regulations require pharmaceutical manufacturers to conduct Phase IV clinical studies and postmarketing drug use surveillance once an investigational new drug has been approved for therapeutic use in the marketplace (see Figure 30.1). The objective of Phase IV study activities is to obtain additional information for maximizing the drug's safe, effective, and rational use. Clinical research conducted during Phase IV includes evaluating the drug for additional uses and indications; assessing its compatibility with other agents; determining its potential for overuse, misuse, and/or abuse; and documenting and reporting drug interactions. Postmarketing surveillance is deemed necessary for identifying adverse drug reactions and product defects not discovered during clinical trials.[19,20]

In addition to the federally mandated drug surveillance activities required of pharmaceutical manufacturers, the FDA conducts MedWatch, a voluntary medical products surveillance program that encourages health professionals to report serious adverse events and product defects.[21] Because pharmacists are a critical link in the drug use process, the ethical principles of beneficence and respect for persons provide ample ethical justification for the pharmacist's voluntary participation in this program. The recent experience with Omniflox® (temafloxacin) demonstrates why drug surveillance is an ethical as well as professional responsibility for pharmacists. Less than four months after its introduction in 1992, serious adverse effects of the drug were being reported to the FDA. In addition to three deaths, significant side effects associated with the drug's use were dangerously low blood sugar levels, excessive destruction of red blood cells, abnormal liver function tests, and impaired blood clotting. None of these serious adverse effects was discovered in the clinical trials of this drug.

MedWatch Medical Products Reporting forms are obtained from the FDA. The FDA may share the identity of the pharmacist-reporter with the manufacturer unless the pharmacist requests otherwise. The identities of patients involved in MedWatch reports is confidential and legally protected.

Pharmacoeconomic Analysis

Pharmacoeconomic analysis is a new area of drug research that has significant ethical implications for the practicing pharmacist. This area of study is receiving increasing attention because of the pressing need to control drug costs while maintaining or improving the quality of pharmaceutical care.

Pharmacoeconomic studies may be conducted during the clinical trials of a new drug and/or in postmarketing Phase IV clinical studies. Postmarketing pharmacoeconomic studies seek to compare approved drugs of various manufacturers within the same therapeutic category. Practical implications of this research for pharmacists are that it may facilitate formulary decisions, may assist patient-specific therapeutic decisions, or may guide pharmacist-physician consultations about a certain therapy.

Pharmacists who conduct or rely on this type of research for therapeutic or product selection decisions must be aware that the ethical norms and expectations underlying pharmacoeconomic research are the same as for new drug therapy research. Like clinical trial investigations, pharmacoeconomic studies demand good research design and procedures. Unfortunately, pharmacoeconomics research is in its infancy, so "little work has been done to define standards and criteria for high-quality pharmacoeconomic research."[22] This lack of standards and criteria, plus the recognition that the complex nature of the research requires highly trained and competent investigators, places an ethical responsibility on pharmacists to assess the quality and validity of a pharmacoeconomic study before its results are accepted for use in practice.

The potential for harm versus benefit is another ethical norm to be considered when conducting or evaluating pharmacoeconomic research. Why was the study conducted? Who sponsored it? Was the study conducted with the subject's, patient's, or society's needs in mind? How will the study's results be used? These are important questions for the pharmacist to consider. Conflict of interest is another ethical consideration for the pharmacist conducting or using the results of pharmacoeconomic research. Is the pharmacist's role that of institution advocate, patient advocate, pharmaceutical company advocate, or self advocate? Finally, was the study based on the therapeutic justification that the outcomes expected were better than the outcomes obtained with the therapy(s) currently being employed?

Pharmacy-Coordinated Investigational Drug Services

The majority of clinical drug studies are conducted in hospitals. This fact, along with the increasing complexity of drug research, prompted the American Society of Hospital Pharmacists to develop a manual for a pharmacy-coordinated investigational drug service (IDS) for hospital pharmacy practice.[2] The goal of this service is to ensure that hospital-based drug research studies are managed and conducted as safely, effectively, and ethically as possible. In meeting this goal, the IDS has the following specific objectives:

- To help the hospital ensure that investigational drug studies conducted by its staff comply with all applicable legal, ethical, and financial requirements
- To help investigators fulfill the administrative, scientific, educational, and clinical obligations of research protocols
- To help assure that the data received by the study sponsors are complete, valid, timely, and produced at the lowest possible cost
- To add to the knowledge of the chemistry, therapeutics, pharmacology, and usage of investigational drugs.

The ASHP has suggested that an investigational drug service is an appropriate professional and ethical role for hospital pharmacists to perform because of the pharmacy's central role and responsibilities in all aspects of hospital drug usage.

CONCLUSION

Although the majority of pharmacists will not conduct clinical research, all will be consumers of its products on a personal and professional basis. When involved in clinical trials, pharmacists have the responsibility to follow ethical guidelines to protect patients. Pharmacists who are not directly involved in research have the important role of identifying ethical concerns for their patients who may be research subjects and critically analyzing research literature for ethical soundness. Thus, it is essential to good pharmacy practice that pharmacists be aware of the ethical norms of clinical research so they can detect lapses in conforming to these norms; understand the ethical principles that serve as rationale for them; and be familiar with the regulations, laws, and institutional groups charged with the protection of the rights of human subjects.

APPENDIX A:
THE NUREMBERG CODE

The great weight of the evidence before us is to the effect that certain types of medical experiments on human beings, when kept within reasonably well-defined bounds, conform to the ethics of the medical profession generally. The protagonists of the practice of human experimentation justify their views on the basis that such experiments yield results for the good of society that are unprocurable by other methods or means of study. All agree, however, that certain basic principles must be observed in order to satisfy moral, ethical and legal concepts.

1. The voluntary consent of the human subject is absolutely essential.

    This means that the person involved should have legal capacity to give consent; should be so situated as to be able to exercise free power of choice, without the intervention of any element of force, fraud, deceit, duress, overreaching, or other ulterior form of constraint or coercion; and should have sufficient knowledge and comprehension of the elements of the subject matter involved as to enable him to make an understanding and enlightened decision. This latter element requires that before the acceptance of an affirmative decision by the experimental subject there should be made known to him the nature, duration, and purpose of the experiment; the method and means by which it is to be conducted; all inconveniences and hazards reasonably to be expected; and the effects upon his health or person which may possibly come from his participation in the experiment.

The duty and responsibility for ascertaining the quality of the consent rests upon each individual who initiates, directs or engages in the experiment. It is a personal duty and responsibility which may not be delegated to another with impunity.

2. The experiment should be such as to yield fruitful results for the good of society, unprocurable by other methods or means of study, and not random and unnecessary in nature.
3. The experiment should be so designed and based on the results of animal experimentation and a knowledge of the natural history of the disease or other problems under study that the anticipated results will justify the performance of the experiment.
4. The experiment should be so conducted as to avoid all unnecessary physical and mental suffering and injury.
5. No experiment should be conducted where there is an *a priori* reason to believe that death or disabling injury will occur; except perhaps, in those experiments where the experimental physicians also serve as subjects.
6. The degree of risk to be taken should never exceed that determined by the humanitarian importance of the problem to be solved by the experiment.
7. Proper preparations should be made and adequate facilities provided to protect the experimental subject against even remote possibilities of injury, disability, or death.
8. The experiment should be conducted only by scientifically qualified persons. The highest degree of skill and care should be required through all stages of the experiment of those who conduct or engage in the experiment.
9. During the course of the experiment the human subject should be at liberty to bring the experiment to an end if he has reached the physical or mental state where continuation of the experiment seems to him to be impossible.
10. During the course of the experiment the scientist in charge must be prepared to terminate the experiment at any stage, if he has probable cause to believe, in the exercise of the good faith, superior skill and careful judgement required of him that a continuation of the experiment is likely to result in injury, disability, or death to the experimental subject.

From *Trials of War Criminals Before the Nuremberg Military Tribunals Under Control Council Law No. 10.* Vol. 11, Nuremberg, October 1946-April 1949.

APPENDIX B:
DECLARATION OF HELSINKI
The International Code of Ethics for Biomedical Research

1. Biomedical research involving human subjects must conform to generally accepted scientific principles and should be based on adequately performed laboratory and animal experimentation and on a thorough knowledge of the scientific literature.
2. The design and performance of each experimental procedure involving human subjects should be clearly formulated in an experimental protocol which should be transmitted to a specially appointed independent committee for consideration, comment, and guidance.
3. Biomedical research involving human subjects should be conducted only by scientifically qualified persons and under the supervision of a clinically competent medical person. The responsibility for the human subject must always rest with a medically qualified person and never rest on the subject of the research even though the subject has given his or her consent.
4. Biomedical research involving human subjects cannot legitimately be carried out unless the importance of the objective is in proportion to the inherent risk to the subject.
5. Every biomedical research project involving human subjects should be preceded by careful assessment of predictable risks in comparison with foreseeable benefits to the subject or to others. Concern for the interest of the subject must always prevail over the interests of science and society.
6. The right of the research subject to safeguard his or her integrity must always be respected. Every precaution should be taken to respect the privacy of the subject and to minimize the impact of the study on the subject's physical and mental integrity and on the personality of the subject.
7. Doctors should abstain from engaging in research projects involving human subjects unless they are satisfied that the hazards involved are believed to be predictable. Doctors should cease any investigation if the hazards are found to outweigh the potential benefits.
8. In the publication of the results of his or her research, the doctor is obliged to preserve the accuracy of the results. Reports on experimentation not in accordance with the principles laid down in this Declaration should not be accepted for publication.
9. In any research on human beings, each potential subject must be adequately informed of the aims, methods, anticipated benefits, and potential hazards of the study and the discomfort it may entail. He or

she should be informed that he or she is at liberty to abstain from participation in the study and that he or she is free to withdraw his or her consent to participation at any time. The doctor should then obtain the subject's freely given informed consent, preferably in writing.

10. When obtaining informed consent for the research project the doctor should be particularly cautious if the subject is in a dependent relationship to him or her or may consent under duress. In that case the informed consent should be obtained by a doctor who is not engaged in the investigation and who is completely independent of this official relationship.

11. In the case of legal incompetence, informed consent should be obtained from the legal guardian in accordance with national legislation. Where physical or mental incapacity makes it impossible to obtain informed consent, or when the subject is a minor, permission from the responsible relative replaces that of the subject in accordance with national legislation.

12. The research protocol should always contain a statement of the ethical considerations involved and should indicate that the principles enunciated in the present Declaration are complied with.

REFERENCES

1. Young JH. The medical messiahs. Princeton, NJ: Princeton University Press, 1967.

2. Stolar MH, ed. Pharmacy coordinated investigational drug services. Bethesda, MD: American Society of Hospital Pharmacists, 1986.

3. Levine RJ, Holder AR. Legal and ethical problems in clinical research. In: Matoren GM, ed. The clinical research process in the pharmaceutical industry. New York: Marcel Dekker, Inc., 1984:124-135.

4. Chinn P. From the editor. Adv Nurs Sci 1979;1(3):v.

5. Beauchamp TL, Childress JF. Principles of biomedical ethics. 4th ed. New York: Oxford University Press, 1994.

6. Beecher HK. Ethics and clinical research. N Engl J Med 1966;274:1354-60.

7. U.S. Office of Science and Technology. Privacy and behavioral research. Washington, DC: Government Printing Office, 1967.

8. Larson E. Exclusion of certain groups from clinical research. Nurs Res 1994;26(3):185-90.

9. Merton V. The exclusion of pregnant, pregnable, and once-pregnable people (a.k.a. women) from biomedical research. Am J Law Med Ethics 1993;19:369-451.

10. Lin AYF. Should women be included in clinical trials? Pharm Times 1994;(Nov):27.

11. President's Commission for the Study of Ethical Problems in Medicine and Biomedical and Behavioral Research. Compensating for research injuries. Vol. 1. Washington, DC: Government Printing Office, 1982.

12. Mondon J. Chance and necessity. New York: Knopf, 1971.

13. Shrader-Frechette K. Ethics of scientific research. Lantham, MD: Rowman and Littlefield, 1994.

14. Spriet A, Simon P. Methodology of clinical drug trials. Edelstein R, Weinstraub M, trans. Switzerland: S. Karger A.G., 1985.

15. Bulger RE, Heitman E, Reiser SJ. The ethical dimensions of the biological sciences. New York: Cambridge University Press, 1993.

16. Hellman S, Hellman DS. Of mice but not men: problems of the randomized clinical trial. N Engl J Med 1991;324:1585-9.

17. Informed consent of human subjects. Federal Register 46:17(27 Jan, 1981) pp. 8951-2.

18. Anema MG. Ethical considerations in conducting clinical research. Dimensions Crit Care Nurs 1989;8:288-96.

19. Johnson JM, Tanner A. Postmarketing surveillance: curriculum for the clinical pharmacologist. Part I: Postmarketing surveillance within the continuum of the drug approval process. J Clin Pharmacol 1993;33:904-11.

20. Faich GA. Adverse-drug-reaction monitoring. N Eng J Med 1986;314:1589-92.

21. Ropp KL. MedWatch, on lookout for medical product problems. FDA Consumer 1993;27:14-7.

22. Bootman JL, Larson LN, McGhan WF, Townsend RJ. Pharmacoeconomic research and clinical trials: concepts and issues. DICP Ann Pharmacother 1989;23:693-7.

ADDITIONAL READINGS

For the Instructor

1. Bernstein JE. Ethical consideration in human experimentation. In: Smith M, Strauss S, Baldwin HJ, Alberts KT, eds. Pharmacy ethics. Binghamton, NY: Pharmaceutical Products Press, 1991:391-408.

2. Bush JK. The industry perspective on the inclusion of women in clinical trials. Acad Med 1994;69:708-15.

3. Caplan AR. How should science handle data from unethical research? J NIH Res 1993;5(May):22+.

4. Clark PI, Leaverton PE. Scientific and ethical issues in the use of placebo controls in clinical trials. Annu Rev Public Health 1994;15:19-38.

5. Forsythe P. Research utilization and the consumers of care: the parents' perspective. J Obstet Gynecol Neonatal Nurs 1994;23:350-1.

6. Geiselmann B, Helmchen H. Demented subjects' competence to consent to participate in field studies: the Berlin Ageing Study. Med Law 1994;13:177-84.

7. Good BAV, Rodrigues-Fisher L. Vulnerability: an ethical consideration in research with older adults. West J Nurs Res 1993;15:780-3.

8. Hansen BC. Ask not what research can do for you . . . issues for the year 2000. J Allied Health 1994;23(2):89-93.

9. Holm S. Moral reasoning in biomedical research protocols. Scand J Soc Med 1994;22(2):81-5.

10. Kalichman MW, Friedman PJ. A pilot study of biomedical trainees' perceptions concerning research ethics. Acad Med 1992;67:769-75.

11. McCarthy CR. Historical background of clinical trials involving women and minorities. Acad Med 1994;69:695-8.

12. O'Rourke K. Informed consent: therapeutic and nontherapeutic trials. Health Care Ethics USA 1994;2(2):6-7.

13. Passamani E. Clinical trials: are they ethical? N Engl J Med 1991;324: 1589-91.

14. Robinson A. Science and scandal: what can be done about scientific misconduct? Can Med Assoc J 1994;151:831-4.

15. Shimm DS, Spece RG. Ethical issues and clinical trials. Drugs 1993;46:579-84.

16. Slatkoff SF, Curtis P, Coker A. Patients as subjects for research: ethical dilemmas for the primary care clinician-investigator. J Am Board Fam Pract 1994;7:196-201.

17. Sutherland HJ, Mesline EM, da Cunha R, Till JE. Judging clinical research questions: what criteria are used? Soc Sci Med 1993;37:1427-30.

For the Student

1. Applebaum PS, Roth LH, Lidz CW, Benson P, Winslade W. False hopes and best data: consent to research and the therapeutic misconception. Hastings Cent Rep 1987;17(Apr):20-4.

2. Bandman EL. Protection of human subjects. Top Clin Nurs 1985;7:15-23.

3. Broad W, Wade N. Betrayers of the truth: fraud and deceit in the halls of science. New York: Simon and Schuster, 1983.

4. Miller B. The ethics of random clinical trials. In: Van de Veer D, Regan T, eds. Health care ethics: an introduction. Philadelphia: Temple University Press, 1987:128-38, 152-7.

5. Pellegrino EM. Beneficence, scientific autonomy, and self-interest: ethical dilemmas in clinical research. Georgetown Med 1991;1(1):21-8.

Epilogue

Albert I. Wertheimer

When all is said and done, the basic principles remain: mental and physical health are related and people need reassurance sometimes. They also need guidance, and often only a friendly greeting or a smile. There are patients who are truly lonely and desperately want to chat. There are also patients who are too timid or too ashamed to ask for help or who have communications difficulties or sensory problems and do not ask for help.

There are people who do not know that they should be patients, and there are countless others who would be much better off if they were compliant with their medication regimens. Due to ignorance or a too casual attitude, there are persons who choose to ignore the progressing severity of an illness or who believe in self-medication or some other unorthodox healing system long after it should be obvious that additional professional help is required.

While a great deal is being said and written about health reform and even more is actually being done about the organization, delivery, and funding of health care delivery in the United States, several constants remain. Some people will become ill from time to time and some portion of those who become ill will seek the assistance of physicians. From past experience, we can extrapolate that perhaps two-thirds of those patients will receive prescriptions. Probably close to 90% of those patients will obtain the prescribed medication, and we can only estimate, based on the conflicting data in the literature, that somewhere between 40% and 75% of those patients will follow their drug regimen with any degree of accuracy and consistency.

So, irrespective of the arrangements for delivery of and payment for services, there will be a need for a system to distribute drugs, maintain records, and provide education and counseling. The bottom line is that the pharmacy profession must be aware of structural plans and systems changes but not be blinded by this domain at the expense of other professional domains.

It would be difficult, if not actually impossible, to predict with any degree of accuracy how pharmacy might be practiced 20 years from now. Actually, it is probably of little consequence whether pharmacy is practiced in huge clinics or in community pharmacies. And it is probably unlikely that the public cares whether pharmacies are individually owned or are units of large corporations. In any case, we will not see the pharmacy of 2020 as we see a pharmacy today.

We do know that much more computerization and automation, including robotics and cognitive services provision areas, will be found. It is a good bet that the pharmacy may have few patrons present on the premises. If current trends continue, prescriptions will be sent electronically from the physician's office to the pharmacy dispensing area. That prescription order may be prepared and then delivered to the patient's residence. Another possibility is that the patient can go to a nearby vending machine, akin to the automatic teller machines used for banking today, to obtain the medication. The other major difference will be the increased use of biotechnology drugs.

Biotechnology drugs, proteins, and peptides require a nonoral route of administration. If our colleagues in drug delivery systems succeed in developing dosage forms that enable oral use of these products, it would be wonderful. Otherwise, we can imagine drug administration taking place at offices with little individual booths where drugs are administered by drug administration aides, a huge group who train only for this role. These drug administration centers could be operated by hospitals, clinics, HMOs, group practices, or even by pharmacies (and what about by technology-based organizations using robot-like apparatus).

It could be possible for a patient to complete a history and a workup questionnaire from home via interactive television. Based upon the answers to the questions, the patient could be instructed to leave 30 milliliters of urine or have a blood sample drawn at a nearby location where laboratory work may be performed or held for pickup by a central laboratory entity. Later that day, the patient could be instructed to go to a pharmacy automated dispensing center and, with a health coverage card, key in 34567. That number would release a prescription specially prepared for the patient, already in a bottle, and including pertinent patient information. How does pharmaceutical care fit into this scenario? Perhaps it does, or perhaps it doesn't. Surely, physicians are asking how conventional medical care and payment for it will be affected, as are all health services providers.

This must be a component of what Alvin Tofler had in mind when he told us that a need for "high touch" would follow the rush to high tech. That makes perfect sense today as we envision individuals using vending machines for banking, airline ticketing, paying tolls, purchasing gasoline, dining, and interfacing with the health care delivery system. It would seem reasonable that people would crave one-on-one interaction with another human being. We cannot tell yet whether this will be the pharmacist or the social worker, bartender, local government, or religion, but we can be sure that increased transaction automation will lead to more market possibilities for low-tech, one-on-one services.

When all is said and done, we can expect to see a health care system in the future that is vastly different from what we are accustomed to today. While the hospital is probably the entity at greatest risk, it is also the one with the greatest extent of opportunity. Only a small fraction of patients using overnight hospital beds today for surgical procedures and for diagnostic tests will use those beds in the future. Only major surgery will require an overnight hospital stay. And, to be sure, the definition of "major" in major surgery will change over time. We should approach a time later this decade when only surgery involving major internal organs will require multiday hospital stays. Physician group practices will operate their own clinical laboratories, imaging centers, and outpatient special procedures units for diagnostic testing such as endoscopic procedures, and surgery (gynecological, plastic, and laparoscopic, etc.). Skimming by other organizations will take away long-term care and hospice patients, patients requiring only bed rest and a minimum of care (post myocardial infarction), psychiatric patients, and others.

Hospitals also see this, and that recognition is probably the single most important motivational force in the current trend today in the development of the physician hospital organization. PHOs are intended as a counterattack to the growing power and influence of HMOs and other administrative health entities. Hospitals are asking, "Why don't we take back control and offer our services directly to employers and other payers?" And that is exactly what they are doing. They are eliminating the middleman (the HMO) and, in doing so, reducing costs while gaining control over quality and service features. This explains the rash of recent and pending purchases by hospitals of local clinical laboratories, group practices, imaging centers, and long-term care facilities. Then these PHOs are learning that there are considerable economies of scale by merging with nearby PHOs in the same community and nearby. These moves create integrated services networks (ISNs), which are also described by several similar names.

Now, overlap and duplication can be eliminated for profit and competitive reasons, a goal never fully achieved even after 40 years of various efforts by governments with certificate of need and health planning legislation. This sounds exciting unless you are an independent pharmacy owner. It only stands to reason that another area of possible integration is in the delivery of pharmaceutical services. Hospitals already have and know how to operate inpatient and outpatient pharmacies. Why not purchase some existing pharmacies at or near their hospital and group practice clinic sites or, better yet, build small professional dispensing pharmacies attached to these same sites? Of course they will, and then, as night follows day, the prescriptions will be routed to and dispensed at the hospitals' own pharmacy operations.

So, the community practice of pharmacy is being assailed by chains, hospital outpatient facilities, HMOs, PHOs, mail-service pharmacies, biotechnology, and even physician dispensing in some areas of the country.

What does all of this mean, one should ask. While there can be multiple interpretations of events, it looks reasonably clear that HMOs, PHOs, and other organized care providers will be operating and building and/or buying more community locations. We probably would do the same thing if we were in the position of HMO administrators. Let us pause and examine the economics and policy implications of such a move.

For our example, let's use a small apothecary-style dispensing pharmacy of about 1,800 square feet, perhaps located in the lobby of a medical building or immediately across the street. Existing physical plant personnel can remodel the space. The pharmacy must purchase fixtures for $50,000 and increase its inventory by another $10,000. Let us assume a $60,000 cost for salary and benefits for a pharmacist and $20,000 for a technician. If that pharmacy dispensed 400 prescriptions per day five days per week, that is 2,000 prescriptions per week or 104,000 prescriptions per year. The economics are compelling. Personnel costs are about $0.78 per prescription. Of course, there are expenses for tape, labels, utilities, etc., but most of these are insignificant additions. Imagine the tiny increase in the clinic's or hospital's electric or telephone bill.

Moreover, such an arrangement enables an organization to purchase drugs at volume discounts. Were that plan to send those prescriptions out to unrelated pharmacies, a dispensing fee of $3.00 times 104,000, plus a cost of 5% of AWP not captured on a $25.00 average prescription amounts to another $130,000. The $130,000 AWP expense plus the $312,000 in dispensing fees equals a cost of $442,000. Subtract not only the $80,000 personnel costs, but also add another $20,000 miscellaneous expense, and the numbers tell it all. Doing pharmacy yourself costs $100,000 and contracting it out costs $442,000. Multiply this times the numbers of PHOs, clinics, and towns, and we can begin to understand the future. Some might argue that this is the death of the profession, but cool heads would argue that this is not the case at all. In fact, historians might point out that this is but one of many, many phases through which the profession has passed and survived.

Pharmacy successfully passed through the phase in the last century when drugs moved from extracts and tinctures made from natural products to synthesized compounds. Later, it made the transition from extemporaneous compounding to industrial manufacturing of drugs. It also survived the move from independent pharmacies to chains and, later, the development of HMOs, clinical pharmacy, managed care, and the information age. While the ownership, financing, and operations of pharmacies will undoubtedly continue to evolve, the need for the functions in the "pharmaceutical services" rubric remains unchanged. The distributive, educational, counseling, and record keeping remain unchanged. Patients could be served from vending machines, trailers, drive-in windows, and mail-service vendors even thousands of miles away, as well as from the more conventional sources of pharmaceutical service, or pharmaceutical care.

For the future, this means that the profession must see where the health care delivery system is headed and be certain that pharmacy's future is likely to be compatible with that vision. This requires that we listen to consumers carefully and search for clues and hints about voids, vacuums, and areas where dissatisfaction exists. There can be a magnificent future if we do this instead of falling into the trap of offering our health professional colleagues and consumers what we want to provide or offer them, rather than what they perceive as valuable and useful.

Perhaps it is appropriate for the profession to take an introspective stance and to consider in an objective fashion what we contribute and what portions of that contribution are unique. Of course, we should divide our services into cognitive and distributive. Our claim to legal, monopoly status here is rather shaky. Technicians can dispense, as can vending machines, robots, and other automated or assembly-line systems, such as those found at mail-service pharmacies. Physicians or persons in their offices can dispense, as could nurses and others as well.

In the domain of cognitive services, we can offer compliance aid; drug history taking; DUR services; counseling; and education regarding dosage, storage, side effects, administration, cautions, etc. But so can others. Pharmacy probably only needs to concern itself with groups or individuals who can perform equally well or better at the same or lower cost. Physicians and nurses become excluded, but not technicians, aides, and others such as patient educators.

As more and more of the population is being spoken for by managed care organizations and as these organizations increase the proportion of capitated contracts, we can see that the hard, cold realities of finance and economics will dictate strategies and policies at the loss of emotion, friendship, loyalty, and tradition.

The profession has an opportunity to demonstrate the cost-effectiveness of its services. If the equation is positive, the only question that will remain is whether some other occupation or profession can offer an equivalent benefit at a lower cost. There is no place for imputed or hypothesized savings. Actual experience-generated numbers are needed using recognized, scientifically sound and accepted methods, sample sizes, and objective researchers. That $2,000 *could* have been saved were a patient taking regimen X to have been hospitalized instead just won't cut it, as they say. A demonstration project using actual controls and blinded to the investigators for actual patient experience is needed.

At one time, a hospital administrator could say that "adding a service made sense and would look good to the public." Today, that service implementation would have had to pass muster that it increased quality, added to patient satisfaction, contained costs, was revenue neutral, and was needed to maintain a competitive state with other providers.

Outside the scope of pharmacy's control, events are happening that could have an enormous impact on pharmacy practice. An example is managed care selective provider contracting. This could lead to tens of thousands of unemployed or underemployed physicians. Such persons might attempt to capture niche markets that do not appear sufficiently attractive to medicine today. Some of these might include pharmacokinetic dosing and kinetics lab services, patient education and counseling with drugs, patient drug histories, discharge instructions, drug dispensing, etc.

Other areas where developments could have a massive impact on pharmacy services include biotechnology; financing of health service; automation; other new technologies; developments within other professions; legislation; OTC area growth; and clinical medicine advances, especially in genetics, immunology, and biochemistry.

While it is difficult to see very clearly into the future for more than six months or so, some factors are relatively certain. Dubos was ever so correct when, in *Mirage of Health*, he told us that fighting disease is like building objects in the sand at the edge of the water. Our interventions, like sand castles, cannot survive in any permanent status. The waves will continue to melt away our construction projects, just as the clearing in the jungle or woods will revert to its original state if not continuously manicured by outsiders.

We cure diseases, and perhaps live a little longer, only to encounter previously unknown threatening pathogens or disease states. If people only lived into their forties, as was the case 100 years ago, we would not yet have seen Alzheimer's disease.

This epilogue does not offer answers. Rather, it is written in the hope that considering pending issues and mentioning some of the more significant trends will and can stimulate the reader to think of individual defensive strategies for progressive change and progress with a focus on the patient.

As is abundantly clear, there are numerous alternatives to how pharmaceutical services are delivered and paid for presently. To be honest, most of these would probably be satisfactory in the distribution of drugs, but the option of using a patient-oriented pharmacy practitioner who has excellent interpersonal communication skills and at least some elementary counseling and/or psychological exposure can prove to be the optimal alternative.

Index